Hydrology and Hydrogeology

Hydrology and Hydrogeology

Edited by William Sobol

SYRAWOOD
PUBLISHING HOUSE
New York

Published by Syrawood Publishing House,
750 Third Avenue, 9ᵗʰ Floor,
New York, NY 10017, USA
www.syrawoodpublishinghouse.com

Hydrology and Hydrogeology
Edited by William Sobol

International Standard Book Number: 978-1-68286-463-0 (Hardback)

Cataloging-in-publication Data

Hydrology and hydrogeology / edited by William Sobol.
    p. cm.
Includes bibliographical references and index.
ISBN 978-1-68286-463-0
1. Hydrology. 2. Hydrogeology. 3. Groundwater. I. Sobol, William.
GB656 .H93 2017
551.48--dc23

Printed in the United States of America.

TABLE OF CONTENTS

PREFACE

This book traces the progress of hydrology and hydrogeology and highlights some of their key concepts and applications. Hydrology refers to the study of the movement, quality and distribution of the water on our planet as well as other celestial bodies. The three main sub-fields of hydrology are ground water hydrology also known as hydrogeology, surface water hydrology and marine hydrology. This text will provide significant information about these rapidly growing fields. It will give in-depth knowledge about the latest advances within this area and its uses in earth sciences. The book is an essential guide for both academicians and those who wish to pursue this discipline further. In this book, using case studies and examples, constant effort has been made to make the understanding of the different concepts of hydrology and hydrogeology as easy and informative as possible, for the readers.

This book has been a concerted effort by a group of academicians, researchers and scientists, who have contributed their research works for the realization of the book. This book has materialized in the wake of emerging advancements and innovations in this field. Therefore, the need of the hour was to compile all the required researches and disseminate the knowledge to a broad spectrum of people comprising of students, researchers and specialists of the field.

At the end of the preface, I would like to thank the authors for their brilliant chapters and the publisher for guiding us all-through the making of the book till its final stage. Also, I would like to thank my family for providing the support and encouragement throughout my academic career and research projects.

Editor

Soil–aquifer phenomena affecting groundwater under vertisols

D. Kurtzman[1], S. Baram[2], and O. Dahan[3]

[1] Institute of Soil, Water and Environmental Sciences, The Volcani Center, Agricultural Research Organization, P.O. Box 6, Bet Dagan 50250, Israel
[2] Dept. of Land, Air and Water Resources, University of California Davis, CA 95616, USA
[3] Dept. of Hydrology and Microbiology, Zuckerberg Institute for Water Research, Blaustein Institutes for Desert Research, Ben Gurion University of the Negev, Sde Boker Campus, Negev 84990, Israel

Correspondence to: D. Kurtzman (daniel@volcani.agri.gov.il)

Abstract. Vertisols are cracking clayey soils that (i) usually form in alluvial lowlands where, normally, groundwater pools into aquifers; (ii) have different types of voids (due to cracking), which make flow and transport of water, solutes and gas complex; and (iii) are regarded as fertile soils in many areas. The combination of these characteristics results in the unique soil–aquifer phenomena that are highlighted and summarized in this review. The review is divided into the following four sections: (1) soil cracks as preferential pathways for water and contaminants: in this section lysimeter-to basin-scale observations that show the significance of cracks as preferential-flow paths in vertisols, which bypass matrix blocks in the unsaturated zone, are summarized. Relatively fresh-water recharge and groundwater contamination from these fluxes and their modeling are reviewed; (2) soil cracks as deep evaporators and unsaturated-zone salinity: deep sediment samples under uncultivated vertisols in semiarid regions reveal a dry (immobile), saline matrix, partly due to enhanced evaporation through soil cracks. Observations of this phenomenon are compiled in this section and the mechanism of evapoconcentration due to air flow in the cracks is discussed; (3) impact of cultivation on flushing of the unsaturated zone and aquifer salinization: the third section examines studies reporting that land-use change of vertisols from native land to cropland promotes greater fluxes through the saline unsaturated-zone matrix, eventually flushing salts to the aquifer. Different degrees of salt flushing are assessed as well as aquifer salinization on different scales, and a comparison is made with aquifers under other soils; (4) relatively little nitrate contamination in aquifers under vertisols: in this section we turn the light on observations showing that aquifers under cultivated vertisols are somewhat resistant to groundwater contamination by nitrate (the major agriculturally related groundwater problem). Denitrification is probably the main mechanism supporting this resistance, whereas a certain degree of anion-exchange capacity may have a retarding effect as well.

1 Introduction

Vertisols can be briefly defined as soils with 30 % or more clay to a depth of 50 cm that have shrinking/swelling properties (Brady and Weil, 2002). More detailed definitions require the existence of a subsurface vertic horizon in which slickenside features are formed by the shrink/swell dynamics (FAO Corporate Document Repository, 2015; IUSS Working Group WRB, 2014). Other names used for these types of soils are vertosols (common in Australian studies, e.g., Radford et al., 2009; Silburn et al., 2009; Gunawardena et al., 2011; Ringrose-Voase and Nadelko, 2013), and the more general *cracking clays* (e.g., Bronswijk, 1991; Liu et al., 2010). This latter generic term emphasizes both the hydrological complexity of these soils due to the inherent discontinuities (cracks) and their relevance for agriculture, being heavy, relatively fertile soils in many semiarid regions (good water-holding capacity, relatively higher organic content, etc.). Vertisols usually form in lowlands (Yaalon, 1997) where, typically, groundwater pools into alluvial aquifers. Hence, the interface between agricultural activity on these soils and the underlying groundwater resources is both complex and relevant. This review focuses on vertisol studies that have implications for the underlying groundwater resources; it does not cover the substantial body of literature concerning shrinking/swelling dynamics and its modeling (e.g., Bronswijk, 1988; Chertkov et al., 2004; te Brake et al., 2013), the purely agricultural and mineralogical aspects of vertisols (e.g., Bhattacharyya et al., 1993; Ahmad and Mermut, 1996; Hati et al., 2007) or environmental topics like the capacity of vertisols to sequester carbon (Hua et al., 2014).

Vertisols cover 335 million hectares out of a total earth land area of 14.8 billion hectares (2.3 %). The largest areas covered with vertisols are in eastern Australia, India, Sudan–Ethiopia and Argentina–Uruguay (FAO Corporate Document Repository, 2015). Smaller areas of vertisols are found in various countries (e.g., China, Israel, Mexico, Spain, Tunisia, USA and many more). Although vertisols are very hard when dry, and very sticky when wet (making them difficult to till), in semiarid regions, irrigated crops such as cotton, corn, wheat, soybeans and others are grown on this soil. We acknowledge the dominant contribution of Australia to the literature on agro-hydrological aspects of vertisols. Conclusions from those studies have strengthened and generalized some of the findings obtained by the authors of this review in vertisol–groundwater studies in Israel, and have motivated this review (e.g., Arnon et al., 2008; Kurtzman and Scanlon, 2011; Baram et al., 2012a, b, 2013; Kurtzman et al., 2013).

The review is divided into four sections, which are partially connected and together deal with major issues concerning aquifers under agricultural land: recharge, salinization and nitrate contamination (other contaminants are mentioned as well). The following four sections cover the most general and relevant issues concerning soil–aquifers phenomena under vertisols:

- soil cracks as preferential pathways for water and contaminants (Sect. 2)

- soil cracks as deep evaporators and unsaturated-zone salinity (Sect. 3)

- impact of cultivation on flushing of the unsaturated zone and aquifer salinization (Sect. 4)

- the relatively little nitrate contamination in aquifers under vertisols (Sect. 5).

2 Soil cracks as preferential pathways for water and contaminants

There are probably hundreds of studies that acknowledge preferential flow and transport through cracks in clays – too many to be mentioned here. This section aims to review works from the soil-column and lysimeter scale to the basin and aquifer scales that show the relations between preferential flow via cracks, deep drainage, and aquifer recharge and contamination. It also provides a short description of the development of models of preferential flow and transport through soil cracks.

2.1 Preferential flow of water in vertisols – evidence from the lysimeter to aquifer scale

On a small scale, Kosmas et al. (1991) observed bypass flow through cracks in clayey soils from Greece using undisturbed soil columns (authors' terminology) with a diameter of 23 cm. Ringrose-Voase and Nadelko (2013) measured flow in preferential paths directly using a field lysimeter that was installed 2 m below the surface of a furrow-irrigated cotton field without disturbing the overlying soil. Significant drainage was collected in this study when the hydraulic gradient in the matrix was in the upward direction, advocating drainage through preferential pathways that bypasses the matrix. In a paragraph on tension-lysimeters measurements, Silburn et al. (2013) acknowledge that "deep drainage measured at 1 m depth was dominated by matrix flow, with only 10 % of drainage attributed to preferential flow (note that the soil was never dry enough to crack)", pointing out that under well-irrigated vertisols matrix deep-drainage and recharge may be of importance despite the low saturated hydraulic conductivity of the clay. A weighing-lysimeter experiment in irrigated vertisols in eastern Australia revealed a complex drainage mechanism following spray irrigation, where only deep parts of the cracks act as preferential pathways for the drainage when the top soil is moist and uncracked (Greve et al., 2010).

On the field scale (~ 100–$1000\,\mathrm{m}^2$), a similar phenomenon – i.e., open cracks at depth when surface cracks are mostly sealed – was reported by Baram et al. (2012b) throughout the rainy season in Israel. These authors compared transient deep (up to 12 m) water-content data collected by vadose-zone-monitoring systems (VMSs; Dahan et al., 2009) at various sites, including very sandy soils; the comparison showed that by far, the fastest propagation of wetting fronts in deep vadose zones is observed in cracking clays.

Ben-Hur and Assouline (2002) conducted measurements of runoff in a vertisol cotton field in Israel that was irrigated with a moving sprinkler irrigation system. They observed that the high infiltration of runoff through soil cracks limited the overall surface runoff from the field. Other field-scale vadose-zone studies reported that preferential flow through cracked clay enhances infiltration from rice paddies (Liu et al., 2004). Losses of up to 83 % of the water to deep drainage (including preferential and/or matrix flows) during furrow irrigation of cotton and sugar cane in vertisols were reported (Raine and Bakker, 1996; Dalton et al., 2001; Moss et al., 2001; Smith et al., 2005). Losses to deep drainage averaged 42.5 mm per irrigation (Smith et al., 2005), ranging from 50 to $300\,\mathrm{mm\,yr^{-1}}$ (Silburn and Montgomery, 2004). Chen et al. (2002) and Bandyopadhyay et al. (2010) showed that the transition from flood to micro-sprinkler irrigation and careful scheduling of water-application rates can dramatically reduce water losses and contaminant transport due to deep drainage. Observation from groundwater supported this phenomenon: Acworth and Timms (2009) used nested piezometers and automated logging of groundwater levels and electrical conductivity to show evidence of shallow-aquifer (16 m depth) freshening (decrease in salinity) due to fast deep drainage of irrigation water during the irrigation season.

At the small-watershed scale ($\sim 10\,000\,\text{m}^2$), Pathak et al. (2013) indicated that runoff from vertisols is smaller than runoff from sandier soils (alfisols) in an agricultural watershed near Hyderabad, India. The smaller runoff from the vertisols was attributed to preferential infiltration of local runoff into the soil cracks. Similar observations of minimal drainage and rapid recharge of shallow groundwater (~ 3 m) below a vertisol–shale watershed in Texas following rainstorms were reported by Allen et al. (2005) and Arnold et al. (2005). This process was most dominant during the first rainstorms when the cracks were fully developed (at the end of the dry season).

On the aquifer scale ($100+\,\text{km}^2$), Kurtzman and Scanlon (2011) concluded that parts of the Israeli coastal aquifer overlaid by vertisols were fresh (before the influence of modern intensive cultivation) only due to recharge flow through preferential paths that bypassed the saline vadose-zone matrix. Dafny and Silburn (2014) reported that following the growing evidence of the feasibility of percolation through cracking clays, several recent studies have included a component of diffuse recharge in their assumptions or models of the Condamine River alluvial aquifer in Australia. This diffused recharge originates in deep drainage flowing through clay matrix and/or preferential paths.

2.2 Preferential transport in vertisols

In the last 2 decades, many transport studies with dyes and/or other conservative tracers (e.g., bromide, Br^-) have indicated the pervasiveness of deep preferential transport through cracks in vertisols. Bronswijk et al. (1995) sprayed a bromide solution on cracking clays in the Netherlands that overlay a shallow water table (~ 1 m from ground surface). The authors reported rapid (on the order of days after rain event) preferential transport of Br^- into the groundwater, and relatively fast (weeks to months) propagation within the vadose zone. Bronswijk et al. (1995) concluded that large cracks control the rapid transport of Br^- to the groundwater, and preferential paths made up of tortuous *mesopores* control transport in the unsaturated zone (suggesting that transport through vertisols could be described as a triple domain medium–macropores, mesopores and matrix). Van Dam (2000) used the Crack module in SWAP to model the aforementioned experiment. This effort improved fits to the observations (relative to a single-domain model), but the variability of Br^- in the unsaturated zone still could not be well reproduced. Lin and McInnes (1995) used dye to study and model flow in vertisols. They showed that infiltrating water passes first through the soil cracks and then into the soil matrix; they concluded that uniform flow through the soil cannot be used to describe the dye transport. A dye experiment in a soil column consisting of a sandy A horizon and a vertic clay B horizon showed preferential downward flow through the cracks in horizon B, bypassing more than 94 % of the matrix (Hardie et al., 2011).

Kelly and Pomes (1998) estimated equivalent hydraulic conductivities from arrival times of Br^- and ^{15}N-labeled nitrate in gravity lysimeters installed above and under a clay pan in Missouri (USA). They reported equivalent conductivities that were 4 orders of magnitude higher than the saturated hydraulic conductivity of the clay matrix.

Unlike tracers used in experiments, fast transport of herbicides and pesticides is of concern in aquifers and drainage systems down gradient from cultivated fields. Graham et al. (1992) reported that in cultivated vertisols in California (USA), herbicides were only found deep below the root zone in samples taken from the cracks' walls and not within the matrix, suggesting rapid transport of herbicides through the cracks, either as solutes or on colloids. Transport of pesticides in preferential-flow paths absorbed on colloids was also suggested for cotton fields on vertisols in Australia (Weaver et al., 2012). Early and deep drainage of herbicides from a lysimeter in cracking clays in the UK (early meaning well before reaching field capacity in the matrix) was reported by Harris et al. (1994). Similarly, fast arrival of herbicides to drains in cultivated clays was observed by Tediosi et al. (2013) on a larger scale (small catchment).

Due to the fact that in semiarid regions vertisols are arable, agriculture-oriented settlements have developed on these soils. In many cases, these settlements include concentrated animal feeding operations (CAFOs), such as dairy farms. Arnon et al. (2008) reported deep transport (> 40 m) of estrogen and testosterone hormones into the unsaturated zone under an unlined dairy-waste lagoon constructed in a 6 m thick vertisol in Israel. They concluded that deep transport of such highly sorptive contaminants can only occur by preferential transport. Baram et al. (2012a, b) reported that preferential infiltration of dairy effluents through the cracks at the same site can transport water and solutes into the deep unsaturated zone. Locally, groundwater under dairy farm areas also shows relatively high concentrations of nitrate (Baram et al., 2014).

Figure 1 provides a visual summary of Sect. 2.1 and 2.2. It shows the potential for matrix-bypassing groundwater recharge and pollution under vertisols. Passing the biogeoactive matrix enables both freshwater recharge and transport of reactive substances.

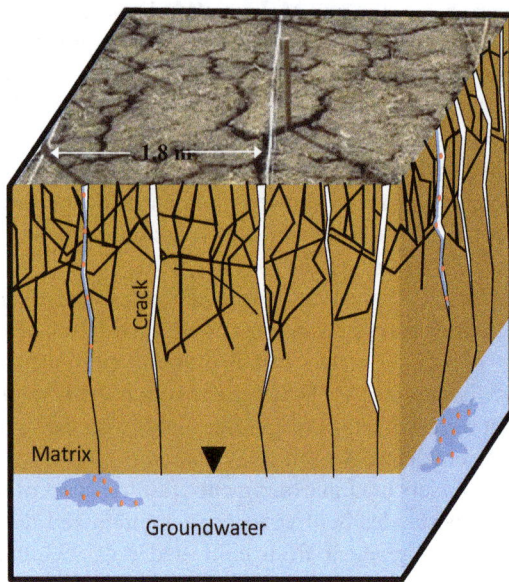

Figure 1. Illustration of potential fluxes of water and pollutants that bypass the matrix, which is typical of vertisols.

2.3 Development of flow and transport models in cracking clays

The field evidence described above motivated the development of quantitative methods to enable better predictions of flow and transport from ground surface to water table under vertisols. Nevertheless, modeling of unsaturated flow and transport as a dual (or multiple) domain in their different variants (e.g., mobile–immobile, dual porosity, dual permeability) did not develop exclusively to deal with cracking clays. Macropores such as voids between aggregates, or worm holes, are the preferential-flow paths of interest in many agricultural problems. Computer codes for modeling unsaturated preferential flow include among others: MACRO (Jarvis et al., 1994) and nonequilibrium flow and transport in HYDRUS (Šimůnek and van Genuchten, 2008). For further information on the kinematic wave approach used in MACRO, the reader is referred to German and Beven (1985); for comparative reviews of the different models and codes see Šimůnek et al. (2003), Gerke (2006), Köhne et al. (2009) and Beven and Germann (2013). The latter is critical of the common use of the Richards (1931) formulation in single and multiple-domain unsaturated-flow simulators.

One of the earlier crack-specific unsaturated-flow models was developed by Hoogmoed and Bouma (1980), who coupled vertical (crack) and horizontal (into the matrix) 1-D models using morphological data for parameterization of the linkage between the two flows. Novák et al. (2000) attached a FRACTURE module to HYDRUS in which a source term was added to the Richards equation accounting for infiltration from the bottom of the fractures, bypassing matrix bulks. Van Dam (2000), added a crack sub-model to SWAP

(van Dam et al., 2008) and Hendriks et al. (1999) used a code called FLOCR/AMINO, to study flow and transport phenomenon in shallow and cracked clayey unsaturated-zones in the Netherlands. A model of herbicide transport through the preferential paths was fitted successfully with the improved MACRO version 5.1 (Larsbo et al., 2005).

A more comprehensive dual-permeability module for 2-D and 3-D variably saturated models was introduced into HY-DRUS much later (Šimůnek et al., 2012) following the formulations of Gerke and van Genuchten (1993). Coppola et al. (2012) took another step forward in modeling flow and transport in cracking clays by also introducing cracking dynamics (adopting formulation of Chertkov, 2005) into a dual-permeability flow and transport model.

3 Soil cracks as deep evaporators and unsaturated-zone salinity

Whereas during rain events or under irrigation, cracks are a concern in terms of loss of water and fertilizers and/or contamination of groundwater (Sect. 2), under dry conditions, deep soil cracks are relevant for their evaporation capacity from deep parts of the soil column. Kurtzman and Scanlon (2011), Baram et al. (2013) and others have reported the low water content and high salinity typical of the sediment matrix under uncultivated vertisols. Deep chloride profiles under native-land vertisols often show an increase in salinity down to 1–3 m and a relatively constant concentration in deeper parts of the vadose zone (e.g., Radford et al., 2009; Kurtzman and Scanlon, 2011; Silburn et al., 2011). In the reported cases from Israel, water uptake by roots was limited to the upper 1 m of the soil profile and to the rainy season, and therefore could not fully explain the increase in salinization in the deeper layers. Deep cracks form an additional mechanism of deep evaporation that supports the chloride profiles and low water content in the matrix under vertisols.

Sun and Cornish (2005) used SWAT to model runoff and groundwater recharge at the catchment scale ($\sim 500\,\text{km}^2$) in a vertisolic catchment in eastern Australia. Considering water balances at this scale, they concluded that recharge models need to have a component that enables taking moisture out of the lower soil profile or groundwater during dry periods. Trees with roots in groundwater and deep soil cracks can maintain deep evaporation in long dry periods. Another, indirect observation that supports evaporation through cracks in vertisol was reported by Liu et al. (2010). In this work discrepancies between satellite and model estimates of soil water content in dry seasons in vertisols are assumed to be related to the extra evaporation through the cracks. Both local- and higher-scale observations and analyses point to a possible significant role of soil cracks as deep evaporators in dry periods.

Baram et al. (2013) suggested a conceptual model termed desiccation-crack-induced salinization (DCIS) based on pre-

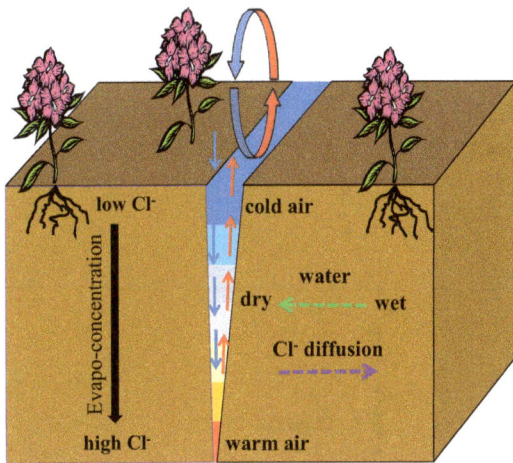

Figure 2. Desiccation-crack-induced salinization (DCIS), Baram et al. (2013). Convective instability of air in soil cracks, occurring mainly at night, leads to drying and salinization of the unsaturated zone.

vious work on subsurface evaporation and salinization in rock fractures (e.g., Weisbrod and Dragila, 2006; Nachshon et al., 2008; Kamai et al., 2009; Weisbrod et al., 2009). In DCIS, vertical convective flow of air in the cracks is driven by instability due to cold (and dense) air in the crack near the surface and warmer air down in deeper parts of the crack at night or other surface-cooling periods. The difference in the relative humidity between the invading surface air (low humidity) and the escaping air (high humidity) leads to subsurface evaporation and salt buildup (Fig. 2). Earlier studies that support the significance of evaporation via cracks in vertisols through field and laboratory observations include: Selim and Kirkham (1970), Chan and Hodgson (1981) and Adams and Hanks (1964). The latter showed enhanced evaporation from crack walls due to increase in surface wind velocity, this is another mechanism (in addition to surface cooling described before) causing instability in the crack's air, hence convection, evaporation and salt build up.

Leaching of salts from horizontal flow through the crack network evident in salinity rise in tail water of furrow-irrigated fields in cracking clays in California was reported by Rhoades et al. (1997). This Californian study acknowledge that this phenomenon was not observed in similar fields (crop and irrigation technique) in lighter soils with no cracks.

In many semiarid regions, deep matrix percolation under non-cultivated vertisols is very small due to the clay's high retention capacity and low hydraulic conductivity, root uptake of the natural vegetation in the rainy season and further evaporation through cracks in dry periods. Low water content in the deeper unsaturated zone results in low hydraulic conductivities and makes aquifer recharge through matrix flow very small year-round. Matrix fluxes on the order of $1 \, \mathrm{mm \, yr^{-1}}$ under the root/crack zone were reported in a number of studies (e.g., Silburn et al., 2009; Kurtzman and Scan-

lon, 2011; Timms et al., 2012). These very low water fluxes contain the conservative ions (e.g., chloride) originating from $200–600 \, \mathrm{mm \, yr^{-1}}$ of precipitation (with salts from wet and dry fallout) that enter the matrix at soil surface. Therefore, a dry (relatively immobile) and salty deep unsaturated matrix, developed for centuries–millennia under these non-cultivated vertisols. Nevertheless, some fresh recharge to the underlying aquifer through preferential paths related to cracks during heavy rain events creates an anomaly whereby relatively fresh water in the aquifer (e.g., $\sim 250 \, \mathrm{mg \, L^{-1}}$ chloride; Kurtzman and Scanlon, 2011) lies beneath a salty and immobile unsaturated zone with porewater chloride concentration of a few thousands of milligrams per liter (O'Leary, 1996; Kurtzman and Scanlon, 2011; Tolmie et al., 2011; Baram et al., 2013). River, mountain-front, paleo- or other types of recharge may contribute, as well, to a situation where a relatively fresh aquifer exists under a saline vadose zone.

4   Impact of cultivation on flushing of the unsaturated zone and aquifer salinization

The anomalous situation of fresh groundwater under a saline unsaturated zone found in some native-land vertisols in semiarid regions exists due to the efficient evapotranspiration by natural vegetation and cracks (making deep unsaturated matrix immobile and saline) and fresh groundwater recharge through preferential flow in cracks or other types of recharge. However, what happens when natural conditions are changed to less favorable for native-vegetation and soil cracks (e.g., cultivated land and more ever irrigated intensive cropping)? The answer is obvious: higher fluxes may develop in the unsaturated matrix, which will flush salts and ultimately cause salinization of the underlying aquifer.

A large bulk of literature from eastern Australia has reported increased deep-drainage and leaching of salts, and in some cases, salinization of aquifers under cultivated vertisols. A typical increase in deep drainage from $< 1 \, \mathrm{mm \, yr^{-1}}$ under native conditions to $10–20 \, \mathrm{mm \, yr^{-1}}$ under rain-fed cropping were reported by Silburn et al. (2009); Timms et al. (2012) and Young et al. (2014); whereas variable deep fluxes often in the 100's $\mathrm{mm \, yr^{-1}}$ range were reported for irrigated fields (mostly furrow-irrigated cotton; Gunwardena et al., 2011; Silburn et al., 2013; Weaver et al., 2013). These deep fluxes desolate salts that accumulated in the vadose zone in the native-vegetation period, moving them down towards the water table (Fig. 3). Earlier studies reporting leaching of salts from the vadose zone after clearing of natural eucalyptus trees for cropping include Allison and Hughes (1983) and Jolly et al. (1989), who worked in semiarid zones in southern Australia. In those studies, neither vertisols nor the role of soil cracks was mentioned; however, deep eucalyptus roots act similar to cracks to form a very saline and immobile deep-unsaturated-zone matrix, which becomes more mobile and less saline after the land-

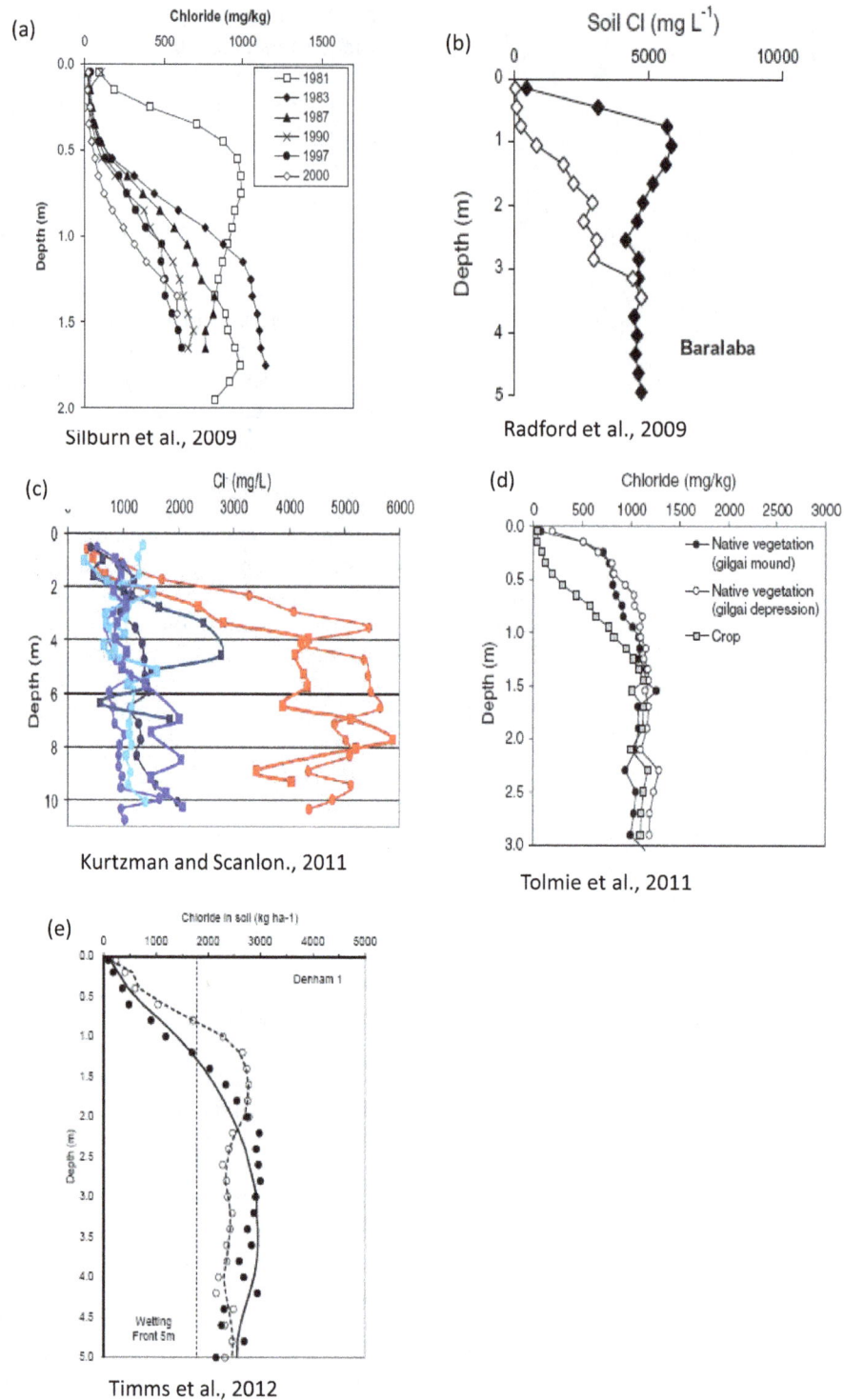

Figure 3. Flushing of chloride down through the unsaturated zone under cultivated vertisols: **(a)** Silburn et al. (2009) – 19 years of flushing; **(b)** Radford et al. (2009) – full diamond, native vegetation; empty, annual cropping; flushing from the top 3 m **(c)** Kurtzman and Scanlon (2011) – red, natural land; blue, irrigated cropping; flushing from 2–10 m depth; **(d)** Tolmie et al. (2011) – flushing from the top 1.5 m. **(e)** Timms et al. (2012) – black, cropping; empty – grass; flushing from the top 2 m.

use change. Timms et al. (2012) inferred, from combined soil and groundwater data, deep drainage and salt leaching after conversion to cropping under gray vertosols in the Murray–Darling Basin. Fresh groundwater was found in that study under shallower saline waters, strengthening the source of groundwater salinity from the vadose zone.

Scanlon et al. (2009) compared mobilization of solutes in the vadose zone after a change in the natural landscape to cultivated fields in three semiarid regions: Amargosa Desert (southwestern USA), southern High Plains (central USA) and Murray Basin (southeast Australia). Flushing of chloride from the top 6–10 m of the vadose zone after cultivation was very clear (e.g., Fig. 3 in Scanlon et al., 2009). Flushing has been observed in many arid and semiarid regions, and not exclusively related to vertisols (e.g., Oren et al., 2004, in the arid Arava Valley, southern Israel). Nevertheless, salinization of an aquifer due to cultivation and salt mobilization may be more pronounced under vertisols due to preferential-flow paths related to soil cracks (enabling the native aquifer to be relatively fresh) and the cracks evaporative capabilities (making the native deep vadose zone more saline).

A good example of an aquifer in which vertisols made a difference is the Mediterranean coastal aquifer in Israel (Fig. 4). Although known as a coastal aquifer, the phenomena discussed here are all a few kilometers inland and are not related to seawater intrusions. The parts of this aquifer overlain by vertisols were salinized a few decades after intensive cultivation, whereas the water in those parts of the aquifer overlain by cultivated loamy sand is still potable (Kurtzman, 2011; Kurtzman and Scanlon, 2011; Fig. 4). Similar to the Murray–Darling Basin (Timms et al., 2012), the upper groundwater under vertisols in this aquifer were more saline than the deep groundwater (e.g., Fig. 1 in Baram et al., 2014). Identification of the source of the salt and the cause of the salinization in the deep unsaturated zone under vertisols and land-use change, respectively, contradicted previous works, which attributed the salinization of these parts of the Israeli coastal aquifer to the intrusion of deep brines and intensive pumping (e.g., Vengosh and Ben-Zvi, 1994; Avisar et al., 2004). A different and shorter temporal trend that might also be interpreted in light of soils covering the recharge area is the response of groundwater salinity of the Israeli coastal aquifer to extreme precipitation (e.g., winter of 1991/1992): under vertisols, freshening of the aquifer (decrease in salinity) was generally observed due to recharge of freshwater through preferential paths, mostly under uncultivated parts; under loamy-sand soils, salinization of the aquifer was observed due to piston-flow recharge pushing relatively saline vadose-zone porewater down to the water table (Goldenberg et al., 1996; interpreted by Kurtzman and Scanlon, 2011).

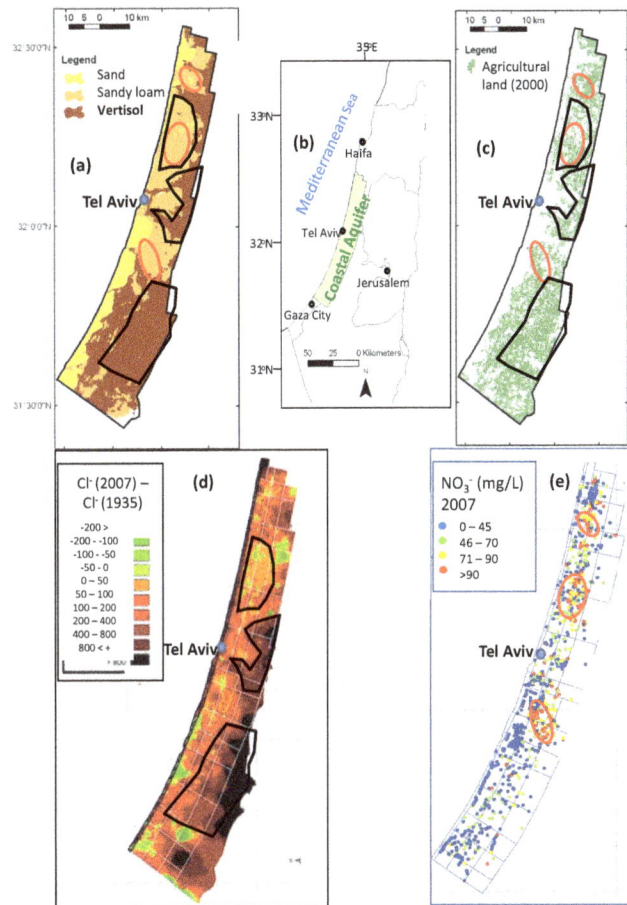

Figure 4. Plan views of the Israeli coastal aquifer. **(a)** Soil type (black polygons and red ellipses for spatial comparisons with panels d and e, respectively). **(b)** Location map. **(c)** Cultivated land in the year 2000. **(d)** Difference in chloride concentrations between 2007 and 1935 (modified from Livshitz and Zentner, 2009). Black polygons are characteristic cultivated areas that were severely salinized (southern polygons) and barely salinized (northern polygon) relative to soil type (panel a). **(e)** Nitrate concentration in groundwater wells in 2007 (modified from Hydrological Service, 2008). Red ellipses – areas with many wells contaminated with nitrate relative to soil type (panel a; modified from Kurtzman et al., 2013).

5　Relatively little nitrate contamination in aquifers under vertisols

Whereas the literature concerning salinization of aquifers and draining of salts from the vadose zone under cultivated vertisols is abundant, much less has been written about the contamination of groundwater by nitrate under these soils. Nitrate is the most problematic groundwater contaminant associated with agriculture worldwide (Jalali, 2005; Erisman et al., 2008; Burow et al., 2010; Vitousek et al., 2010; Kourakos et al., 2012). Both mineral nitrogen fertilizers (e.g., Kurtzman et al., 2013) and organic forms of nitrogen (e.g., Dahan et al., 2014) are often applied in excess with respect to the plants' ability to take up the nitrogen, leaving significant quantities of nitrate as a potential groundwater contaminant.

While in the previous sections aquifers under vertisols were shown to be vulnerable to salinization, due to the agricultural practice above, there is an increasing number of observations that indicate lesser nitrate contamination in groundwater under cultivated vertisols relative to groundwater of the same aquifer located under cultivated land of lighter soils. Kurtzman et al. (2013), dealing with nitrate contamination problems of the Israeli coastal aquifer, showed that at the groundwater basin scale ($\sim 2000\,\mathrm{km}^2$) the contamination plumes of nitrate are present in the aquifer only under cultivated sandy loams, whereas under cultivated vertisols sporadic wells seldom produce water with nitrate concentration above the drinking-water standard (Fig. 4). Dafny (personal communication, 2014) revealed, by chi-square analysis, that groundwater under cultivated vertisols and a thick clayey-alluvial unsaturated zone is less likely than groundwater, under coarser sediments, to get contaminated by nitrate in the Condamine floodplain aquifer in Australia.

In contrast to the relatively high capability of vertisols to reduce nitrate leaching from cultivated land, both Baram et al. (2014; Israel, Coastal Aquifer) and Dafny (personal communication, 2014; Condamine floodplain, eastern Australia) acknowledge that CAFOs can be significant point sources of nitrate in vertisols as well. This might be due to incidental percolation of CAFO wastewater through the crack systems.

Silburn et al. (2013) indicated that modern deep drainage and solutes are still migrating down through the unsaturated zone in vertisol–alluvial systems in Australia and the nitrate is accumulating in the unsaturated zone. Nevertheless in vertisols areas overlaying the Israeli coastal aquifer, the rise in salinity and unsaturated-flow and transport models, indicate that the cultivation effects reached the water table, yet nitrate contamination is not severe, suggesting other mechanism are responsible for the low levels of nitrate contamination.

Denitrification in clayey soils is thought to be the major reason for the reduced deep leaching of nitrate in semiarid climates; this reduction of nitrate to gaseous nitrogen is less likely to be significant in lighter soils (Sigunga et al., 2002; Baram et al., 2012b; Boy-Roura et al., 2013; He et al., 2013).

Jahangir et al. (2012) found that adding carbon to deeper soil horizons significantly enhances denitrification in those layers. Profiles of dissolved organic carbon (DOC) in deep vadose zones (down to 9 m below ground) under citrus orchards on thick vertisols versus sandy-loam in Israel were compared. Whereas DOC in the lighter soils was higher than $15\,\mathrm{mg\,kg^{-1}_{dry\,soil}}$ only in the top 1 m, in the vertisols it was above $30\,\mathrm{mg\,kg^{-1}_{dry\,oil}}$ in the entire 9 m profile (Shapira, 2012). These latter two studies support the notion that denitrification in the root zone, and perhaps beyond, results in less nitrate problems in aquifers under cultivated vertisols than under lighter soils. Thayalakumaran et al. (2014) reported high DOC in shallow groundwater overlain by irrigated sug-

arcane corresponds with the absence of nitrate in this aquifer in northeast Australia.

Denitrification in the root zone and deeper in the soil profile explains the small amount of nitrate leached to the groundwater under rice fields in clayey soils in California, USA (Liang et al., 2014). Shallow groundwater (< 1.5 m) under cultivated vertisols (e.g., the Netherlands) showed large variability (spatial and temporal) in nitrate concentration, probably due to the highly variable oxygen concentrations and therefore variability in nitrogen transformations in these systems (Hendriks et al., 1999).

A more speculative mechanism that might explain the relatively lower occurrence of groundwater nitrate contamination involves the anion-exchange capacity of the clay. Harmand et al. (2010) observed very significant adsorption of nitrate to kaolinite and oxyhydroxides under a fertilized coffee plantation growing on an acrisol in Costa Rica. In vertisols, montmorillonite is usually the dominant clay mineral; nevertheless, some kaolinite is found in most vertisols (e.g., Singh and Heffernan, 2002; Krull and Skjemstad, 2003; Baram et al., 2012b). Another drawback of this mechanism as dominant in vertisols is the adsorption of anions to a positively charged surface is more efficient at low pH, while vertisols in semiarid regions are usually neutral to alkaline. Retardation of nitrate in the vadose zone due to adsorption to positively charged sites within the clay might slow down groundwater contamination under cultivated vertisols. Nevertheless, if significant, this mechanism would only retard groundwater contamination, whereas denitrification removes the nitrogen from the soil – unsaturated zone – aquifer system. The idea of nitrate adsorption has been tested as an engineered solution for reducing deep nitrate percolation. Artificially synthesized materials that have nitrate-sorption capacity (e.g., $[\mathrm{Mg}^{2+}_{0.82}\,\mathrm{Al}^{3+}_{0.18}\,(\mathrm{OH})_2]^{0.18+}[(\mathrm{Cl}-)_{0.18}\,0.5(\mathrm{H2O})]^{0.18-}$) are being tested as soil additives to buffer nitrate leaching (Torres-Dorante et al., 2009).

6 Conclusions

Vertisols are considered arable soils in semiarid climates, and are intensively cultivated. Located in lowlands, vertisols often overlie aquifers. Flow and transport through the cracking clays is complex and results in unique land–aquifer phenomena. Observations from the lysimeter to basin scale have shown (directly and indirectly) the significance of cracks as preferential-flow paths in vertisols that bypass matrix blocks in the unsaturated zone. These preferential paths support recharge with relatively fresh water in uncultivated vertisols, and groundwater contamination from point sources such as CAFOs and under some conditions, from crop fields. Deep soil samples under uncultivated vertisols in semiarid regions reveal a dry (immobile), saline matrix, partly due to enhanced evaporation through the soil cracks. This evaporation is related to convective instability due to colder air at ground surface and warmer air deep in the crack during the night. In some aquifers lying beneath vertisols in these regions, relatively fresh groundwater exists under the saline unsaturated zone. Land-use change to cropland promotes greater fluxes through the saline matrix, which flush salts into the aquifer and eventually cause groundwater salinization. In contrast to the vulnerability of groundwater under vertisols to salinization, observations show that this soil–aquifer setting has some resistance to groundwater contamination by nitrate (the major agriculturally related groundwater contamination). Denitrification is probably the main mechanism supporting this resistance, whereas anion-exchange capacity may have a retarding effect as well.

Acknowledgements. The study was supported by the Agricultural Research Organization (ARO), Israel.

References

Acworth, R. I. and Timms, W. A.: Evidence for connected water processes through smectite-dominated clays at Breeza, New South Wales, Aust. J. Earth Sci., 56, 81–96, 2009.

Adams, J. E. and Hanks, R. J.: Evaporation from Soil Shrinkage Cracks, Soil Sci. Soc. Am. J., 28, 281–284, 1964.

Ahmad, N. and Mermut, A.: Vertisols and technologies for their management, Developments in Soil Science, Volume 24, Elsevier Science, Netherlands, 1996.

Allen, P. M., Harmel, R. D., Arnold, J., Plant, B., Yelderman, J., and King, K.: Field data and flow system response in clay (Vertisol) shale terrain, north central Texas, USA, Hydrol. Process., 19, 2719–2736, doi:10.1002/hyp.5782, 2005.

Allison, G. B. and Hughes, M. W.: The use of natural tracers as indicators of soil-water movement in a temperate semi-arid region, J. Hydrol., 60, 157–173, 1983.

Arnold, J. G., Potter, K. N., King, K. W., and Allen, P. M.: Estimation of soil cracking and the effect on the surface runoff in a Texas Blackland Prairie watershed, Hydrol. Process., 19, 589–603, doi:10.1002/hyp.5609, 2005.

Arnon, S., Dahan, O., Elhanany, S., Cohen, K., Pankratov, I., Gross, A., Ronen, Z., Baram, S., and Shore, L.S.: Transport of testosterone and estrogen from dairy farm waste lagoons to groundwater, Environ. Sci. Technol., 42, 5521–5526, 2008.

Avisar, D., Rosenthal, E., Shulman, H., Zilberbrand, M., Flexer, A., Kronfeld, J., Ben Avraham, Z., and Fleischer, L.: The Pliocene Yafo Fm in Israel: hydrogeologically inert or active?, Hydrogeol. J., 12, 291–304, 2004.

Bandyopadhyay, K. K., Misra, A. K., Ghosh, P. K., Hati, K. M., Mandal, K. G., and Moahnty, M.: Effect of irrigation and nitrogen application methods on input use efficiency of wheat under limited water supply in a Vertisol of Central India, Irrig. Sci., 28, 285–299, 2010.

Baram, S., Arnon, S., Ronen, Z., Kurtzman, D., and Dahan, O.: Infiltration mechanism controls nitrification and denitrification processes under dairy waste lagoons, J. Environ. Qual., 41, 1623–1632, 2012a.

Baram, S., Kurtzman, D., and Dahan, O.: Water percolation through a clayey vadose zone, J. Hydrol., 424–425, 165–171, 2012b.

Baram, S., Ronen, Z., Kurtzman, D., Külls, C., and Dahan, O.: Desiccation-crack-induced salinization in deep clay sediment, Hydrol. Earth Syst. Sci., 17, 1533–1545, doi:10.5194/hess-17-1533-2013, 2013.

Baram, S., Kurtzman, D., Ronen, Z., Peeters, A., and Dahan, O.: Assessing the impact of dairy waste lagoons on groundwater quality using a spatial analysis of vadose zone and groundwater information in a coastal phreatic aquifer, J. Environ. Manage., 132, 135–144, 2014.

Ben-Hur, M. and Assouline S.: Tillage effects on water and salt distribution in a vertisol during effluent irrigation and rainfall, Agron. J., 94, 1295–1304, 2002.

Beven, K. and Germann, P.: Macropores and water flow in soils revisited, Water Resour. Res., 49, 3071–3092, doi:10.1002/wrcr.20156, 2013.

Bhattacharyya, T., Pal, D. K., and Deshpande, S. B.: Genesis and transformation of minerals in the formation of red (alfisols) and black (inceptisols and vertisols) soils on Deccan Basalt in the Western Ghats, India, J. Soil Sci., 44, 159–171, 1993.

Boy-Roura, M., Nolan, B. T., Menció, A., and Mas-Pla, J.: Regression model for aquifer vulnerability assessment of nitrate pollution in the Osona region (NE Spain), J. Hydrol., 505, 150–162, 2013.

Brady, N. C. and Weil, R. R.: The Nature and Properties of Soils, thirteenth ed., Pearson Education, New Jersey, 2002.

Bronswijk, J. J. B.: Modeling of water balance, cracking and subsidence of clay soils, J. Hydrol., 97, 199–212, 1988.

Bronswijk, J. J. B.: Relation between vertical soil movements and water-content changes in cracking clays, Soil Sci. Soc. Am. J., 55, 1220–1226, 1991.

Bronswijk, J. J. B., Hamminga, W., and Oostindie, K.: Field-scale solute transport in a heavy clay soil, Water Resour. Res., 31, 517–526, 1995.

Burow, K. R., Nolan, B. T., Rupert, M. G., and Dubrovsky, N. M.: Nitrate in groundwater of the United States, 1991–2003, Environ. Sci. Technol., 44, 4988–4997, doi:10.1021/es100546y, 2010.

Chan, K. Y. and Hodgson, A. S.: Moisture regimes of a cracking clay soil under furrow irrigated cotton, Aust. J. Exp. Agr., 21, 538–542, 1981.

Chen, C. C., Roseberg, R. J., and Selker, J. S.: Using microsprinkler irrigation to reduce leaching in a shrink/swell clay soil, Agric. Water Manage., 54, 159–171, 2002.

Chertkov, V. Y.: The shrinkage geometry factor of a soil layer, Soil Sci. Soc. Am. J., 69, 1671–1683, doi:10.2136/sssaj2004.0343, 2005.

Chertkov, V. Y., Ravina, I., and Zadoenko, V.: An approach for estimating the shrinkage geometry factor at a moisture content, Soil Sci. Soc. Am. J., 68, 1807–1817, 2004.

Coppola, A., Gerke, H. H., Comegna, A., Basile, A., and Comegna, V.: Dual-permeability model for flow in shrinking soil with dominant horizontal deformation, Water Resour. Res., 48, W08527, doi:10.1029/2011WR011376, 2012.

Dafny E. and Silburn D. M.: The hydrogeology of the Condamine River Alluvial Aquifer, Australia: a critical assessment, Hydrogeol. J., 22, 705–727, 2014.

Dahan, O., Talby, R., Yechieli, Y., Adar, E., Lazarovitch, N., and Enzel, Y.: In situ monitoring of water percolation and solute transport using a vadose zone monitoring system, Vadose Zone J., 8, 916–925, doi:10.2136/vzj2008.0134, 2009.

Dahan, O., Babad, A., Lazarovitch, N., Russak, E. E., and Kurtzman, D.: Nitrate leaching from intensive organic farms to groundwater, Hydrol. Earth Syst. Sci., 18, 333–341, doi:10.5194/hess-18-333-2014, 2014.

Dalton, P., Raine, S. R., and Broadfoot, K.: Best management practices for maximising whole farm irrigation efficiency in the Australian cotton industry, Final report to the Cotton Research and Development Corporation, National Centre for Engineering in Agriculture Report, 179707/2, USQ, Toowoomba, 2001.

Erisman, J. W., Sutton, M. A., Galloway, J. N., Klimont, Z., and Winiwarter, W.: How a century of ammonia synthesis changed the world, Nat. Geosci., 1, 636–639, 2008.

FAO Corporate Document Repository: Lecture notes on the major soils of the world, available at: http://www.fao.org/docrep/003/y1899e/y1899e06.htm, last access: July 2015.

Gerke, H. H.: Preferential flow descriptions for structured soils, J. Plant Nutr. Soil Sci., 169, 382–400, doi:10.1002/jpln.200521955, 2006.

Gerke, H. H. and van Genuchten, M. Th.: A dual-porosity model for simulating the preferential movement of water and solutes in structured porous media, Water Resour. Res., 29, 305–319, 1993.

Germann, P. F. and Beven, K.: Kinematic wave approximation to infiltration into soils with sorbing macropores, Water Resour. Res., 21, 990–996, 1985.

Goldenberg, L. C., Melloul, A. J., and Zoller, U.: The "short cut" approach for the reality of enhanced groundwater contamination, J. Environ. Manage., 46, 311–326, doi:10.1006/jema.1996.0024, 1996.

Graham, R. C., Ulery, A. L., Neal, R. H., and Teso, R .R.: Herbicide residue distributions in relation to soil morphology in two California vertisols, Soil Sci., 153, 115–121, 1992.

Greve, A., Andersen, M. S., and Acworth, R. I.: Investigations of soil cracking and preferential flow in a weighing lysimeter filled with cracking clay soil, J. Hydrol., 393, 105–113, 2010.

Gunawardena, T. A., McGarry, D., Robinson, J. B., and Silburn, D. M.: Deep drainage through Vertosols in irrigated fields measured with drainage lysimeters, Soil Res., 49, 343–354, 2011.

Hardie, M. A., Cotching, W. E., Doyle, R. B., Holz, G., Lisson, S., and Mattern, K.: Effect of antecedent soil moisture on preferential flow in a texture-contrast soil, J. Hydrol., 398, 191–201, 2011.

Harmand, J. M., Avila, H., Oliver, R., Saint-Andre, L., and Dambrine, E.: The impact of kaolinite and oxi-hydroxides on nitrate adsorption in deep layers of a Costarican Acrisol under coffee cultivation, Geoderma, 158, 216–224, 2010.

Harris, G. L., Nicholls, P. H., Bailey, S. W., Howse, K. R., and Mason, D. J.: Factors influencing the loss of pesticides in drainage from a cracking clay soil, J. Hydrol., 159, 235–253, 1994.

Hati, K. M., Swarup, A., Dwivedi, A. K., Misra, A. K., and Bandyopadhyay, K. K.: Changes in soil physical properties and organic carbon status at the topsoil horizon of a vertisol of central India after 28 years of continuous cropping, fertilization and manuring, Agric. Ecosyst. Environ., 119, 127–134, 2007.

He, J., Dougherty, M., and AbdelGadir, A.: Numerical assisted assessment of vadose-zone nitrogen transport under a soil moisture controlled wastewater SDI dispersal system in a Vertisol, Ecol. Eng., 53, 228–234, 2013.

Hendriks, R. F .A., Oostindie, K., and Hamminga, P.: Simulation of bromide tracer and nitrogen transport in a cracked clay soil with the FLOCR/ANIMO model combination, J. Hydrol., 215, 94–115, 1999.

Hoogmoed, W. B. and Bouma, J.: A simulation model for predicting infiltration into cracked clay soil, Soil Sci. Soc. Am. J., 44, 458–461, doi:10.2136/sssaj1980.03615995004400030003x, 1980.

Hua, K., Wang, D., Guo, X., and Guo, Z.: Carbon Sequestration Efficiency of Organic Amendments in a Long-Term Experiment on a Vertisol in Huang-Huai-Hai Plain, China, PLoS ONE, 9, e108594, doi:10.1371/journal.pone.0108594, 2014.

Hydrological Service: Current condition and development of water resources in Israel until autumn 2007, Annual report, Israel Water Authority, Jerusalem, 2008 (in Hebrew).

IUSS Working Group WRB: World Reference Base for Soil Resources, 2014, International soil classification system for naming soils and creating legends for soil maps, World Soil Resources Reports No. 106, FAO, Rome, 2014.

Jahangir, M. M. R., Khalil, M. I., Johnston, P., Cardenas, L. M., Hatch, D. J., Butler, M., Barrett, M., O'flaherty, V., and Richards, K. G.: Denitrification potential in subsoils: a mechanism to reduce nitrate leaching to groundwater, Agric. Ecosyst. Environ., 147, 13–23, 2012.

Jalali, M.: Nitrates leaching from agricultural land in Hamadan, western Iran, Agric. Ecosyst. Environ., 110, 210–218, 2005.

Jarvis, N. J., Stähli, M., Bergström, L., and Johnsson, H.: Simulation of dichlorprop and bentazon leaching in soils of contrasting texture using the MACRO model, J. Environ. Sci. Health, A29, 1255–1277, 1994.

Jolly, I. D., Cook, P. G., Allison, G. B., and Hughes, M. W.: Simultaneous water and solute movement through unsaturated soil following an increase in recharge, J. Hydrol., 111, 391–396, doi:10.1016/0022-1694(89)90270-9, 1989.

Kamai, T., Weisbrod, N., and Dragila, M. I.: Impact of ambient temperature on evaporation from surface-exposed fractures, Water Resour. Res., 45, W02417, doi:10.1029/2008WR007354, 2009.

Kelly, B. P. and Pomes, M. L.: Preferential flow and transport of nitrate and bromide in claypan soil, Ground Water, 36, 484–494, doi:10.1111/j.1745-6584.1998.tb02820.x, 1998.

Köhne, J. M., Köhne, S., and Simunek, J.: A review of model applications for structured soils: a) water flow and tracer transport, J. Contam. Hydrol., 104, 4–35, 2009.

Kosmas, C., Moustakas, N., Kallianou, C., and Yassoglou, N.: Cracking patterns, bypass flow and nitrate leaching in Greek irrigated soils, Geoderma, 49, 139–152, 1991.

Kourakos, G., Klein, F., Cortis, A., and Harter, T.: A groundwater nonpoint source pollution modeling framework to evaluate long-term dynamics of pollutant exceedance probabilities in wells and other discharge locations, Water Resour. Res., 48, W00L13, doi:10.1029/2011WR010813, 2012.

Krull, E. S. and Skjemstad, J.O.: Delta(13)C and delta(15)N profiles in (14)C-dated Oxisol and Vertisols as a function of soil chemistry a mineralogy, Geoderma, 112, 1–29, 2003.

Kurtzman, D.: The soil, the unsaturated zone, and groundwater-salinization in the south-eastern part of the (Israeli) Coastal Aquifer Agricultural Research Organization publication no. 603/11 (in Hebrew), available upon request: daniel@volcani.agri.gov.il, 2011.

Kurtzman, D. and Scanlon, B. R.: Groundwater recharge through vertisols: irrigated cropland vs. natural land, Israel, Vadose Zone J., 10, 662–674, 2011.

Kurtzman, D., Shapira, R. H., Bar-Tal, A., Fine, P., and Russo, D.: Nitrate fluxes to groundwater under citrus orchards in a Mediterranean climate: observations, calibrated models, simulations and agro-hydrological conclusions, J. Contam. Hydrol., 151, 93–104, 2013.

Larsbo, M., Roulier, S., Stenemo, F., Kastreel, R., and Jarvis, N.: An improved dual-permeability model of water flow and solute transport in the vadose zone, Vadose Zone J., 4, 398–406, 2005.

Liang, X. Q., Harter, T., Porta, L., van Kessel, C., and Linquist, B. A.: Nitrate leaching in Californian rice fields: a field and regional-scale assessment, J. Environ. Qual., 43, 881–894, 2014.

Lin, H. and McInnes, K.: Water-flow in clay soil beneath a tension infiltrometer, Soil Sci. 159, 375–382, 1995.

Liu, C. W., Chen, S. K., and Jang, C. S.: Modelling water infiltration in cracked paddy field soil, Hydrol. Process., 18, 2503–2513, doi:10.1002/hyp.1478, 2004.

Liu, Y. Y., Evans, J. P., McCabe, M. F., de Jeu, R. A. M., van Dijk, A. I. J. M., and Su, H.: Influence of cracking clays on satellite estimated and model simulated soil moisture, Hydrol. Earth Syst. Sci., 14, 979–990, doi:10.5194/hess-14-979-2010, 2010.

Livshitz, Y. and Zentner, E.: Changes in water and salt storage of the Coastal Aquifer during the years 1933–2007: development and application of a method and comparison to previous methods, A Hydrological Service Report, Israel Water Authority, 2009 (in Hebrew).

Moss, J., Gordon, I. J., and Zischke, R.: Best management practices to minimise below root zone impacts of irrigated cotton, Final report to the Murray–Darling Basin Commission (Project I6064), Department of Natural Resources and Mines, Queensland, 2001.

Nachshon, U., Weisbrod, N., and Dragila, M. I.: Quantifying air convection through surface-exposed fractures: a laboratory study, Vadose Zone J., 7, 948–956, 2008.

Novák, V., Šimůnek , J., and van Genuchten, M. Th. Infiltration of water into soils with cracks, ASCE J. Irrig. Drain. Eng., 126, 41–47, 2000.

O'Leary, G. J.: The effects of conservation tillage on potential groundwater recharge, Agric. Water Manage., 31, 65–73, 1996.

Oren, O., Yechieli, Y., Bohlke, J. K., and Dody, A.: Contamination of groundwater under cultivated fields in an arid environment central Arava Valley, Israel, J. Hydrol., 290, 312–328, 2004.

Pathak, P., Sudi, R., Wani, S. P., and Sahrawat, K. L.: Hydrological behavior of Alfisols and Vertisols in the semi-arid zone: implications for soil and water management, Agric. Water Manage., 118, 12–21, 2013.

Radford, B. J., Silburn, D. M., and Forster, B. A.: Soil chloride and deep drainage responses to land clearing for cropping at seven sites in central Queensland, northern Australia, J. Hydrol., 379, 20–29. doi:10.1016/j.jhydrol.2009.09.040, 2009.

Raine, S. R. and Bakker, D.: Increased furrow irrigation efficiency through better design and management of cane fields, in: Proceedings of the Australian Society Sugar Cane Technologists, 119–124, 1996.

Rhoades, J. D., Lesch, S. M., Burch, S. L., Letey, J., LeMert, R. D. Shouse, P. J., Oster, J. D., and O'Halloran T.: Salt distributions in cracking soils and salt pickup by runoff waters, J. Irrig. Drain. Eng., 123, 323–328, 1997.

Richards, L. A.: Capillary conduction of liquids through porous mediums, Physics, 1, 318–333, 1931.

Ringrose-Voase, A. J. and Nadelko, A. J.: Deep drainage in a Grey Vertosol under furrow-irrigated cotton, Crop Pasture Sci., 64, 1155–1170, 2013.

Scanlon, B. R., Stonestrom, D. A., Reedy, R. C., Leaney, F. W., Gates, J., and Cresswell, R. G.: Inventories and mobilization of unsaturated zone sulfate, fluoride, and chloride related to land use change in semiarid regions, southwestern United States and Australia, Water Resour. Res., 45, W00A18, doi:10.1029/2008WR006963, 2009.

Selim, H. W. and Kirkham, D.: Soil temperature and water content changes during drying as influenced by cracks: a laboratory experiment, Soil Sci. Soc. Am. J., 34, 565–569, 1970.

Shapira, R. H.: Nitrate flux to groundwater under citrus orchards: observations, modeling and simulating different nitrogen application rates, M.Sc. Thesis, The Hebrew University of Jerusalem, 2012 (in Hebrew with English abstract).

Sigunga, D. O., Janssen, B. H., and Oenema, O.: Denitrification risks in relation to fertilizer nitrogen losses from Vertisols and Phaoezems, Commun. Soil Sci. Plant Anal., 33, 561–578, 2002.

Silburn D. M. and Montgomery J.: Deep drainage under irrigated cotton in Australia – A review, WATERpak a guide for irrigation management in cotton, Cotton Research and Development Corporation/Australian Cotton Cooperative Research Centre, Narrabri, 29–40, 2004.

Silburn, D. M., Cowie, B. A., and Thornton, C. M.: The Brigalow Catchment Study revisited: effects of land development on deep drainage determined from non-steady chloride profiles, J. Hydrol., 373, 487–498, doi:10.1016/j.jhydrol.2009.05.012, 2009.

Silburn, D. M., Tolmie, P. E., Biggs, A. J. W., Whish, J. P. M., and French, V.: Deep drainage rates of Grey Vertosols depend on land use in semi-arid subtropical regions of Queensland, Soil Res., 49, 424–438, doi:10.1071/SR10216, 2011.

Silburn, D. M., Foley, J. L., Biggs, A. J. W., Montgomery, A. J., and Gunawardena, T. A.: The Australian Cotton Industry and four decades of deep drainage research: a review, Crop Pasture Sci., 64, 1049–1075, 2013.

Šimůnek, J. and van Genuchten, M. T.: Modeling nonequilibrium flow and transport processes using HYDRUS, Vadose Zone J., 7, 782–797, 2008.

Šimůnek, J., Jarvis, N. J., van Genuchten, M. T., and Gärdenäs, A.: Review and comparison of models for describing nonequilibrium and preferential flow and transport in the vadose zone, J. Hydrol., 272, 14–35, doi:10.1016/S0022-1694(02)00252-4, 2003.

Šimůnek, J., Šejna, M., and van Genuchten M. T.: The DualPerm Module for HYDRUS (2-D/3-D) Simulating Two-Dimensional Water Movement and Solute Transport in Dual-Permeability Porous Media, Version 1.0, PC Progress, Prague, Czech Republic, 2012.

Singh, B. and Heffernan, S.: Layer charge characteristics of smectites from vertosols (vertisols) of New South Wales, Aust. J. Soil Res., 40, 1159–1170, doi:10.1071/SR02017, 2002.

Smith, R., Raine, S., and Minkevich, J.: Irrigation application efficiency and deep drainage potential under surface irrigated cotton, Agric. Water Manage., 71, 117–130, 2005.

Sun, H. and Cornish, P. S.: Estimating shallow groundwater recharge in the headwaters of the Liverpool Plains using SWAT, Hydrol. Process., 19, 795–807, 2005.

te Brake, B., van der Ploeg, M. J., and de Rooij, G. H.: Water storage change estimation from in situ shrinkage measurements of clay soils, Hydrol. Earth Syst. Sci., 17, 1933–1949, doi:10.5194/hess-17-1933-2013, 2013.

Tediosi, A., Whelan, M. J., Rushton, K. R., and Gandolfi, C.: Predicting rapid herbicide leaching to surface waters from an artificially drained headwater catchment using a one dimensional two-domain model coupled with a simple groundwater model, J. Contam. Hydrol., 145, 67–81, 2013.

Thayalakumaran T., Lenahan M. J., and Bristow K. L.: Dissolved organic carbon in groundwater overlain by irrigated sugarcane, Groundwater, 53, 525–530, 2014.

Timms, W. A., Young, R. R., and Huth, N.: Implications of deep drainage through saline clay for groundwater recharge and sustainable cropping in a semi-arid catchment, Australia, Hydrol. Earth Syst. Sci., 16, 1203–1219, doi:10.5194/hess-16-1203-2012, 2012.

Tolmie, P. E., Silburn, D. M., and Biggs, A. J. W.: Deep drainage and soil salt loads in the Queensland Murray–Darling Basin using soil chloride: comparison of land uses, Soil Res., 49, 408–423, 2011.

Torres-Dorante, L. O., Lammel, J., and Kuhlmann, H.: Use of a layered double hydroxide (LDH) to buffer nitrate in soil: long-term nitrate exchange properties under cropping and fallow conditions, Plant Soil, 315, 257–272, 2009.

Van Dam, J. C.: Simulation of field-scale water flow and bromide transport in a cracked clay soil, Hydrol. Process., 14, 1101–1117, 2000.

Van Dam, J. C., Groenendijk, P., Hendriks, R. F. A., and Kroes, J. G.: Advances of modeling water flow in variably saturated soils with SWAP, Vadose Zone J., 7, 640–665, 2008.

Vengosh, A. and Ben-Zvi, A.: Formation of a salt plume in the Coastal Plain aquifer of Israel: the Be'er Toviyya region, J. Hydrol., 160, 21–52, doi:10.1016/0022-1694(94)90032-9, 1994.

Vitousek, P. M., Naylor, R., Crews, T., David, M. B., Drinkwater, L. E., Holland, E., Johnes, P. J., Katzenberger, J., Martinelli, L. A., Matson, P. A., Nziguheba, G., Ojima, D., Palm, C. A., Robertson, G. P., Sanchez, P. A., Townsend, A. R., and Zhang, F. S.: Nutrient imbalances in agricultural development, Science, 324, 1519–1520. doi:10.1126/science.1170261, 2010.

Weaver, T. B., Ghadiri, H., Hulugalle, N. R., and Harden, S.: Organochlorine pesticides in soil under irrigated cotton farming systems in Vertisols of the Namoi Valley, north-western New South Wales, Australia, Chemosphere, 88, 336–343, 2012.

Weaver, T. B., Hulugalle, N. R., Ghadiri, H., and Harden, S.: Quality of drainage water under irrigated cotton in Vertisol of the lower Namoi Valley, New South Wales, Australia, Irrig. Drainage, 62, 107–114, doi:10.1002/ird.1706, 2013.

Weisbrod, N. and Dragila, M. I.: Potential impact of convective fracture venting on salt-crust buildup and ground-water salinization in arid environments, J. Arid Environ., 65, 386–399, 2006.

Weisbrod, N., Dragila, M. I., Nachshon, U., and Pillersdorf, M.: Falling through the cracks: the role of fractures in Earth-atmosphere gas exchange, Geophys. Res. Lett., 36, L02401, doi:10.1029/2008GL036096, 2009.

Yaalon, D. H.: Soils in the Mediterranean region: what makes them different?, Catena, 28, 157–169, 1997.

Young, R., Huth, N., Harden, S., and McLeod, R.: Impact of rainfed cropping on the hydrology and fertility of alluvial clays in the more arid areas of the upper Darling Basin, eastern Australia, Soil Res., 52, 388–408, 2014.

Differences in the water-balance components of four lakes in the southern-central Tibetan Plateau

S. Biskop[1], F. Maussion[2], P. Krause[3], and M. Fink[1]

[1]Department of Geography, Friedrich Schiller University Jena, Germany
[2]Institute of Atmospheric and Cryospheric Sciences, University of Innsbruck, Austria
[3]Thuringian State Institute for Environment and Geology, Jena, Germany

Correspondence to: S. Biskop (sophie.biskop@uni-jena.de)

Abstract. The contrasting patterns of lake-level fluctuations across the Tibetan Plateau (TP) are indicators of differences in the water balance over the TP. However, little is known about the key hydrological factors controlling this variability. The purpose of this study is to contribute to a more quantitative understanding of these factors for four selected lakes in the southern-central part of the TP: Nam Co and Tangra Yumco (increasing water levels), and Mapam Yumco and Paiku Co (stable or slightly decreasing water levels). We present the results of an integrated approach combining hydrological modeling, atmospheric-model output and remote-sensing data. The J2000g hydrological model was adapted and extended according to the specific characteristics of closed-lake basins on the TP and driven with High Asia Refined analysis (HAR) data at 10 km resolution for the period 2001–2010. Differences in the mean annual water balances among the four basins are primarily related to higher precipitation totals and attributed runoff generation in the Nam Co and Tangra Yumco basins. Precipitation and associated runoff are the main driving forces for inter-annual lake variations. The glacier-meltwater contribution to the total basin runoff volume (between 14 and 30 % averaged over the 10-year period) plays a less important role compared to runoff generation from rainfall and snowmelt in non-glacierized land areas. Nevertheless, using a hypothetical ice-free scenario in the hydrological model, we indicate that ice-melt water constitutes an important water-supply component for Mapam Yumco and Paiku Co, in order to maintain a state close to equilibrium, whereas the water balance in the Nam Co and Tangra Yumco basins remains positive under ice-free conditions. These results highlight the benefits of

linking hydrological modeling with atmospheric-model output and satellite-derived data, and the presented approach can be readily transferred to other data-scarce closed lake basins, opening new directions of research. Future work should go towards a better assessment of the model-chain uncertainties, especially in this region where observation data are scarce.

1 Introduction

The drainage system of the interior Tibetan Plateau (TP) is characterized by numerous closed-lake (endorheic) basins. Because an endorheic lake basin integrates all hydrological processes in a catchment, lake-level or volume changes provide a cumulative indicator of the basin-scale water balance. While most of the lakes located in the central part of the TP are characterized by a water-level increase over recent decades (e.g., Zhang, et al., 2011; Phan et al., 2012), there are also several lakes with nearly stable or slightly decreasing water levels in the southern part of the TP. These high-elevation lakes are therefore considered to be one of the most sensitive indicators for regional differences in the water balance in the TP region (e.g., B. Zhang et al., 2013; G. Zhang et al., 2013; Song et al., 2014).

Neglecting the influence of long-term storage changes such as deep groundwater and lake–groundwater exchange, the net water balance of an endorheic lake basin with water supply from glaciers can be expressed as $\Delta V_{lake} = P_{lake} - E_{lake} + R_{land} + R_{glacier}$, where ΔV_{lake} is the lake-volume change (net annual lake-water storage), P_{lake} the on-lake precipitation, E_{lake} the evaporation rate from the lake, and R_{land}

and R_{glacier} are the runoff from non-glacierized land surfaces and from glaciers (in units of volume per unit time). Under constant climatic conditions, endorheic lakes will eventually tend towards a stable equilibrium ($\Delta V_{\text{lake}} = 0$), where the several water-balance terms are balanced (Mason, 1994). Lake-level changes thus result from a shift in the water input or output.

Due to the accelerated glacier mass loss, it has been hypothesized that lake-level increases are primarily due to an increased inflow of glacier meltwater (e.g., Yao et al., 2007; Zhu et al., 2010; Meng et al., 2012). Nevertheless, glacier runoff into lakes itself should not increase the overall water-volume mass on the TP, as indicated by GRACE satellite gravimetry data (G. Zhang et al., 2013). Furthermore, numerous lakes of the TP are not linked to glaciers (Phan et al., 2013), and the water-level changes of lakes without glacier meltwater supply in the 2000s were as high as those of glacier-fed lakes (Song et al., 2014). In other studies, increased precipitation and decreased evaporation were generally considered to be the principal factors causing the rapid lake-level increases (e.g., Morrill, 2004; Lei et al., 2013, 2014). Y. Li et al. (2014) argued for the importance of permafrost degradation on recent lake-level changes. Thus, recent studies addressing the controlling mechanism of lake-level fluctuations remain controversial.

In order to explore differences in the water balance of endorheic lake basins in the TP region, recent studies emphasize the urgency of the quantification of water-balance components by using hydrological models (e.g., Cuo et al., 2014; Lei et al., 2014; Song et al., 2014). Hydrological modeling studies of endorheic lake basins in the TP region are rare (e.g., Krause et al., 2010), principally due to a lack of hydro-climatological observations and limitations in spatial and temporal coverage of available gridded climate data (Biskop et al., 2012). The paucity of spatial information of climatological variables was addressed by Maussion et al. (2014), who developed a high-resolution (up to $10\,\text{km} \times 10\,\text{km}$) atmospheric data set for the 2001–2011 period, the High Asia Refined analysis (HAR). The HAR10 data set was successfully applied in surface energy balance/mass balance (SEB/MB) modeling studies (Mölg et al., 2014; Huintjes et al., 2015) and in a hydrological modeling study in the Pamir Mountains (Pohl et al., 2015), but has not yet been used as input for catchment-scale hydrological modeling studies in the central TP. The objective of this study is the hydrological modeling of endorheic lake basins across the southern-central part of the TP in order to

- analyze spatiotemporal patterns of water-balance components and to contribute to a better understanding of their controlling factors, and

- quantify single water-balance components and their contribution to the water balance, and obtain a quantitative knowledge of the key factors governing the water

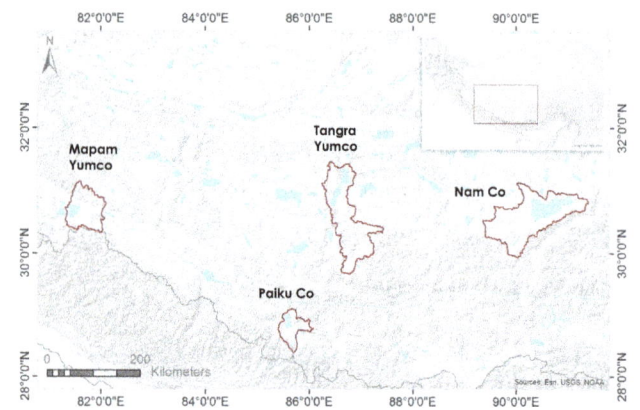

Figure 1. Location of the study region comprising four selected endorheic lake basins.

balance and lake-level variability during the 2001–2010 period.

Lakes Nam Co and Tangra Yumco with increasing water levels (i.e., positive water balance) and lakes Mapam Yumco and Paiku Co with stable or slightly decreasing water levels (i.e., stable or slightly negative water balance, respectively) were selected to investigate differences in the water-balance components. The paper is organized as follows. In Sect. 2, we describe the study area and the data used. Section 3 gives details of the hydrological modeling approach and, in Sect. 4, we present the modeling results and assess similarities and differences among the basins; in Sect. 5, the results, limitations and uncertainties of this study are discussed with respect to findings from other studies. Finally, Sect. 6 highlights the principal results and concludes with remarks on future research needs and potential future model applications.

2 Study area and data

2.1 Description of the study area

The study region comprises four endorheic lake basins along a west–east (W–E) lake transect in the southern-central part of the TP between 28–32° N and 81–92° E (Fig. 1). Basic characteristics of the selected lake basins are summarized in Table 1. Climatologically, the study region encompasses a semi-arid zone and is characterized by two distinct seasons: a temperate-wet summer season dominated by the Indian Monsoon and a cold-dry winter season determined by the westerlies. The mean annual air temperature (MAAT) lies between 0 and −3 °C and the mean annual precipitation ranges between 150 and 500 mm, with 60–80 % of this total occurring between June and September (Leber et al., 1995). The study region features a climate gradient, with increasingly cooler and drier conditions in a westward direction.

Due to the semi-arid and cold climate conditions as well as the complex topography, soils in the study area in general are

Table 1. Basic information of selected basins in the study region. Data sources are described in Sect. 2.2.

Lake name	Elev. (m a.s.l.)	Lake center Lat	Lake center Long	Basin area (km^2)	Lake area (km^2)	Land cover (%) Lake	Land cover (%) Glacier	Land cover (%) Grassland	Land cover (%) Wetland	Land cover (%) Barren land
Nam Co	4725	30°42	90°33	10 760	1950	18	2	39	8	33
Tangra Yumco	4540	31°00	86°34	9010	830	9	0.96	31	0.04	59
Paiku Co	4585	28°55	85°35	2380	270	10	6.5	43	0.5	40
Mapam Yumco	4580	30°42	81°28	4440	420	10	1.5	64	2.5	22

poorly developed and vegetation throughout the study area is generally sparse. The growing period lasts approximately 5 months, from late April or early May to late September or mid-October (B. Zhang et al., 2013). The highest mountain regions are covered by glaciers and permanent snow. Among all basins, the Paiku Co catchment exhibits the largest glacier coverage (6.5 % of the basin area). The areas covered by glaciers in the Nam Co, Tangra Yumco and Mapam Yumco basins accounts for 2, 1 and 1.5 % of catchment area. The lake area in the several basins corresponds to 18 % (Nam Co), 11 % (Mapam Yumco), 9.5 % (Paiku Co) and 9 % (Tangra Yumco). Based on GLAS/ICESat data, the lake levels for Nam Co and Tangra Yumco rose by approximately $0.25 \, m \, yr^{-1}$ between 2003 and 2009, whereas the lake levels for Paiku Co and Mapam Yumco slightly decreased by around $-0.05 \, m \, yr^{-1}$ (Zhang et al., 2011; Phan et al, 2012).

2.2 Data used

Because of limited availability of climatological data in the TP region, we used a new atmospheric data set for the TP, the HAR (Maussion et al., 2014) as input for the hydrological model. The HAR data sets were generated by dynamical downscaling of global-analysis data (final analysis data from the Global Forecasting System; data set ds083.2), using the Weather Research and Forecasting (WRF) model (Skamarock and Klemp, 2008). A detailed description of this procedure is given in Maussion et al. (2014). HAR products are freely available (http://www.klima.tu-berlin.de/HAR) in different spatial (30 km × 30 km and 10 km × 10 km) and temporal (hourly, daily, monthly and yearly) resolutions. In this study, we used the daily HAR10 data. In the WRF model version 3.3.1, which was used for the generation of the HAR10 data, the lake-surface temperature is initialized by averaging the surrounding land-surface temperatures. By analyzing the influence of the assimilation of satellite-derived lake-surface temperatures, Maussion (2014) found that the standard method of WRF leads to a much cooler lake than observed, which in turn has a strong influence on local climate. Therefore, the HAR10 data points over water surfaces were not included for hydrological modeling purposes.

The HAR10 precipitation output was compared to rain-gauge data and to Tropical Rainfall Measuring Mission (TRMM) satellite precipitation estimates by Maussion et al. (2014). They concluded that HAR10 accuracy in compar-

ison to rain gauges was slightly less than TRMM; however, orographic precipitation patterns and snowfall were more realistically simulated by the WRF model. HAR10 temperatures in the summer months are closer to ground observations than in winter (Maussion, 2014). Despite the winter cold bias, the overall seasonality is well reproduced (Maussion, 2014). The cold bias effect on the accuracy of the hydrologic-modeling results is assumed to be low, because hydrological processes governing lake-level changes are more critical during the other three seasons of the year.

Lake-surface water temperature (LSWT) estimates from the ARC-Lake v2.0 data products (MacCallum and Merchant, 2012) served as additional input for the hydrological modeling in the Nam Co and Tangra Yumco basins. ARC-Lake v2.0 data products contain daytime and nighttime LSWT observations from the series of (advanced) along-track scanning radiometers for the period 1991–2011. Daytime and nighttime MODIS land-surface temperature (LST) 8-day data at 1 km spatial resolution (MOD11A2) were averaged after a plausibility check to obtain mean daily LSWT time series for Paiku Co and Mapam Yumco, where no ARC-Lake v2.0 data were available.

Shuttle Radar Topography Mission (SRTM) 90 m digital elevation model (DEM) data (Farr et al., 2007) were retrieved from the Consortium for Spatial Information (CGAIR-CSI) Geoportal (http://srtm.csi.cgiar.org). We used the SRTM Version 4 data for derivations of catchment-related information such as catchment boundary, river network, flow accumulation and flow direction, as well as terrain attributes (slope and aspect).

For the Nam Co and Tangra Yumco basins, land-cover classifications were generated using Landsat TM/ETM+ satellite imagery. The land-cover classifications consist of five classes used for this analysis: water, wetland, grassland, barren land and glacier. For the Paiku Co and Mapam Yumco basins, land-cover information could be obtained from the Himalaya Regional Land Cover database (http://www.glcn.org/databases/hima_landcover_en.jsp). The Himalaya land-cover map was produced as part of the Global Land Cover Network – Regional Harmonization Program, an initiative to compile land-cover information for the Hindu Kush–Karakorum–Himalaya mountain range using a combination of visual and automatic interpretation of recent Landsat ETM+ data. The land-cover classes were reclassified according to the five classes mentioned above. Classes with similar characteristics (e.g., vegetation type, degree of vegetation cover) were consolidated into a single class.

Lake-level observations from 2006 to 2010 for Nam Co were provided by the Institute of Tibetan Plateau Research (ITP), Chinese Academy of Sciences (CAS), and used for model validation. However, lake-level values during the freezing (wintertime) periods are missing, because the lake-level gauge was destroyed by lake ice, and therefore rendered inoperable each winter. Thus, data are only available for the ice-free period (May/June–November/December). Unfortunately, the lake-level observation data contain an unknown shift between the consecutive years.

Due to the absence of continuous lake-level measurements, we obtained satellite-based lake-level and water-volume data for the four studied basins from the HydroWeb database (http://www.legos.obs-mip.fr/en/soa/hydrologie/hydroweb/) provided by LEGOS/OHS (Laboratoire d'Etudes en Geodesie et Oceanographie Spatiales (LEGOS) from the Oceanographie, et Hydrologie Spatiales, OHS) (Crétaux et al., 2011). LEGOS lake-level and water-volume data for the lakes included in this study were available for different time spans (see Table 2). The start and end date of each time series were taken from the same season (as far as available) in order to make lake levels or volumes comparable. Water-volume data calculated through a combination of satellite images (e.g., MODIS, Landsat) and various altimetric height level data (e.g., Topex/Poseidon, Jason-1) (Crétaux et al., 2011) were used for model calibration (see Sect. 3.3). The mean annual lake-level changes derived from LEGOS data for Nam Co, Tangra Yumco, Paiku Co and Mapam Yumco (0.25, 0.26, −0.07, −0.05 m yr^{-1}) are close to the change rates estimated by Zhang et al. (2011) (0.22, 0.25, −0.04, −0.02 m yr^{-1}) and Phan et al. (2012) (0.23, 0.29, −0.12, −0.04 m yr^{-1}) using GLAS/ICESat data (2003–2009) (Table 4, lower part).

MODIS snow-cover 8-day data of Terra (MOD10A2) and Aqua (MYD10A2) satellites at a spatial resolution of 500 m served for validation of the snow modeling. As proposed in the literature (e.g., Parajka and Blöschl, 2008; Gao et al., 2010; Zhang et al., 2012), we combined Terra and Aqua data on a pixel basis to reduce cloud-contaminated pixels. The cloud pixels in the Terra images were replaced by the corresponding Aqua pixel. For the period of time before the Aqua satellite was launched (May 2002), this combination procedure was not possible, and we used the original MODIS/Terra snow-cover data. After the combination procedure the cloud-cover percentage was on average less than 1–2 % for all basins.

3   Methods

3.1   Hydrological model concept and implementation

The challenge for hydrological modelers is to balance the wish to adequately represent complex processes with the need to simplify models for regions with limited data availability (Wagener and Kollat, 2007). Therefore, we selected a semi-distributed conceptual model structure, primarily following the J2000g model (Krause and Hanisch, 2009). The J2000g model is a simplified version of the fully distributed J2000 model (Krause, 2002). The main differences with J2000 are that complex process descriptions (e.g., soil-water dynamics) are simplified, leading to a reduced number of land-surface and calibration parameters in the J2000g model, and lateral flow processes between spatial model units and streamflow routing are not accounted for by the J2000g model. The J2000g was successfully applied for hydrological predictions in data-scarce basins (e.g., Deus et al., 2013; Knoche et al., 2014; Rödiger et al., 2014; Pohl et al., 2015), including a previous modeling study in the Nam Co basin (Krause et al., 2010).

The conceptual model presented here was realized within the Jena Adaptable Modelling System (JAMS) framework (http://jams.uni-jena.de/). An overview of JAMS, especially the JAMS software architecture and common structure of JAMS models, is given in Kralisch and Fischer (2012). Primarily, JAMS was developed as a JAVA-based framework for the implementation of model components of the J2000 model. During recent years, a solid library of single easily manageable components has been developed by implementing a wide range of existent hydrological-process concepts as encapsulated process modules and developing new model modules as needed. Due to the modular structure, the J2000g model could be easily adapted and extended according to the specific characteristics of endorheic lake basins in the TP region.

Meteorological data requirements for this study were daily times series of precipitation, minimum, maximum and average air temperature, solar radiation, wind speed, relative humidity and cloud fraction obtained from daily HAR10 data. Daily LSWT data served as additional input for the calculation of the long-wave radiation term over the lake surface. Process simulations were grouped into the following categories: (i) lake, (ii) land (non-glacierized) and (iii) glacier. A schematic illustration of the model structure and a detailed description of the model components are given in the Supplement.

In brief, we used the regionalization procedure implemented in J2000g for the interpolation of the HAR10 raster points (centroid of the raster cell) to each hydrological response unit (HRU). This combines inverse distance weighting (IDW) with an optional elevation correction. Net radiation was calculated following the Food and Agriculture Organization of the United Nations (FAO) proposed use of the

Table 2. Lake-level and water-volume changes derived from LEGOS data for the four studied lakes.

Lake name	Start date	Start volume (km^3)	Start level (m)	End date	End volume (km^3)	End level (m)	Δ Lake volume (km^3 yr^{-1})	Δ Lake level (m yr^{-1})
Nam Co	27 Sep 2001	1.3	4722.683	1 Oct 2010	5.3	4724.697	0.44	0.22
Tangra Yumco	7 Oct 2001	0	4533.997	25 Oct 2009	1.7	4535.987	0.21	0.25
Paiku Co	2 Jun 2004	0	4578.067	4 Mar 2008	−0.08	4577.768	−0.02	−0.07
Mapam Yumco	30 Oct 2003	0.02	4585.551	21 Nov 2009	−0.1	4585.231	−0.01	−0.05

Penman–Monteith model (Allen et al., 1998). We adapted the long-wave radiation part of the FAO56 calculation to the special high-altitude conditions on the TP, according to the recommendations of Yin et al. (2008), and implemented the commonly used approach for calculating net long-wave radiation over water surfaces (e.g., Jensen, 2010).

Potential evapotranspiration (PET) from land and snow surfaces (sublimation) is calculated based on Penman–Monteith (Allen et al., 1998). For the estimation of open-water evaporation rates from large lakes, we modified the Penman equation through the addition of an empirical estimation of the lake heat storage (Jensen et al., 2005). As suggested by Valiantzas (2006), we used the reduced wind function proposed by Linacre (1993) for the estimation of evaporation from large open-water body surfaces.

The simple degree-day snow modeling approach of the standard J2000g model version was replaced by the J2000 snow module that combines empirical or conceptual approaches with more physically based routines. This module takes into account the phases of snow accumulation and the compaction of the snow pack caused by snowmelt or rain on the snow pack. For a detailed description, see Nepal et al. (2014). The glacier module calculates ice melt according to an extended temperature-index approach (Hock, 1999).

Soil-water budget and runoff processes are simulated using a simple water storage approach (Krause and Hanisch, 2009). The storage capacity is defined from the field capacity of the specific soil type within the respective modeling unit. Actual evapotranspiration (AET) is calculated depending on the saturation of the soil-water storage, PET and a calibration parameter. The J2000g model generates runoff only when the soil-water storage is saturated. The partition into surface runoff and percolation depends on the slope and the maximum percolation rate of the respective modeling unit that can be adapted by a calibration parameter. The percolation component is transferred to the groundwater storage component. The groundwater module calculates base flow using a linear outflow routine and a recession parameter (Krause and Hanisch, 2009).

The lake module calculates the net evaporation (lake evaporation minus precipitation over the lake's surface area). The lake-water storage change is the sum of (i) direct runoff and base flow from each modeling unit of the non-glacierized areas and (ii) glacier runoff (snowmelt and ice melt, and rainfall over glaciers) from each glacier HRU minus lake net evaporation. For simplicity, the terms land runoff, glacier runoff and net evaporation are used to refer to several water-balance components. Because the J2000g model does not account for water routing and thus time delay of the discharge, the model is not fully suited for providing continuous and precise estimates of lake-water storage changes.

3.2 Delineation of spatial model entities

In order to provide spatially distributed information on landscape characteristics for the hydrological modeling, we applied the hydrological response unit (HRU) approach (Flügel, 1995). Using ArcGIS software, HRUs with similar hydrological behavior were delineated by overlaying topographic-related and land-cover information. Soil and hydro-geology information were not included in the overlay analysis, due to a lack of detailed data. The distribution concept applied represents the landscape heterogeneity with a higher spatial resolution in the complex high mountain areas (a large number of small polygons) than in the relatively flat terrains at the lower elevations (smaller number of large polygons). The total number of HRUs varies between 1928 (Paiku Co) and 8058 (Nam Co).

3.3 Model-parameter estimation and model evaluation

The J2000g model requires the definition of spatially distributed land-surface parameters describing the heterogenic land surface and the estimation of calibration parameters. Land-surface parameters were derived from field studies or literature values. The field capacity was derived as a function of the soil types obtained from our own field surveys. Due to the limited availability of soil information for the TP, soil parameters were distributed according to different land-cover and slope classes (Table 3).

Parameter-optimization procedures are difficult to apply in data-scarce regions such as the TP (e.g., Winsemius et al., 2009). Moreover, various parameter set combinations may yield equally acceptable representations of the (often limited) calibration data, which are referred to as the equifinality problem (e.g., Beven, 2001; Beven and Freer, 2001). Due to a lack of calibration data, we used default settings or parameter values given in the literature (see Table S1 in the Supplement).

Table 3. Soil parameters used as input for hydrological modeling.

Combination land cover–slope	Soil depth (cm)	Field capacity							
		Total (mm)	0–1 dm (mm dm^{-1})	1–2 dm (mm dm^{-1})	2–3 dm (mm dm^{-1})	3–4 dm (mm dm^{-1})	4–5 dm (mm dm^{-1})	5–6 dm (mm dm^{-1})	6–7 dm (mm dm^{-1})
Wetland	70	236	60	60	60	14	14	14	14
Grassland < 15°	70	120	18	18	18	18	16	16	16
Grassland > 15°	40	68	18	18	16	16	–	–	–
Barren land < 5°	20	14	7	7	–	–	–	–	–
Barren land > 5°	10	7	7	–	–	–	–	–	–

Following Mölg et al. (2014), we implemented a precipitation-scaling factor as an additional model parameter to account for (i) HAR10 precipitation overestimation related to atmospheric-model errors and/or (ii) sublimation of blowing or drifting snow that was neglected in the model. Due to the high uncertainty in the range of the precipitation-scaling factor in various regions of the TP (Huintjes, 2014; Mölg et al., 2014; Pohl et al., 2015), we performed model runs with precipitation-scaling factors varying between 0.3 and 1.0 with a 0.05 increment. Because the precipitation-scaling factor was judged to be the parameter that contributes the most to uncertainties in model results, all other climate forcing variables and model parameters were held constant. We compared simulated mean annual lake-volume changes of each model run with water-volume changes derived from remote-sensing data (Fig. 2). The dotted line in Fig. 2 indicates the lake-volume changes derived from LEGOS data (see Table 2). The model run with the minimum difference between modeled and satellite-derived lake-volume change was defined as a reference run and thereby was used for an assessment of model results. The "best" match between simulated and satellite-derived lake-volume change was achieved by applying the following precipitation-scaling factors: 0.80 (Nam Co), 0.75 (Tangra Yumco), 0.85 (Paiku Co) and 0.50 (Mapam Yumco). We discuss the possible reasons for the lower parameter value for the Mapam Yumco basin in Sect. 5.2.

Similar to the calibration process, data scarcity limited the establishment of rigorous and systematic validation tests. Because water-level measurements from Nam Co provide consistent time series between the months of June through November for the years 2006–2010, we chose this period for validation. Given the fact that water routing is not considered in the model, we compared mean monthly, instead of daily, water-level simulations and measurements. For the calculation of monthly average lake levels, the lake-level value of 1 June was set to zero in each year and the subsequent values were adjusted accordingly to make the lake-level changes during the June–November period of the years 2006–2010 comparable.

For an independent assessment of the snow model capabilities, we compared modeled snow-water equivalent (SWE) simulations with MODIS snow-cover data (see Sect. 2.2). Because MODIS data provide no information about the

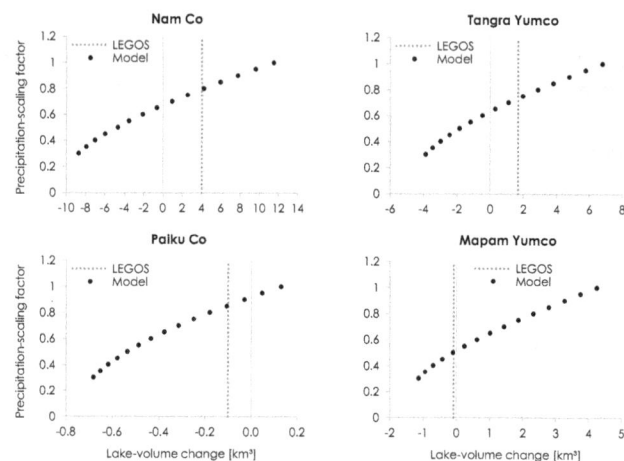

Figure 2. Model-simulated lake-volume changes for Nam Co, Tangra Yumo, Paiku Co and Mapam Yumco for the time periods given in Table 2 using precipitation-scaling factors varying between 0.3 and 1.0. The dotted line indicates lake-volume changes derived from remote-sensing data provided by LEGOS. The point where model dots are closest to the dotted line was taken as the precipitation-scaling factor for each basin.

amount of water stored as snow (i.e., SWE), this comparison was only possible in an indirect way by comparing the percent or fraction of snow-covered area (SCAF) derived from the model simulation and MODIS data. Any given spatial model unit was considered to be snow-covered on days when the amount of SWE was larger than a specific threshold (i.e., 1, 10, and 50 mm).

4 Results

Section 4.1 contains the comparison of simulated and measured water levels of lake Nam Co (Sect. 4.1.1) and of simulated snow-cover dynamics with MODIS for all four study basins (Sect. 4.1.2). Section 4.2 deals with the assessment of the modeling results regarding spatiotemporal variations of water-balance components (Sect. 4.2.1) and their contributions to each basin's water balance (Sect. 4.2.2) during the 2001–2010 period.

4.1 Model evaluation

4.1.1 Comparison of simulated and measured water levels of the Nam Co lake

Lake-level observations of Nam Co indicate a distinct seasonal dynamic with continuously increasing lake levels during the months of June through September caused by runoff from the non-glacierized land surface and glacier areas, a lake-level peak in September and decreasing lake levels from October on primarily caused by lake evaporation. The overall seasonal dynamic during the June–November period is well represented by the J2000g model ($r = 0.81$) (Fig. 3). However, the model overestimates the lake level for the month of November, except for the year 2006.

In general, the magnitude of the lake-level evolution is less well simulated than its timing. The comparison reveals a non-systematic pattern (Fig. 3). In 2006, the model is not able to reproduce the observed increase in lake levels. The substantial lake-level rise of Nam Co in 2008 simulated by the model compares well with observed data. However, the lake-level increase in 2009 is slightly overestimated. The absolute deviation between observed and simulated relative changes of monthly averaged lake levels during the June–November period ranges between −0.31 m (2006) and 0.30 m (2009). The simulated relative lake-level change during the June–November period averaged over the years 2006 to 2010 is 0.41 m, which is approximately 0.05 m higher than the measured one (0.36 m).

4.1.2 Comparison of the simulated snow-cover dynamics with MODIS

The comparison of mean monthly values of modeled snow-covered area fraction (SCAF) (SWE > 1 mm) and MODIS indicates that the model captures seasonal variability quite well. However, there are large deviations in the magnitude. The modeled SCAF (SWE > 1 mm) is generally greater than the MODIS SCAF, with higher deviations in the Mapam Yumco and Paiku Co basins (Nam Co: 30 % vs. 22 %; Tangra Yumco: 17 % vs. 8 %; Mapam Yumco: 54 % vs. 28 %; Paiku Co: 49 % vs. 20 %; Fig. 4). During the winter months November through April, the overestimation by the model (up to a factor of 2) is generally higher than during the summer season. During the months May through October, the modeled SCAF (SWE > 1 mm) in the Nam Co and Tangra Yumco basins is even approximately 50 % lower compared to MODIS. Figure 4 indicates how sensitive the results are by using different thresholds for the amount of SWE to depict an area as snow-covered in the model. The use of higher thresholds (SWE > 10, 50 mm) for derivations of SCAF from the model reduces the overestimation, but also leads to an underestimation of the SCAF in early winter in most basins (Fig. 4). A threshold larger than 10 mm seems to be inappropriate for deriving SCAF from the SWE simulations. It is

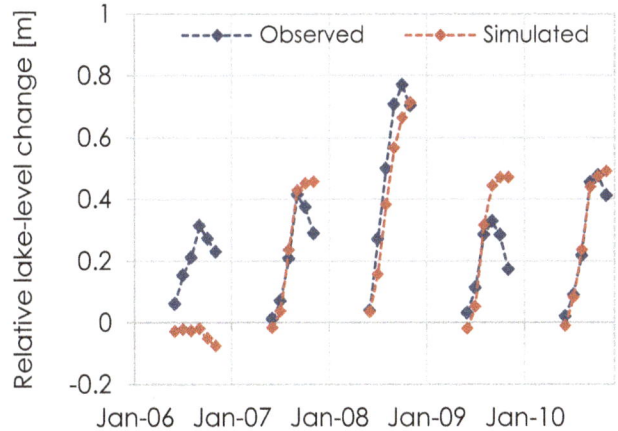

Figure 3. Monthly averaged lake-level observations from Nam Co (blue) vs. simulated lake levels (red) for the June–November period of the years 2006 through 2010.

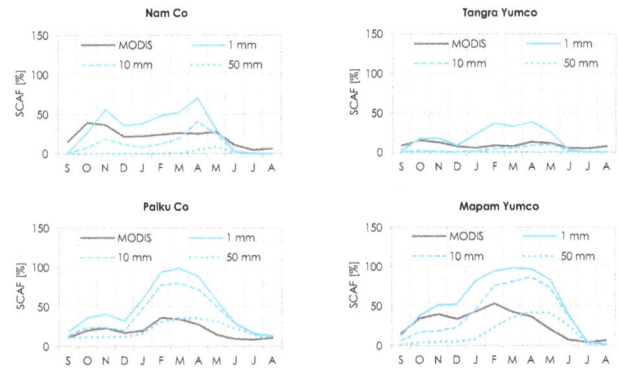

Figure 4. Mean monthly modeled–derived SCAF (blue) using SWE > 1 mm (solid line), > 10 mm (dashed line), and > 50 mm (dotted line) vs. SCAF derived from MODIS (black) for the four study basins.

more likely that the J2000g model overestimates SCAF. This will be discussed later in Sect. 5.2.

4.2 Comparative analysis of the four selected lake basins

4.2.1 Spatiotemporal patterns of hydrological components

The percentage of the precipitation occurring during the wet season (June through September) is more than half of the annual precipitation in all basins. Specifically, June-through-September precipitation is approximately 80 % of the annual total in the Nam Co and Tangra Yumco basins and around 60 % in the Paiku Co and Mapam Yumco basins (Fig. 5a). This indicates a higher influence of the westerlies in the Paiku Co and Mapam Yumco basins. As simulated by the model, snow accumulation in the basins generally occurs beginning in mid-September, reaching a first smaller peak be-

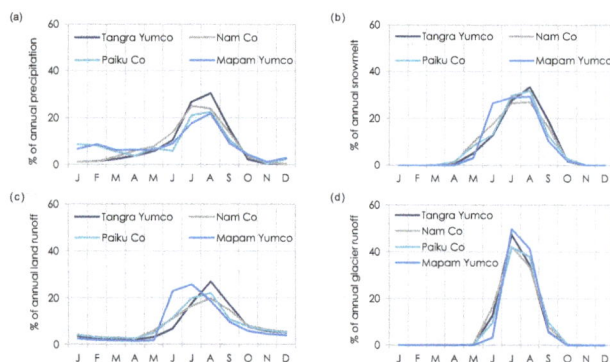

Figure 5. (a) Monthly percentage of annual precipitation, (b) snowmelt from non-glacierized land areas, (c) runoff from non-glacierized land areas, and (d) glacier runoff for the four studied basins.

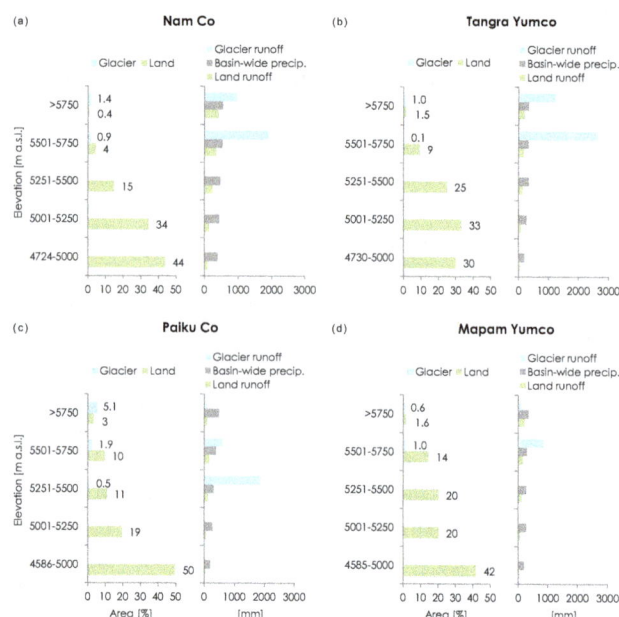

Figure 6. (a–d, left panels) Hypsometry of glacier and non-glacierized areas based on mean elevations of respective model entities for the four studied basins. (a–d, right panels) Variability of precipitation and runoff from glacier and non-glacierized areas related to altitude for the four studied basins.

tween October and November and the maximum peak between April and May, followed by a rapid decrease in snow between May and June and a slower rate of decrease until September. In the Mapam Yumco basin, simulated snowmelt starts later and occurs over a shorter time period compared to the other basins (Fig. 5b). This can be explained by lower air temperatures in this basin.

About 80 % of simulated annual terrestrial actual evapotranspiration (AET) occurs during the growing season (May–October). Modeled AET has its maximum in July when the availability of soil water and energy is highest. The seasonal cycle of modeled lake evaporation is influenced by seasonal heat-storage changes in the lakes. The released heat in fall acts as an energy source for evaporation. Thus, the evaporation is higher in fall than in spring.

Approximately 70 % of annual precipitation is released to the atmosphere through AET and does not contribute to the runoff in all basins. Discharge from non-glacierized land areas is concentrated during the wet season (~ 80 % of annual runoff occurs during May through October). Runoff starts to increase in spring with the beginning of snowmelt. The land runoff peak in the Mapam Yumco basin occurs 1 month earlier (between June and July) compared to the other basins (Fig. 5c), because of a higher contribution of snowmelt to the discharge. Glacier runoff occurs during June through September in all basins (Fig. 5d), but with a later beginning and a shorter duration of the melt season in the Mapam Yumco basin due to the colder climate conditions.

Table 4 (upper part) summarizes annual means of modeled water-balance components for the 2001–2010 period for each basin. The annual mean of the model-simulated lake evaporation rates varies between 700 and 900 mm yr^{-1} for the four basins averaged over the 10-year study period. Because of unlimited water availability, the modeled mean annual lake evaporation is substantially higher than the land AET (see Table 4, upper part). Due to higher precipitation amounts in the eastern part of the study region, the simulated mean an-

nual AET is higher in the east (~ 290 mm in the Nam Co basin) than in the west (~ 170 mm in the Mapam Yumco basin) (Table 4, upper part).

Impacted by the decreasing precipitation gradient spatially from east to west, the model-simulated mean annual land runoff in the Nam Co basin (~ 130 mm) is estimated to be more than twice that in the Mapam Yumco basin (~ 60 mm) during the study period (Table 4, upper part). The combination of various influencing variables such as local climate, topography, land cover, soil and hydro-geological properties results in a spatially heterogeneous pattern of runoff generation within the catchments. Figure 6 illustrates the altitudinal dependence of the mean annual basin-wide precipitation total and runoff from glaciers and non-glacierized land areas, as computed by the J2000g model. The area–altitude relation (hypsometry) for glacier and non-glacierized areas is based on mean elevations of the respective model entities. Larger precipitation amounts in the high mountainous and hilly headwater areas result in higher land runoff estimates compared to lower elevation areas (Fig. 6). Indeed, the increase in land runoff with altitude is higher than the elevation-dependent increase in precipitation. The non-glacierized high-elevation areas characterized by sparse vegetation, poorly developed soils, steep topography and lower air temperatures indicate smaller soil-water contents and lower AET rates compared to lower elevation bands, resulting in higher runoff rates.

Table 4. Mean annual water-balance components, water budget and lake-level changes for the four studied lake basins for the study period 2001–2010 derived from the reference run. The variation ranges of the mean annual water-balance components correspond to model runs with precipitation-scaling factors ±0.05.

		Western basin → eastern basin			
		Mapam Yumco	Paiku Co	Tangra Yumco	Nam Co
Water-balance components (mm yr^{-1})					
Land	Precipitation	230 (±24)	250 (±15)	300 (±20)	420 (±27)
	AET	170 (±9)	180 (±5)	210 (±7)	290 (±8)
	Land runoff	60 (±14)	70 (±8)	90 (±12)	130 (±18)
Glacier	Precipitation	330 (±33)	480 (±28)	330 (±22)	560 (±35)
	Glacier runoff	600 (±8)	320 (±4)	1320 (±12)	1320 (±4)
Lake	On-lake precipitation	90 (±9)	140 (±8)	150 (±10)	290 (±18)
	Lake evaporation	710 (−)	910 (−)	840 (−)	770 (−)
	Net evaporation	620 (±9)	770 (±8)	690 (±10)	580 (±18)
Water budget (km^3 yr^{-1})					
Water gain	Land runoff (% of total basin runoff)	0.23 (85)	0.14 (70)	0.70 (86)	1.15 (81)
	Glacier runoff (% of total basin runoff)	0.04 (15)	0.06 (30)	0.11 (14)	0.27 (19)
Water loss	Net evaporation	−0.26	−0.22	−0.57	−0.95
Net water budget	Lake-volume change	0.01	−0.02	0.24	0.47
Lake level (m yr^{-1})	Simulated	0.02	−0.07	0.29	0.24
	Zhang et al. (2011) (GLAS/ICESat 2003–2009)	−0.02	−0.04	0.26	0.25
	Phan et al. (2012) (GLAS/ICESat 2003–2009)	−0.043	−0.118	0.291	0.230
	LEGOS*	−0.05	−0.07	0.25	0.22

* Mean annual lake-level rates for the studied basins correspond to the following time periods: Nam Co – 2001–2010; Tangra Yumco – 2001–2009; Paiku Co – 2004–2008; Mapam Yumco – 2003–2009.

In all studied basins, the runoff from glacier areas located in lower-elevation zones (< 5750 m a.s.l.) significantly exceeds the land runoff in the same elevation zones (Fig. 6), due to high ice-melt rates in the ablation areas. Because of lower temperatures and higher snowfall rates at higher elevations, the modeled glacier runoff decreases with altitude. The modeled mean annual glacier runoff averaged over all glacier HRUs in the Nam Co and Tangra Yumco basins (∼ 1300 mm) is considerably higher than in the Paiku Co (∼ 300 mm) and Mapam Yumco (∼ 600 mm) basins. This is judged to be caused by lower air temperatures (∼ 2 °C less) in the glacier areas of the Paiku Co and Mapam Yumco basins.

4.2.2 Contributions of the individual hydrological components to the water balance

Table 4 (lower part) summarizes the model-simulated mean annual lake-volume and level changes and the contribution of non-glacierized land, glacier, and lake areas to the total water budget during the 2001–2010 study period. Comparative values for the mean annual lake-level changes derived from remote-sensing data are also given in Table 4 (lower part). The contribution of glacier runoff to the total basin runoff volume in the Nam Co (19 %), Tangra Yumco (14 %) and Mapam Yumco (15 %) basins is relatively low compared to the runoff contribution from non-glacierized land areas. The glacierization in the Paiku Co basin is about 2 to 5 times larger than in the other three basins, but the glacier-melt contribution to the total basin runoff volume is only around twice

as high (30 %) due to lower glacier-melt rates. Despite the generally higher glacier contribution in Paiku Co, the water balance is slightly negative during the study period (Table 4, lower part). The water loss for Paiku Co exceeds the water gain by 10 %. In contrast, the total water inflow in the Tangra Yumco and Nam Co basins exceeds the water loss by factors of 1.4 or 1.5, respectively. In the Mapam Yumco basin the water gain and loss terms tend to balance each other out (Table 4, lower part), based upon the model simulation.

In order to better predict and understand the role of glaciers in the mean annual water balance, a hypothetical scenario with ice-free conditions was evaluated through model simulations for each lake basin. Therefore, the land-cover class of all glacier HRUs was changed to barren land. In the absence of glaciers, the total runoff volumes in the Nam Co and Tangra Yumco basins would be about 13 % lower than with an ice-melt water contribution during the 2001–2010 period (compared to the reference run). Thus, the mean annual lake-level increases of Nam Co and Tangra Yumco would be reduced from 0.24 to 0.15 m and from 0.29 to 0.17 m, respectively. In the Mapam Yumco and Paiku Co basins, the total runoff volumes would decrease by approximately 30 % and the resulting mean annual lake-level changes would change from 0.02 to −0.18 m and from −0.07 to −0.25 m, respectively, under ice-free conditions. From this latter evaluation, it can be concluded that the mean annual net water budget would noticeably change without ice-melt water contribution; however, the water balance in Nam Co and Tangra Yumco remains positive.

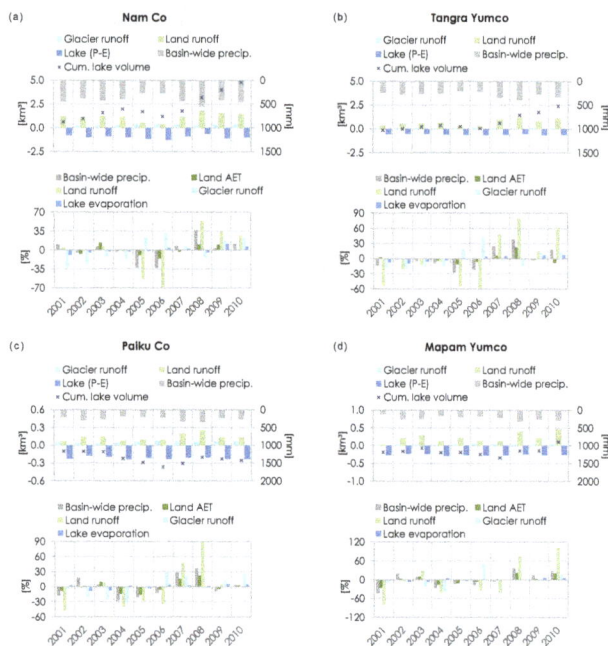

Figure 7. (**a–d**, upper panels) Cumulative lake-volume change (km^3), contribution of several water-balance components (km^3) to lake-volume change and annual basin-wide precipitation amounts (mm yr^{-1}) for the four studied basins. (**a–d**, lower panels) Annual percentage deviations from the 10-year average of several water-balance components for the four studied basins.

Based upon the J2000g modeling results, the differences in the water balance among the four studied lakes are primarily caused by relatively higher land runoff contributions in the Nam Co and Tangra Yumco basins compared to the Paiku Co and Mapam Yumco basins. This is related to relatively higher precipitation totals in the Nam Co and Tangra Yumco basins compared to the other two basins during the 2001–2010 period.

Figure 7a–d (upper panels) illustrates the yearly water contribution in km^3 of each land-cover type (land, glacier, and lake) for the 2001–2010 period. The annual percentage deviations from the 10-year average of several hydrological system components are presented in Fig. 7a–d (lower panels). Over the study period, annual relative lake-volume changes in the four basins indicate similar patterns. A relatively high correlation of lake-volume changes is found between the Nam Co and Tangra Yumco basins ($r = 0.82$). These are the basins with a higher proportion of June-through-September precipitation compared to the Paiku Co and Mapam Yumco basins.

The modeled annual lake-volume changes of all four lakes are highly correlated with inter-annual variations of land runoff ($r \approx 0.99$). The year-to-year variability of runoff from non-glacierized land surfaces, in turn, is strongly related to inter-annual variations of precipitation ($r \approx 0.92$). Inter-annual variability of lake evaporation is low in all four stud-

ied basins and not correlated with lake-level changes. Thus, lake evaporation seems to have a minor impact on inter-annual lake-level variations during the study period. There is also no correlation between annual glacier-melt amounts and lake-volume changes for the four basins. This suggests that glacier-melt runoff is not the main driving force for inter-annual lake variations during the last decade. Although the modeled annual glacier runoff is greater than the 10-year average in the year 2006 in all basins, lower precipitation amounts lead to less land runoff, causing a lake-volume decrease in this year in all basins. In contrast, the year 2008 is judged as having anomalous conditions, with modeled precipitation and land runoff substantially above average and with below-average glacier melt, resulting in a lake-volume increase in all basins. Differences in annual lake-volume changes among the basins are caused principally by regional differences in the inter-annual variations of precipitation.

5 Discussion

5.1 Comparison with other studies

5.1.1 Estimation of the water-balance components

Due to the scarcity of field measurements, model simulations of water-balance components are limited in the TP region. Evaporation over lake surfaces has been estimated for only a few lakes on the TP, based on model simulations (e.g., Morrill, 2004; Haginoya et al., 2009; Xu et al., 2009; Yu et al., 2011). Mean annual lake-evaporation estimates vary between 700 and 1200 mm. The lake-evaporation rates simulated with the J2000g model (between 710 and 910 mm yr^{-1}) are within this range (Table 4). There are only a few studies for the TP for assessing the actual evapotranspiration over alpine grassland, based on measurements and model simulations (e.g., Gu et al., 2008; Yin et al., 2013; Zhu et al., 2014). Yin et al. (2013) estimated AET over the entire TP using meteorological data available between 1981 and 2010 from 80 weather stations as model input for the Lund-Potsdam-Jena dynamic vegetation model (Sitch et al., 2003). For the southern-central TP, the simulated mean annual AET ranges from 100 to 300 mm, with generally higher values in the east and lower values in the drier regions in the west. Our simulated AET estimates for the four basins vary between 170 and 290 mm yr^{-1}, decreasing from east (Nam Co basin) to west (Mapam Yumco basin) (Table 4). This compares favorably with the study reported by Yin et al. (2013).

Using a simplified procedure, Yin et al. (2013) developed spatial patterns of the surface-water budget over the entire TP for the 1981–2010 period by estimating the difference between precipitation and AET (P-AET). The results revealed that P-AET depends on climate regimes and gradually decreases from the east (~ 150 mm yr^{-1}) to the west (~ 50 mm yr^{-1}) in the study region. Our model simulations indicate quite similar runoff patterns compared to the findings of Yin et al. (2013), with decreasing annual means from the east (~ 130 mm in the Nam Co basin) to the west (~ 60 mm in the Mapam Yumco basin). The calculated AET / precipitation ratio of around 0.7 in all basins agrees well with study results from Gu et al. (2008).

The mean annual glacier runoff of 1320 mm, simulated with the J2000g model for the Nam Co basin, compares quite well with estimated glacier-melt quantities for the Zhadang glacier in the Nam Co basin using more complex SEB/MB models (Mölg et al., 2014: 1375 mm yr^{-1}; Huintjes et al., 2015: 1325 mm yr^{-1}).

5.1.2 Factors controlling the water balance and lake-level variability

Many studies emphasize the importance of glacier-meltwater contribution to the water budget of Tibetan lakes (e.g., Zhu et al., 2010). However, only a few studies have quantitatively estimated glacier-meltwater contribution to total runoff in the TP region, due to the difficulty in estimating glacier-volume changes (B. Li et al., 2014). B. Li et al. (2014) estimated a glacier-runoff contribution of 15 % of the total runoff (during 2006–2011) in a sub-basin of the Nam Co basin, the Qugaqie basin (8.4 % glacier coverage), using an energy-balance-based glacier-melt model and the Gridded Subsurface Hydrologic Analysis (GSSHA) model (Downer and Ogden, 2004). Based upon the J2000g model results, the glacier contribution ranges between 14 and 30 % in the four studied basins (1–6 % glacier coverage) during the 2001–2010 period. This range of values is higher than that computed by B. Li et al. (2014) considering also the lower percentage of the basin area covered by glaciers in our study basins.

Simulated glacier-meltwater contribution is generally lower compared to the runoff contribution from non-glacierized areas. However, glaciers make an important contribution to the water budget during the 10-year period considering the small extent of ice-covered areas in the four studied lake basins. Indeed, the water balance in the Nam Co and Tangra Yumco basins would also be positive without ice-meltwater contribution during the study period, based on the results of the ice-free scenarios (Sect. 4.2.2). Thus, the question arises why Nam Co and Tangra Yumco indicate a non-equilibrium state, whereas Paiku Co and Mapam Yumco are at a state close to the hydrologic equilibrium.

Endorheic lakes respond to climatic changes to maintain equilibrium between input and output, and to reach steady state. Due to the time lag of lakes in responding to climatic changes, this modeling study cannot confirm whether or not the shift towards a positive water balance in the Nam Co and Tangra Yumco basins or the negative shift in the water balance of the Paiku Co basin, respectively, was primarily caused by changes in precipitation, glacier melt, evapotranspiration, etc. However, inter-annual lake-level variations are highly positively correlated with precipitation and land

runoff. This supports the assumption of other studies (e.g., Lei et al., 2014; Y. Li et al., 2014; Song et al., 2014) that increasing precipitation is the primary factor causing lake-level increases in the central TP (where Nam Co and Tangra Yumco are located). The relative stability or slight lake-level declines in the marginal region of the TP (where Paiku Co and Mapam Yumco are located) seem to be related to relatively stable or slightly decreased precipitation (e.g., Lei et al., 2014). Both changes in large-scale circulation systems and local circulation are assumed to be responsible for spatially varying changes in moisture flux over the TP (e.g., Gao et al., 2014, 2015). However, these factors are still under debate and further research is needed (Gao et al., 2015).

The potential evaporation decreased in most areas on the TP during 1961–2000, primarily caused by decreasing wind speeds (Xie and Zhu, 2013). A decreasing trend in potential evaporation before 2000 might have resulted in rising lake levels in the central TP. However, this factor did not prevent the lake shrinkage along the southwestern periphery of the TP, indicating that lake evaporation is not a primary factor for explaining the spatial differences of lake-level changes between the central and southern TP (Lei et al., 2014). In addition, Y. Li et al. (2014) argued that recent rapid lake expansion in the central TP cannot be explained by changes in potential evaporation, because the overall increasing tendency of potential evaporation in the TP region after 2000 (Yin et al., 2013; Y. Li et al., 2014) would negate the effect of increasing precipitation on lake levels.

Under the assumption that glacier-melt runoff increased during the last decades due to climate warming, it is very likely that glacier-meltwater supply augmented the precipitation-driven lake areal expansion in the central TP region (Song et al., 2014). In the Mapam Yumco and Paiku Co basins, glacier-meltwater discharge might have mitigated lake-level decline and acted as a regulating factor (Ye et al., 2008; Nie et al., 2012).

Y. Li et al. (2014) suggest that spatial variations in lake-level changes might be related to different distributions and types of permafrost. Most lakes in the central-northern TP with continuous permafrost are rapidly expanding, whereas lakes in the southern region with isolated permafrost are relatively stable or slightly decreasing. Thus, accelerated permafrost melting might have contributed to the rapid lake expansion in the central and northern TP subregions (Y. Li et al., 2014). The water contribution from permafrost will become limited when ground temperature remains above the melting point of the frozen soil and the water held in the frozen soil has been released (Y. Li et al., 2014). Y. Li et al. (2014) suggest that the permafrost-meltwater contribution may have already become limited in the southern TP. However, this is difficult to corroborate given the absence of observational data for the studied lake basins. The questions remain (i) how large the volume of water released due to thinning and thawing of permafrost is, and (ii) to what extent it can modulate basin runoff. These cannot be answered without adequate information about permafrost occurrence, thickness and ice content in the studied basins.

Differences in the response time or, in other words, the time required to reach an equilibrium state, could also be a reason for observed differences in lake-level changes. Based upon remote-sensing data, the lake areas of Nam Co and Tangra Yumco expanded by 4.6 and 1.8 %, respectively, between 1970 and 2008 (Liao et al., 2013). The lakeshore slopes of Tangra Yumco are steeper compared to Nam Co. Steep-sided lakes have a longer equilibrium response time, because of a lower rate of change of the lake area with volume (Mason, 1994). Based upon paleo-shorelines, the post-glacial lake-level high of Nam Co and Tangra Yumco was about 29 m (Schütt et al., 2008) and 185 m (Rades et al., 2013), respectively, above the present-day lake level, supporting the assumption that Nam Co has a shorter response time to compensate for the increment in net inflow (i.e., faster and stronger reaction of its lake area). Moreover, the water supply coefficient (basin area / lake area ratio) for Nam Co is smaller than for the Tangra Yumco basin (5.5 vs. 11.0).

The lake extent of Paiku Co and Mapam Yumco decreased by 3.7 and 0.8 %, respectively, between 1970 and 2008 (Liao et al., 2013). The lakeshores of Mapam Yumco are generally flatter compared to Paiku Co. However, due to only small lake-area variations of Mapam Yumco during the last decades (Liao et al., 2013), differences in lake morphology seem not to be the reason for the relative stability of the recent water levels of Mapam Yumco. Differences in the water-supply coefficient of the Paiku Co and Mapam Yumco basin are quite low (8.8 vs. 10.6).

Based upon results of other studies in the region (e.g., Lei et al., 2013), the effects of upwelling and downwelling groundwater related to fault zones and lake–groundwater exchanges on the water balance were assumed to be negligible. However, Zhou et al. (2013) suggest that water leakage related to seepage might play an essential role in the hydrological cycle of the TP, due to the large numbers of lakes and the sub-surface fault system in the TP region. Groundwater outflow from Mapam Yumco to Langa Co (located only a few kilometers to the west about 15 m below Mapam Yumco) cannot be excluded and could be a reason for the relatively stable water levels of Mapam Yumco. In the more recent past, Mapam Yumco and Langa Co were connected by the natural river Ganga Chu having an extent of about 10 km. However, currently there is no surface outflow (Liao et al., 2013).

5.2 Limitations and uncertainties

Hydrological modeling is hindered by systematic or random model-input errors, model-parameter uncertainty and model-structure inadequacies (Sivapalan, 2003). As stated in many studies (e.g., Knoche et al., 2014; Pohl et al., 2015), precipitation input is the primary source of uncertainty in hydrological modeling studies in data-scarce regions. HAR10 data have been successfully used as modeling input in vari-

ous studies (Huintjes, 2014; Mölg et al., 2014; Huintjes et al., 2015; Pohl et al., 2015). However, these studies also needed to apply a precipitation-scaling factor of less than 1. Maussion et al. (2014) could not find a systematic bias in comparison with station observations, but it is probable that overestimation of precipitation amounts occurs at high altitudes.

The precipitation-scaling factors were kept constant for the entire 10-year period, because there is no opportunity to derive varying scaling factors for individual years, due to a lack of observations in the lake basins included in this study. This may have an impact on inter-annual variations of modeling results. The non-systematic deviations between simulated and measured lake levels of Nam Co (Fig. 3) might be related to a non-systematic error pattern in the HAR10 precipitation data. The primary issue is that HAR10 precipitation cannot be validated to a sufficient degree, because available data are for stations that are located at lower elevations, and no accuracy assessment can be done for the higher-elevation zones where the study basins are located. The comparison with available station data suggests that the accuracy of the precipitation data is probably regionally dependent (Maussion et al., 2014). This makes it difficult to find a fixed precipitation-scaling factor that is applicable for different regions of the TP.

As described in Sect. 3.3, we conducted multiple model runs using precipitation-scaling factors between 0.3 and 1.0, seeking a precipitation-scaling factor that best simulates satellite-derived lake-volume changes. There may be errors in the satellite-derived water-volume data, which in turn might have affected the estimation of the precipitation-scaling factor and thereby the accuracy of model results. However, the precipitation-scaling factors obtained for the Nam Co (0.80), Tangra Yumco (0.75) and Paiku Co (0.85) basins are relatively close to the scaling factor used for the Zhadang glacier in the Nam Co basin in the study of Mölg et al. (2014). They found very good agreement between glacier mass-balance model calculations and available in situ measurements by applying a precipitation-scaling factor of 0.79. This gives us confidence that the scaling factors used in our study seemed to be within an acceptable range.

The relatively low precipitation-scaling factor of 0.50 obtained for the Mapam Yumco basin seems to be plausible when comparing HAR10 precipitation with weather station data of Burang (30°17' N, 81°15' E, ∼ 30 km to the south, the closest station with available data) published in Liao et al. (2013). The mean annual precipitation total of Burang is 150 mm yr^{-1} for the period 2001–2009, whereas the nearest HAR10 point gives a mean annual precipitation amount of 330 mm yr^{-1}. Huintjes (2014) also found that a reduction of the precipitation by more than 50 % leads to more reliable mass-balance results for the Naimona'nyi glacier (Gurla Mandhata, southwestern TP), which is located close to the Mapam Yumco basin.

Uncertainty also arises from the fact that the precipitation-scaling factor can compensate for not only input data errors, but also model-structure inadequacies. Blowing-snow sublimation was neglected in our modeling approach, due to the complexity of this process in complex terrain (Vionnet et al., 2014). However, wind-induced sublimation of suspended snow above the snow pack can be a significant water loss to the atmosphere (e.g., Bowling et al., 2004; Strasser et al., 2008; Vionnet et al., 2014). Vionnet et al. (2014) simulated total sublimation (surface + blowing snow) in Alpine terrain (French Alps) using a fully coupled snowpack–atmosphere model. They estimated that blowing-snow sublimation is two-thirds of total sublimation. This process is judged to be important in the study area, due to the relatively dry near-surface conditions and relatively higher wind speeds occurring during the winter months. Thus, the low values of the scaling factor applied in the Mapam Yumco basin (0.5) and in the study of Pohl et al. (2015) (0.37) might be an indication that drifting-snow sublimation plays a greater role in regions that are more strongly influenced by westerlies.

The omission of processes such as snow redistribution by wind and avalanches and snow loss by blowing-snow sublimation may affect snow-cover patterns as well as the magnitude and timing of melt runoff (Pellicciotti et al., 2014). This could also be a reason for the larger areal snow-cover extent in the model simulation during the winter season compared to MODIS (Sect. 4.1.2). Explanations for lower SCAF values of the model during the summer period could be related to the fact that the MODIS/Terra data are collected only in the morning (10:30) rather than at several times during the day. That means that MODIS indicates snow cover on days when snow was accumulated during the previous night or early morning but which might be sublimated or melted later during the day (Kropacek et al., 2010).

Given the limited data availability, further assumptions and simplifications in the model were required. The currently implemented glacier-melt model component according to Hock (1999) is a simple, robust and easy-to-use methodology that does not account for the transformation of snow into ice. Thus, simulated snowmelt amounts on glacier surfaces might be overestimated. Because glacier-volume changes are not considered in J2000g, unrealistic amounts of glacier-meltwater could be generated. However, the impact of this effect on model results is assumed to be small over the 10-year period. The consideration of glacier-volume changes would be of higher importance for long-term model simulations.

Effects of lake–groundwater interactions were neglected in the model, because the quantification of flow between aquifer systems and a deep lake is difficult (Rosenberry et al., 2014). However, it is unclear whether and to what extent intermittent (at irregular time intervals) exfiltration and infiltration processes might occur, thereby impacting water-level fluctuations. The stated values of lake–groundwater exchange rates do strongly vary within the literature by more than 5 orders of magnitude (Rosenberry et al., 2014). The lack of consideration of lake–groundwater interactions could be the reason that the observed lake-level decrease of Nam Co during the

months of October and November is not well represented by the model. If lake levels rise higher than adjacent groundwater levels, lake water may move into the adjacent lakeshores' subsurface. This additional storage factor would basically have a dampening effect on lake-level dynamics. However, in view of multi-annual lake changes, lake–groundwater exchanges are assumed to be negligible.

6   Conclusions and outlook

Hydrological modeling is required to allow for a quantitative assessment of differences in the water balance and thus a better understanding of the factors affecting water balance in the TP region. Addressing this research need, we developed a modeling framework integrating atmospheric-model output and satellite-based data, and applied it to four selected endorheic lakes across the southern-central part of the TP. The J2000g hydrological model was adapted to the specific characteristics of endorheic lake basins in the TP region. The model-derived HAR10 atmospheric data and satellite-derived lake-water surface temperature (Sect. 2.2) served as input for the modeling period 2001–2010. Due to missing continuous lake-level in situ data, we used satellite-derived lake-volume changes as a model-performance criterion.

The adapted J2000g model version reasonably captured seasonal dynamics of relevant hydrological processes. Water-balance estimates of individual years should be interpreted with care, due to possible unsystematic error patterns in HAR10 precipitation. Nevertheless, uncertainties that appear to be related to the precipitation-scaling factor should not affect the overall conclusions drawn from model-application results, as discussed in Sect. 5.2.

The major outcomes can be summarized as follows.

The seasonal hydrological dynamics and spatial variations of runoff generation within the basins are similar for all lake basins; however, the several water-balance components vary quantitatively among the four basins.

Differences in the mean annual water balances among the four basins are primarily related to higher precipitation totals and attributed runoff generation in the basins with a higher monsoon influence (Nam Co and Tangra Yumco).

The glacier-meltwater contribution to the total basin runoff volume (between 14 and 30 % averaged over the 10-year period) plays a less important role compared to runoff generation from rainfall and snowmelt in non-glacierized land areas. However, considering the small part of glacier areas in the study basins (1–6 %), glaciers make an important contribution to the water balance.

Based upon hypothetical ice-free scenarios in the hydrological model, ice-melt water constitutes an important water-supply component for basins with lower precipitation (Mapam Yumco and Paiku Co) in order to maintain a state close to equilibrium, whereas the water balance in the basins with higher precipitation (Nam Co and Tangra Yumco) would still be positive under ice-free conditions.

Precipitation and associated runoff are the main driving forces for inter-annual lake-level variations during the 2001–2010 period. Both are highly positively correlated with annual lake-level changes, whereas no correlation is found between inter-annual variability of lake levels and glacier runoff or lake evaporation.

For the 10-year modeling period used in this study, it is not possible to draw definitive conclusions about the hydrological changes that might have led to imbalances in the water budgets of the four studied lakes. However, the model results support the assumption of other studies that contrasting patterns in lake-level fluctuations across the TP are closely linked to spatial differences in precipitation.

This study demonstrates the feasibility of a methodological approach combining distributed hydrological modeling with atmospheric-model output and various satellite-based data to overcome the data-scarcity problem in the TP region. The integration of readily available model-derived atmospheric and remote-sensing data with hydrological modeling has the potential to improve our understanding of spatiotemporal hydrological patterns and to quantify water-balance components, even in ungauged or poorly gauged basins. The modeling framework presented in this study provides a useful basis for future regionally focused investigations on the space–time transition of lake changes in the TP region.

Model applications in such a data-scarce region have inherent uncertainty that should be perceived as useful information rather than a lack of basic knowledge or understanding (Blöschl and Montanari, 2010). An uncertainty and sensitivity analysis that includes the assessment of spatially and temporally variable effects on model outputs will allow specific and detailed recommendations on the timing and locations of future field measurements (e.g., Ragettli et al., 2013). There is an urgent need in such studies for meteorological observations (particularly precipitation in high mountain regions) and monitoring of land-surface characteristics (vegetation, soil and hydrogeological properties), in order to reduce the model uncertainties arising from input data and land-surface parameterization.

Overall, future research should focus on model-independent data describing hydrological system components that can be used for multi-response calibration and validation purposes. Water-level and volume estimations with a higher temporal resolution are expected to be produced from new satellite-altimetry data, such as from Cryosat (data continuously available since 2012, planned until 2017), Sentinel-3 (2015) and Jason-CS (2017) (Kleinherenbrink et al., 2015), which could be used as calibration or validation data in further model applications in the future.

Author contributions. S. Biskop designed the study, extended the J2000g model, performed modeling studies, analyzed data and wrote the main paper and the supplementary information. F. Maussion developed HAR and analyzed HAR data. P. Krause developed the original J2000g and helped to enhance the model. F. Maussion and M. Fink participated in field work. M. Fink carried out soil analysis. All authors continuously discussed the results and developed the analysis further. F. Maussion, M. Fink and P. Krause commented on and/or edited the manuscript.

Acknowledgements. This work is supported by the German Federal Ministry of Education and Research (BMBF) Central Asia-Monsoon Dynamics and Geo-Ecosystems (CAME) program through the WET project (Variability and Trends in Water Balance Components of Benchmark Drainage Basins on the Tibetan Plateau) under the code 03G0804C. F. Maussion acknowledges support by the Austrian Science Fund (FWF project P22443-N21). We sincerely thank the ITP/CAS for providing lake-level data from Nam Co and Jan Kropacek for proving land-cover data for the Nam Co basin. Finally, we are grateful to Timothy D. Steele, Alexander Brenning and four anonymous reviewers for their valuable comments and suggestions on the manuscript.

References

Allen, R. G., Pereira, L. S., Raes, D., and Smith, M.: Crop evapotranspiration: Guidelines for computing crop water requirements, FAO Irrigation and drainage paper 56, Rome, Italy, 1998.

Beven, K.: How far can we go in distributed hydrological modelling?, Hydrol. Earth Syst. Sci., 5, 1-12, doi:10.5194/hess-5-1-2001, 2001.

Beven, K. and Freer, J.: Equifinality, data assimilation, and uncertainty estimation in mechanistic modelling of complex environmental systems using the GLUE methodology, J. Hydrol., 249, 11–29, 2001.

Biskop, S., Krause, P., Helmschrot, J., Fink, M., and Flügel, W.-A.: Assessment of data uncertainty and plausibility over the Nam Co Region, Tibet, Adv. Geosci., 31, 57–65, doi:10.5194/adgeo-31-57-2012, 2012.

Blöschl, G. and Montanari, A.: Climate change impacts-throwing the dice?, Hydrol. Process., 24, 374–381, 2010.

Bowling, L. C., Pomeroy, J. W., and Lettenmaier, D. P.: Parameterization of Blowing-Snow Sublimation in a Macroscale Hydrology Model, J. Hydrometeorol., 5, 745–762, 2004.

Crétaux, J.-F., Jelinski, W., Calmant, S., Kouraev, A., Vuglinski, V., Bergé-Nguyen, M., Gennero, M.-C., Nino, F., Abarca Del Rio, R., Cazenave, A., and Maisongrande, P.: SOLS: A lake database to monitor in the Near Real Time water level and storage variations from remote sensing data, Adv. Space Res., 47, 1497–1507, 2011.

Cuo, L., Zhang, Y., Zhu, F., and Liang, L.: Characteristics and changes of streamflow on the Tibetan Plateau: A review, J. Hydrol. Reg. Stud., 2, 49–68, 2014.

Deus, D., Gloaguen, R., and Krause, P.: Water Balance Modeling in a Semi-Arid Environment with Limited in situ Data Using Remote Sensing in Lake Manyara, East African Rift, Tanzania, Remote Sens., 5, 1651–1680, 2013.

Downer, C. and Ogden, F.: GSSHA: Model To Simulate Diverse Stream Flow Producing Processes, J. Hydrol. Eng., 9, 161–174, 2004.

Farr, T. G., Rosen, P. A., Caro, E., Crippen, R., Duren, R., Hensley, S., Kobrick, M., Paller, M., Rodriguez, E., Roth, L., Seal, D., Shaffer, S., Shimada, J., Umland, J., Werner, M., Oskin, M., Burbank, D., and Alsdorf, D. E.: The shuttle radar topography mission, Rev. Geophys., 45, RG2004, doi:10.1029/2005RG000183, 2007.

Flügel, W.-A.: Delineating Hydrological Response Units by Geographical Information System analyses for regional hydrological modelling using PRMS/MMS in the drainage basin of the river Bröl, Germany, Hydrol. Process., 9, 423–436, 1995.

Gao, Y., Xie, H., Yao, T., and Xue, C.: Integrated assessment on multi-temporal and multi-sensor combinations for reducing cloud obscuration of MODIS snow cover products of the Pacific Northwest USA, Remote Sens. Environ., 114, 1662–1675, 2010.

Gao, Y., Cuo, L., and Zhang, Y.: Changes in moisture flux over the tibetan plateau during 1979-2011 and possible mechanisms, J. Climate, 27, 1876–1893, 2014.

Gao, Y., Leung, L. R., Zhang, Y., and Cuo, L.: Changes in Moisture Flux over the Tibetan Plateau during 1979–2011: Insights from a High-Resolution Simulation, J. Climate, 28, 4185–4197, 2015.

Gu, S., Tang, Y., Cui, X., Du, M., Zhao, L., Li, Y., Xu, S., Zhou, H., Kato, T., Qi, P., and Zhao, X.: Characterizing evapotranspiration over a meadow ecosystem on the Qinghai-Tibetan Plateau, J. Geophys. Res., 113, D08118, doi:10.1029/2007JD009173, 2008.

Haginoya, S., Fujii, H., Kuwagata, T., Xu, J., Ishigooka, Y., Kang, S., and Zhang, Y.: Air Lake Interaction Features Found in Heat and Water Exchanges over Nam Co on the Tibetan Plateau, Sci. Online Lett. Atmos., 5, 172–175, 2009.

Hock, R.: A distributed temperature-index ice- and snowmelt model including potential direct solar radiation, J. Glaciol., 45, 101–111, 1999.

Huintjes, E.: Energy and mass balance modelling for glaciers on the Tibetan Plateau – Extension, validation and application of a coupled snow and energy balance model, Dissertation, RWTH Aachen, 2014.

Huintjes, E., Sauter, T., Schröter, B., Maussion, F., Yang, W., Kropáček, J., Buchroithner, M., Scherer, D., Kang, S., and Schneider, C.: Evaluation of a Coupled Snow and Energy Balance Model for Zhadang Glacier, Tibetan Plateau, Using Glaciological Measurements and Time-Lapse Photography, Arctic, Antarct. Alp. Res., 47, 573–590, 2015.

Jensen, M., Dotan, A., and Sanford, R.: Penman-Monteith Estimates of Reservoir Evaporation, in: Impacts of Global Climate Change, Proceedings of World Water and Environmental Resources Congress, edited by: Raymond Walton, P. E., American Society of Civil Engineers, Anchoraga, Alaska, USA, 15–19 March 2005, 1–24, doi:10.1061/40792(173)548, 2005.

Jensen, M. E.: Estimating evaporation from water surfaces, presented at the CSU/ARS Evapotranspiration Workshop, Fort Collins, Colorado, USA, 12 March 2010, available at: http://ccc.atmos.colostate.edu/ET_Workshop/ET_Jensen/ET_water_surf.pdf (last access: 12 January 2013), 2010.

Kleinherenbrink, M., Lindenbergh, R. C., and Ditmar, P. G.: Monitoring of lake level changes on the Tibetan Plateau and Tian Shan by retracking Cryosat SARIn waveforms, J. Hydrol., 521, 119–131, 2015.

Knoche, M., Fischer, C., Pohl, E., Krause, P., and Merz, R.: Combined uncertainty of hydrological model complexity and satellite-based forcing data evaluated in two data-scarce semi-arid catchments in Ethiopia, J. Hydrol., 519, 2049–2066, 2014.

Kralisch, S. and Fischer, C.: Model representation, parameter calibration and parallel computing – the JAMS approach, in: Proceedings of the International Congress on Environmental Modelling and Software, Sixth Biennial Meeting, edited by: Seppelt, R. Voinov, A. A., Lange, S., and Bankamp, D., International Environmental Modelling and Software Society (iEMSs), Leipzig, Germany, 1–5 July 2012, 1177–1184, 2012.

Krause, P.: Quantifying the impact of land use changes on the water balance of large catchments using the J2000 model, Phys. Chem. Earth, 27, 663–673, 2002.

Krause, P. and Hanisch, S.: Simulation and analysis of the impact of projected climate change on the spatially distributed waterbalance in Thuringia, Germany, Adv. Geosci., 21, 33–48, doi:10.5194/adgeo-21-33-2009, 2009.

Krause, P., Bäse, F., Bende-Michl, U., Fink, M., Flügel, W.-A., and Pfennig, B.: Multiscale investigations in a mesoscale catchment – hydrological modelling in the Gera catchment project, Adv. Geosci., 9, 53–61, doi:10.5194/adgeo-9-53-2006, 2006.

Krause, P., Biskop, S., Helmschrot, J., Flügel, W.-A., Kang, S., and Gao, T.: Hydrological system analysis and modelling of the Nam Co basin in Tibet, Adv. Geosci., 27, 29–36, doi:10.5194/adgeo-27-29-2010, 2010.

Kropacek, J., Feng, C., Alle, M., Kang, S., and Hochschild, V.: Temporal and Spatial Aspects of Snow Distribution in the Nam Co Basin on the Tibetan Plateau from MODIS Data, Remote Sens., 2, 2700–2712, 2010.

Leber, D., Holawe, F., and Häusler, H.: Climatic Classification of the Tibet Autonomous Region Using Multivariate Statistical Methods, GeoJournal, 37, 451–472, 1995.

Lei, Y., Yao, T., Bird, B. W., Yang, K., Zhai, J., and Sheng, Y.: Coherent lake growth on the central Tibetan Plateau since the 1970s: Characterization and attribution, J. Hydrol., 483, 61–67, 2013.

Lei, Y., Yang, K., Wang, B., Sheng, Y., Bird, B. W., Zhang, G., and Tian, L.: Response of inland lake dynamics over the Tibetan Plateau to climate change, Clim. Change, 125, 281–290, 2014.

Li, B., Yu, Z., Liang, Z., and Acharya, K.: Hydrologic response of a high altitude glacierized basin in the central Tibetan Plateau, Glob. Planet. Change, 118, 69–84, 2014.

Li, Y., Liao, J., Guo, H., Liu, Z., and Shen, G.: Patterns and Potential Drivers of Dramatic Changes in Tibetan Lakes, 1972–2010, PLoS One, 9, e111890, doi:10.1371/journal.pone.0111890, 2014.

Liao, J., Shen, G., and Li, Y.: Lake variations in response to climate change in the Tibetan Plateau in the past 40 years, Int. J. Digit. Earth, 6, 534–549, 2013.

Linacre, E. T.: Data-sparse estimation of lake evaporation, using a simplified Penman equation, Agr. Forest Meteorol., 64, 237–256, 1993.

MacCallum, S. N. and Merchant, C. J.: Surface water temperature observations of large lakes by optimal estimation, Can. J. Remote Sens., 38, 25–45, 2012.

Mason, I. M.: The response of lake levels and areas to climatic change, Clim. Change, 27, 161–197, 1994.

Maussion, F.: A new atmospheric dataset for High Asia: Development, validation and applications in climatology and in glaciology, Dissertation, TU Berlin, 2014.

Maussion, F., Scherer, D., Mölg, T., Collier, E., Curio, J., and Finkelnburg, R.: Precipitation Seasonality and Variability over the Tibetan Plateau as Resolved by the High Asia Reanalysis, J. Climate, 27, 1910–1927, 2014.

Meng, K., Shi, X., Wang, E., and Liu, F.: High-altitude salt lake elevation changes and glacial ablation in Central Tibet, 2000–2010, Chinese Sci. Bull., 57, 525–534, 2012.

Mölg, T., Maussion, F., and Scherer, D.: Mid-latitude westerlies as a driver of glacier variability in monsoonal High Asia, Nat. Clim. Chang., 4, 68–73, 2014.

Morrill, C.: The influence of Asian summer monsoon variability on the water balance of a Tibetan lake, J. Paleolimnol., 32, 273–286, 2004.

Nie, Y., Zhang, Y., Ding, M., Liu, L., and Wang, Z.: Lake change and its implication in the vicinity of Mt. Qomolangma (Everest), central high Himalayas, 1970–2009, Environ. Earth Sci., 68, 251–265, 2012.

Parajka, J. and Blöschl, G.: Spatio-temporal combination of MODIS images – potential for snow cover mapping, Water Resour. Res., 44, W03406, doi:10.1029/2007WR006204, 2008.

Pellicciotti, F., Ragettli, S., Carenzo, M., and McPhee, J.: Changes of glaciers in the Andes of Chile and priorities for future work., Sci. Total Environ., 493, 1197–1210, 2014.

Phan, V. H., Lindenbergh, R. C., and Menenti, M.: Geometric dependency of Tibetan lakes on glacial runoff, Hydrol. Earth Syst. Sci., 17, 4061–4077, doi:10.5194/hess-17-4061-2013, 2013.

Phan, V. H., Lindenbergh, R., and Menenti, M.: ICESat derived elevation changes of Tibetan lakes between 2003 and 2009, Int. J. Appl. Earth Obs. Geoinf., 17, 12–22, 2012.

Pohl, E., Knoche, M., Gloaguen, R., Andermann, C., and Krause, P.: Sensitivity analysis and implications for surface processes from a hydrological modelling approach in the Gunt catchment, high Pamir Mountains, Earth Surf. Dyn., 3, 333–362, 2015.

Rades, E. F., Hetzel, R., Xu, Q., and Ding, L.: Constraining Holocene lake-level highstands on the Tibetan Plateau by 10Be exposure dating: a case study at Tangra Yumco, southern Tibet, Quaternary Sci. Rev., 82, 68–77, 2013.

Ragettli, S., Pellicciotti, F., Bordoy, R., and Immerzeel, W. W.: Sources of uncertainty in modeling the glaciohydrological response of a Karakoram watershed to climate change, Water Resour. Res., 49, 6048–6066, 2013.

Rödiger, T., Geyer, S., Mallast, U., Merz, R., Krause, P., Fischer, C., and Siebert, C.: Multi-response calibration of a conceptual hydrological model in the semiarid catchment of Wadi al Arab, Jordan, J. Hydrol., 509, 193–206, 2014.

Rosenberry, D. O., Lewandowski, J., Meinikmann, K., and Nützmann, G.: Groundwater – the disregarded component in lake water and nutrient budgets. Part 1: effects of groundwater on lake hydrology, Hydrol. Process., doi:10.1002/hyp.10403, 2014.

Schütt, B., Berking, J., Frenchen, M., and Yi, C.: Late Pleistocene Lake Level fluctuations of the Nam Co, Tibetan Plateau, China, Z. Geomorph. N. F., 52, 57–74, 2008.

Sitch, S., Smith, B., Prentice, I. C., Arneth, A., Bondeau, A., Cramer, W., Kaplan, J. O., Levis, S., Lucht, W., Sykes, M. T.,

Thonicke, K., and Venevsky, S.: Evaluation of ecosystem dynamics, plant geography and terrestrial carbon cycling in the LPJ dynamic global vegetation model, Glob. Chang. Biol., 9, 161–185, 2003.

Sivapalan, M.: Prediction in ungauged basins: a grand challenge for theoretical hydrology, Hydrol. Process., 17, 3163–3170, 2003.

Skamarock, W. C. and Klemp, J. B.: A time-split nonhydrostatic atmospheric model for weather research and forecasting applications, J. Comput. Phys., 227, 3465–3485, 2008.

Song, C., Huang, B., Richards, K., Ke, L., and Hien, V. P.: Accelerated lake expansion on the Tibetan Plateau in the 2000s: Induced by glacial melting or other processes?, Water Resour. Res., 50, 3170–3186, 2014.

Strasser, U., Bernhardt, M., Weber, M., Liston, G. E., and Mauser, W.: Is snow sublimation important in the alpine water balance?, The Cryosphere, 2, 53–66, doi:10.5194/tc-2-53-2008, 2008.

Valiantzas, J. D.: Simplified versions for the Penman evaporation equation using routine weather data, J. Hydrol., 331, 690–702, 2006.

Vionnet, V., Martin, E., Masson, V., Guyomarc'h, G., Naaim-Bouvet, F., Prokop, A., Durand, Y., and Lac, C.: Simulation of wind-induced snow transport and sublimation in alpine terrain using a fully coupled snowpack/atmosphere model, The Cryosphere, 8, 395–415, doi:10.5194/tc-8-395-2014, 2014.

Wagener, T. and Kollat, J.: Numerical and visual evaluation of hydrological and environmental models using the Monte Carlo analysis toolbox, Environ. Model. Softw., 22, 1021–1033, 2007.

Winsemius, H. C., Schaefli, B., Montanari, A., and Savenije, H. H. G.: On the calibration of hydrological models in ungauged basins: A framework for integrating hard and soft hydrological information, Water Resour. Res., 45, W12422, doi:10.1029/2009WR007706, 2009.

Xie, H. and Zhu, X.: Reference evapotranspiration trends and their sensitivity to climatic change on the Tibetan Plateau (1970–2009), Hydrol. Process., 27, 3685–3693, 2013.

Xu, J., Yu, S., Liu, J., Haginoya, S., Ishigooka, Y., Kuwagata, T., Hara, M., and Yasunari, T.: The Implication of Heat and Water Balance Changes in a Lake Basin on the Tibetan Plateau, Hydrol. Res. Lett., 5, 1–5, 2009.

Yao, T., Pu, J., Lu, A., Wang, Y., and Yu, W.: Recent Glacial Retreat and Its Impact on Hydrological Processes on the Tibetan Plateau, China, and Surrounding Regions, Arctic, Antarct. Alp. Res., 39, 642–650, 2007.

Ye, Q., Yao, T., Chen, F., Kang, S., Zhang, X., and Wang, Y.: Response of Glacier and Lake Covariations to Climate Change in Mapam Yumco Basin on Tibetan Plateau during 1974–2003, J. China Univ. Geosci., 19, 135–145, 2008.

Yin, Y., Wu, S., Zheng, D., and Yang, Q.: Radiation calibration of FAO56 Penman-Monteith model to estimate reference crop evapotranspiration in China, Agric. Water Manag., 95, 77–84, 2008.

Yin, Y., Wu, S., Zhao, D., Zheng, D., and Pan, T.: Modeled effects of climate change on actual evapotranspiration in different ecogeographical regions in the Tibetan Plateau, J. Geogr. Sci., 23, 195–207, 2013.

Yu, S., Liu, J., Xu, J., and Wang, H.: Evaporation and energy balance estimates over a large inland lake in the Tibet-Himalaya, Environ. Earth Sci., 64, 1169–1176, 2011.

Zhang, B., Wu, Y., Lei, L., Li, J., Liu, L., Chen, D., and Wang, J.: Monitoring changes of snow cover, lake and vegetation phenology in Nam Co Lake Basin (Tibetan Plateau) using remote sensing (2000–2009), J. Great Lakes Res., 39, 224–233, 2013.

Zhang, G., Xie, H., Kang, S., Yi, D., and Ackley, S. F.: Monitoring lake level changes on the Tibetan Plateau using ICESat altimetry data (2003–2009), Remote Sens. Environ., 115, 1733–1742, 2011.

Zhang, G., Xie, H., Yao, T., Liang, T., and Kang, S.: Snow cover dynamics of four lake basins over Tibetan Plateau using time series MODIS data (2001–2010), Water Resour. Res., 48, W10529, doi:10.1029/2012WR011971, 2012.

Zhang, G., Xie, H., Yao, T., and Kang, S.: Water balance estimates of ten greatest lakes in China using ICESat and Landsat data, Chinese Sci. Bull., 58, 3815–3829, 2013.

Zhou, S., Kang, S., Chen, F., and Joswiak, D. R.: Water balance observations reveal significant subsurface water seepage from Lake Nam Co, south-central Tibetan Plateau, J. Hydrol., 491, 89–99, 2013.

Zhu, G., Su, Y., Li, X., Zhang, K., Li, C., and Ning, N.: Modelling evapotranspiration in an alpine grassland ecosystem on Qinghai-Tibetan plateau, Hydrol. Process., 28, 610–619, 2014.

Zhu, L., Xie, M., and Wu, Y.: Quantitative analysis of lake area variations and the influence factors from 1971 to 2004 in the Nam Co basin of the Tibetan Plateau, Chinese Sci. Bull., 55, 1294–1303, 2010.

3

Carbon and nitrogen dynamics and greenhouse gas emissions in constructed wetlands treating wastewater

M. M. R. Jahangir[1,2], **K. G. Richards**[2], **M. G. Healy**[3], **L. Gill**[1], **C. Müller**[4,5], **P. Johnston**[1], and **O. Fenton**[2]

[1]Department of Civil, Structural & Environmental Engineering, Trinity College Dublin, Dublin 2, Ireland

[2]Department of Environment, Soils & Land Use, Teagasc Environment Research Centre, Johnstown Castle, Co. Wexford, Ireland

[3]Civil Engineering, National University of Ireland, Galway, Co. Galway, Ireland

[4]School of Biology and Environmental Science, University College Dublin, Belfield, Dublin, Ireland

[5]Department of Plant Ecology (IFZ), Justus-Liebig University Giessen, Giessen, Germany

Correspondence to: M. M. R. Jahangir (jahangim@tcd.ie)

Abstract. The removal efficiency of carbon (C) and nitrogen (N) in constructed wetlands (CWs) is very inconsistent and frequently does not reveal whether the removal processes are due to physical attenuation or whether the different species have been transformed to other reactive forms. Previous research on nutrient removal in CWs did not consider the dynamics of *pollution swapping* (the increase of one pollutant as a result of a measure introduced to reduce a different pollutant) driven by transformational processes within and around the system. This paper aims to address this knowledge gap by reviewing the biogeochemical dynamics and fate of C and N in CWs and their potential impact on the environment, and by presenting novel ways in which these knowledge gaps may be eliminated. Nutrient removal in CWs varies with the type of CW, vegetation, climate, season, geographical region, and management practices. Horizontal flow CWs tend to have good nitrate (NO_3^-) removal, as they provide good conditions for denitrification, but cannot remove ammonium (NH_4^+) due to limited ability to nitrify NH_4^+. Vertical flow CWs have good NH_4^+ removal, but their denitrification ability is low. Surface flow CWs decrease nitrous oxide (N_2O) emissions but increase methane (CH_4) emissions; subsurface flow CWs increase N_2O and carbon dioxide (CO_2) emissions, but decrease CH_4 emissions. Mixed species of vegetation perform better than monocultures in increasing C and N removal and decreasing greenhouse gas (GHG) emissions, but empirical evidence is still scarce. Lower hydraulic loadings with higher hydraulic retention times enhance nutrient removal, but more empirical evidence is required to determine an optimum design. A conceptual model highlighting the current state of knowledge is presented and experimental work that should be undertaken to address knowledge gaps across CWs, vegetation and wastewater types, hydraulic loading rates and regimes, and retention times, is suggested. We recommend that further research on process-based C and N removal and on the balancing of end products into reactive and benign forms is critical to the assessment of the environmental performance of CWs.

1 Introduction

Increasing anthropogenic loading of reactive nitrogen (Nr; all forms of nitrogen (N) except di-nitrogen gas, N_2) along the N cascade in the environment raises many critical concerns for human health, drinking water quality (Gray, 2008), coastal and marine water degradation as well as algal blooms and hypoxia (Conley et al., 2009; Rabalais et al., 2010). Constructed wetlands (CWs) are artificial sinks for Nr (Galloway et al, 2003; Tanner et al., 2005), and have been successfully used to treat domestic sewage, urban runoff and storm water, industrial and agricultural wastewater, and leachate. While the biogeochemistry of wetlands in general has been discussed in the literature (Whalen, 2005; Reddy and Delaune, 2008), less is known about the delivery pathways of the transformation products of carbon (C) and N from CWs treating wastewater. Although CWs have a proven potential for organic C and N removal, with few exceptions (Dzakpasu et al., 2014), studies have rarely quantified all relevant pathways. This has meant that reported removal efficiencies have been variable (Seitzinger et al., 2002). If the fate of C and N is accurately quantified, appropriate design and management strategies may be adopted.

Constructed wetlands are complex bioreactors that facilitate a number of physical, chemical, and biological processes, but are frequently evaluated as a *black box* in terms of process understanding (Langergraber, 2008). Many investigations target single contaminant remediation and disregard the reality of mixed contaminants entering and leaving CWs. They do not consider the dynamics of *pollution swapping* (the increase in one pollutant as a result of a measure introduced to reduce a different pollutant) driven by transformational processes within and around the system. This means that potential negative impacts that CWs may have on the environment, such as greenhouse gas (GHG) emissions (IPCC, 2013; Clair et al., 2002; Mander et al., 2008; Mitsch and Gosselink, 2000) or enhancement of pollution swapping (Reay, 2004), are not accounted for in analyses. There are many pathways by which the removed N can contribute to water and air pollution: accumulation and adsorption in soils, leaching of nitrate (NO_3^-) and ammonium (NH_4^+) to groundwater, emissions of nitrous oxide (N_2O) and ammonia (NH_3^+) to the atmosphere, and/or conversion to N_2 gas. Constructed wetlands significantly contribute to atmospheric N_2O emissions either directly to the atmosphere from the surface of the wetland (IPCC, 2013; Søvik et al., 2006; Ström et al., 2007; Elberling et al., 2011) or indirectly via dissolved N_2O in the effluent or groundwater upon discharge to surface waters. The IPCC (2013) has recognized the significance of indirect N_2O emissions from CW effluent that is discharged to aquatic environments, and estimate emission factors (EF) ranging from 0.0005 to 0.25. Production and reduction processes of N_2O in the environment are not yet fully understood.

Constructed wetlands receive organic C from the influent wastewater and from fixation by the photosynthetic hydrophytes, which are incorporated into soil as organic C. Soil organic C undergoes the biogeochemical processes that regulate C accretion in soil and microbial respiration, producing carbon dioxide (CO_2). Anaerobic mineralization of organic C by methanogenic archaea can produce methane (CH_4) (Laanbroek, 2010; Ström et al., 2007; Søvik et al., 2006; Pangala et al., 2010). Constructed wetlands can also contribute to the dissolved organic carbon (DOC) load transfer to ground- and surface waters, which may produce and exchange substantial amounts of CO_2 and CH_4 with the atmosphere (Clair et al., 2002; Elberling et al., 2011). Therefore, CWs can diminish the environmental benefits of wastewater treatment. The dynamics of dissolved N_2O, CO_2, and CH_4 in CWs is a key knowledge gap in global GHG budgets.

Surface emissions of GHG from CWs have been commonly measured by the closed chamber method (Johansson et al., 2003, 2004; Mander et al., 2005, 2008), but have rarely been measured by ebullition and diffusion methods (Søvik et al., 2006). The measured rates have shown high spatial, temporal, and diurnal variations due to the change in biogeochemistry of C and N and plant–microbe–soil interaction over time and space. Surface emissions cannot explain the kinetics of production and consumption rates of GHG, which we need to know in order to adopt better management practices to mitigate emissions. In addition, subsurface export of dissolved nutrients and GHG, an important pathway of nutrient loss (Riya et al., 2010), is frequently ignored. Mass balance analysis of the different components of the N cycle and kinetics of their transformation processes occurring within the treatment cells using the isotope-tracing ^{15}N technique can provide mechanistic information for N transformation products (Lee et al., 2009; O'Luanaigh et al., 2010) and may be used to start to answer such questions. Similarly, ^{14}C application and measurement of C species (e.g. CO_2, CH_4, and DOC) may elucidate the C mineralization and CO_2 and CH_4 production and consumption. Used in combination, these methods may provide a comparative analysis of the rates of C and N transformation processes and the role of these processes in delivering NO_3^-, NH_4^+, and DOC to ground/surface waters and N_2O, CO_2, and CH_4 to the atmosphere.

Past reviews on CWs, though very limited, summarize the performance of different types of CWs on C and N removal (Vymazal, 2007) and surface emissions of GHG (Mander et al., 2014), but have not discussed the mechanisms of nutrient removal and the fate of the nutrients delivered and removed to and from CWs. Therefore, the objectives of this review are to (i) understand the biogeochemical dynamics of C and N in CWs, (ii) better understand the fate of various C and N species in a holistic manner, in addition to the conventional influent/effluent balance for nutrient removal, (iii) identify the research gaps that need to be addressed to optimize nutrient removal and mitigate GHG emissions, and (iv) discuss emerging measurement techniques that may give insights into the production and reduction of GHG.

2 Removal efficiency, hydraulic loading, and retention time

In CWs, the efficiency of C and N removal is generally limited and highly variable over CW types, plant types, seasons, climatic regions, and management practices. On average, it appears that 50 and 56 % of the influent total nitrogen (TN) and total organic carbon (TOC), respectively, can be removed, but the removal rates are very inconsistent. Mean (± standard error) TN removals, obtained from the literature cited in this paper, ranged from 31.3 ± 6.3 % in surface flow (SF) CWs to 40.4 ± 4.4 % in subsurface flow (SSF) CWs,

Table 1. TN input (mg N L^{-1}), TN output (TN, mg N L^{-1}), and TN removal (%) in various CWs treating wastewater; average standard error (± SE) is presented for TN removal; NA – data not available.

CW type	Treatment	N input (mg N L^{-1})			N output (mg N L^{-1})			N removal (%)			References
		TN	NH$_4^+$	NO$_3^-$	TN	NH$_4^+$	NO$_3^-$	TN	NH$_4^+$	NO$_3^-$	
SF_Finland	Municipal	1.4±150	0.03±5.8	0.3±95	1.1±48	0.01±3.0	0.02±6.7	21.4	66.7	93.3	Søvik et al. (2006)
SF_Finland	Agril. runoff	66.1±1.9	63.5±1.3	0.7±0.13	64.7±1.7	61.2±1.7	0.3±0.09	2.1	3.6	57.1	Søvik et al. (2006)
SF_Norway	Municipal	43.4±3.6	41.5±3.0	0.0±0.0	36.7±2.7	32.6±1.9	0.9±0.4	15.4	21.4	−800	Søvik et al. (2006)
SF	Municipal	NA	4.5	15.5	NA	NA	NA	61	NA	NA	Song et al. (2011)
SF	Domestic	NA	40	5	NA	NA	NA	97–98	NA	NA	Dzakpasu et al. (2011)
SF	Various	NA	39	4.4	NA	NA	NA	39–48	NA	NA	Vymazal (2007)
SF	Municipal	NA	36		NA	NA	NA	39	NA	NA	Vymazal (2010)
SF	Municipal	NA	196	<2	NA	NA	NA	35	NA	NA	Shamir et al. (2001)
SF	various	NA	80	< 1	NA	NA	NA	>90	NA	NA	Harrington et al. (2007)
SF	Municipal	NA	0.95	1.54	NA	NA	NA	45	NA	NA	Toet et al. (2005)
SF	Dairy washout	227	NA	NA	NA	NA	NA	40	NA	NA	Van der Zaag et al. (2010)
All SF								31.3±6.3			
HSSF_Estonia	Municipal	96.5±3.0	83.9±2.7	0.2±0.02	46.2±1.5	36.2±1.4	5.9±0.65	52.1	56.9	−2850	Søvik et al. (2006)
HSSF_Norway	Municipal	53.4±4.3	38.4±7.7	14.1±7.5	45.0±4.1	43.1±4.7	1.0±0.8	15.7	−12.2	92.9	Søvik et al. (2006)
HSSF	Dairy washout	306±101*	NA	NA	177±58*	NA	NA	42.2	NA	NA	Van der Zaag et al. (2010)
HSSF	Domestic	NA	74.9	3.9	NA	NA	NA	29	NA	NA	O'Luanaigh et al. (2010)
HSSF	Domestic	87						46–48			Mander et al. (2008)
HSSF	Dairy washout	227						28			Van der Zaag et al. (2010)
HSSF	Milk parlour	112	22	NA	24	11	NA	78	50	NA	Kato et al. (2006)
HSSF	Agriculture	67	40	0.85	27	11	1.1	47	39	−29	Vymazal and Kröpfelova (2010)
HSSF	Industry	124	65	8.5	103	31	7.4	20	20	8	Vymazal and Kröpfelova (2010)
HSSF	Landfill	157	149	1.5	147	98	1.3	30	33	31	Vymazal and Kröpfelova (2010)
HSSF	Municipal	43	24	2	24	14	1.2	40	30	33	Vymazal and Kröpfelova (2010)
All HSSF								40.4±4.4			
VSSF_Estonia	Municipal	50.9±9.2	35.7±6.2	1.1±0.32	43.1±7.6	31.7±5.5	1.7±0.84	15.3	11.2	−54.5	Søvik et al. (2006)
VSSF_Norway	Municipal	52.6±5.2	49.6±4.0	0.0±0.0	47.8±6.9	21.4±6.9	25.5±1.3	9.1	56.9	−25 400	Søvik et al. (2006)
VSSF	Municipal	41.0±0.5	NA	NA	20.7±0.8	NA	NA	49.3±1.8	NA	NA	Yan et al. (2012)
VSSF	Municipal	46±13	NA	NA	NA	NA	NA	74±3	NA	NA	Zhao et al. (2014)
All VSSF								37.0±10.9			

SF – surface flow; HSSF – horizontal subsurface flow; VSSF – vertical subsurface flow; * mg N m^{-2} h^{-1}.

Table 2. Total organic C (TOC) removal (%) in various CWs treating wastewater; average standard error (± SE) is presented for TOC removal; NA – data not available.

CWs type	Treatment	C input (TOC; mg C L^{-1})	C output (TOC; mg C L^{-1})	TOC Removal (%)	References
SF_Finland	Municipal	13.0±0.3	14.0±0.5	−7.7	Søvik et al. (2006)
SF_Finland	Agril runoff	25.0±3.4	20.0±3.4	20.0	Søvik et al. (2006)
SF_Norway	Municipal	26.7±2.9	17.1±1.8	36.0	Søvik et al. (2006)
SF	Dairy wash out	186[a]	136[a]	27	Van der Zaag et al. (2010)
All SF				18.8±9.4	
HSSF	Domestic	150[b]	NA	NA	Garcia et al. (2007)
HSSF	Dairy wash out	186[a]	107.9[a]	42	Van der Zaag et al. (2010)
HSSF_Estonia	Municipal	62.8±16.6[a]	41.0±11.3[a]	34.7	Søvik et al. (2006)
HSSF_Norway	Municipal	40.5±11.3	15.0±2.4	63.0	Søvik et al. (2006)
All HSSF				46.6±7.3	
VSSF_Estonia	Municipal	132.2±32.2[a]	62.8±16.6[a]	52.5	Søvik et al. (2006)
VSSF_Norway	Municipal	40.5±11.3	15.0±2.4	63.0	Søvik et al. (2006)
VSSF	Municipal	106±35	74±21	26±4.6	Yan et al. (2012)
VSSF	Municipal	249±49	NA	83±1.0	Zhao et al. (2014)
All VSSF				56.2±9.5	

SF – surface flow; HSSF – horizontal subsurface flow; VSSF – vertical subsurface flow; [a] BOD; [b] mg m^{-2} h^{-1}.

whereas TOC removal ranged from 18.8 ± 9.4 % in SF CWs to 56.2 ± 9.5 % in vertical subsurface flow CWs (Tables 1 and 2). In European systems, for example, typical removals of ammoniacal N in long-term operation are around 35 %, but can be enhanced if some pre-treatment procedures are followed (Verhoeven and Meuleman, 1999; Luederitz et al., 2001). Generally, TN removal is higher in SF CWs than SSF CWs (Table 1), but studies differ. For example, Van der Zaag et al. (2010) showed higher N removal in SF CWs than SSF CWs, but Søvik et al. (2006) and Gui et al. (2007) showed the opposite. In SSF CWs, limited removal can be caused by a reduced environment that enhances NH$_4^+$ accumulation and limits NH_4^+ oxidation. In SF CWs, denitrification rates can be limited due to lack of NO_3^-. In vertical subsurface flow (VSSF) CWs, aeration can increase NH_4^+ oxidation to NO_3^-, which can be denitrified or converted to NH_4^+ by dissimilatory NO_3^- reduction to NH_4^+ (DNRA).

Plant species are important components of CWs, and affect C and N removals. Optimal species selection for best removal is difficult because some species are efficient in removing one pollutant but not the other (Bachand and Horne, 2000; Bojcevska and Tonderski, 2007; da Motta Marques et al., 2001). In some studies there are no inter-species differences at all (Calheiros et al., 2007). Mixed species perform better than monocultures to remove C and N pollutants because they increase microbial biomass and diversity. Payne et al. (2014a) discussed the role of plants in nutrient removal. Plants regulate CW hydrology (evaporation and transpiration) and temperature (insulating CWs from seasonal temperature change, trapping falling and drifting snow, and heat loss of wind). Some species can create a large surface area for microbial attachment and enhance microbial diversity, but experimental evidence is still scarce.

Soil physico-chemical properties, such as permeability (Dzakpasu et al., 2014) and cation exchange capacity (Drizo et al., 1999) are important factors controlling the purification capacity in CWs. Microbial activities and growth depend on substrate C quality and C : N ratios, which affect nutrient removal. Growth of heterotrophic microorganisms is a function of the wastewater C : N (Makino et al., 2003). High C : N ratios can enhance denitrification by providing electron donors for denitrifiers, but the opposite can increase nitrification. High C : N ratios can also encourage DNRA over denitrification. Yan et al. (2012) measured a high TN removal but low TOC removals at a C : N ratio 2.5 : 1, which indicates that removal of one parameter might lead to a problem with a different one. The uncertainty in the conditions for achievement of optimum removal suggests that the rates of C and N transformations and the fate of the removed nutrients within CWs should be investigated. However, to our knowledge, no study has provided a holistic evaluation of C and N attenuation and transformation.

The removal of pollutants in CWs depends on hydraulic loading rates (HLR) and hydraulic retention time (HRT) (Toet et al., 2005). The HLR and HRT are considered to be significant design parameters determining the nutrient removal efficiencies (Weerakoon et al., 2013). Longer HRTs of wastewater in CWs increase the removal of C and N (Wang et al., 2014) by increasing sedimentation and duration of contact between nutrients and the CWs. The effects of HLR and HRT can vary with the nature of the use of CWs, e.g. whether they are used for treating single or mixed pollutants. To reduce Nr delivery to the receiving waters or to the atmosphere, CWs need to be optimally designed with respect to HLR and HRT.

Fluctuating hydraulic loading influences all biotic and abiotic processes in CWs (Mander et al., 2011). For example, if the groundwater table is lowered through changes in hydraulic loading, soil aeration can increase or decrease. Ammonification and nitrification rates increase with increased soil aeration and this enhances C utilization by bacteria and, therefore, can stimulate the removal of C and N. Investigation into the effects of fluctuating hydraulic loadings (hydraulic pulsing) on C and N removals and their transformation products will provide information about the fate of the added nutrients in terms of their environmental benefits and/or pollution swapping potential. For example, if the dominant product is N_2, the system will be relatively benign in terms of its impact on the environment, but if it is NH_4^+, it can be fixed in the soils or transported to ground- and surface waters connected to CWs if the cation exchange sites become saturated. Several authors have used a wide range of HLRs and HRTs to measure nutrient removal efficiency, but experimental evidence linking HLR and HRT to removal efficiency is scarce (Toet et al., 2005). Luo et al. (2005) reported that low HLR results in incomplete denitrification, whereas Zhang et al. (2006) argued that low HLR increases NH_4^+ and chemical oxygen demand oxidation. The way in which the performance of a CW is assessed can lead to different conclusions regarding the removal of Nr. For future studies, evaluation of CWs in a holistic manner, which includes pollution swapping at different HLRs and HRTs, is important, particularly within the context of the changing hydrologic cycle in a changing climate. In addition, local legislative targets should be considered and weighting factors (e.g. the relative importance of, say, GHG over water quality targets) should be developed to evaluate the overall performance of CWs. In addition to the estimation of nutrient removal rates, investigation of the effect of HLR and HRT on the different forms of nutrients in the final effluent and their fate in the natural environment may help elucidate the pollution swapping potential of CWs.

3 Accumulation of C and N in CWs soils

The soil in CWs is a major sink for C and N (Mustafa and Scholz, 2011). However, although data on the influent and effluent N concentrations are available, data on N accumulation (dissolved organic nitrogen (DON), TN, NH_4^+, or NO_3^- – N) within the soil profile of various CWs are scarce. The wide range of N accumulation reported in the literature (e.g. 30–40 %, Shamir et al., 2001; 39 %, Harrington et al., 2007; 9 %, Mander et al., 2008; 2.5 %, Obarska-Pempkowiak and Gajewska, 2003) may be due to the variations in CW types and management strategies. The accumulated species of N are reactive unless they have been transformed to N_2 by biogeochemical processes. However, there is a dearth of information on the extent of Nr accumulation in soils and discharge to surface waters and air (Shamir et al., 2001). Accumulated organic N could be mineralized to NH_4^+ and NO_3^-, depending on the physico-chemical properties of soil. The Nr could be assimilated by plants and microbes, which are recycled in a soil–plant–soil continuum. Nitrogen spiralling occurs from NH_4^+ to organic N and back to NH_4^+ within the CW (O'Luanaigh et al., 2010). Typically, N accumulation has been found to decrease with soil depth (Shamir et al., 2001). In terms of the conventional input–output balance, these are considered as removed N, but may, in fact, remain in such a biogeochemically active system. In addition to N, organic C accumulation occurs in CW soils (Nguyen, 2000).

Soils of CWs represent organic C and Nr-rich systems, where the products of continuously occurring biogeochemical processes, such as accumulation in soil and transportation to fresh waters and to the atmosphere, need to be quantified. Such an approach will show the shortcomings of conventional removal efficiency estimation methods and will also demonstrate how the apparently removed C and N species can become a source of contamination. Estimation of the rates of nutrient accumulation in soils in various types of CWs under different management systems is important. The stability of the accumulated C and N under changing climatic scenarios also needs to be addressed to consider the long-term sustainability of CWs.

4   C and N dynamics and greenhouse gas emissions

Increased nutrient input to CWs increases the productivity of wetland ecosystems and the production of GHG. As CWs are designed to remove pollutants in an anaerobic/suboxic environment, they change the C and N biogeochemistry and contribute significantly to CH_4 and N_2O emissions (Johansson, 2002, Johansson et al., 2003; Mander et al., 2005, 2008; Stadmark and Leonardson, 2005; Liikanen et al., 2006). Søvic et al. (2006) measured N_2O, CH_4, and CO_2 emissions in various CWs in different European countries, and suggested that the potential atmospheric impacts of CWs should be examined as their development is increasing globally. Management of CWs must consider the negative climatic aspects of increased emissions of GHG in addition to their primary functions (Ström et al., 2007). Therefore, estimation of the contribution of CWs to global warming is required. In this regard, measurement of spatial and temporal variations (seasonal and diurnal) of GHG emissions is necessary to accurately estimate CW-derived GHG emissions. A holistic assessment of ecologically engineered systems has been outlined by Healy et al. (2011, 2014) and developed further by Fenton et al. (2014). Such assessments can be applied in evaluating nutrient dynamics in CWs. Moreover, plant mediated GHG emissions could be an important component of total emissions, but again research in this area is very limited. Effective modelling or up-scaling of GHG emissions from watershed to regional/national scales is important for the improvement of global GHG budgets. Such up-scaling needs an accurate estimation of C and N inputs and outputs, i.e. a

balance coupled with net GHG emissions, while considering all possible processes and pathways involved. A study of the dynamics of C and N in CWs is crucial, as the forms of removed C and N are particularly pertinent to their potential for pollution swapping, global warming, and water pollution.

Processes involved in N removal and N transformations in wetlands include sedimentation of particulates (Koskiaho, 2003), nitrification, denitrification, DNRA (Poach et al., 2003; Burgin et al., 2013), microbial assimilation and plant uptake–release (Findlay et al., 2003), and anammox (anaerobic ammonium oxidation) and deamox (DEnitrifying AMmonium OXidation). Constructed wetlands are complex systems that facilitate aerobic and anaerobic microsites (Wynn and Liehr, 2001). Nitrification, denitrification, and nitrifier denitrification are the processes responsible for the production of N_2O. Depending on the environmental conditions or management practices prevailing, a certain process will dominate; e.g., denitrification is the dominant process in SF CWs (Beaulieu et al., 2011), but nitrifier denitrification is dominant in VSSF CWs (Wunderlin et al., 2013). Generally, CWs are anaerobic but aquatic macrophytes can transport oxygen from the atmosphere to the rooting zone, where it can sustain nitrification. The existence of microsites that support high activity and promote denitrification has been shown in soils (Parkin, 1987). Such conditions are also likely to occur in CWs, which have patchy distributions of organic material (e.g. particulate organic carbon), due to rhizodepositions (Minett et al., 2013; Hamersley and Howes, 2002). Minett et al. (2013) found that simultaneous oxygenation of the rhizosphere, through radial oxygen loss, and enhanced oxygen consumption by the soil occurs in the area immediately surrounding the roots. Nitrate produced in the rooting zone can be taken up by plants or denitrified and/or converted back to NH_4^+ by DNRA.

Competition for NO_3^- may occur between denitrification and biotic assimilation. This is likely governed by the prevailing aerobic/anaerobic conditions and therefore dependent on the type of wetland. For instance, in storm water biofiltration systems, prolonged periods of inundation and dry periods may support bio-assimilation over denitrificaton (Payne et al., 2014a, b).

The conditions that favour the occurrence of either denitrification or DNRA are still in debate (Rütting et al., 2011). DNRA is thought to be favoured by a $C:NO_3^-$ ratio of > 12 (Rütting et al., 2011) and occurs at low levels of oxidation-reduction potential (Thayalakumaran et al., 2008). The differences between denitrification and DNRA may be due to the availability of organic matter, because DNRA is favoured at a high $C:NO_3^-$ ratio and denitrification is favoured when carbon supplies are limiting (Korom, 2002; Kelso et al., 1997). The fermentative bacteria that carry out DNRA are obligate anaerobes, and so cannot occupy all the niches that denitrifiers can (Buss et al., 2005). Takaya (2002) stated that a more reducing state favours DNRA over denitrification.

Pett-Ridge et al. (2006) showed that DNRA is less sensitive to dissolved oxygen (DO) than denitrification. Fazzolari et al. (1998) showed that the partitioning between DNRA and denitrification depends on the $C:NO_3^-$ ratio and C rather than DO.. Significant DNRA may occur only at a $C:NO_3^-$ ratio above 12 (Yin et al., 1998). Different numbers of electrons are required in the reduction of each NO_3^- molecule: five for denitrification and eight for DNRA. Therefore, more organic matter can be oxidized for each molecule of NO_3^- by DNRA than by denitrification. In addition, NO_3^- reduction is generally performed by fermentative bacteria that are not dependent on the presence of NO_3^- for growth under anaerobic conditions. Therefore, DNRA bacteria may be favoured in NO_3^--limited conditions (Laanbroek, 1990). Recent studies have suggested that DNRA may be an important process compared to denitrification in wetland sediments (Burgin and Hamilton, 2008). Van Oostrom and Russell (1994) found a 5 % contribution of DNRA to NO_3^- removal in CWs. Little is known about the eventual fate of the NO_3^- that is converted to NH_4^+ via DNRA pathways. In recent years, N-cycling studies have increasingly investigated DNRA in various ecosystems to explore its importance in N cycling (Rütting et al., 2011), but controls on DNRA are relatively unknown (Burgin et al., 2013), DNRA being probably the least studied process of N transformation in wetlands (Vymazal, 2007). However, DNRA can be a significant pathway of NO_3^- reduction that impacts on the CW ecosystem services and so should therefore be evaluated.

Denitrification has been estimated to be a significant N removal process, but actual quantification data are scarce. Few studies have estimated N losses by denitrification, e.g. 19 % (Mander et al., 2008) and 86 % (Obarska-Pempkowiak and Gajewska, 2003) of the total N input based on the mass balance study. To our knowledge, no data are available on denitrification measurements in soil/subsoils of surface flow CWs. While many of these pathways transfer Nr (mainly NH_4^+ and N_2O) to the environment, other pathways can convert Nr to N_2 (e.g. denitrification, anammox, and deamox). Anammox can remove NO_2^- and NH_4^+ as N_2 when the existing environment is hypoxic. Deamox can remove NO_3^- and NH_4^+ as N_2, where NO_3^- is converted to NO_2^- by autotrophic denitrification with sulfide (Kalyuzhnyi et al., 2006). In CWs, anammox and deamox are not well understood, so it is crucial to identify which of the processes are occurring in a specific type of CW and the rate at which they occur. Once a process that provides N_2 as the end product is determined, then the management of the CW can be directed towards enhancement of that process. Hence, quantifying the rates of these processes for various types of CW is required for improved N management towards lowering Nr in the environment.

The various components of the C cycle include: fixation of C by photosysnthesis, respiration, fermentation, methanogenesis ,and CH_4 oxidation with reduction of sulfur, iron, and NO_3^-. Anaerobic methane oxidation coupled with denitrification, a recently proposed pathway of the C cycle (á Norði and Thamdrup, 2014; Haroon et al., 2013; Islas-Lima et al., 2004), can reduce CH_4 emissions in CWs. The C removal processes are sedimentation, microbial assimilation, gaseous emissions, dissolved C losses through water to ground- and surface water bodies, and chemical fixation (bonding with chemical ions). Net primary productivity of wetland hydrophytes varies across CW type, season, climatic region, and local environmental conditions. For example, results can vary remarkably for CWs containing the same plant species in different geographical regions (Brix et al., 2001). The rate of carbon mineralization in CW sediments depends on the redox chemistry of soil, the bio-availability of organic C and temperature. In particular, areas of sediment subjected to prolonged low redox conditions (e.g, $-150\,mV$) are conducive to methanogens and rates of CH_4 emissions exceeding $132\,mg\,m^{-2}\,d^{-1}$ (Brix et al., 2001), but this is highly variable depending on C:N ratio of the influent water and wetland seasonality. In summer, oxygen diffusion to the topsoil can reduce methanogenesis and stimulate CH_4 oxidation (Grünfeld and Brix, 1999). However, an increase in temperature can decrease DO in deeper subsoil layers, which can enhance CH_4 production. As in all biochemical reactions, temperature increases C and N turnover in CWs, causing high variations in GHG emissions in different types of CWs in different regions (temperate/tropical/arctic). This warrants the acquisition of more measurement data across CW types and regions for the better extrapolation of GHG emissions. The C:N ratios of wastewater affect microbial growth and development that, in turn, affect their response to C and N cycles and GHG emissions. Previous research on the effects of C:N ratios on nutrient removal and GHG emissions is limited. A few examples include Yan et al. (2012) and Zhao et al. (2014), who measured lower CO_2 and CH_4 emissions at C:N ratios of between 2.5:1 and 5:1, but this lower range of C:N ratios decreased TOC removal. Hence, investigation of the influence of C:N ratio on nutrient removal efficiencies and GHG emissions across CW and management types is crucial.

Emissions of GHG in CWs can vary across CW typologies, e.g. surface flow or subsurface flow (Tables 3 and 4). Generally, CH_4 emissions are higher in SF CWs than in SSF CWs (Table 3), but may vary with season, which requires investigation. Nitrous oxide and CO_2 emissions are higher in VSSF CWs than horizontal subsurface flow (HSSF) and SF CWs. The N_2O EF (N_2O / TN input $\times$ 100) ranged from 0.61 ± 0.21 % in SF CWs to 1.01 ± 0.48 % in VSSF CWs. The EF for CH_4 emissions ranged from 1.27 ± 0.31 % in VSSF CWs to 16.8 ± 3.8 % in SF CWs. The GHG from CWs can vary between vegetated and non-vegetated systems (Table 5).

Aquatic plants play an important role in GHG production and transport to the atmosphere by releasing GHG through their interconnected internal gas lacunas (Laanbroek, 2010). Emergent plants can transport atmospheric oxygen to the

Table 3. Nitrous oxide (N$_2$O) emissions (mg N m^{-2} d^{-1}); N$_2$ emissions (mg N m^{-2} d^{-1}) and N$_2$O emission factor (N$_2$O / TN input $\times$ 100) in various type of CWs; mean standard error ($\pm$ SE) was presented for N$_2$O emission factor; NA – data not available.

CW type	Treatment	Denitrification		N$_2$O-N / TN (%)	N$_2$-N / TN (%)	References
		N$_2$O emissions (mg N m^{-2} d^{-1})	N$_2$ emissions (mg N m^{-2} d^{-1})			
HSF	Agril. tile drainage	0.01–0.12	NA	0.19–1.4	NA	Xue et al. (1999)
HSF	Treated municipal	2.0 ± 3.3	NA	0.02–0.27	NA	Johansson et al. (2003)
HSF	Agril. drainage	−0.2–1.9	NA	−0.14–0.52	NA	Wild et al. (2002)
HSF	Dairy wash out	16.8 ± 7.0	NA	0.33 ± 0.12	NA	Van der Zaag et al. (2010)
HSF_Finland	Municipal	0.01 ± 0.01	NA	1.6 ± 1.3	NA	Søvik et al. (2006)
HSF_Finland	Agril. runoff	0.40 ± 0.25	NA	0.37 ± 0.18	NA	Søvik et al. (2006)
HSF_Norway	Municipal	4.0 ± 1.6	NA	1.5 ± 4.4	NA	Søvik et al. (2006)
All SF		2.78 ± 1.72		0.61 ± 0.21		
HSSF	Domestic	0.2–17.0	NA	0.06–3.8	NA	Mander et al. (2005)
HSSF_Estonia	Municipal	7.1 ± 1.2	NA	0.05 ± 0.31	NA	Søvik et al. (2006)
HSSF_Norway	Municipal	6.9 ± 4.3	NA	0.24 ± 0.53	NA	Søvik et al. (2006)
HSSF	Domestic	1.3–1.4	160–170	0.37–0.60	15.2–22.7	Mander et al. (2008)
HSSF	Domestic	0.003–0.001	0.01–5.42	NA	NA	Teiter and Mander (2005)
HSSF	Domestic	0.13	NA	0.008	NA	Fey et al. (1999)
HSSF	Dairy wash out	9.5 ± 1.5	NA	0.18 ± 0.12	NA	Van der Zaag et al. (2010)
HSSF	Domestic	0.17	NA	0.23	NA	Liu et al. (2009)
VSSF	Domestic	0.17	NA	0.01		Mander et al. (2011)
All HSSF		4.23 ± 1.87		0.62 ± 0.38		
VSSF	Domestic	0.001–0.002	0.01–5.0	NA	NA	Teiter and Mander (2005)
VSSF	Domestic	4.6	150	0.45–0.50	NA	Mander et al. (2008)
VSSF	Domestic	11.0	NA	0.29	NA	Mander et al. (2005)
VSSF	Domestic	1.44	NA	0.03		Mander et al. (2011)
VSSF	Domestic	0.005	NA	0.09	NA	Gui et al. (2007)
VSSF	Domestic	0.003	NA	0.04	NA	Liu et al. (2009)
VSSF_Estonia	Municipal	15 ± 3.9	NA	04.3 ± 0.95		Søvik et al. (2006)
VSSF_Norway	Municipal	960 ± 40	NA	1.4 ± 0.72		Søvik et al. (2006)
All VSSF		123.8 ± 106		1.01 ± 0.48		

SF – surface flow; HSSF – horizontal subsurface flow; VSSF – vertical subsurface flow.

rooting zone and contribute to increased N$_2$O and CO$_2$ production and CH$_4$ consumption (Brix, 1997). Vascular plants can exchange GHG between the rooting zone and atmosphere (Yavitt and Knapp, 1998). Vegetation and its composition affect the nutrient dynamics and the production, consumption, and transport of GHG and hence their exchange between wetlands and atmosphere (Ström et al., 2003, 2005; Søvic et al., 2006; Johansson et al., 2003). They can also affect the biogeochemistry of CWs due to the differences in their growth and development, longevity, root systems, root density, root depth, and microbial ecology in the rhizosphere. As some plant litter decomposes, organic matter with lignocellulose and humic compounds may be released that are more or less labile or stable in nature than others. Release of low molecular weight organic matter that is labile in nature is more likely to produce GHGs than stable forms. For example, *Z. latifolia* showed higher nutrient removal and CH$_4$ fluxes than *P. australis* (Inamori et al., 2007). The *Z. lotifolia* root system is shallow and the activity of methanotrophs is primarily confined to the top soil. The root systems of *P. australis* are deeper, which is more favourable for the oxidization of CH$_4$.

A fluctuating water table in CWs has significant impacts on GHG dynamics. Pulsing hydrologic regimes decreases CH$_4$ but increases N$_2$O emissions (Mander et al., 2011). In aerobic and anaerobic conditions caused by pulsing hydrology, incomplete nitrification and denitrification increase N$_2$O emissions (Healy et al. 2007). However, the effects of pulsing hydrologic regimes on GHG emissions are contradictory. For example, intermittent hydrologic regimes decrease both N$_2$O (Sha et al., 2011) and CH$_4$ emissions (Song et al., 2010). Highly contrasting results on gas emissions with fluctuating water levels have been reported and the controlling mechanisms are unclear (Elberling et al., 2011).

Therefore, the assessment of GHG emissions in various types of CWs (surface flow, subsurface flow, vertical and horizontal), vegetation cover (vegetated, non-vegetated) and species type, and management system employed (HLR, HRT, soil used, and water table), is necessary in light of the national and global GHG budgets. In addition, such measurements will help scientists, environmental managers, and policy makers adopt environmentally friendly construction and management of CWs. The enhanced reduction of N$_2$O to N$_2$ needs further elucidation.

Table 4. Carbon dioxide (CO_2; $mg\,C\,m^{-2}\,d^{-1}$), CH_4 ($mg\,C\,m^{-2}\,d^{-1}$), and CH_4 emission factor (CH_4–C / TOC input $\times$ 100) in various types of CWs; mean standard error ($\pm$ SE) was presented for CH_4 emission factor; NA – data not available.

CWs type	Treatment	CO_2 emissions ($mg\,C\,m^{-2}\,d^{-1}$)	CH_4 emissions ($mg\,C\,m^{-2}\,d^{-1}$)	CH_4 / TC (%)	References
SF	Municipal	NA	5.4	NA	Tai et al. (2002)
SF	Domestic	0.19	NA	26	Gui et al. (2007)
SF	Domestic	1.13	NA	16	Liu et al. (2009)
SF	Agril. drainage	NA	0.88	31	Wild et al. (2002)
SF	Dairy wash out	4250 ± 550	223 ± 35	9.45	Van der Zaag et al. (2010)
SF_Finland	Municipal	1200 ± 420	29 ± 6.4	19 ± 4.3	Søvik et al. (2006)
SF_Finland	Agril runoff	3200 ± 560	350 ± 180	11 ± 5.5	Søvik et al. (2006)
SF_Norway	Municipal	1400 ± 250	72 ± 28	4.8 ± 2.2	Søvik et al. (2006)
All SF		1675 ± 703	113 ± 58	16.8 ± 3.8	
HSSF	Domestic	NA	1.7–528	NA	Mander et al. (2005)
HSSF	Domestic	2.54–5.83	0.03–0.40	NA	Teiter and Mander (2005)
HSSF	Domestic	5.33	0.001	0.03	Garcia et al. (2007)
HSSF	Domestic	NA	0.03	4.3	Gui et al. (2007)
HSSF	Domestic	NA	0.29	4.0	Liu et al. (2009)
HSSF	Dairy wash out	3475 ± 375	118 ± 9.0	4.4	Van der Zaag et al. (2010)
HSSF	Domestic	0.6–1.7	1.4–4.1	0.12–0.23	Søvik et al. (2006)
HSSF	Domestic	600	0.48	0.02	Mander et al. (2011)
HSSF_Estonia	Municipal	3800 ± 210	340 ± 240	NA	Søvik et al. (2006)
HSSF_Norway	Municipal	790 ± 170	130 ± 43	9.5 ± 3.3	Søvik et al. (2006)
All HSSF		1010 ± 672	112 ± 74	3.23 ± 1.4	
VSSF	Domestic	5.83–12.13	0.60–5.70		Teiter and Mander (2005)
VSSF	Domestic	NA	16.4	NA	Mander et al. (2005)
VSSF	Domestic	NA	0.013	1.68	Gui et al. (2007)
VSSF	Domestic	NA	0.13	1.73	Liu et al. (2009)
VSSF	Municipal	2662 ± 175	33.5 ± 3.2	NA	Mander et al. (2008)
VSSF	Domestic	1080	3.36	0.05	Mander et al. (2011)
VSSF_Estonia	Municipal	8400 ± 2100	110 ± 35	NA	Søvik et al. (2006)
VSSF_Norway	Municipal	22000 ± 5000	140 ± 160	0.39 ± 0.27	Søvik et al. (2006)
All VSSF		6616 ± 3779	42.9 ± 23.7	1.27 ± 0.31	

SF – surface flow; HSSF – horizontal subsurface flow; VSSF – vertical subsurface flow.

5 Surface emissions vs. subsurface export of C and N

Dissolved GHG produced in soils and subsoils can be emitted to the atmosphere by transpiration of vascular plants (from within the rooting zone), ebullition, and diffusion from soils. Elberling et al. (2011) reported that in wetlands, the transport of gases through subsoil occurs both via diffusive transport in the pores and through the vascular plants. Surface emissions of GHG from CWs have recently been recognized and have been commonly measured by chamber methods (Mander et al., 2008, 2011). As is the case with other dissolved pollutants (Dzakpasu et al., 2014), the GHG produced in CWs can also be transported to the groundwater with the percolating water and emitted to the atmosphere upon discharge to surface waters (Riya et al., 2010). It can also flow towards surface waters by advective transport and/or by dispersion of groundwater. Dissolved nutrients can be preferentially leached down into deeper soil layers and groundwater via different pathways (e.g. root channels). The Nr delivered to groundwater can be transformed in situ to other reactive or benign forms. Hence, quantification of such Nr loadings to groundwater and their in situ consumption (e.g. N_2O to N_2 or CH_4 to CO_2) is necessary to understand their environmental consequences. In addition, DON, NO_3^-, NH_4^+, and DOC delivered to surface waters can undergo biochemical reactions and produce N_2O, CO_2, and CH_4 in streams and estuaries. Ström et al. (2007) measured a considerable quantity of CH_4 in porewater and found a correlation between the surface emissions and porewater CH_4 concentrations in vegetated wetlands. Measuring only the surface emissions of GHG can omit substantial quantities of GHG released from CWs. For example, Riya et al. (2010) measured emissions of CH_4 and N_2O, accounting for 2.9 and 87 % of the total emissions. Measuring porewater GHG and linking these to the surface emissions and subsurface export to groundwater below CWs will help to estimate a better GHG balance from both a national and global context. Elberling et al. (2011)

Table 5. Nitrous oxide (N_2O; mg N m^{-2} d^{-1}), CO_2, and CH_4 emissions (mg C m^{-2} d^{-1}) in various type of CWs under different plant types; NA – data not available.

CW type	Wastewater type	Plant type	N_2O (mg N m^{-2} d^{-1})	CH_4 (mg C m^{-2} d^{-1})	CO_2 (mg m^{-2} d^{-1})	Reference
HSF	Secondary treated	No plant	3.79 ± 2.64	163 ± 209		Johansson et al. (2003); Johansson et al. (2004)
	municipal	*Typha lotifolia*	2.64 ± 4.09	109 ± 185	NA	
		Phalaris arundinacea	3.79 ± 3.44	212 ± 151	NA	
		Glyceria maxima	0.76 ± 1.01	112 ± 178	NA	
		Lemna minor	1.45 ± 1.18	450 ± 182	NA	
		Spirogyra sp.	0.98 ± 1.25	107 ± 135	NA	
HSF	Sewage treatment water	No plant	-0.26 ± 2.53	-4.76 ± 61.8	4.32 ± 0.73	Ström et al. (2007)
		Typha atifolia	4.94 ± 2.00	225 ± 47.7	25.3 ± 4.08	
		Phragmites australis	7.80 ± 2.53	333 ± 76.6	25.1 ± 4.74	
		Juncus effusus	3.87 ± 1.86	489 ± 46.3	26.1 ± 3.00	
HSSF	Domestic	No plant	0.04 ± 0.02	87 ± 6.3	80 ± 6.3	Maltais-Landry et al. (2009)
		Phragmites	0.06 ± 0.03	50 ± 7.5	200 ± 35	
		Typha	0.03 ± 0.01	28 ± 3.0	235 ± 32	
		Phalaris	0.01 ± 0.01	45 ± 6.0	195 ± 31	
VSSF	Municipal	*Phragmites australis*	15 ± 3.9	110 ± 35	8400 ± 2100	Søvik et al. (2006)
VSSF	Municipal	*Phragmites australis*	264	384		Mander et al. (2005)

SF – surface flow; HSSF – horizontal subsurface flow; VSSF – vertical subsurface flow.

linked subsurface gas concentrations in wetlands to the surface fluxes using a diffusion model. This demonstrates the need for future studies on subsurface GHG production, consumption and net GHG emissions in CWs within a climate change context.

It is important to characterize soils and subsoils' physical (e.g. texture, bulk density) and hydraulic (development of a soil water characteristic curve) properties and to assess their potential to percolate dissolved nutrients and gases in the solute phase to the underlying groundwater. To our knowledge, the indirect pathway of GHG emissions from CWs has never been reported, despite the fact that this would appear to have a high biogeochemical potential to produce and exchange GHG. The balance between N and C input and output flows between CWs and aquatic and atmospheric environments, together with the direct and indirect emissions of C and N species, could be an important input to global C and N budgets.

6   Hydrogeochemistry below CWs

Constructed wetlands can be designed with or without a clay liner or a compacted soil bed at the base, which can lead to large differences in permeability of the underlying layers. The variation in permeability of a CW soil bed will affect solute, nutrient, and GHG flows, and their interactions with the underlying groundwater (Dzakpasu et al., 2012, 2014). Groundwater hydrogeochemistry below CWs can therefore provide a unique insight into such interactions. An example of such interactions would be between nutrient-rich water discharging from CW cells mixing with laterally moving regional groundwater. It should be noted that groundwater can also discharge into CWs depending on the hydraulic gradi-

ents. This means that fully screened, multi-level piezometers or boreholes should be installed at such sites to elucidate groundwater flow direction, hydraulic gradients, and conductivities. Such monitoring networks allow water samples to be collected and the sources of nutrients in groundwater bodies below CWs to be identified. The local site hydrology (precipitation, groundwater table fluctuations, and evapotranspiration) has a large impact on the pollutant removal. Hydrogeochemical studies at an accurate spatial and temporal resolution should explain the effects of precipitation on nutrient removal by dilution as well in situ nutrient turnover. Effective CW management requires an understanding of the effects of wetland hydrology on the physical and biochemical attenuation of nutrients in order to assess their impacts on the surface emissions and subsurface export of nutrients and GHG. Data on the species of N in groundwater below the CWs are required to provide an in-depth understanding of wetland ecosystem services, particularly if CWs have the potential to leak pollutants down into the groundwater (Dzakpasu et al., 2014). Higher NH_4^+ concentrations in groundwater below the CW than the effluent are often reported (Harrington et al., 2007; Dzakpasu et al., 2012). Therefore, questions arise with respect to NH_4^+ concentrations in groundwater below the CWs if they have been transported from CWs. Linking geochemistry of groundwater below CWs to site hydrology, water table fluctuations, and soil/subsoil physico-chemical properties is required to elucidate the major environmental drivers of C and N removal, and/or pollution swapping. The quality of groundwater underlying CWs with regards to the Nr species is largely unknown.

7 Methodological developments

To improve the ecosystem services and to minimize the pollution swapping of CWs, quantification of N cycling is crucial. Measurement of GHG using the closed chamber method is widely used, but has large uncertainty in estimating the diurnal variability due to internal changes in temperature and physical access to the chambers over a 24 h time period. Gas ebullition and diffusion measurements are quite challenging in CWs covered by vegetation, because of the difficulties in estimation of gas transfer velocity. Application of the eddy-covariance method is not appropriate for most CWs, as it requires a large surface area (> several ha) to avoid contribution of surrounding area and complication of GHG foot printing. A combination of chamber, ebullition, and diffusion methods in a single system could minimize the uncertainly in GHG estimation. The methane ebullition measurement was found to be similar to surface emissions by the chamber method, but N_2O and CO_2 ebullition measurements were lower than the surface emissions (Søvik et al., 2006).

The use of in situ microcosm studies and soil core incubation methods may give a better estimation of N_2O, CO_2, and CH_4 production and consumption than existing methods. With the recent advancement of isotope pairing and dilution techniques, single or simultaneously occurring C and N transformation processes can be quantified in laboratory or in situ conditions (Huygens et al., 2013; Müller et al., 2014). The isotope technique relies on the introduction of a known amount of ^{14}C and or ^{15}N into the CW and then quantification of C and N concentrations and isotopic compositions through different C and N pools after incubation for a specific period. Laboratory methods involve collection of intact soil/sediment cores, with subsequent incubation in the laboratory. In situ field techniques involve the release of a $^{14}C / ^{15}N$ solution in the CW soils. Incubation of intact soil cores with differentially labelled $^{15}NH_4^{14}NO_3$ and $^{14}NH_4^{15}NO_3$ can be used to quantify the rates of different N transformation processes (Rütting and Müller, 2008). The quantification of simultaneously occurring N transformation rates rely on the analysis with appropriate ^{15}N-tracing models. In recent years, ^{15}N-tracing techniques have evolved, and are now able to identify process-specific NO_2^- pools (Rütting and Müller 2008), pathway-specific N_2O production, and emission, as well as $N_2O : N_2$ ratios (Müller et al., 2014). Traditional techniques for investigation of gross N dynamics in sediments (Blackburn, 1979) may be combined with the latest ^{15}N-tracing techniques, where all N transformation rates are included (Huygens et al., 2013). Thus, current models should consider processes such as anammox and/or deamox, and then be tested in CWs under various operational conditions. Denitrification in porewater samples can be measured by analysing samples for dissolved N_2 in a membrane inlet mass spectrometer (MIMS; Kana et al., 1994) and N_2O in a gas chromatograph (GC; Jahangir et al., 2012). The studies of natural abundance of ^{15}N and ^{18}O ($\delta^{15}N$ and

$\delta^{18}O$) in NO_3^- is an insightful tool for the investigation of the sources, fate, and transformational processes of N in a system (e.g. in shallow groundwater; Baily et al., 2011). The in situ NO_3^- push–pull method has been used to determine denitrification in shallow groundwater (< 3 m) in riparian wetlands (Addy et al., 2002; Kellogg et al., 2005) and in deep groundwater in arable/grassland (Jahangir et al., 2013).

Isotope-based techniques can also be extended to other elements; e.g., a ^{33}P-tracing model has been developed recently to study phosphorus (P) cycle in soil (Müller and Bünemann, 2014). These techniques can be applied in the study of C, N, and P biogeochemistry in aquatic environments. In addition, measurements of DOC and gases (CO_2 and CH_4) will provide insights into the C consumption and transformation associated with the N transformations. Carbon and N dynamics are influenced by the interacting effects of soil conditions with microbial community structure and functioning. Microbial functioning involves transcription of genes, translation of messenger RNA, and activity of enzymes (Firestone et al., 2012). As such, activities of microbial communities under various environmental conditions and how these contribute to C and N dynamics is a very important area of future research (Müller and Clough, 2014). Molecular approaches can be important tools for identifying and quantifying the genes that code for enzyme-mediating C and N cycles (Peterson et al., 2012). These tools help assess the relationships among genes, environmental controllers, and the rates of C and N processes. The scientific tools and multi-disciplinary techniques are now available to better understand C and N transformation rates, processes, and factors controlling the unwanted emission of N and C products to the environment.

8 Conclusions and recommendations

The transformational processes on a mixture of contaminants within and below CWs can cause pollution swapping. A holistic assessment of C and N dynamics in CWs is needed to fully understand their removal, transport, and impact on water quality and emissions to atmosphere. Mixed contaminants entering CWs and those formed within and underneath CWs during transformational processes must be considered in future studies. The overall balance of these constituents will determine whether a CW is a pollution source or a sink. This will necessitate a higher degree of multi-level spatial and temporal monitoring and the use of multi-disciplinary in and ex situ techniques to fully characterize all pathways of C and N loss. At this time we cannot suggest any design optima in terms of nutrient removal and GHG mitigation because empirical information is not yet abundant. To do this, transformation kinetics of C and N and net GHG emissions through all possible pathways are required to provide a holistic assessment. However, a combination of various types of CW and plant types could provide higher removals and lower

CW type	N$_2$O	CH$_4$	CO$_2$	C-cycling	N-cycling	fate of C&N	Research needed
SF	decreases	increases	decreases	?	?	?	Effects of combination of surface and subsurface flow CWs on C and N cycling processes. Measurement of C and N cycle processes and estimation of their individual rate kinetics in the combined CW system. Effects of CW type and management on microbial community structure and functional gene abundance and diversity in CW soil and water
HSSF	decreases	increases	decreases	?	?	?	
VSSF	increases	decreases	increases	?	?	?	
Vegetation type							
Shallow rooted	?	increases	?	?	?	?	Effects of vegetation type on the fractionation of SOC (labile vs. stable), nutrient removal and GHG emissions. Investigate in situ C, N transformations and GHG production and consumption in soil, subsoil and water column of CWs
Deep rooted	?	decreases	?	?	?	?	
Wastewater quality							
C:N ratio	decreases	?	?	?	?	?	Comparative study of denitrification and DNRA in various C:N ratios in different CW types and management regimes. Investigation of soil infiltration rates and their contribution to dissolved C, N and GHG losses to groundwater below Cws. Impact of combined engineering of CW types on NH4+ oxidation, TN removal and N$_2$O reduction to N$_2$. Towards a complete C, N and GHG balance in Cws by integrating all possible pathways of C and N transformation and movement. Link groundwater quality to CW hydrogeochemistry under various CW types and management regimes
NH4+ pollution	decreases	?	?	?	?	?	
NO3- pollution	increases	?	?	?	?	?	
TOC pollution	?	increases	increases	?	?	?	
Soil properties	?	?	?	?	?	?	
HLR	increases	?	?	?	?	?	Investigate the impact of various HLR and HRT on C and N transformation and GHG emissions. Investigate the impact of hydrologic pulsing on C and N removal and GHG emissions across CW type and management regime.
HRT	decreases	?	?	?	?	?	
Hydologic pulsing	increases	increase	?	?	?		

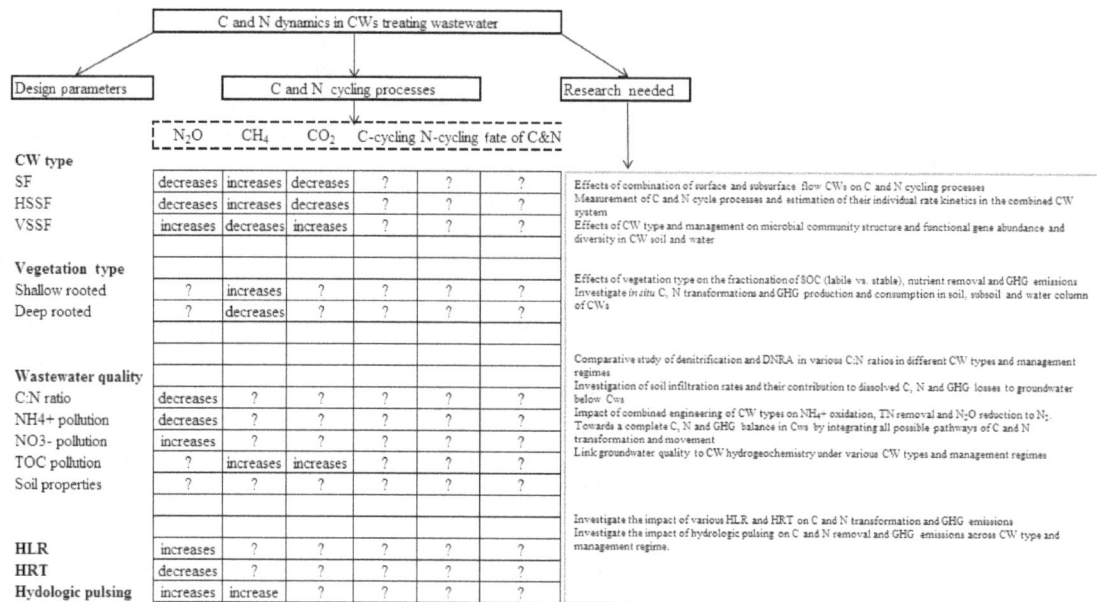

Figure 1. Conceptual model showing the current state of knowledge of C and N dynamics in constructed wetlands treating wastewater and the specific experimental work that needs to be undertaken in the future; SF – surface flow; HSSF – horizontal subsurface flow; VSSF – vertical subsurface flow; HLR – hydraulic loading rate; HTR – hydraulic retention time; ? – not known or very little known.

GHG emissions. A conceptual model highlighting the current state of knowledge in this area and the research gaps is presented in Fig. 1.

Subsurface export of nutrients and GHG to groundwater should be accounted for in CW management. Reducing the saturated hydraulic conductivity below the wetland bed will help reduce nutrients leaching to groundwater. The reactive versus the benign forms of the N transformation products should be evaluated. Data on when, where, and the rates at which denitrification, deamox, and anammox occur in CWs are needed, as well as identification of the key factors that control such processes. The provenance of NH$_4^+$ in groundwater below CW cells and its impact on down-gradient receptors needs further elucidation. Constructed wetlands have the potential to produce N$_2$O, DON, DOC, dissolved inorganic C (DIC), CO$_2$, and CH$_4$, which may be exported to fresh waters via groundwater and degassed upon discharge to surface waters. Moreover, the DOC and DIC transferred to the fresh water sediments (rivers and lakes) can produce GHG that, in turn, emit to the atmosphere. The amount of C and N exported from terrestrial ecosystems via the subsurface pathway to fresh waters has been the missing piece of our understanding of global C and N budgets. It is clear that data on the various C and N species, along with the GHG emissions, are crucial to make a robust input–output balance of C and N in CWs. Spatial and temporal variations of GHG emissions in CWs under different management systems are also critical to get much more rigorous estimates of emission factors. These data will reduce the existing uncertainties in global C and N budgets.

Managing wetting and drying spells (pulsing hydrology) in CWs can enhance NH$_4^+$ removal. Similarly, oxidation of organic C will increase CO$_2$ production and, in anaerobic conditions, may be reduced to CH$_4$. This requires more research into the C and N cycle processes over the wetting and drying spells, which is now possible with the advancement in ^{14}C / ^{15}N-tracing and modelling techniques. The selection of appropriate plant species is important to optimize nutrient removal, sequester C, and decrease GHG emissions, but more research is needed across species and geographical locations. Further research is also needed to investigate the impacts of hydraulic retention time on nutrient dynamics. Rates of nutrient accumulation or fixation in soils and their in situ transformation in CWs need to be quantified to evaluate their contribution to C sequestration and GHG emissions.

Author contributions. The first author, M. M. R. Jahangir has reviewed articles in the relevant area, analysed results, identified knowledge gaps, and prepared the draft paper. All co-authors were directly involved in preparation of the paper and edited the paper for its improvement.

Acknowledgements. The research was funded by Irish Research Council and Department of Agriculture, Food and Marine in Association with The University of Dublin, Trinity College.

References

á Norði, K. and Thamdrup, B.: Nitrate-dependent anaerobic methane oxidation in a freshwater sediment, GeochiM. Cosmochim. Acta, 132, 141–150, 2014.

Addy, K., Kellogg, D. Q., Gold, A. J., Groffman, P. M., Ferendo, G., and Sawyer, C.: *In situ* push-pull method to determine groundwater denitrification in riparian zones, J. Environ. Qual., 31, 1017–1024, 2002.

Bachand, P. A. M. and Horne, A. J.: Denitrification in constructed free-water surface wetlands: II. Effects of vegetation and temperature, Ecol. Eng., 14, 17–32, 2000.

Baily, A., Rock, L., Watson, C. J., and Fenton, O.: Spatial and temporal variations in groundwater nitrate at an intensive dairy farm in south-east Ireland: Insights from stable isotope data, Agril. Ecosysts. Environ., 308–318, 2011.

Beaulieu, J. J., Tank, J. L., Hamilton, S. K., Wollheim, W. M., Hall, R. O., and Mulholland, P. J.: Nitrous oxide emission from denitrification in stream and river networks, P. Natl. Acad. Sci., 108, 214–219, 2011.

Blackburn, T. H.: Methods for measuring rates of NH_4^+ turnover in anoxic marine sediments, using a ^{15}N-NH_4^+ dilution technique, Appl. Environ. Microbiol., 37, 760–765, 1979.

Bojcevska, H. and Tonderski, K.: Impact of loads, season, and plant species on the performance of a tropical constructed wetland polishing effluent from sugar factory stabilization ponds, Ecol. Eng., 29, 66–76, 2007.

Brix, H.: Do macrophytes play a role in constructed treatment wetlands, Water Sci. Technol., 35, 11–17, 1997.

Brix, H., Sorrell, B. K., and Lorenzen, B.: Are Phragmites-dominated wetlands a net source or net sink of greenhouse gases?, Aquat. Bot., 69, 313–324, 2001.

Burgin, A. J., Hamilton, S. K., Gardner, W. S., and McCarthy, M. J.: Nitrate reduction, denitrification, and dissimilatory nitrate reduction to ammonium in wetland sediments, in: Methods in Biogeochemistry of Wetlands, edited by: DeLaune, R. D., Reddy, K. R., Richardson, C. J., and Megonigal, J. P., SSSA Book Series, no. 10, Madison, USA, 307–325, 2013.

Calheiros, C. S. C., Rangel, A. O. S. S., and Castro, P. M. L.: Constructed wetland systems vegetated with different plants applied to the treatment of tannery wastewater, Water Res., 41, 1790–1798, 2007.

Clair, T. A., Arp, P., Moore, T. R., Dalva, M., and Meng, F. R.: Gaseous carbon dioxide and methane, as well as dissolved organic carbon losses from a small temperate wetland under a changing climate, Environ. Pollut., 116, S143–S148, 2002.

Conley, D. J., Paerl, H. W., Howarth, R. W. Boesch, D. F., Seitzinger, S. P., Havens, K. E., Lancelot, C., and Likens, G. E.: Controlling eutrophication: Nitrogen and phosphorus, Science, 323, 1014–1015, 2009.

da Motta Marques D. M. L., Leite, G. R., and Giovannini, S. G. T.: Performance of two macrophyte species in experimental wetlands receiving variable loads of anaerobically treated municipal wastewater, Water Sci. Technol., 44, 311–316, 2001.

Drizo, A., Frost, C. A., Grace, J., and Smith, K. A.: Physico-chemical screening of phosphate-removing substrates for use in constructed wetland systems, Wat. Res., 33, 3595–3602, 1999.

Dzakpasu, M., Hofmann, O., Scholz, M., Harrington, R., Jordan, S. N., and McCarthy, V.: Nitrogen removal in an integrated constructed wetland treating domestic wastewater, J. Environ. Sci. Health, Part A, 46, 742–750, 2011.

Dzakpasua, M., Scholz, M., Harrington, R., Jordan, S. N., and McCarthy, V.: Characterising infiltration and contaminant migration beneath earthen-lined integrated constructed wetlands, Ecol. Eng., 41, 41–51, 2012.

Dzakpasu, M., Scholz, M., Harrington, R., McCarthy, V., and Jordan, S.: Groundwater Quality Impacts from a Full-Scale Integrated Constructed Wetland, Groundw. Monit. Remote, 34, 51–64, 2014.

Elberling, B., Louise A., Christian, J. J., Hans, P. J., Michael K., Ronnie N. G., and Frants, R. L.: Linking soil O_2, CO_2, and CH_4 concentrations in a wetland soil: implications for CO_2 and CH_4 fluxes, Environ. Sci. Technol., 45, 3393–3399, 2011.

Fazzolari, C. E., Nicolardot, B., and Germon, J. C.: Simultaneous effects of increasing levels of glucose and oxygen partial pressures on denitrification and dissimilatory reduction to ammonium in repacked soil cores, Eur. J. Soil Biol., 34, 47–52, 1998.

Fenton, O., Healy, M. G., Brennan, F., Jahangir, M. M. R., Lanigan, G. J., Richards, K. G., Thornton, S. F., and Ibrahim, T. G.: Permeable reactive interceptors – blocking diffuse nutrient and greenhouse gas losses in key areas of the farming landscape, J. Agric. Sci., 152, S71–S81, 2014.

Fey, A., Benckiser, G., and Ottow, J. C. G.: Emissions of nitrous oxide from a constructed wetland using a ground filter and macrophytes in waste-water purification of a dairy farm, Biol. Fertil. Soils, 29, 354–359, 1999.

Findlay, S., Groffman, P., and Dye, S.: Effects of *Phragmites australis* removal on marsh nutrient cycling, Wetlands Ecol. Manage., 11, 157–165, 2003.

Firestone, M., Blazewicz, S., Peterson, D. G., and Placella, S.: Can molecular microbial ecology provide new understanding of soil nitrogen dynamics, edited by: Richards, K. G., Fenton, O., Watson, C. J., in: Proceedings of the 17th Nitrogen Workshop, 11–14, 2012.

Galloway, J. N., Aber, J. D., Erisman, J. W., Seitzinger, S. P., Howarth, R. W., Cowling, E. B., and Cosby, B. J.: The nitrogen cascade, BioSci., 53, 341–356, 2003.

Garcia, J., Capel, V., Castro, A., Ruiz, I., and Soto, M.: Anaerobic biodegradation tests and gas emissions from subsurface flow constructed wetlands, Bioresour. Technol., 98, 3044–3052, 2007.

Gray, N. F.: Drinking water quality: problems and solutions, 2nd edn., Cambridge Univ. Press, Cambridge, UK, 116–134, 2008

Grünfeld, S. and Brix, H.: Methanogenesis and methane emissions: effects of water table, substrate type and presence of *Phragmites australis*, Aquat,. Bot., 64, 63–75, 1999.

Gui, P., Inamori, R., Matsumura, M., and Inamori, Y.: Evaluation of constructedwetlands by waste water purification ability and greenhouse gas emissions, Water Sci. Technol., 56, 49–55, 2007.

Haroon, M. F., Hu, S., Shi, Y., Imelfort, M., Keller, J., Hugenholtz, P., Yuan, Z., and Tyson, G. W.: Anaerobic oxidation of methane coupled to nitrate reduction in a novel archaeal lineage, Nature, 500, 567–570, 2013.

Harrington, R., Carroll, P., Carty, A. H., Keohane, J., and Ryder, C.: Integrated constructed wetlands: concept, design, site evaluation and performance, Int. J. Water, 3, 243–255, 2007.

Healy, M. G., Rodgers, M., and Mulqueen, J.: Treatment of dairy wastewater using constructed wetlands and intermittent sand filters, Bioresour. Technol., 98, 2268–2281, 2007.

Healy, M. G., Ibrahim, T. G., Lanigan, G., Serrenho, A. J., and Fenton, O.: Nitrate removal rate, efficiency and pollution swapping potential of different organic carbon media in laboratory denitrification bioreactors, Ecol. Eng., 40, 198–209, 2011.

Healy, M. G., Barrett, M., Lanigan, G. J., João Serrenho, A., Ibrahim, T. G., Thornton, S. F., Rolfe, S. A., Huang, W. E., and Fenton, O.: Optimizing nitrate removal and evaluating pollution swapping trade-offs from laboratory denitrification bioreactors, Ecol. Eng., 74, 290–301, 2014

Huygens, D., Trimmer, M., Rütting, T., Müller, C., Heppell, C. M., Lansdown, K., and Boeckx, P.: Biogeochemical N cycling in wetland ecosystems: ^{15}N isotope techniques, Methods in biogeochemistry of wetlands, edited by: Reddy, K. R., Megonigal, J. P., and Delaune, R. D., Soil Sci. Soc. Amer., 30, 553–591, 2013.

Inamori, R., Gui, P., Dass, P., Matsumura, M., Xu, K.Q., Kondo, K., Ebie, Y., and Inamori, Y.: Investigating CH_4 and N_2O emissions from eco-engineering wastewater treatment processes using constructed wetland microcosms, Proc. Biochem, 42, 363–373, 2007.

IPCC: 2013 Supplement to the 2006 IPCC guidelines for national greenhouse gas inventories: wetlands, edited by: Hirashi, T., Krug, T., Tanabe, K., Srivastava, N., Baasansuren, J., Fukuda, M., Troxler, T. G., Switzerland, IPCC Task Force on National Greenhouse Gas Inventories, 354, 2013.

Islas-Lima, S., Thalasso, F., and Gomez-Hernandez, J.: Evidence of anoxic methane oxidation coupled to denitrification, Water Res., 38, 13–16, 2004.

Jahangir, M. M. R., Johnston, P., Grant, J., Somers, C., Khalil, M. I., and Richards, K. G.: Evaluation of headspace equilibration methods for measuring greenhouse gases in groundwater, J. Environ. Manage., 111, 208–212, 2012.

Jahangir, M. M. R., Johnston, P., Addy, K., Khalil, M. I., Groffman, P., and Richards, K. G.: Quantification of in situ denitrification rates in groundwater below an arable and a grassland system, Water Air Soil Pollut., 224, 1693, doi:10.1007/s11270-013-1693-z, 2013.

Johansson, A. E., Kasimir-Klemedtsson, A., Klemedtsson, L., and Svensson, B. H.: Nitrous oxide exchanges with the atmosphere of a constructed wetland treating wastewater, Tellus, 55B, 737–750, 2003.

Johansson, A. E., Gustavsson, A. M., Öquist, M. G., and Svensson, B. H.: Methane emissions from a constructed wetland treating wastewater – seasonal and spatial distribution and dependence on edaphic factors, Water Res., 38, 3960–3970, 2004.

Johansson, E.: Constructed wetlands and deconstructed discorses–greenhouse gas fluxes and discorses on purifying capacities, Dept. Water Environ. Studies, Linköping Univ., Sweden, 2002.

Kalyuzhnyi, S., Gladchenko, M., Mulder, A., and Versprille, B.: DEAMOX-new biological nitrogen, Water Res., 40, 3637–3645, 2006.

Kana, T. M., Darkangelo, C., Hunt, M. D., Oldham, J. B., Bennett, G. E, and Cornwell, J. C.: Membrane inlet mass spectrometer for rapid high precision determination N_2, O_2 and Ar in environmental water samples, Anal. Chem., 66, 4166–4170, 1994.

Kato, K., Koba, T., Ietsugu, H., Saigusa, T., Nozoe, T., Kobayashi, S., Kitagawa, K., and Yanagiya, S.: Early performance of hybrid reed bed system to treat milking parlour wastewater in cold climate in Japan, in: 10th international conference wetland systems

for water pollution control, MAOTDR, Lisbon, Portugal, 1111–1118, 2006.

Kellogg, D. Q., Gold, A. J., Groffman, P. M., Addy, K., Stolt, M. H., and Blazejewski, G.: In situ groundwater denitrification in stratified, permeable soils underlying riparian wetlands, J. Environ. Qual., 34, 524–533, 2005.

Kelso, B. H. L., Smith, R. V., Laughlin, R. J., and Lennox, S. D.: Dissmilatory nitrate reduction in anaerobic sediments leading to river nitrate accumulation, Appl. Enviro. Microbiol., 63, 4679–4685, 1997.

Korom, S. F.: Natural denitrification in the saturated zone: a review, Water Resour. Res., 28, 1657–1668, 2002.

Koskiaho, J., Ekholm, P., Räty, M., Riihimäki, J., and Puustinen, M.: Retaining agricultural nutrients in constructed wetlands- experiences under boreal conditions, Ecol. Eng., 20, 89–103, 2003.

Laanbroek, H. J.: Bacterial cycling of minerals that affect plant growth in waterlogged soils: a review, Aquat. Bot., 38, 109–125, 1990.

Laanbroek, H. J.: Methane emission from natural wetlands: interplay between emergent macrophytes and soil microbial processes, A mini-review, Ann. Bot., 105, 141–153, 2010.

Langergraber, G.: Modeling of processes in subsurface flow constructed wetlands: A review. Vadose Zone, J. 7, 830-842, 2008.

Lee, C., Fletcher, T. D., and Sun, G.: Nitrogen removal in constructed wetland systems, Eng. Life Sci., 9, 11–22, 2009.

Liikanen, A., Huttunen, J. T., Kaijalainen, S. M., Heikkinen, K. Vdisdinen, T. S., Nykiinen, H., and Martikainen, P. J.: Temporal and seasonal changes in greenhouse gas emissions from a constructed wetland purifying peat mining runoff water, Ecol. Eng., 26, 241–251, 2006.

Liu, C., Xu, K., Inamori, R., Ebie, Y., Liao, J., and Inamori, Y.: Pilot-scale studies of domestic wastewater treatment by typical constructed wetlands and their greenhouse gas emissions, Front. Environ. Sci. Eng. China, 3, 477–482, 2009.

Luederitz, V., Eckert, E., Lange-Weber, M., Lange, A., and Gersberg, R. M.: Nutrient removal efficiency and resource economics of vertical flow and horizontal flow constructed wetlands, Ecol. Eng., 18, 157–171, 2001.

Luo, W. G., Wang, S. H., Huang, J., and Qian, W. Y.: Denitrification by using subsurface constructed wetland in low temperature, China Water Wastewater, 21, 37–40, 2005.

Makino, W., Cotner, J. B., Sterner, R. W., and Elser, J. J.: Are bacteria more like plants or animals? Growth rate and resource dependence of bacterial C:N:P stoichiometry, Funct. Ecol., 17, 121–130, 2003.

Maltais-Landry, G., Maranger, R., Brisson, J., and Chazarenc, F.: Greenhouse gas production and efficiency of planted and artificially aerated constructed wetlands, Environ. Pollut., 157, 748–754, 2009.

Mander, Ü., Lõhmus, K., Teiter, S., Nurk, K., Mauring, T., and Augustin, J.: Gaseous fluxes from subsurface flow constructed wetlands for wastewater treatment, J. Environ. Sci. Health A 40, 1215-1226, 2005.

Mander, U., Lõhmus, K., Teiter, S., Mauring, T., Nurk, K., and Augustin, J.: Gaseous fluxes in the nitrogen and carbon budgets of subsurface flow constructed wetlands, Sci. Total Environ., 404, 343–353, 2008.

Mander, Ü., Maddison, M., Soosaar, K., and Karabelnik, K.: The impact of intermit-tent hydrology and fluctuating water table on greenhouse gas emissions from subsurface flow constructed wetlands for wastewater treatment, Wetlands, 31, 1023–1032, 2011.

Mander, Ü., Dotro, G., Ebie, Y., Towprayoon, S., Chiemchaisri, C., Nogueira, S. F., Jamsranjav, B., Kasak, K, Truu, J., Tournebize, J., and Mitsch, W. J.: Greenhouse gas emission in constructed wetlands for wastewatertreatment: A review, Ecol. Eng., 66, 19–35, 2014.

Minett, D. A., Cook, P. L. M., Kessler, A. J., and Cavagnaro, T. R.: Root effects on the spatial and temporal dynamics of oxygen in sand-based laboratory-scale constructed biofilters, Ecol. Eng., 58, 414–422, 2013.

Mitsch, W. J. and Gosselink, J. G.: Wetlands, 3rd Edn. New York, John Wiley & Sons, p. 936, 2000.

Müller, C. and Bünemann, E. K.: A ^{33}P tracing model for quantifying gross P transformation rates in soil, Soil Biol. Biochem., 76, 218–226, 2014.

Müller, C. and Clough, T. J.: Advances in understanding nitrogen flows and transformations: gaps and research pathways, J. Agril. Sci., 152, S1, 34–44, 2014.

Müller, C., Laughlin, R. J., Spott, O., and Rütting, T.: Quanti?cation of N$_2$O emission pathways via a ^{15}N tracing model, Soil Biol. Biochem., 72, 44–54, 2014.

Mustafa, A. and Scholz, M.: Nutrient accumulation in *Typha latifolia* L. and sediment of a representative integrated constructed wetland, Water Air Soil Pollut., 219, 329–341, 2011.

Nguyen, L. M.: Organic matter composition, microbial biomass and microbial activity in gravel-bed constructed wetlands treating farm dairy wastewaters, Ecol. Eng., 16, 199–221, 2000.

O'Luanaigh, N. D., Goodhue, R., and Gill, L. W.: Nutrient removal from on-site domestic wastewater in horizontal subsurface flow reed beds in Ireland, Ecol. Eng., 36, 1266–1276, 2010.

Obarska-Pempkowiak, H. and Gajewska, M.: The dynamics of processes responsible for transformation of nitrogen compounds in hybrid wetlands systems in a temperate climate, in: Wetlands-Nutrients, Metals and Mass Cycling, edited by: Vymazal, J., Backhuys Publishers, Leiden, The Netherlands, 129–142, 2003.

Pangala, S. R., Reay, D. S., and Heal, K. V.: Mitigation of methane emissions from constructed farm wetlands, Chemosphere, 78, 493–499, 2010.

Parkin, T. B.: Soil microsites as a source of denitrification variability, Soil Sci. Am. J., 51, 1194–1199, 1987.

Payne, E. G. I., Fletcher, T. D., Cook, P. L. M., Deletic, A., and Hatt, B. E.: Processes and drivers of nitrogen removal in stormwater biofiltration, Crit. Rev. Environ. Sci. Technol., 44, 796–846, 2014a.

Payne, E. G. I., Fletcher, T. D., Russell, D. G., Grace, M. R., Cavagnaro, T. R., Evrard, V., Deletic, A., Hatt, B. E., and Cook, P. L. M.: Temporary storage or permanent removal? The division of nitrogen between biotic assimilation and denitrification in stormwater biofiltration systems, PLoS ONE, 9, e90890, 2014b.

Peterson, D. G., Blazewicz, S., Herman, D. J., Firestone, M., Turetsky, M., and Waldrop, M.: Abundance of microbial genes associated with nitrogen cycling as indices of biogeochemical process rates across a vegetation gradient in Alaska, Environ. Microbiol., 14, 993–1008, 2012.

Pett-Ridge, J., Silver, W. L., and Firestone, M. K.: Redox fluctuations frame microbial community impacts on N-cycling rates in humid tropical forest soil, Biogeochem., 81, 95–110, 2006.

Poach, M. E., Hunt, P. G., Vanotti, M. B., Stone, K. C., Matheny, T. A., Johnson, M. H., and Sadler, E. J.: Improved nitrogen treatment by constructed wetlands receiving partially nitrified liquid swine manure, Ecol. Eng., 20, 183–197, 2003.

Rabalais, N. N., Díaz, R. J., Levin, L. A., Turner, R. E., Gilbert, D., and Zhang, J.: Dynamics and distribution of natural and human-caused hypoxia, Biogeosciences, 7, 585–619, doi:10.5194/bg-7-585-2010, 2010.

Reay, D. S.: Fertilizer 'solution' could turn local problem global – Protecting soil and water from pollution may mean releasing more greenhouse gas, Nature, 427, 485–485, 2004.

Reddy, K. R. and Delaune, R. D.: Biogeochemistry of Wetlands: Science and Applications, CRC Press, 800 pp., 2008.

Riya, S., Zhou, S., Nakashima, Y., Terada, A., and Hosomi, M.: Direct and indirect greenhouse gas emissions from vertical flow constructed wetland planted with forage rice, Kagaku Kogaku Ronbunshu, 36, 229–236, 2010.

Rütting, T. and Müller, C.: Process-specific analysis of nitrite dynamics in a permanent grassland soil by using a Monte Carlo sampling technique, Euro. J. Soil Sci., 59, 208–215, 2008.

Rütting, T., Boeckx, P., Müller, C., and Klemedtsson, L.: Assessment of the importance of dissimilatory nitrate reduction to ammonium for the terrestrial nitrogen cycle, Biogeosci., 8, 1169–1196, 2011.

Seitzinger, S. P., Sanders, R. W., and Styles, R.: Bioavailability of DON from natural and anthropogenic sources to estuarine plankton, Limnol. Oceanogr., 47, 353–366, 2002.

Sha, C. Y., Mitsch, W. J., Mander, Ü., Lu, J. J., Batson, J., Zhang, L., and He, W. S.: Methane emissions from freshwater riverine wetlands, Ecol. Eng., 37, 16–24, 2011.

Shamir, E., Thompson, T. L., Karpisak, M. M., Freitas, R. J., and Zauderer, J.: Nitrogen accumulation in a constructed wetland for dairy wastewater treatment, J. Amer. Water Resour. Assoc., 37, 315–325, 2001.

Song, K., Lee, S. H., Mitsch, W. J., and Kang, H.: Different responses of denitrification rates and denitrifying bacterial communities to hydrological pulses in created wetlands, Soil Biol. Biochem., 42, 1721–1727, 2010.

Song, K., Lee, S. H., and Kang, H.: Denitrification rates and community structure of denitrifying bacteria in newly constructed wetland, Eur. J. Soil Biol., 47, 24–29, 2011.

Søvik, A. K., Augustin, J., Heikkinen, K., Huttunen, J. T., Necki, J. M., Karjalainen, S. M., Klove, B., Liikanen, A., Mander, U., Puustinen, M., Teiter, S., and Wachniew, P.: Emission of the greenhouse gases nitrous oxide and methane from constructed wetlands in Europe, J. Environ. Qual., 35, 2360–2373, 2006.

Stadmark, J. and Leonardson, L.: Emissions of greenhouse gases from ponds constructed for nitrogen removal, Ecol. Eng., 25, 542–551, 2005.

Ström, L., Ekberg, A., Mastepanov, M., and Christensen, T. R.: The effect of vascular plants on carbon turnover and methane emissions from a tundra wetland, Global Change Biol. 9, 1185–1192, 2003.

Ström, L., Mastepanov, M., and Christensen, T. R.: Species specific effects of vascular plants on carbon turnover and methane emissions from wetlands, Biogeochem., 75, 65–82, 2005.

Ström, L., Lamppa, A., and Christensen, T. R.: Greenhouse gas emissions from a constructed wetland in southern Sweden, Wetlands Ecol. Manage., 15, 43–50, 2007.

Tai, P. D., Li, P. J., Sun, T. H., He, Y. W., Zhou, Q. X., Gong, Z. Q., Mizuochi, M., and Inamori, Y.: Greenhouse gas emissions from a constructed wetland for municipal sewage treatment, J. Environ. Sci. China, 14, 27–33, 2002.

Takaya, N.: Dissimilatory nitrate reduction metabolisms and their control in fungi, J. Biosci. Bioeng., 94, 506–510, 2002.

Tanner, C. C., Nguyen, M. L., and Sukias, J. P. S.: Nutrient removal by a constructed wetland treating subsurface drainage from grazed dairy pasture, Agric. Ecosysts. Environ., 105, 145–162, 2005.

Teiter, S. and Mander, Ü.: Emission of N_2O, N_2, CH_4 and CO_2 from constructed wetlands for wastewater treatment and from riparian buffer zones, Ecol. Eng., 25, 528–541, 2005.

Thayalakumaran, T., Bristow, K. L., Charlesworth, P. B., and Fass, T.: Geochemical conditions in groundwater systems: Implications for the attenuation of agricultural nitrate, Agril. Water Manage., 95, 103–115, 2008.

Toet, S., Richards, S. P., van Logtestijn, Kamp, R., Schreijer, M., and Verhoeven, J. T. A.: The effect of hydraulic retention time on the performance of pollutants from sewage treatment plant effluent in a surface flow constructed wetland system, Wetlands, 25, 375–391, 2005.

Van der Zaag, A. C., Gordon, R. J., Burton, D. L., Jamieson, R. C., and Stratton, G. W.: Greenhouse gas emissions from surface flow and subsurface flow constructed wetlands treating dairy wastewater, J. Environ. Qual., 39, 460–471, 2010.

Van Oostrom, A. J. and Russell, J. M.: Denitrification in constructed wastewater wetlands receiving high concentrations of nitrate, Water Sci. Technol., 29, 7–14, 1994.

Verhoeven, J. T. A. and Meuleman, A. F. M.: Wetlands for wastewater treatment: Opportunities and limitations, Ecol. Eng., 12, 5–12, 1999.

Vymazal, J.: Removal of nutrients in various types of constructed wetlands, Sci. Total Environ., 380, 48–65, 2007.

Vymazal, J. and Kröpfelová, L.: Types of constructed wetlands for wastewater treatment in: Wastewater Treatment in Constructed Wetlands with Horizontal Sub-Surface Flow, Environmental Pollution Springer, Heidelberg, 121–202, 2010.

Wang, H., Huang, C., Ge, Y., zhi Wu, J., and Chang, J.: The Performance of Species Mixtures in Nitrogen and Phosphorus Removal at Different Hydraulic Retention Times, Pol. J. Environ. Stud., 23, 917–922, 2014.

Weerakoon, G. M .P. R., Jinadasa, K. B. S. N., Herath, G. B. B., Mowjood, M. I. M., and van Bruggen, J. J. A.: Impact of the hydraulic loading rate on pollutants removal in tropical horizontal subsurface flow constructed wetlands, Ecol. Eng., 61, 154–160, 2013.

Whalen, S. C.: Biogeochemistry of methane exchange between natural wetlands and the atmosphere, Environ. Eng. Sci., 22, 73–94, 2005.

Wild, U., Lenz, A., Kamp, T., Heinz, S., and Pfadenhauer, J.: Vegetation development, nutrient removal and trace gas fluxes in constructed Typha wetlands, in: Natural wetlands for wastewater treatment in cold climates, edited by: Mander, Ü., Jenssen P. D., Adv. Ecol. Sci., Southampton, Boston, WIT Press, 101–126, 2002.

Wunderlin, P., Lehmann, M. F., Siegrist, H., Tuzson, B., Joss, A., Emmenegger, L., and Mohn, J.: Isotope signatures of N_2O in a mixed microbial population system: constraints on N_2O producing pathways in wastewater treatment. Environ. Sci. Technol., 47, 1339–1348, 2013.

Wynn, T. M. and Liehr, S. K.: Development of a constructed subsurface-flow wetland simulation model, Ecol. Engg., 16, 519–536, 2001.

Xue, Y., Kovacic, D. A., David, M. B., Gentry, L. E., Mulvaney, R. L., and Lindau, C. W.: In situ measurements of denitrification in constructed wetlands, J. Environ. Qual., 28, 263–269, 1999.

Yan, C., Zhang, H., Li, B., Wang, D., Zhao, Y., and Zheng, Z.: Effects of influent C/N ratios on CO_2 and CH_4 emissions from vertical subsurface flow constructed wetlands treating synthetic municipal wastewater, J. Hazardous Mat., 203–204, 188–194, 2012.

Yavitt, J. B. and Knapp, A. K.: Aspects of methane flow from sediment through emergent cattail (Typha latifolia) plants, New Phytol., 139, 495–503, 1998.

Yin, S., Shen, Q., Tang, Y., and Cheng, L.: Reduction of nitrate to ammonium in selected paddy soils in China, Pedosphere, 8, 221–228, 1998.

Zhang, J., Shao, W. S., He, M., Hu, H. Y., and Gao, B. Y.: Treatment performance and enhancement of subsurface constructed wetland treating polluted river water in winter, Environ. Sci., 27, 1560–1564, 2006.

Zhao, Y., Zhang, Y., Ge, Z., Hu, C., and Zhang, H.: Effects of influent C/N ratios on wastewater nutrient removal and simultaneous greenhouse gas emission from the combinations of vertical subsurface flow constructed wetlands and earthworm eco-filters for treating synthetic wastewater, Environ. Sci. Proc. Impacts, 16, 567–575, 2014.

4

Assessing changes in urban flood vulnerability through mapping land use from historical information

M. Boudou, B. Danière, and M. Lang

Irstea, UR HHLY, Hydrology-Hydraulics, 5 rue de la Doua, 69626 Villeurbanne, France

Correspondence to: M. Boudou (martin.boudou@gmail.com)

Abstract. This paper presents an appraisal of the temporal evolution of flood vulnerability of two French cities, Besançon and Moissac, which were largely impacted by floods in January 1910 and March 1930, respectively. Both flood events figure among the most significant events recorded in France during the 20th century, in terms of certain parameters such as the intensity and severity of the flood and spatial extension of the damage. An analysis of historical sources allows the mapping of land use and occupation within the areas affected by the two floods, both in past and present contexts, providing an insight of the complexity of flood risk evolution at a local scale.

1 Introduction

Directive 2007/60/EC on the assessment and management of flood risks draws up a new framework for the promotion of historical information. It aims to reduce and manage the risks that floods pose to human health, the environment, cultural heritage and economic activity. The directive requires member states to first carry out a preliminary assessment by 2011 to identify the river basins and then the associated coastal areas which are at risk of flooding. For such zones, subsequent steps would involve drawing up flood risk maps by 2013 and establishing flood risk management plans focused on prevention, protection and preparedness by 2015. The directive applies to inland waters as well as all coastal waters across the whole territory of the EU. In France, a national historical database (http://bdhi.fr/), based on the inventory of major floods, was produced in 2011 within the framework of the EU Flood Directive (Lang and Coeur, 2014; Lang et al., 2012) and was made available to the public in 2015. It contains a description of 176 "remarkable" flood events from 1770 to 2011.

A key issue of the Flood Directive is the accurate assessment of flood risk. A commonly accepted definition of flood risk is the combination of a flood hazard and the vulnerability of the assets that are exposed (de Bruijn, 2005; Schanze, 2006; Cardona et al., 2012). Following this definition, the French Government distinguished two main steps for flood risk assessment. A first step consists of mapping the potential flood extent to evaluate the number of infrastructure assets exposed. Starting from this data, a second step consists of determining the exposure and vulnerability of the asset. For this purpose, some indicators have been adopted, according to the potential impacts on human health, economic activity, the environment, and cultural heritage within the potential flood extent. To mention just a few, these indicators include the number of inhabitants affected, the number of single-storey buildings, the number of employed persons, the number of nuclear power stations, and the area of remarkable built heritage. Following this approach, flood risk assessment leads to a contrasted overview of the actual flood risk. The results indicate a strong and unequal exposure of assets over the French territory and raise some concerns in a context of increasing flood damage (SwissRe, 2015) and global change.

The term "vulnerability" has long been a subject of debate in the scientific literature, being covered by several definitions (Birkmann, 2006; Wisner et al., 1994). A commonly used definition of vulnerability is the likelihood of the elements at risk to produce damage. Based on that definition, assessing the vulnerability and its evolution can be broken down into two main steps: firstly, appraising the exposure by listing the elements at risk and, secondly, by evaluating the susceptibility of the elements at risk (Merz et al., 2007). To carry out these two steps, we identify a series of indicators adapted for a retrospective analysis.

On the one hand, the exposure analysis is supported by quantifying the number of buildings and inhabitants at risk. On the other hand, the susceptibility analysis is based on identifying the building use type, providing some keys for understanding the kind of damage to be expected during floods (Barroca et al., 2006). For example, some building types are especially likely to trigger major damage (industrial or commercial activities) or cause disturbances for society (e.g. public infrastructures such as hospitals or schools), thus requiring special attention from risk managers (Merz et al., 2007).

Many authors have already highlighted the importance of historical data as a tool for risk assessment (Glade et al., 2001; Brazdil et al., 2006; Coeur and Lang, 2008; Kjeldsen et al., 2014). A general survey of flood mapping techniques in Europe by de Moel et al. (2009) provides evidence that flood maps are available in almost all countries, based on historical floods or design-basis floods. As an example, Barnikel (2004), Tropeano and Turconi (2004), and Luino et al. (2012) reported past flood extents in relation to present-day land use, which allows developing the prospective analysis of flood risk.

Assessing flood impacts and understanding the past vulnerability of a territory is an essential step towards a long-term mitigation strategy (Changnon et al., 2000). Firstly, it allows for a better understanding of the circumstances that lead to a disaster. Secondly, it helps to shed light on the actual state of vulnerability within a territory. This vulnerability (especially visible through the exposure of assets) should be seen as the result of a complex historical evolution, partly related to the occurrence of damaging flood events in the past (Barrera et al., 2006).

To take account of a potential increase in flood risk, the Flood Directive assessment has to be considered in terms of a long timescale. The indicators developed during the preliminary phase are in fact closely correlated with the present-day situation and raise some questions about the past situation of vulnerability. How do we assess the vulnerability and exposure situations for past flood events based on uncertain and sparse historical sources? Can we validate an increase in the exposure and vulnerability of stakeholders based on a temporal analysis of past disasters? Are these disasters still relevant and easily integrated into risk management policies as indicated in the Flood Directive text?

To address these issues, the present study sets out to highlight the importance of historical information by applying a multidisciplinary and mapping approach (Danière, 2014). Our study is based on the set of 176 major floods in France, which offers an opportunity to explore the vulnerability associated with past flood events. We apply this methodology to two case studies selected for their "remarkability": the Jan-

uary 1910 flood event (generalized over all the north-east of France) and the March 1930 flood event (concentrated on the Tarn River valley). We focus our analysis on two cities, Besançon and Moissac, which were largely affected by the floods of 1910 and 1930, respectively. After a brief presentation of the two flood events (Sect. 2), we present the methodological framework used for mapping the vulnerability (Sect. 3). This approach is applied to the two case studies (Sect. 4), illustrating the past and present vulnerability situations in the two cities. Finally, some key points are given (Sect. 5) concerning the importance of historical information for assessing vulnerability changes during the 20th century.

2 Case studies

2.1 Selection of two remarkable flood events

During the inventory work carried out for the Flood Directive in 2011, we selected a total of 176 major floods in France since 1770 (see Lang and Coeur, 2014) based on the following considerations: diversity of flood types, strong flood hazard or spatial extent, and important socio-economic impacts, in addition to reference events used in planning documents (flood mapping area) or last significant flood in living memory. Using a multidisciplinary methodology, we established an evaluation grid based on three main features (Boudou et al., 2015): (1) flood intensity (score between 3.5 and 14) according to several criteria (return period of maximum peak discharge; duration of submersion; dyke breaches or log jams); (2) flood severity (score between 3 and 12), with two main indicators: flood damage (number of fatalities, economic loss) and social, media or political impacts of the event (establishing a new risk policy, calling for international solidarity to face the crisis, etc.); and (3) spatial extent of damage (score between 2 and 8). This grid allowed us to rank the 176 major floods (Boudou, 2015). Then, a second level of selection led us to focus on the nine events shown in Fig. 1 (January 1910, March 1930, October 1940, December 1947/January 1948, December 1959, January 1980, November 1999, December 2000/April 2001 and February 2010). These flood events cover all flood typologies (oceanic/snowmelt/Mediterranean floods, storm surges, cyclones, dam breaching) and are considered as some of the most remarkable in accordance with the evaluation grid. Lang et al. (2012) presented the main characteristics of these nine events (except for the 1947–48 flood).

In this study, we investigate the two oldest selected events, which took place in January 1910 and March 1930, focusing on the urban situation in Besançon and Moissac (Fig. 2). The aim is to focus on two cities that have been significantly flooded in the past and to understand how their vulnerability to flooding has changed up to the present day. A detailed inventory of documentary sources for these two events can be found in the Supplement.

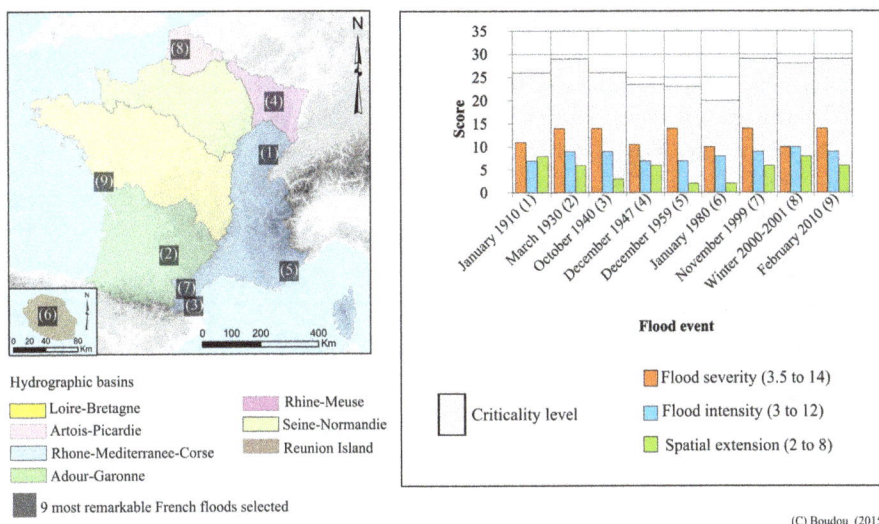

Figure 1. Location map of the nine most remarkable French flood events selected in this study and chart showing their related remarkability scores (Boudou, 2015).

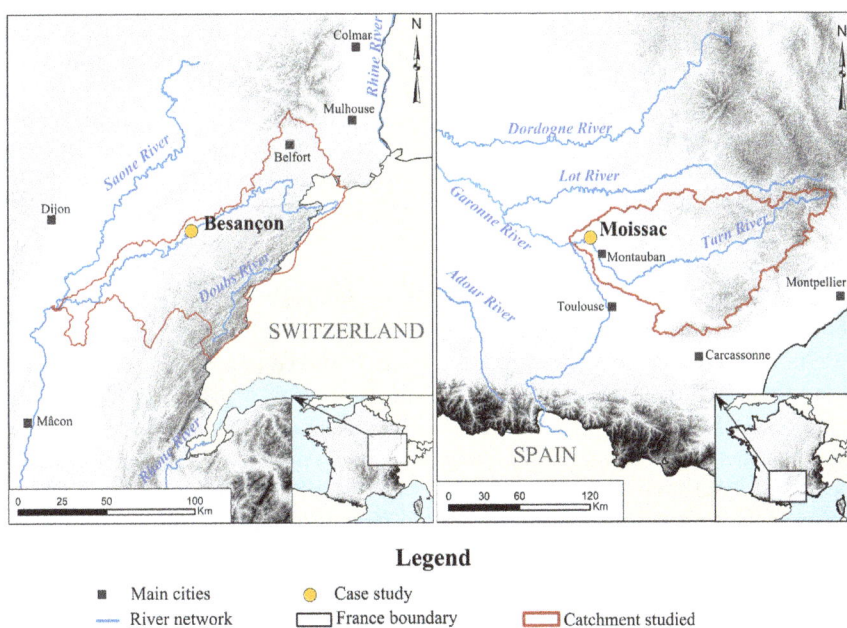

Figure 2. Location of the case studies: Doubs Basin and Besançon (left panel), Tarn Basin and Moissac (right panel).

2.2 The January 1910 flood event in Besançon (Doubs River catchment)

The flood of January 1910 ranks fifth among the nine floods selected as remarkable according to the evaluation grid (Fig. 1). This flood event is mostly known for being the most significant flood affecting the city of Paris, with a return period of about 100 years for several rivers of the Seine Basin. After a very wet end during the year 1909 (450 mm of rainfall in 3 months), the Seine Basin received a large amount of rain and snow in January 1910 (about 300 mm in the upper part, 110 mm in the central part, and 280 mm in the downstream part). The water level at Paris Austerlitz was 8.66 m, the second highest historical level after the flood of February 1658 (8.80 m) (Champion, 1858–1864; Goubet, 1997). There was a relatively small number of direct fatalities (seven deaths) plus nine indirect deaths (collapsing of several cavities), but the impact within the Paris region was extremely high, with 150 000 persons affected and economic losses of about FRF 400 million (EUR 1.5 billion, 2015) (Picard, 1910). Despite the fact that a large part of northern France was also affected, most of the attention of society and recollections of this event have been focused on Paris. To demonstrate the remarkability of this event, not only for the Seine catchment area but also for more rural regions, we concentrate our study on the Doubs Basin where the flood of January 1910 remains one of the most significant historical floods, with the highest water level being recorded in the city of Besançon (see Fig. 3, e.g. $Z = 245.55$ m at Poterne, Place la Revolution). While the flood event across the Seine Basin was characterized by a clustering of several oceanic rainfall events, the flood event in the Doubs Basin was triggered by an episode of heavy rainfall from 18 to 21 January (between 150 and 250 mm), plus the presence of extensive snow cover after a wet winter which led to significant snow melting. A large part of the old city of Besançon was flooded, with huge damage. Many shops, houses and their basements were inundated, causing important losses of furniture. The streets were also particularly badly affected due to the high flow velocity. In total, the cost of the flooding at Besançon is estimated at around FRF 2 million (DREAL Franche-Comté, 2010), corresponding to EUR 7.7 million in the present day.

According to several documentary sources (Allard, 1910; Ministère de l'Ecologie, 2011), it appears that the hydrometeorological conditions of the event (peak discharge at Besançon of about 1750 m^3 s^{-1}, with a return period of about 100 years; catchment area of 4379 km^2) cannot explain why the flood level was so high throughout the old city. Such exceptional water levels in the city centre were the consequence of energy losses at the bridges of the town. These energy losses were larger than usual (cf. Fig. 3, in comparison with the 1882 and 1896 flood events) due to a log jam (about 35 000 m^3), resulting from the inundation of a paper factory a few kilometres upstream of Besançon, contributing significantly to a raising of the water level.

Archive sources (especially administrative reports produced by the chief engineer of the Ponts-et-Chaussées, Serial S, Doubs departmental archives) also reveal some major failures of flood warning during the event. Surprised both by the arrival and the intensity of the flood, the local authorities did not succeed in setting up temporary protective structures at the different open city gates ("postern gates"), which directly contributed to the inundation of the city (Fig. 4).

2.3 The March 1930 flood in Moissac (Tarn River catchment)

At the end of February 1930, an intense Mediterranean rainfall event occurred in the south-west of France, with hot and moist air from the Mediterranean Sea penetrating deep into the Massif Central highlands. From 25 February to 4 March, a large area was affected by heavy rainfall (e.g. more than 200 mm over 6000 km^2 during 4 days), with a maximum of 694 mm in 7 days at Saint-Gervais-sur-Mare (spring of the Orb river). The very serious adverse consequences of

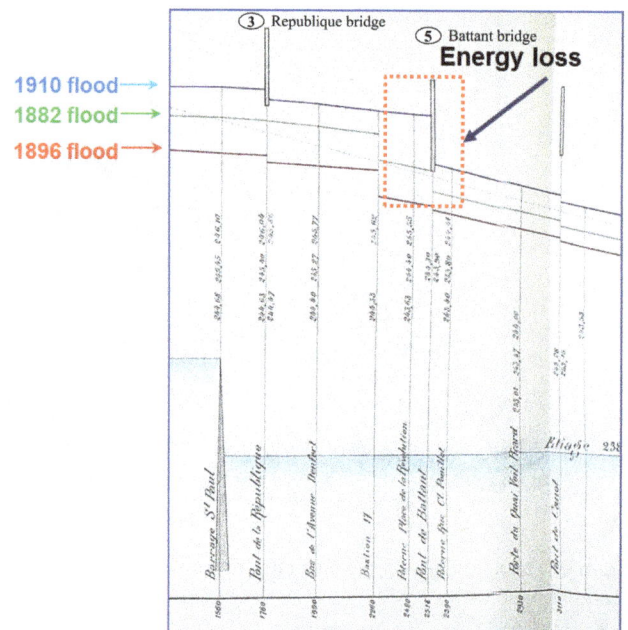

Figure 3. Longitudinal profile of the Doubs River within the old city of Besançon and inter-comparison of floods (sources: Ville de Besançon – Service de la voirie et des eaux: Profil en long des crues du Doubs du 21 janvier 1910, 28 décembre 1882 et 10 mars 1896, 10 mars 1910, Bibliothèque et archives municipales de Besançon, série 0). Locations of the Republique and Battant bridges are shown in Fig. 4.

this rainfall event can be explained by at least two factors. From October 1929 to February 1930, high rainfall totals were observed (e.g. 1177 mm at Lodève, 840 mm at Florac), thus favouring a strong reaction of the basins which were already saturated. Moreover, a warming in temperature associated with intense rainfall was causing a large amount of snow melting (20–100 cm) above 600 m.

Due to its intensity and unusual date of occurrence (at the end of a wet winter) the rainfall event triggered an exceptional flood event (Pardé, 1930). The following flood hazard intensity can be judged exceptional for the downstream part of the Tarn catchment (8000 m^3 s^{-1} at Moissac, 15 400 km^2; mean annual discharge 230 m^3 s^{-1}), with a return period of about 250–300 years (Dreal Midi-Pyrénées, 2014). Between 210 and 230 fatalities were recorded during this Tarn River flood (resp. Bichambis, 1930, and Boudou, 2015), which represents one of the most destructive flood events ever recorded in France and surely the most significant during the 20th century. The economic loss for the entire surrounding region was estimated at around FRF 1 billion, which corresponds to EUR 570 million in 2015 (Journal Officiel de la République Française, 1930).

One of the striking features of the disaster can be found in the concentration of damage in the town of Moissac (120 deaths out of a total of 210). Reconstructing and mapping the flood chronology using historical sources provides

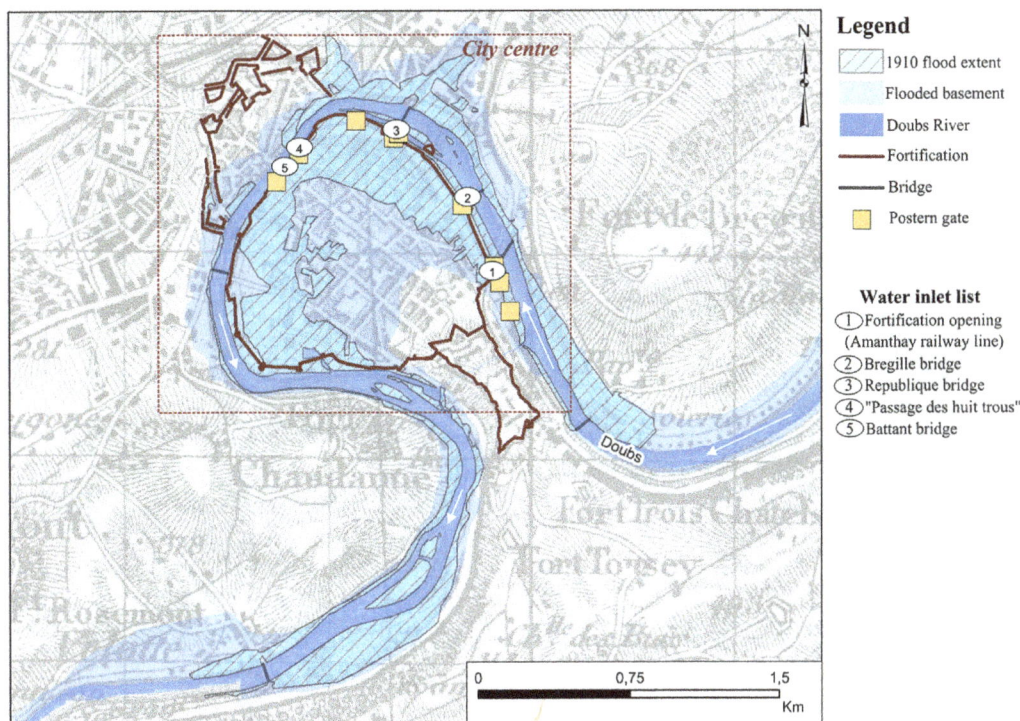

Source : Departemental Archives of the Doubs departement, IGN. (C) Danière, 2014

Figure 4. Old Besançon city centre with characteristic water inlets during the flood event on 17–21 February 1910.

us with a better understanding of the circumstances of the disaster (Fig. 5). On 3 March 1930, the flood arrived in the town. Before 18:30 LT (local time) the Tarn River was already overflowing the main channel, on both the south and north banks. Fortunately, the town centre was protected by three main dykes and the railway line embankment. From 18:30 to 23:00 LT, the water level rose and the flood extent covered the area between the main dikes at the eastern part of the town. Around 23:00 LT, at the time of maximum discharge (estimated at around 8000 $m^3 s^{-1}$), three breaches suddenly appeared along the railway embankment. These breaches led to a sudden outburst of the dykes and final inundation of the town.

According to the locations of fatalities and the feedback of information on the disaster, the explanation of the high death toll is twofold. Firstly, the rapid influx of water into the city due to the flash flood and dyke failures induced a surprise effect on the inhabitants of Moissac. Secondly, the collapse of more than 600 houses was related to the typical kind of housing in this region, being built of raw bricks especially vulnerable to flooding and sustained contact with water.

3 Methodology for monitoring changes in flood vulnerability

3.1 Relevance of historical events in the present context?

One of the main requirements of the Flood Directive is to identify areas with a potential high level of flood risk, based on historical floods that would have significant adverse consequences if they occurred again. As the consequences are dependent on the flood hazard as well as the personal, social and economic assets located in the flood risk zones, one of the main concerns is to assess the changes in local vulnerability of city centres as a function of time. In both case studies, the main casualties and/or economic losses within the catchment were located in a single municipal area. But some aggravating factors are time dependent, such as woody debris upstream of bridges at Besançon or dyke failures to the east of Moissac. Other aggravating factors are related to social vulnerability, such as failure of flood warnings at Besançon or vulnerable building materials at Moissac.

To obtain a better understanding of the local disaster process, our study aims to monitor changes in flood vulnerability, comparing the past and present situations. Several questions have to be addressed. Is it possible to assess correctly the changes in vulnerability over time according to the available sources? Does the mapping of land use provide enough information to identify indicators of vulnerability? Can we

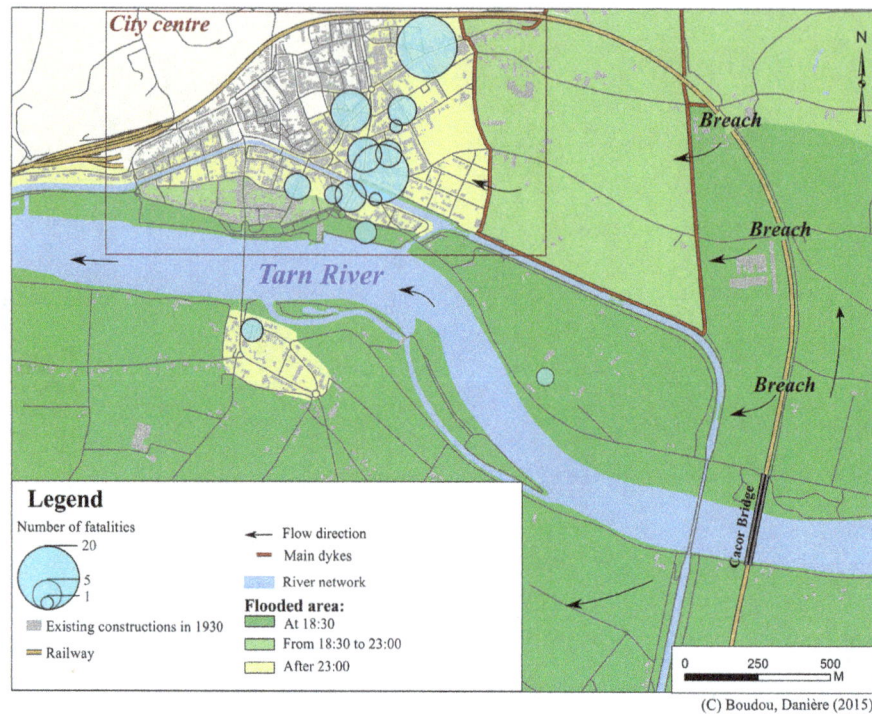

Figure 5. Flood chronology and location of fatalities during the flood event in Moissac on 3 March 1930.

establish scenarios concerning the impact of a future flood based on a historical flood?

After a preliminary analysis that involves georeferencing historical information in the present-day context, we then consider the mapping of land use and estimating the population at risk, while comparing the past and the present situations.

3.2 Dynamic mapping to locate historical information

A preliminary step of this study consists of carrying out dynamic mapping with a spatial display of the previously collected historical information. The historical corpus made up of various document formats and sources is included in a GIS by locating the information available. However, some place names have changed since the date of the flood event, thus requiring supplementary treatment of the data.

The dynamic consultation of historical information is not only of interest for correctly locating the various sources of information on flood vulnerability, it can also be used to develop risk awareness and risk culture on an exposed territory. As an example, the high-water mark inventory developed for the Seine River catchment (www.reperesdecrues-seine.fr/carte.php) provides dynamic mapping which is easily understandable and interactive for the general public, in contrast to the maps resulting from hydraulic or hydromorphogenic modelling (de Moel et al., 2009).

3.3 Evolution of land use

In this section, we address the exposure and susceptibility to flood risk (Fig. 6) using simplified descriptors which remain consistent with the level of data availability and accuracy of historical information (Barnikel and Becht, 2003; Barnikel, 2004).

Firstly, the exposure analysis is based on the changes in the population living per building and provides information about the evolution of built-up areas. Secondly, susceptibility analysis based on land-use classification provides relevant information to evaluate the nature of buildings affected during flooding. Historical information is required which at least describes the land cover on different dates. For example, historical maps and aerial photos often depict the built-up territory for a specific year.

To perform a spatial analysis of historical maps, it is necessary to integrate them into a GIS. Three steps are executed: scanning, georeferencing, and digitization supported by a spatial reference system (Fig. 6a) (Rumsey and Williams, 2002; Levin et al., 2010). A set of historical maps and aerial photographs produced by the French National Institute of Geographic and Forest Information (IGN) are used to depict the extent of built-up areas at the scale of a block of houses. A total of seven topographic maps (from 1911 to 1988) are used for Besançon and 26 aerial photographs for Moissac (from 1947 to 1983). Aerial photographs are favoured in the case of Moissac because of the inconvenient representation of the town on topographic maps, which is split between four

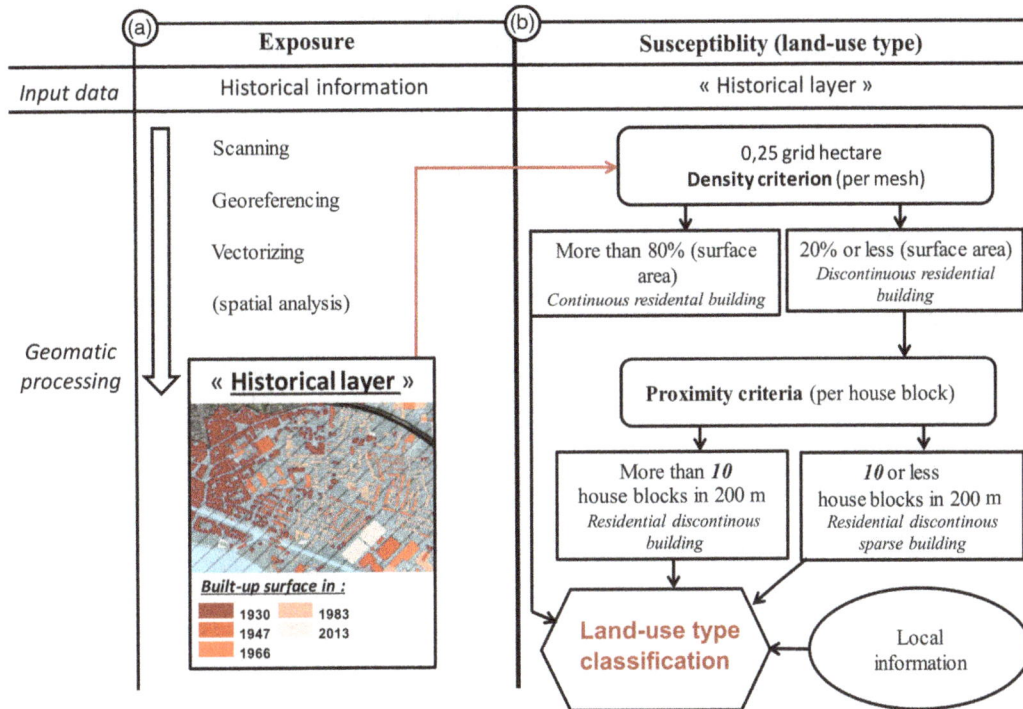

Figure 6. Evolution of vulnerability: **(a)** exposure and **(b)** susceptibility (building use type).

map plates. These raster data are then imported and georeferenced. A spatial database (BD TOPO) produced by the IGN, describing the present French territory and its infrastructures, is used to select control points and evaluate distortions during the digitizing step. During this last step, information from topographic maps is vectorized into a unique "historical layer". In this way, each object is given a spatial reality (via the GIS representation) and a temporal reality (by associating a temporal field to indicate its existence for a specific year). Consequently, the "historical layer" allows us to obtain "temporal snapshots" (Langran and Chrisman, 1988; Gregory and Healey, 2007) of the urban fabric: the space is discretized based on information available at the time of the event.

Subsequently, the description of "historical layer" objects provides information on the nature of building exposure. A land-use classification is drawn up based on a nomenclature adapted from the Urban Atlas of the European Environment Agency (http://www.eea.europa.eu/data-and-maps/data/urban-atlas), according to historical information constraints (Fig. 6b). A first geomatic processing step is performed to discretize the residential buildings on a 0.25 ha grid. A density criterion is applied in each grid cell, based on the percentage contribution to the building footprint, leading to a distinction between dense and sparse areas. To enhance the classification, a second processing step is carried out using a proximity criterion for each building based on the number of buildings within a 200 m radius (continuous and discontinuous buildings). Local information is then added re-

lated to the location and nature of non-residential constructions. BD TOPO data are used to describe the current situation, and a point-in-time layer is built with our "historical corpus" information for earlier historical periods.

3.4 Census of the exposed population within the flood extent

General information is provided by the evolution of population at the scale of the municipality. Figure 7 presents the data derived from several population censuses during the 20th century. It shows than the number of inhabitants has grown by about +100 % at Besançon (from 57 978 to 116 914, between 1911 and 2010) and +60 % at Moissac (from 7814 to 12 354, between 1911 and 2006). As only part of the built-up area was affected by floods, especially in the case of Besançon, it is necessary to cross two layers of information: the number of inhabitants per small block and the spatial extent of the historical flood (1910 or 1930 floods at Besançon and Moissac, respectively).

Human exposure is taken into account by census or an estimation of the resident population. The aim here is to distribute the raw demographic data throughout the blocks of houses by following its evolution at different scales (Wu et al., 2008). The maps so produced can shed light on the evolution of human exposure within the area affected by the flood.

To assess the current population living within the flood extent, we make use of two demographic data sets produced by the French National Institute for Statistics and Economic

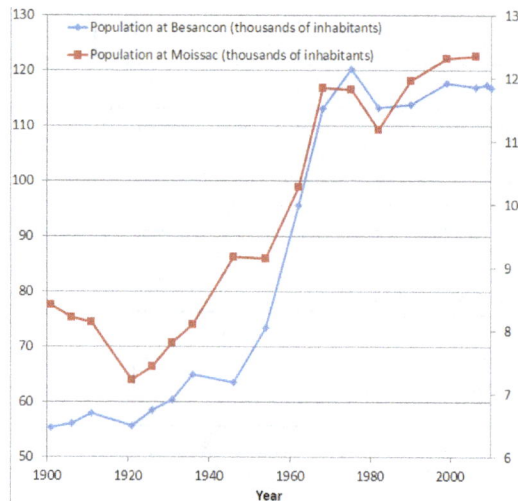

Figure 7. Evolution of the number of inhabitants during the 20th century at Besançon and Moissac. Source: EHESS-Cassini before 1962, INSEE from 1968.

Table 1. Exposed population in 1930 and 2013 for each flooded area (cf. Fig. 11) in Moissac.

Flooded area (Fig. 11)	1930	2013
(1)	4089	1160
(2)	1044	2880
(3)	2267	2000
Total	7400	6040

Studies (INSEE), applying Eq. (1) to redistribute the population data at the scale of blocks of houses. The first data set is defined at infra-municipal scale with IRIS (infra-urban statistical area) data. The second data set is based on an estimation of the fiscal population within a 200 m × 200 m grid. These data sets are distributed at the scale of residential blocks of houses, based on a volumetric method (Lwin and Murayama, 2009), in proportion to the building footprint area multiplied by the vertical density, using the building height provided by BD TOPO:

$$\text{developed area} = \frac{\text{building height} \times \text{building floor area}}{\text{average storey height}}. \quad (1)$$

Historical information, in the form of a census or raw demographic data, is required to estimate (Ekamper, 2010) the numbers of the population exposed at the time of the disaster. General census reports are available for every French municipality (sometimes online), generally compiled every 5 years up until 1946, with some exceptions. These documents contain nominative information about the municipal population, grouped by building and street, at different dates. The comparison between past and present exposed population within the flood extent should take account of possible changes of census methodology over time.

4 Change of vulnerability based on two case studies

We now consider the changes of vulnerability in the two case studies, from past to present, using historical sources and current information.

4.1 Changes in vulnerability of Besançon with respect to the January 1910 flood

Figure 8 displays the land use within the area affected by the 1910 flood in Besançon, based on the situations in 1911 and 2013 (resp. dates of two censuses). No significant change can be seen in terms of vulnerability, according to the spatial extent of the built-up area. Since the centre of Besançon is located within a meander of the Doubs River, with no opportunity for spatial expansion or urban densification, there has been no increase of exposure, apart from the hospital area. Although the city has experienced a spatial expansion towards the north, on the right bank, this area is located outside our zoning at a larger scale.

According to the land-use classification, we can note significant changes in the various activities. There has been a fall in military employment, in favour of an increase in administrative and public facilities. While military areas have decreased by 74 % between 1911 and 2013, administrative areas have grown by a factor of 12. A reduction of human exposure is noticeable between 1911 (the census year closest to the 1910 flood) and 2013, with a 24 % decrease in the city-centre population.

The demographic evolution is represented on Fig. 9 at the scale of a block of houses, reflecting the decrease in household size (decline in the number of inhabitants per building) and a decline in residential function (reduction of inhabited buildings within the city centre).

4.2 Changes in vulnerability of Moissac with respect to the March 1930 flood

The flood risk mapping of Moissac yields an opposite diagnosis, with a major increase of vulnerability within the area affected by the 1930 flood (Fig. 10). Built-up areas have expanded by 122 % between 1930 and 2013. Such spatial extension is explained by new residential development (mainly housing estates) and economic buildings east of the city centre and by a progressive densification of the low-density area on the south bank flood plain.

Despite a new distribution of the population (Table 1), the human exposure has not changed significantly. The reduction of population density in the city centre is compensated by a spatial expansion (Fig. 11). The human expo-

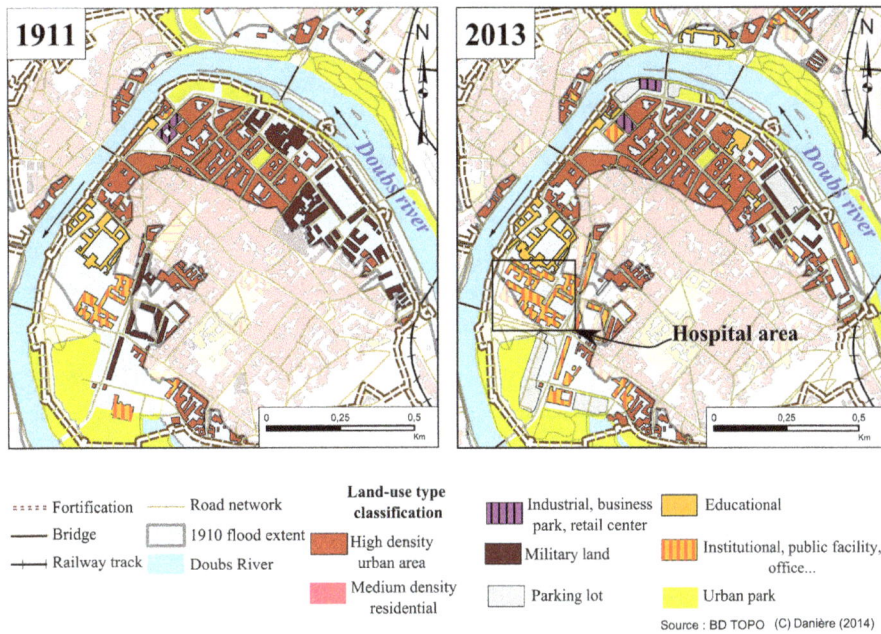

Figure 8. Land-use types and soil occupation within the area affected by the 1910 flood in Besançon: **(a)** in 1911 and **(b)** in 2013.

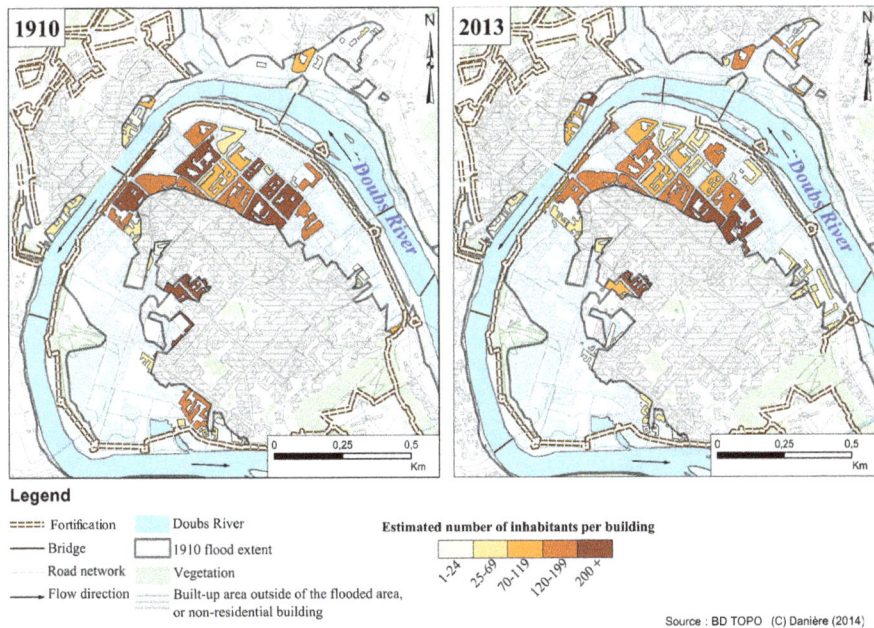

Figure 9. Estimated number of inhabitants per building within the area affected by the 1910 flood in Besançon: **(a)** in 1911 and **(b)** in 2013. Some blocks of houses are depicted only on one of the maps, because of land-use changes. Non-residential blocks of houses are not taken into account here.

sure has mainly increased on the east side of the city centre, especially in the area located between the two levees. It should be noted that no general census report is available for Moissac in the 1930s. Therefore, the population exposed to flood risk in 1930 was estimated from a raw demographic data set, obtained from an internet database containing a historical population census at the municipality scale (http://cassini.ehess.fr/), which was then distributed according to the volume-based method.

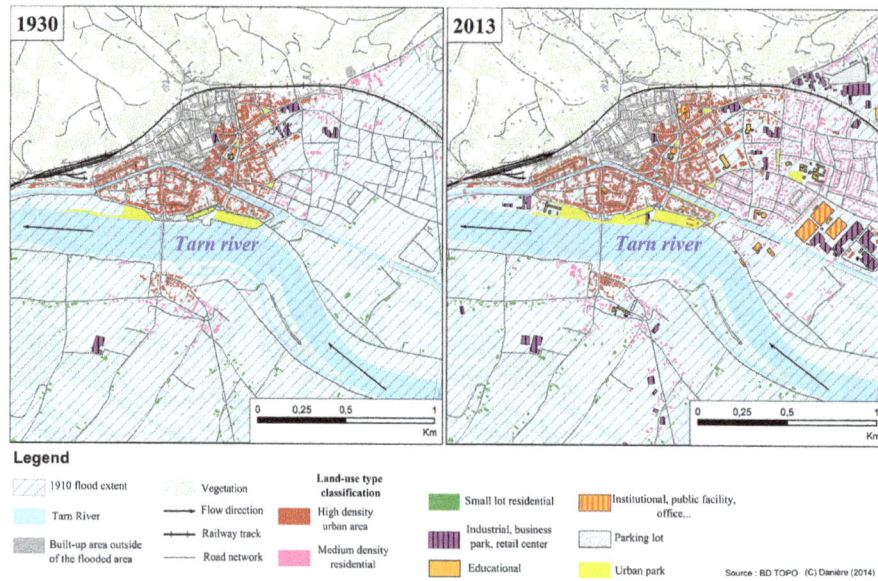

Figure 10. Land-use types and soil occupation within the area affected by the 1930 flood in Moissac: **(a)** in 1930 and **(b)** in 2013.

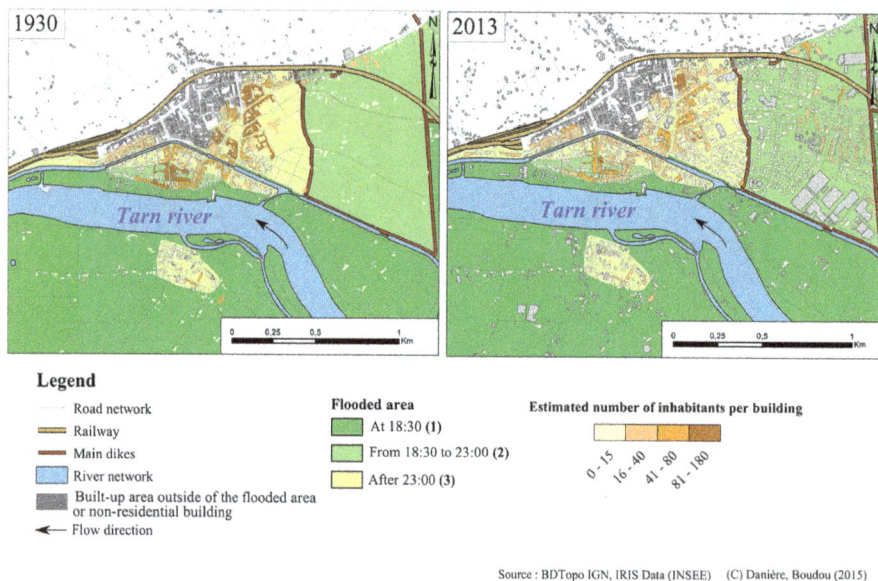

Figure 11. Estimated number of inhabitants per building within the area affected by the 1930 flood in Moissac: **(a)** in 1930 and **(b)** in 2013.

4.3 Appraisal of the temporal evolution of flood risk

These two case studies shed light on the complexity of flood-risk evolution. At the nationwide scale, it is clearly acknowledged that the increase of flood damage over the last few decades is induced by a general increase in flood vulnerability (Kron, 2002; Luino et al., 2012; Kundzewicz et al., 2014; Smith et al., 2014). At a local scale, where topographic, social, and economic contexts are crucial, it is necessary to have a more detailed analysis.

In Besançon, there has been no extension of the urban area within the old city since 1910, but significant land-use changes have led to a decrease of flood vulnerability as some previously residential areas are now used as administrative buildings. The frequency of flooding has changed in the historical centre, due to the establishment of safety measures, especially with the construction of mitigation structures such as cofferdams to close the postern gates. Some uncertainties remain for determining the flooded area in the case of an event comparable to the 1910 reference flood, since opposite effects come into play. The log jams at the bridges are not expected to be repeated, but additional hydraulic losses have been introduced by new hydraulic structures since 1910.

Nowadays, the reference flood selected in the regulatory documents is a simulated flood larger than the January 1910 flood.

In Moissac, the changes in vulnerability show a more contrasted pattern. As in various other French regions, the built-up areas have grown in spatial extent since 1930, characterized by an important development of housing estates. One critical point is the development of one-storey buildings, leading to a higher human vulnerability due to the lack of a refuge floor. On the other hand, building quality has improved. During the 1930 flood, the house collapses in Moissac and the consequent fatalities were closely related to the construction materials used. To increase the resistance of the structures, new materials and building techniques were used during the reconstruction stage. Another positive change is related to the improvement of safety measures, due to progress in flood-warning decision-making as well as in regards to emergency population evacuation schemes implemented by the civil protection services. The 1930 flood in Moissac, with a return period estimated at around 250 years, is nowadays considered as the reference flood hazard for the local flood risk management strategy as well as for planning and development documents. This territory appears to remain vulnerable, especially to risks of dyke failure.

5 Conclusion and perspectives

This paper presents a case study on the urban vulnerability of two French cities that were largely involved in floods occurring in January 1910 and March 1930. This approach gives an insight of the complexity of flood risk evolution, not ignoring the local characteristics. Old maps can provide reliable information on the flood vulnerability in the past, but this requires a necessary evaluation of the modifications occurred in the examined area. A first step is necessary to locate and georeference the historical information within the present geographical reference system. Qualitative information (images, technical reports, national and local newspaper articles, paintings, marble plaques, etc.) can be interpreted as a complement to historical maps on land use. An assessment of the population exposed at risk within spatial units can be inferred from technical documents with nominative lists of people (or inhabitants) as well from old censuses. Historical information on past floods can therefore be useful when building scenarios on future possible floods, providing a reliable reference of what might be possible in terms of water depth, flow velocity and flood extent. Additional work is needed to account for possible changes both in vulnerability and flood hazard over the past several decades (from historical floods to the present day) and for future decades (prospective studies). It is also important to consider the uncertainties associated with historical data and to use relevant scales when mapping vulnerability indicators.

As usual, the temporal analysis of flood risk evolution at a local scale implies a good knowledge of the general context of the socio-economic development of territories, as well as changes in the recollection and perception of risk. According to data availability, this study focuses on a small component of vulnerability only. However, to carry out a comprehensive flood vulnerability analysis, other indicators should be taken into account. After the Xynthia storm surged in 2010 (41 fatalities due to floods in France), Vinet et al. (2012) showed that the age of the population is a key component of local vulnerability. It is clear that the insurance system may benefit from similar analyses on urban flood vulnerability over the last few decades in order to better evaluate the future damages of remarkable floods. Depending on the analysis results, some vulnerability scenarios could be carried out by the risk managers, allowing for the identification of risky areas on which prospective mitigation strategies would be established. Such measures could be realized and financially supported by public authorities, following the example of the experience ALABRI (2012), which led to setting up individual flood protection in the houses preliminarily identified as exposed in the Gard department (http://www.les-gardons.com/alabri/).

This study addresses the issue of flood vulnerability, which is an important component of the flood risk. In parallel, research on flood hazard is also necessary to simulate past floods in a present-day context, taking into account modifications of the river (morphological changes and river engineering) and new settlements on the flood plain.

Author contributions. M. Boudou established the evaluation grid used for the selection of "remarkable" flood events. He collected data on the two historical floods and produced thematic maps on flood hazard. B. Danière carried out dynamic mapping to locate historical information and thematic maps on flood vulnerability. M. Lang supervised the drafting of the paper.

Acknowledgements. The authors especially thank the DREAL of Besançon, the DDT of Moissac, and the IGN for providing data. We are also grateful to Freddy Vinet and Denis Cœur for their advice. Maria-Carmen Llasat and two anonymous referees are acknowledged for their useful comments. Finally, the authors would like to thank the French Minister of Ecology, Sustainable Development and Energy (MEDDE) for the financial support of Martin Boudou's PhD. Michael Carpenter post-edited the English style and grammar.

References

Allard, M.: Les récentes inondations à Besançon, Bibliothèque et archives municipales de la ville de Besançon, Besançon, 1910.

Barnikel, F.: The value of historical documents for hazard zone mapping, Nat. Hazards Earth Syst. Sci., 4, 599–613, doi:10.5194/nhess-4-599-2004, 2004.

Barnikel, F. and Becht, M.: A historical analysis of hazardous events in the Alps – the case of Hindelang (Bavaria, Germany), Nat. Hazards Earth Syst. Sci., 3, 625–635, doi:10.5194/nhess-3-625-2003, 2003.

Barrera, A., Llasat, M. C., and Barriendos, M.: Estimation of extreme flash flood evolution in Barcelona County from 1351 to 2005, Nat. Hazards Earth Syst. Sci., 6, 505–518, doi:10.5194/nhess-6-505-2006, 2006.

Barroca, B., Bernardara, P., Mouchel, J. M., and Hubert, G.: Indicators for identification of urban flooding vulnerability, Nat. Hazards Earth Syst. Sci., 6, 553–561, doi:10.5194/nhess-6-553-2006, 2006.

Bichambis, P.: Inondations du midi en mars 1930: les paisibles rivières devenues torrents de ruine et de mort. Les deuils, les ruines, les héros, Toulouse, 128 pp., 1930.

Birkmann, J.: Measuring vulnerability to promote disaster-resilient societies: Conceptual frameworks and definitions, in: Measuring vulnerability to natural hazards: Towards disaster resilient societies, United Nations Univ. Press, New York, 9–54, 2006.

Boudou, M.: Approche multidisciplinaire pour la caractérisation d'inondations remarquables: enseignements tirés de de neufs évènements en France (1910–2010), PhD, Univ. Montpellier, Montpellier, 463 pp., 2015.

Boudou, M., Coeur, D., Lang, M., and Vinet, F.: Grille de lecture pour la caractérisation d'événements remarquables d'inondation en France: exemple d'application pour la crue de mars 1930, Environnement, politiques publiques et pratiques locales, Toulouse, 2015.

Brazdil, R., Kundzewicz, Z. W., and Benito, G.: Historical hydrology for studying flood risk in Europe, Hydrolog. Sci. J., 51, 739–764, 2006.

Cardona, O. D., Van Alast, M. K., Birkmann, M., Fordham, M., McGregor, G., Perez, R., Pulwarty, R. S., Schipper, E. L. F., and Sinh, B. T.: Determinants of risk: exposure and vulnerability, in: Managing the Risks of Extreme Events and Disasters to Advance Climate Change Adaptation, A Special Report of Working Groups I and II of the Intergovernmental Panel on Climate Change (IPCC), edited by: Field, C. B., Barros, V., Stocker, T. F., Qin, D., Dokken, D. J., Ebi, K. L., Mastrandrea, M. D., Mach, K. J., Plattner, G.-K., Allen, S. K., Tignor, M., andMidgley, P. M., Cambridge University Press, Cambridge, UK, and New York, NY, USA, 65–108, 2012.

Champion, M.: Les inondations en France depuis le VIe siècle jusqu'à nos jours, 6 volumes, Re-édition Cemagref Editions, Paris, 1858–1864.

Changnon, S. A., Pielke, R. A., Changnon, D., Sylves, R. T., and Pulwarty, R.: Human Factors Explain the Increased Losses from Weather and Climate Extremes, B. Am. Meteorol. Soc., 81, 437–442, 2000.

Coeur, D. and Lang, M.: Use of documentary sources on past flood events for flood risk management and land planning, C. R. Geosci., 340, 644–650, 2008.

Danière, B.: Analyse cartographique de l'évolution de la vulnérabilité en zone urbaine face aux inondations dites remarquables, Master 2 Univ. J. Monet Saint-Etienne, Irstea, Lyon, 111 pp., 2014.

de Bruijn, K. M.: Resilience and flood risk management: a systems approach applied to lowland rivers, PhD dissertation, Delft Univ., Delft, p. 210, 2005.

de Moel, H., van Alphen, J., and Aerts, J. C. J. H.: Flood maps in Europe – methods, availability and use, Nat. Hazards Earth Syst. Sci., 9, 289–301, doi:10.5194/nhess-9-289-2009, 2009.

DREAL Franche-Comté, EPTB Saône-et-Doubs, Ville de Besançon: 1910: la Crue du siècle à Besançon – Dossier de Presse, p. 9, www.franche-comte.developpement-durable.gouv.fr (last access: 1 December 2015), 2010.

Dreal Midi-Pyrénées: Mise en œuvre de la Directive Inondation. Rapport d'accompagnement des cartographies du TRI Montauban Moissac, Toulouse, p. 29 + annexes, 2014.

Ekamper, P.: Using cadastral maps in historical demographic research: Some examples from the Netherlands, Hist. Family, 15, 1–12, 2010.

Glade, T., Albini,, P., and Frances, F.: The use of historical data in natural hazard assessments Advances in Natural and Technological Hazards Research, Kluwer Academic Publishers, Dordrecht, p. 220, 2001.

Goubet, A.: Les crues historiques de la Seine à Paris, La Houille Blanche, 8, 23–27, 1997.

Gregory, I. N. and Healey, R. G.: Historical GIS: structuring, mapping and analysing geographies of the past, Prog. Human Geogr., 31, 638–653, 2007.

Journal Officiel de la République Française: Loi portant création d'un fonds provisionnel d'un milliard de francs, en vue de la réparation des dommages de caractère exceptionnel causés par les orages et les crues du 1er au 30 mars 1930, 88, 3970, Paris, 11 avril 1930.

Kjeldsen, T. R., Macdonald, N., Lang, M., Mediero, L., Albuquerque, T., Bogdanowicz, E., Brazdil, R., Castellarin, A., David, V., Fleig, A., Gül, G. O., Kriauciuniene, J., Kohnova, S., Merz, B., Nicholson, O., Roald, L. A., Salinas, J. L., Sarauskienel, D., Sraj, M., Strupczewski, W., Szolgay, J., Toumazis, A., Vanneuville, W., Veijalainen, N., and Wilson, D.: Documentary evidence of past floods in Europe and their utility in flood frequency estimation, J. Hydrol., 517, 963–973, doi:10.1016/j.jhydrol.2014.06.038, 2014.

Kron, W.: Keynote lecture: Flood risk = hazard × exposure × vulnerability, Proceedings of the Flood Defence, Science Press, New York,, 82–97, 2002.

Kundzewicz, Z. W., Kanae, S., Seneviratne, S. I., Handmer, J., Nicholls, N., Peduzzi, P., Mechler, R., Bouwer, L. M., Arnell, N., Mach, K., Muir-Wood, R., Brakenridge, G. R., Kron, W., Benito, G., Honda, Y., Takahashi, K., and Sherstyukov, B.: Flood risk and climate change: global and regional perspectives, Hydrolog. Sci. J., 59, 1–28, 2014.

Lang, M. and Coeur, D., 2014. Les inondations remarquables en France, Inventaire 2011 pour la directive Inondation, Quae, Versailles, p. 640, 2014.

Lang, M., Coeur, C., Bacq, B., Bard, A., Becker, T., Bignon, E., Blanchard, R., Bruckmann, L., Delserieys, M., Edelblutte, C., and Merle, C.: Preliminary Flood Risk Assessment for the European Directive: inventory of French past floods, in: Comprehensive Flood Risk Management, edited by: Kjlin, F. and Schweckendiek, T., Taylor and Francis group, Rotterdam, 1211–1217, 2012.

Langran, G. and Chrisman, N. R.: A framework for temporal geographic information. Cartographica, Int. J. Geogr. Inf. Geovisual., 25, 1–14, 1988.

Levin, N., Kark, R., and Galilee, E.: Maps and the settlement of southern Palestine, 1799–1948: an historical/GIS analysis, J. Hist. Geogr., 36, 1–18, 2010.

Luino, F., Turconi, L., Petrea, C., and Nigrelli, G.: Uncorrected land-use planning highlighted by flooding: the Alba case study (Piedmont, Italy), Nat. Hazards Earth Syst. Sci., 12, 2329–2346, doi:10.5194/nhess-12-2329-2012, 2012.

Lwin, K. and Murayama, Y.: A GIS Approach to Estimation of Building Population for Micro spatial Analysis, T. GIS, 13, 401–414, 2009.

Merz, B., Thieken, A., and Gocht, M.: Flood risk mapping at the local scale: concepts and challenges, in: Flood risk management in Europe, Springer, New York, 231–251, 2007.

Ministère de l'Ecologie: L'évaluation préliminaire des risques d'inondation 2001, Bassin Rhône Méditerranée – Partie III Unité de présentation du Doubs, Lyon, 159–177, 2011.

Pardé, M.: La crue de mars 1930 dans le sud et le sud-ouest de la France: Genèse de la catastrophe, Revue Géographique des Pyrénées et du sud-ouest, 1, 3–99, 1930.

Picard, A.: Rapport de la commission chargée d'analyser les inondations sur le bassin de la Seine de janvier 1910, Rapport au président du Conseil et au ministère de l'Intérieur, Paris, IN, 1910.

Rumsey, D. and Williams, M.: Historical maps in GIS, in: Past time, past place: GIS for history, edited by: Knowles, A. K., ESRI Press, Redlands, CA, 1–18, 2002.

Schanze, J.: Flood risk management – A basic framework, in: Flood Risk Management: Hazards, Vulnerability and Mitigation Measures, Chap. I, Springer, Dordrecht, 1–20, 2006.

Smith, A., Martin, D., and Cockings, S.: Spatio-Temporal Population Modelling for Enhanced Assessment of Urban Exposure to Flood Risk, Appl. Spat. Anal. Policy, 10, 1–19, 2014.

SwissRe: Natural catastrophes and man-made disaster in 2014: convective and winter storms generate most losses, Sigma, 2, 52, 2015.

Tropeano, D. and Turconi, L.: Using Historical Documents for Landslide, Debris Flow and Stream Flood Prevention. Applications in Northern Italy, Nat. Hazards, 31, 663–679, 2004.

Vinet, F., Lumbroso, D., Defossez, S., and Boissier, L.: A comparative analysis of the loss of life during two recent floods in France: the sea surge caused by the storm Xynthia and the flash flood in Var, Nat. Hazards, 61, 1179–1201, 2012.

Wisner, B., Blaikie, P., Cannon, T., and Davis, I: At risk: natural hazards, people's vulnerability and disasters, Routledge, London, p. 284, 1994.

Wu, S. S., Wang, L., and Qiu, X.: Incorporating GIS building data and census housing statistics for sub-block-level population estimation, Profess. Geogr., 60, 121–135, 2008.

5

Climatological characteristics of raindrop size distributions in Busan, Republic of Korea

S.-H. Suh[1]**, C.-H. You**[2]**, and D.-I. Lee**[1,2]

[1]Department of Environmental Atmospheric Sciences, Pukyong National University, Daeyeon campus 45, Yongso-ro, Namgu, Busan 608-737, Republic of Korea
[2]Atmospheric Environmental Research Institute, Daeyeon campus 45, Yongso-ro, Namgu, Busan 608-737, Republic of Korea

Correspondence to: C.-H. You (youch@pknu.ac.kr)

Abstract. Raindrop size distribution (DSD) characteristics within the complex area of Busan, Republic of Korea (35.12° N, 129.10° E), were studied using a Precipitation Occurrence Sensor System (POSS) disdrometer over a 4-year period from 24 February 2001 to 24 December 2004. Also, to find the dominant characteristics of polarized radar parameters, which are differential radar reflectivity (Z_{dr}), specific differential phase (K_{dp}) and specific attenuation (A_h), **T**-matrix scattering simulation was applied in the present study. To analyze the climatological DSD characteristics in more detail, the entire period of recorded rainfall was divided into 10 categories not only covering different temporal and spatial scales, but also different rainfall types. When only convective rainfall was considered, mean values of mass-weighted mean diameter (D_m) and normalized number concentration (N_w) values for all these categories converged around a maritime cluster, except for rainfall associated with typhoons. The convective rainfall of a typhoon showed much smaller D_m and larger N_w compared with the other rainfall categories.

In terms of diurnal DSD variability, we analyzed maritime (continental) precipitation during the daytime (DT) (nighttime, NT), which likely results from sea (land) wind identified through wind direction analysis. These features also appeared in the seasonal diurnal distribution. The DT and NT probability density function (PDF) during the summer was similar to the PDF of the entire study period. However, the DT and NT PDF during the winter season displayed an inverse distribution due to seasonal differences in wind direction.

1 Introduction

Raindrop size distribution (DSD) is controlled by the microphysical processes of rainfall, and therefore it plays an important role in development of the quantitative precipitation estimation (QPE) algorithms based on forward scattering simulations of radar measurements (Seliga and Bringi, 1976). DSD data accurately reflect local rainfall characteristics within an observation area (You et al., 2014). Many DSD models have been developed to characterize spatial–temporal differences in DSDs under various atmospheric conditions (Ulbrich, 1983). Marshall and Palmer (1948) developed an exponential DSD model using DSD data collected by a filter paper technique ($N(D) = 8 \times 10^3 \exp\left(-4.1R^{-0.21}D\right)$ in $\mathrm{m^{-3}\,mm^{-1}}$ D in mm and R in $\mathrm{mm\,h^{-1}}$). In subsequent studies, a lognormal distribution was assumed to overcome the problem of exponential DSD mismatching with real data (Mueller and Sims, 1966; Levin, 1971; Markowitz, 1976; Feingold and Levin, 1986).

To further investigate natural DSD variations, Ulbrich (1983) developed a gamma DSD that permitted one to change the dimension of the intercept parameter (N_0 in $\mathrm{m^{-3}\,mm^{-1-\mu}}$) with $N(D) = N_0 D^{\mu} \exp(-\Lambda D)$. In addition, to enable the quantitative analysis of different rainfall events, the development of a normalized gamma DSD model that accounted for the independent distribution of DSD from the disdrometer channel interval enabled a better representation of the actual DSD (Willis, 1984; Dou et al., 1999; Testud et al., 2001).

DSDs depend on the rainfall type, geographical and atmospheric conditions, and observation time. Also, these are closely linked to microphysical characteristics that control rainfall development mechanisms. In the case of stratiform rainfall, raindrops grow by the accretion mechanism because of the relatively long residence time in weak updraft condition, in which almost all water droplets are changed to ice particles. With time, the ice particles grow sufficiently and fall to the ground. The raindrop size of stratiform rainfall observed at the ground level is larger than that of convective rainfall for a same rainfall intensity due to the resistance of the ice particles to break-up mechanisms. In contrast to stratiform rainfall, convective rainfall raindrops grow by the collision–coalescence mechanism associated with relatively strong vertical wind speeds and short residence time in the cloud. Fully grown raindrops of maritime precipitation are smaller in diameter than those in stratiform rainfall due to the break-up mechanism in the case of the same rainfall rate (Mapes and Houze Jr., 1993; Tokay and Short, 1996). Convective rainfall can be classified into two types based on the origin and direction of movement. Rainfall systems occurring over ocean and land are referred to as maritime and continental rainfall, respectively (Göke et al., 2007). Continental rainfall is related to a cold-rain mechanism whereby raindrops grow in the form of ice particles. In contrast, maritime rainfall is related to a warm-rain mechanism whereby raindrops grow by the collision–coalescence mechanism. Therefore, the mass-weighted drop diameter (D_m) of continental rainfall observed on the ground is larger than that of maritime rainfall, and a smaller normalized intercept parameter (N_w) is observed in continental rainfall (Bringi et al., 2003).

Specific heat is a major climatological feature that creates differences between DSDs in maritime and continental regions. These two regions have different thermal capacities, and thus different temperature variations have occurred with time. The surface temperature of the ocean changes slowly because of the higher thermal capacity compared with land, while the continental regions that have a comparatively lower thermal capacity show greater diurnal temperature variability. Sea winds generally occur from afternoon to early evening when the temperature gradient between the sea and land becomes negative, which is the opposite gradient in the daytime (DT). In coastal regions, the land and sea wind effect causes a pronounced difference between the DT and nighttime (NT) DSD characteristics. Also, when mountains are located near the coast, the difference is intensified by the effect of mountain and valley winds (Qian, 2008).

In the present study, we analyzed a 4-year data set spanning from 2001 to 2004, collected from Busan, Republic of Korea (35.12° N, 129.10° E), using a Precipitation Occurrence Sensor System (POSS) disdrometer, to investigate the characteristics of DSDs in Busan, Republic of Korea, which consist of a complex mid-latitude region comprising both land and ocean. To quantify the effect of land and sea wind on these characteristics, we also analyzed diurnal variations in DSDs. The remainder of the paper is organized as follows. In Sect. 2 we review the normalized gamma model

and explain the DSD quality control method and the classification of rainfall. In Sect. 3 we report the results of DSD analysis with respect to stratiform/convective and continental/maritime rainfall, and discuss diurnal variations. Finally, a summary of the results and the main conclusions are presented in Sect. 4.

2 Data and methods

2.1 Normalized gamma DSD

DSDs are defined by $N(D) = N_0 \exp(-\Lambda D)\,(\mathrm{m^{-3}\,mm^{-1}})$ and reflect the microphysical characteristics of rainfall using the number concentration of raindrops ($N(D)$). Also, DSDs are able to calculate the many kinds of parameters that show the dominant feature of raindrops. Normalization is used to define the DSD and to solve the non-independence of each DSD parameter (Willis, 1984; Dou et al., 1999; Testud et al., 2001). Furthermore, a normalized gamma DSD enables the quantitative comparison for rainfall cases regardless of timescale and rain rate. Here, we use the DSD model designed by Testud et al. (2001):

$$N(D) = N_w f(\mu)\left(\frac{D}{D_m}\right)^{\mu} \exp\left[-(4+\mu)\frac{D}{D_m}\right], \quad (1)$$

where D is the volume-equivalent spherical raindrop diameter (mm), and $f(\mu)$ is defined using the DSD model shape parameter (μ) and gamma function (Γ) as follows:

$$f(\mu) = \frac{6}{4^4}\frac{(4+\mu)^{\mu+4}}{\Gamma(\mu+4)}. \quad (2)$$

From the value of $N(D)$, the median volume diameter (D_0 in mm) can be obtained as follows:

$$\int_0^{D_0} D^3 N(D)\,dD = \frac{1}{2}\int_0^{D_{max}} D^3 N(D)\,dD. \quad (3)$$

Mass-weighted mean diameter (D_m in mm) is calculated as the ratio of the fourth to the third moment of the DSD:

$$D_m = \frac{\int_0^{D_{max}} D^4 N(D)\,dD}{\int_0^{D_{max}} D^3 N(D)\,dD}. \quad (4)$$

The normalized intercept parameter (N_w in $\mathrm{m^{-3}\,mm^{-1}}$) is calculated as follows:

$$N_w = \frac{4^4}{\pi \rho_w}\left(\frac{\mathrm{LWC}}{D_m^4}\right). \quad (5)$$

The shape of the DSD is calculated as the ratio of D_m to the standard deviation (SD) of D_m (σ_m in mm) (Ulbrich and Atlas, 1998; Bringi et al., 2003; Leinonen et al., 2012):

$$\sigma_m = \left[\frac{\int_0^{D_{max}} D^3 (D - D_m)^2 N(D)\,dD}{\int_0^{D_{max}} D^3 N(D)\,dD}\right]^{\frac{1}{2}}. \quad (6)$$

In addition, σ_m/D_m is related to μ as follows:

$$\frac{\sigma_m}{D_m} = \frac{1}{(4+\mu)^{1/2}}. \quad (7)$$

Liquid water content (LWC in $g\,m^{-3}$) can be defined from the estimated DSD:

$$LWC = \frac{\pi}{6}\rho_w \int_0^{D_{max}} D^3 N(D)\,dD, \quad (8)$$

where ρ_w is the water density ($g\,m^{-3}$), and it is assumed to be $1 \times 10^6 g\,m^{-3}$ for a liquid. Similarly, the rainfall rate (R in $mm\,h^{-1}$) can be defined as follows:

$$R = \frac{3.6}{10^3}\frac{\pi}{6} \int_0^{D_{max}} v(D)D^3 N(D)\,dD, \quad (9)$$

where the value of factor 3.6×10^{-3} is the unit conversion that converts the mass flux unit ($mg\,m^{-2}\,s^{-1}$) to the common unit ($mm\,h^{-1}$) for convenience. $v(D)$ ($m\,s^{-1}$) is the terminal velocity for each raindrop size. The relationship between $v(D)$ and D (mm) is given by Atlas et al. (1973), who developed an empirical formula based on the data reported by Gunn and Kinzer (1949):

$$v(D) = 9.65 - 10.3\exp[-0.6D]. \quad (10)$$

2.2 Quality control of POSS data

POSS is used to measure the number of raindrops within the diameter range of 0.34–5.34 mm, using bistatic, continuous-wave X-band Doppler radar (10.525 GHz) across 34 channels (Fig. 1; Sheppard and Joe, 2008). To estimate DSDs, the Doppler power density spectrum is calculated as follows:

$$S(f) =$$
$$\int_{D_{min}}^{D_{max}} N(D_m)V(D_m, \rho, h, w)\bar{S}(f, D_m, \rho, h, w)dD_m, \quad (11)$$

where $S(f)$ means Doppler spectrum power density, $V(D_m \rho h, w)\bar{S}(f, D_m \rho h, w)$ means weighting function of $S(f)$, $\bar{S}$ is the mean of $S(f)$, ρ is density of precipitation distribution, h is the shape of precipitation distribution, w ($m\,s^{-1}$) is wind speed, $V(x)$ is sample volume, and the symbol "x" means arbitrary parameters that affect the sampling volume. The Doppler power density spectrum has a resolution of 16 Hz, and terminal velocity (v_t) has a resolution of $0.24\,m\,s^{-1}$. Transmitter and receiver skewed about 20° toward each other, and the cross point of the signal is located over 34 cm from the transmitter–receiver. The transmitter–receiver toward the upper side detects $N(D)$ in $V(x)$ (Sheppard, 1990). Also, Sheppard (1990) and Sheppard and Joe (1994) noted some shortcomings as the overestimation of small drops at horizontal wind was larger than $6\,m\,s^{-1}$. However, in the present study, the quality control of POSS for wind effects was not considered, because it lies

Figure 1. Photograph of the POSS instrument used in this research.

Table 1. Specification of the POSS disdrometer.

Specifications	Detail
Manufacturer	ANDREW CANADA INC
Module	PROCESSOR
Model number	POSS-F01
Nominal power	100 mW
Bandwidth	Single frequency
Emission	43 mW
Pointing direction	20° (to the vertical side)
Antenna	Rectangular pyramidal horns
Range of sample area	< 2 m
Wavelength	10.525 GHz ± 15 GHz
Physical dimension	$277 \times 200 \times 200\,cm^3$
Net weight	Approximately 110 kg

beyond the scope of this work. Detailed specifications and measurement ranges and raindrop sizes for each observation channel of the POSS disdrometer are summarized in Table 1.

A POSS disdrometer has been operating in Busan, Republic of Korea (35.12° N, 129.10° E), along with other atmospheric instruments, the locations of which are shown in Fig. 2. Estimating raindrop diameter correctly is challenging,

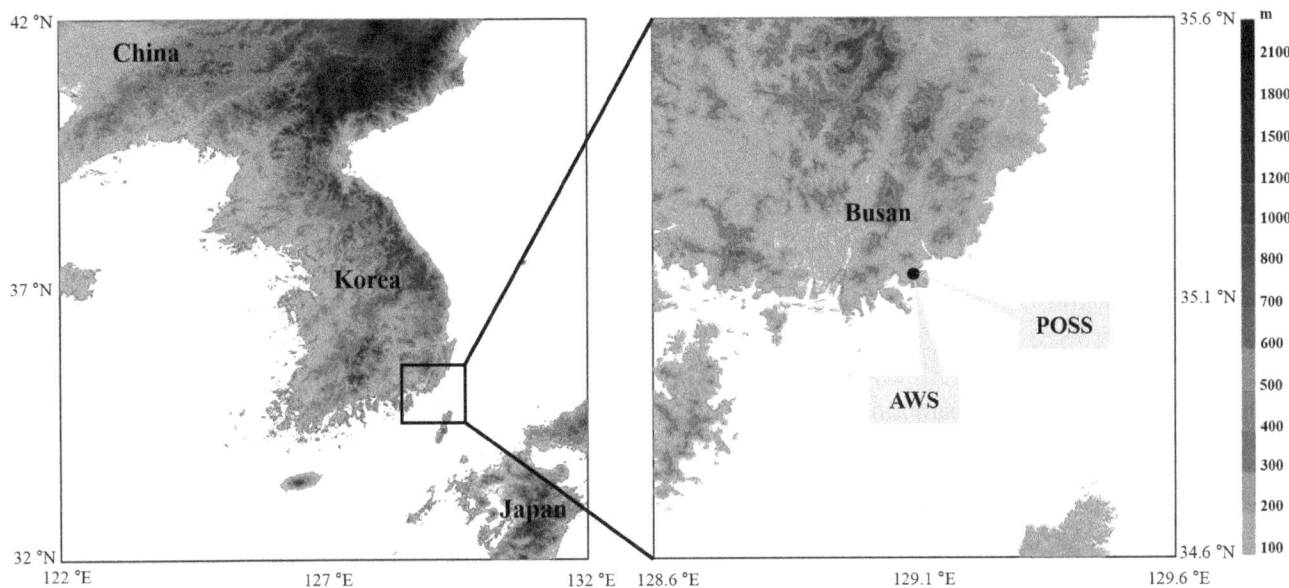

Figure 2. Locations of the POSS and the AWS rain gauge installed in Busan, Republic of Korea.

and care should be taken to ensure reliable data are collected. We performed the following quality controls to optimize the accuracy of the disdrometer estimates. (i) Non-liquid type event data (e.g., snow, hail) detected by POSS were excluded by routine observation and the surface weather chart provided by the Korea Meteorological Administration (KMA). (ii) DSD spectra in which drops were not found in at least five consecutive channels were removed as non-atmospheric. (iii) Only data recorded in more than 10 complete channels were considered. (iv) To compensate for the reduced capability to detect raindrops smaller than 1 mm when $R > 200 \, \mathrm{mm \, h^{-1}}$ (as recorded by the disdrometer), data for $R > 200 \, \mathrm{mm \, h^{-1}}$ were not included in the analyses, even though the number of samples was only 64 for the entire period. (v) To eliminate wind and acoustic noise, data collected when $R < 0.1 \, \mathrm{mm \, h^{-1}}$ are removed (Tokay and Short, 1996).

After performing all quality control procedures, 99 388 spectra were left from the original data (166 682) for 1 min temporal resolution. The accumulated rainfall amount from POSS during the entire period was 4261.49 mm. To verify the reliability of the POSS data, they were compared with data collected by a 0.5 mm tipping bucket rain gauge at an automatic weather system (AWS) located ∼ 368 m from the POSS (Fig. 3).

2.3 Radar parameters

First, the radar reflectivity factor (z in $\mathrm{mm^6 \, m^{-3}}$) and non-polarized radar reflectivity (Z in dBZ) were computed using the DSD data collected by POSS, as follows:

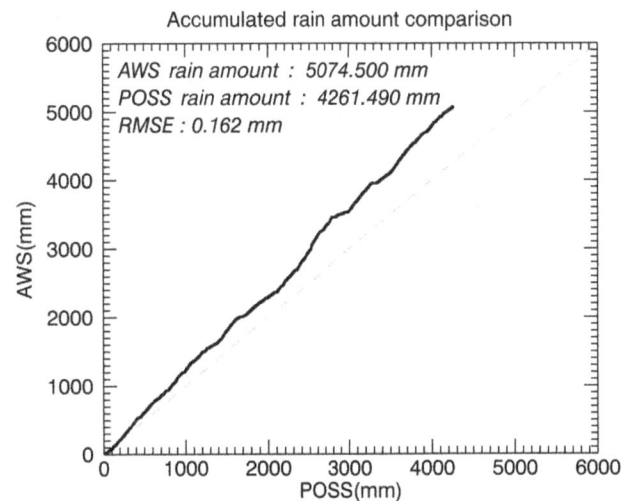

Figure 3. Comparison of the recorded rainfall amounts between the POSS and AWS instruments.

$$z = \int_0^{D_{\max}} D^6 N(D) \, \mathrm{d}D, \qquad (12)$$

$$Z = 10 \log_{10}(z). \qquad (13)$$

The **T**-matrix method used in this study was initially proposed by Waterman (1965, 1971) to calculate electromagnetic scattering by single non-spherical raindrops. The adaptable parameters for this calculation are frequency, temperature, hydrometeor types, raindrop canting angle, and axis ratio (γ), and explained the following sentences. Axis ratios of raindrops differ with atmospheric conditions and rainfall type. To derive the drop shape relation from the drop diameter, we applied the results of numerical simulations and wind tunnel tests employing a fourth-order polynomial equation, as in many previous studies (Beard and Chuang, 1987; Pruppacher and Beard, 1970; Andsager et al., 1999; Brandes et al., 2002). The axis ratio relation used in the present study is a combination of those from Andsager et al. (1999) and Beard and Chuang (1987) for three raindrop size ranges (Bringi et al., 2003).

The raindrop axis ratio relation of Andsager et al. (1999) is applied in the range of $1 < D\,(\mathrm{mm}) < 4$, as follows:

$$\gamma = 1.012 + 0.01445 \times 10^{-1}D - 0.01028 \times 10^{-2}D^2. \quad (14)$$

The drop-shaped relation of Beard and Chuang (1987) is applied in the range of $D < 1\,\mathrm{mm}$ and $D > 4\,\mathrm{mm}$, as follows:

$$\gamma = 1.0048 + 0.0057 \times 10^{-1}D - 2.628 \times 10^{-2}$$
$$\times D^2 + 3.682 \times 10^{-3}D^3 - 1.677 \times 10^{-4}D^4. \quad (15)$$

We assumed SD and the mean canting angle of raindrops to be 7 and $0°$, respectively. The refractive indices of liquid water at $20\,°\mathrm{C}$ were used (Ray, 1972). Also, the condition of frequency for electromagnetic wave of radar is $2.85\,\mathrm{GHz}$ (S-band). We calculated dual polarized radar parameters based on these conditions. The parameters of differential reflectivity (Z_{dr} in dB), specific differential phase (K_{dp} in deg km^{-1}), and attenuation (A_{h} in dB km^{-1}) using DSD data were calculated and analyzed.

2.4 Classification of rainfall types and rainfall events

Rainfall systems can be classified as stratiform or convective in nature, via analysis of the following microphysical characteristics: (i) DSD, using relationships between N_0 and R ($N_0 > 4 \times 10^9 R^{-4.3}$ in m^{-3} mm^{-1} is considered to be convective rainfall, Tokay and Short, 1996; Testud et al., 2001); (ii) Z, where, according to Gamache and Houze Jr. (1982), a rainfall system that displays radar reflectivity larger than 38 dBZ is considered to be convective; and (iii) R, where an average value larger than 0.5 mm per 5 min is considered to be convective rainfall (Johnson and Hamilton, 1988). Alternatively, rainfall that has 1 min $R > 5$ (0.5) mm h^{-1} and a SD of R > (<) 1.5 mm h^{-1} is considered to be of the convective (stratiform) type (Bringi et al., 2003). The rainfall classification method proposed by Bringi et al. (2003) is applied in the present study.

It is necessary to categorize different rainfall systems because their microphysical characteristics show great variation depending on the type of rainfall, as well as the type of rainfall event, e.g., typhoon, Changma, heavy rainfall and seasonally discrete rainfall. To investigate the temporal variation in DSDs, we analyzed daily and seasonal DSDs. Likewise, to investigate diurnal variability in DSD, DT, and NT data were considered by using the sunrise and sunset time in Busan (provided by the Korea Astronomy and Space Science Institute (KASI)). In the middle latitudes, and including Busan, the timings of sunrise and sunset vary due to solar culminating height. The earliest and latest sunrise (sunset) time of the entire period is 05:09 KST (17:12 KST) and 07:33 KST (19:42 KST), respectively. DT (NT) is defined as the period from the latest sunrise (sunset) time to the earliest sunset (sunrise) time for the unity of classification of each time group (Table 2).

To analyze the predominant characteristics of DSDs for typhoon rainfall, nine typhoon events were selected from throughout the entire study period, which is summarized in Table 2.

This study utilizes KMA rainfall warning regulations to identify heavy rainfall events. The KMA issues a warning if the accumulated rain amount is expected to be > 70 mm within a 6 h period, or > 110 mm within a 12 h period. Rainfall events classified as Changma and typhoon were not included in the classification "heavy rainfall".

Changma is the localized rainfall system or rainy season that is usually present over the Korean Peninsula between mid-June and mid-July, which is similar to the Meiyu (China) or Baiu (Japan). The selected dates and periods of each rainfall category are summarized in Table 2.

3 Results

3.1 DSD and radar parameters

Figure 4 shows the probability density function (PDF) and cumulative distribution function (CDF) of DSDs and radar parameters with respect to the entire, stratiform and convective rainfall. The PDFs of DSD and radar parameters were calculated using the non-parameterization kernel estimation to identify the dominant distribution of each parameter recorded in Busan. Non-parameterization kernel estimation was also used to identify continuous distributions of DSDs. The PDF of stratiform rainfall is more similar to that of the data set for the entire analysis period due to the dominant contribution of stratiform rainfall (about 62.93 %) to the overall rainfall than that of convective rainfall. However, the PDF for convective rainfall is significantly different from that of the entire analysis period, and as the convective rainfall contributes only 6.11 % of the overall rainfall (Table 3). When $\mu < 0$, the distribution of μ for convective rain has more values of PDF than that for stratiform rain (Fig. 4a). Alternatively, the frequency of μ for stratiform rainfall is higher than that of convective rainfall when $0 < \mu < 5$. The value of μ for convective rainfall is higher than that for stratiform rainfall because the break-up mechanism would increase the number concentration of small raindrops. The number concentrations of mid-size raindrops increased due to the decrease in the number concentration of relatively large raindrops (Hu and Srivastava, 1995; Sauvageot and Lacaux, 1995). However, we observed a higher frequency of

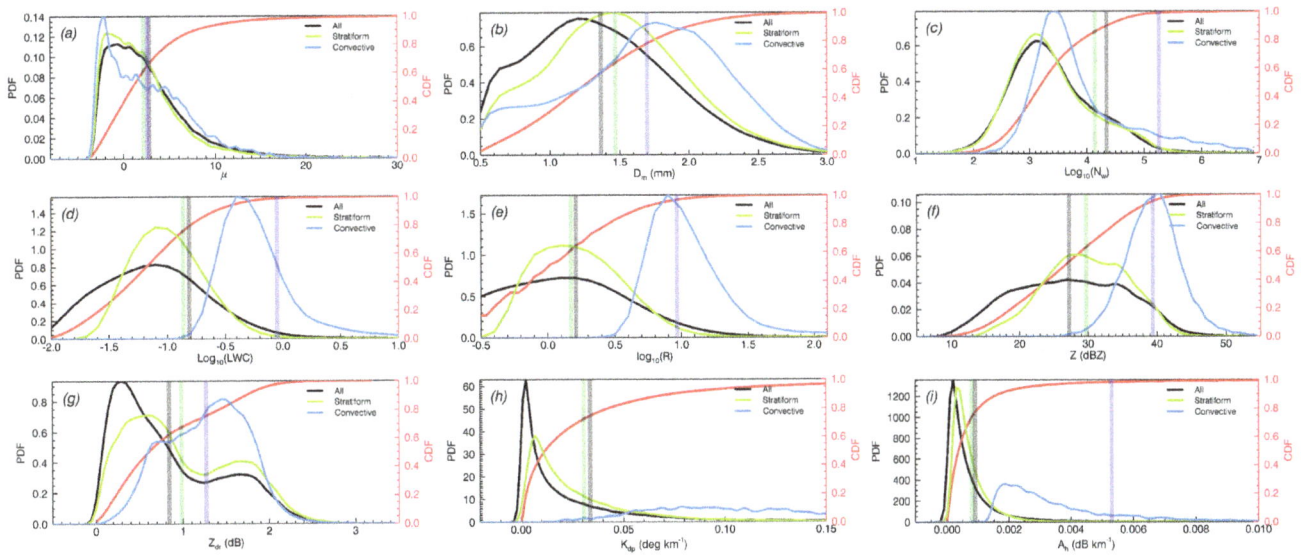

Figure 4. PDF and CDF for **(a)** μ, **(b)** D_m, **(c)** $\log_{10}(N_w)$, **(d)** $\log_{10}(R)$, **(e)** $\log_{10}(\text{LWC})$, **(f)** Z, **(g)** Z_{dr}, **(h)** K_{dp}, and **(i)** A_h for the entire rainfall data set (solid black line), stratiform rainfall (solid green line), and convective rainfall (solid blue line). The solid red line represents the CDF for the entire rainfall data set. The solid vertical line represents the mean value of each type.

Table 2. Designated date with respect to the source of rainfall.

Rainfall category	Period			
Typhoon	2001 –	2002 5–6 Jul, 31 Aug	2003 29 May, 19 Jun, 7 Aug, 11–12 Sep	2004 20 Jun, 19 Aug, 6 Sep
Changma	18–19 Jun, 23–26 Jun, 29–30 Jun, 1 Jul, 5–6 Jul, 11–14 Jul	23–25 Jun, 30 Jun, 1–2 Jul	12–14 Jun, 23 Jun, 27 Jun, 30 Jun, 1 Jul, 3–15 Jul	11–14 Jul
Heavy rainfall	15 Apr 2002, 20:13 KST to 16 Apr 2002, 06:29 KST			
Seasonal	Spring Mar to May	Summer Jun to Aug	Autumn Sep to Nov	Winter Dec to Feb
Diurnal	DT (KST) 07:33–17:12		NT (KST) 19:42–05:09	

convective rainfall than stratiform rainfall in the negative μ range.

The PDF of D_m displays a peak around 1.2 and 1.4 mm for stratiform rainfall and the entire rainfall data set, respectively. We note that a gentle peak exists around 0.7 mm for both stratiform and convective rainfall data sets (Fig. 4b). These features are similar to the distribution of D_m observed in a high-latitude region at Järvenpää, Finland (Fig. 4 of Leinonen et al., 2012). For D_m values > 1.7 mm, the PDF for convective rainfall is higher than stratiform rainfall. Accordingly, the value of DSD for stratiform rainfall is higher than that of convective rainfall when $D_m < 1.7$ mm. Generally, stratiform rainfall that develops by the cold rain process displays weaker upward winds and less efficient break-

up of raindrops. Therefore, in the same rainfall rate, stratiform rainfall tends to produce larger raindrops than convective rainfall, which develops by the warm rain process. However, the average D_m values for convective and stratiform rain for the entire period are approximately 1.45 and 1.7 mm, respectively. In short, D_m is proportional to R regardless of rainfall type. This finding is consistent with the results of Atlas et al. (1999), who found that the D_m of convective rainfall is larger than that of stratiform rainfall on Kapingamarangi, Micronesia.

The PDF of $\log_{10}(N_w)$ for the entire rainfall data set was evenly distributed between 1.5 and 5.5, with a peak at $N_w = 3.3$ (Fig. 4c). The PDF of $\log_{10}(N_w)$ for stratiform rainfall is rarely > 5.5, while for convective rainfall it

64

Hydrology and Hydrogeology

Table 3. Rainfall rate for each rainfall category and the number of sample sizes for 1 min data.

Rainfall category	Total precipitation	Stratiform precipitation (%)	Convective precipitation (%)
Typhoon	5095	3118 (61.19)	652 (12.79)
Changma	18 526	11 099 (59.91)	1611 (8.69)
Heavy rainfall	359	153 (42.61)	150 (41.78)
Spring	30 703	20 370 (66.34)	1478 (4.81)
Summer	37 187	22 566 (60.68)	3409 (9.16)
Autumn	19 809	12 033 (60.74)	850 (4.29)
Winter	11 689	7582 (64.86)	339 (2.90)
Daytime	41 328	26 373 (63.81)	2539 (6.14)
Nighttime	37 455	23 063 (84.00)	2242 (5.89)
Entire	99 388	62 551 (62.93)	6076 (6.11)

is higher at >5.5 than that of stratiform. There is a similar frequency in the stratiform and convective rainfall at 4.4.

The PDF distributions for $\log_{10}(R)$ and $\log_{10}(\text{LWC})$ are similar to each other (Fig. 4d and e). It is inferred that the similar results come from the use of similar moments of DSD such as 3.67 and 3 for R and LWC, respectively. The PDF of $\log_{10}(R)$ for the entire rainfall data set ranged between -0.5 and 2. A peak exists at 0.3 and the PDF rapidly decreases from the peak value as R increases. The PDF for stratiform rainfall has a higher frequency than that of the entire rainfall when $-0.3 < \log_{10}(R) < 0.7$, while the PDF for convective rainfall is denser between 0.4 and 2. Furthermore, the frequency of the PDF for convective rainfall was higher than that of stratiform rainfall in the case of $\log_{10}(R) > 0.65$ and the peak value shown as 0.9.

The PDF and CDF for Z, Z_{dr}, K_{dp} and A_h are shown in Fig. 4f–i. The PDF of Z for stratiform rainfall (Fig. 4f) is widely distributed between 10 and 50 dBZ, with the peak at approximately 27 dBZ. Conversely, for convective rainfall, the values of PDF lie between 27 and 55 dBZ, and the peak frequency value at approximately 41 dBZ. The frequency value of reflectivity is higher for convective rainfall than for stratiform rainfall in the range of $\sim > 35$ dBZ. Furthermore, the shape of the PDF for convective rainfall is similar to that reported for Darwin, Australia (Steiner et al., 1995); however, for stratiform rainfall, there are significant differences between Busan and Darwin in terms of the shape of the frequency distribution. The PDF of Z_{dr} for the entire rainfall primarily exists between 0 and 2.5 dB, and the peaks are at 0.3 and 1.8 dB (Fig. 4g). The distribution of Z_{dr} for convective and stratiform rainfall is concentrated between 0.6 and 1.6 dB, and between 0.3 and 2 dB, respectively. The frequency of Z_{dr} for convective (stratiform) rainfall exists in ranges higher (lower) than stratiform (convective) at 0.9 dB.

The dominant distribution of K_{dp} for the entire data set and for stratiform rainfall lies between 0 and 0.14 deg km^{-1}, with a peak value of 0.03 and 0.08 deg km^{-1}. However, for convective rainfall, the PDF of K_{dp} evenly exists between 0.01 and 0.15 deg km^{-1}. Furthermore, when

$K_{dp} > 0.056$ deg km^{-1}, the frequency of the PDF for convective rainfall is higher than that of stratiform rainfall (Fig. 4h).

The PDF of A_h is similar to that of K_{dp} and exists between 0 and 0.01 dB km^{-1}. For the case of the entire rainfall data set and for stratiform rainfall, the PDF of A_h is concentrated between 0 and 2.0×10^{-3} dB km^{-1} and that of convective rainfall is strongly concentrated between 1.0×10^{-3} and 8.0×10^{-3} dB km^{-1} (Fig. 4i). Unlike the PDF of A_h for convective rainfall, the PDF for stratiform rainfall shows a strong peak at about 7.0×10^{-4} dB km^{-1}.

3.2 Climatological characteristics of DSD in Busan

The climatological characteristics of DSDs for 10 rainfall categories are analyzed in this study. Sample size and ratio rainfall for each category are summarized in Table 3. Figure 5a illustrates the distribution of all 1 min stratiform rainfall data, and Fig. 5b shows scatter plots of averaged D_m and $\log_{10}(N_w)$ for all 10 rainfall categories for stratiform rainfall data. Figure 5a displays a remarkable clear boundary in the bottom sector and shows that most of the data lie below the reference line used by Bringi et al. (2003) to classify convective and stratiform rainfall. The average values of D_m and $\log_{10}(N_w)$ for all rainfall categories, except for heavy rainfall, exist between 1.4 and 1.6 mm and between 3.15 and 3.5, respectively (Fig. 5b). These values are relatively small compared with the reference line presented by Bringi et al. (2003).

The distribution of 1 min convective rainfall data is displayed in Fig. 6a, and the distribution of average values of D_m and N_w for the 10 rainfall categories in the case of convective rainfall in Fig. 6b. The blue and red plus symbols represent maritime and continental rainfall, respectively, as defined by Bringi et al. (2003). The scatter plot of 1 min convective rainfall data shows more in the continental cluster than the maritime cluster; however, the average values for the 10 rainfall categories are all located around the maritime cluster, except for the typhoon category. By considering the entire average values including typhoon events (Fig. 6b), we can induce the simple linear equation using D_m and $\log_{10}(N_w)$ as follows:

$$\log_{10}(N_w) = -1.8D_m + 6.9. \tag{16}$$

Even the coefficients in Eq. (16) might be changed slightly with the typhoon values; this result is not present in $D_m < 1.2$ mm and $D_m > 1.9$ mm. The D_m (N_w) value for the typhoon category was considerably smaller (larger) than that of the other categories as well as that of the stratiform type of typhoon. This result does not agree with that reported by Chang et al. (2009), who noted that the D_m of convective rainfall typhoon showed a large value compared with that associated with stratiform rainfall.

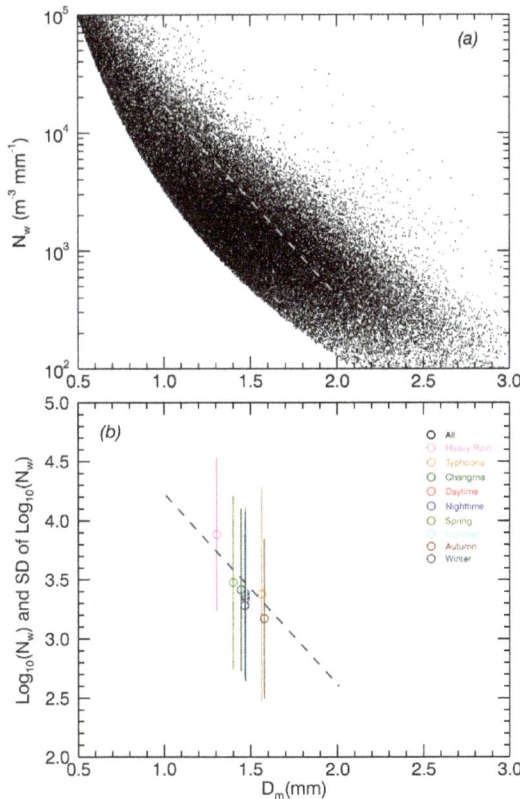

Figure 5. (a) Scatter plot of 1 min D_m and N_w for the 10 rainfall categories with respect to stratiform rainfall data. The broken grey line represents the average line as defined by Bringi et al. (2003). **(b)** Scatter plot of mean D_m and $\log_{10}(N_w)$ values of the 10 rainfall categories with respect to stratiform rainfall; these mean values for each rainfall type are shown as circle symbols. The vertical line represents $\pm 1\sigma$ of $\log_{10}(N_w)$ for each category.

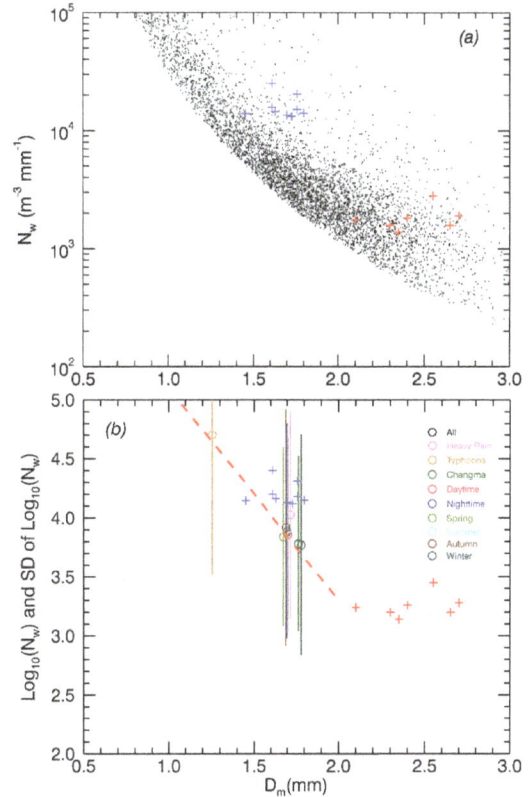

Figure 6. (a) As in Fig. 5a but for convective rainfall. The blue and red plus symbols represent maritime and continental rainfall, respectively, as defined by Bringi et al. (2003). **(b)** As in Fig. 5b but for convective rainfall. The broken red line represents the mathematical expression described in Eq. (16).

3.3 Diurnal variation in raindrop size distributions

3.3.1 Diurnal variations in DSDs

Figure 7a shows a histogram of normalized frequency of 16 wind directions recorded by the AWS, which is the same instrument as that used to collect the data shown in Fig. 3. To establish the existence of a land and sea wind, the difference in wind direction frequencies between DT and NT were analyzed. Figure 7b shows the difference between DT and NT; difference frequency means normalized frequency of wind direction for DT subtracted from that of NT for each direction, in terms of the normalized frequency of 16 wind directions. In other words, positive (negative) values indicate that the frequency of wind is more often observed during DT (NT). Also, land (sea) wind was defined in the present study from 225° (45°) to 45° (225°) according to the geographical condition in Busan. The predominant frequency of wind direction in DT (NT), between 205° (22.5°) and 22.5° (205.5°), is higher than that in NT (DT) (Fig. 7b). The ob-

servation site distance where the POSS was installed at the western side from the closest coastline is about 611 m, suggesting that the effect of the land and sea wind would have been recorded. To understand the effects of the land and sea wind on DSD characteristics, we analyzed the PDF and 2 h averaged DSD parameters for DT and NT. Figure 8 illustrates the distributions of μ, D_m, $\log_{10}(N_w)$, $\log_{10}(\text{LWC})$, $\log_{10}(R)$, and Z. There were large variations of μ with time. The μ values varied from 2.41 to 3.17, and the minimum and maximum μ values occurred at 08:00 and 12:00 KST, respectively (Fig. 8a). A D_m larger than 1.3 mm dominated from 00:00 to 12:00 KST, before decreasing remarkably between 12:00 and 14:00 KST. The minimum and maximum D_m appeared at 14:00 and 08:00 KST, respectively (Fig. 8b).

The N_w distribution showed inversely to D_m; however, no inverse relationship was identified between D_m and N_w in the case of the time series (Fig. 8c). The maximum and minimum values of N_w were found at 06:00 and 22:00 KST.

Variability through time was similar for R, LWC, and Z_h like D_m. There was an increasing trend from 00:00 to 08:00 KST followed by a remarkably decreasing trend from 08:00 to 14:00 KST (Figs. 8d, 11e and 11f). Note that the

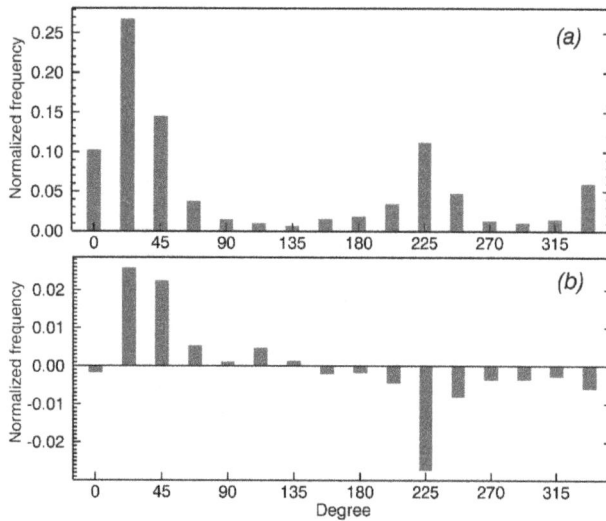

Figure 7. (a) Histogram of normalized frequency of 16 wind directions for the entire study period. (b) Difference values of wind direction frequencies between DT and NT (DT–NT).

Table 4. DT and NT (KST) in the summer and winter seasons.

Rainfall category	Period	Beginning time (KST)	Finishing time (KST)
Summer	DT	05:33	19:27
	NT	19:42	05:09
Winter	DT	07:33	17:12
	NT	18:19	06:54

In the present study, the shape of the PDF of LWC and R for DT and NT are similar, which is the same reason with the results of Fig. 4e–f. LWC and R distributions during DT (NT) are higher (lower) than in NT (DT) when $\log_{10}(\text{LWC})$ and $\log_{10}(R)$ are larger (smaller) than -1.2 and 0, respectively (Fig. 9d and e). Z has a similar pattern with LWC, and R during DT (NT) was higher (lower) than in NT (DT) in the range below (above) about 27 dBZ (Fig. 9f).

3.3.2 Diurnal variations of DSDs with respect to season

Busan experiences distinct atmospheric conditions that are caused by the different frequencies and magnitudes of land and sea winds in response to variable sunrise and sunset times. To identify seasonal variations of DSDs with respect to the effect of the land and sea wind, we analyzed the DT and NT PDF of D_m and N_w in the summer and winter. The start and end times of DT (NT) were sorted using the latest sunrise (sunset) and the earliest sunset (sunrise) time for each season (Table 4), which is the same method as that of the entire period of classification.

Figure 10a shows a histogram of wind directions in summer (light grey) and winter (dark grey). The frequencies of summer and winter wind directions are similar to each other. However, in Fig. 10b, the DT and NT distributions of winter wind direction display opposite frequencies.

Note that the winter season shows remarkable frequency of land (sea) wind between 0° (157.5°) and 45° (202.5°) at DT (NT) compared with results of those for the summer season. The accumulated value of normalized wind frequencies at the sea and land wind show different features between the summer and winter seasons (Table 5).

To identify the variability of DSDs caused by the land and sea wind in summer and winter, a 2 h interval time series of D_m, N_w and R was analyzed. In the summer, the time series of D_m displays considerably large values between 00:00 and 12:00 KST, compared with the period between 14:00 and 22:00 KST (Fig. 11a). The mean value of D_m decreases dramatically between 12:00 and 14:00 KST. $\log_{10}(N_w)$ has a negative relationship with D_m (Fig. 11b). However, the inverse relation between $\log_{10}(N_w)$ and D_m is not remarkable. $\log_{10}(R)$ tends to increase gradually from 00:00 to 08:00 KST and decrease from 08:00 to 14:00 KST, which is similar to the pattern of that of the entire period (Fig. 11c). Kozu et al. (2006) analyzed the diurnal variation in R at

time of the sharp decline for R between 12:00 and 14:00 KST is simultaneous with a D_m decrease. Larger (smaller) drops would contribute to higher R in the morning (afternoon). These variations considerably matched with the diurnal sea wind time series (Fig. 8g). Sea wind is the sum value of normalized wind frequency between 45 and 225°. From 02:00 (14:00) to 12:00 (20:00) KST, Fig. 8g shows a smaller (larger) value of sea wind frequency, which is opposite to the relatively larger (smaller) parts of each parameter (D_m, R, LWC and Z_h).

The PDF distribution of μ between -2 and 0 is more concentrated for NT than for DT. Furthermore, when $\mu > 0$, DT and NT frequency distributions are similar (Fig. 9a). A larger $N(D)$ of small or large raindrops would be expected in NT than in DT.

The distribution of DT $D_m < 0.7$ mm is wider than that of NT. However, between 0.7 and 1.5 mm, the frequency for NT is higher than that for DT, whereas the distribution in the range $D_m > 1.5$ mm is similar for both DT and NT (Fig. 9b). We note that the smaller peak of D_m around 0.6 mm for the entire rainfall data set (Fig. 4b) was observed only in DT.

The distribution of $\log_{10}(N_w)$ for DT has a higher value of PDF at larger $\log_{10}(N_w)$ than that of NT at $\log_{10}(N_w) > 4$ (Fig. 9c).

Bringi et al. (2003) noted that the maritime climatology displayed larger N_w and smaller D_m values than the continental climatology, based on observed DSDs in the low and middle latitudes. Also, Göke et al. (2007) emphasized that rainfall type can be defined by the origin location and movement direction. In accordance with these previous results, we consider NT rainfall in the Busan region to be more likely caused by a continental convective system.

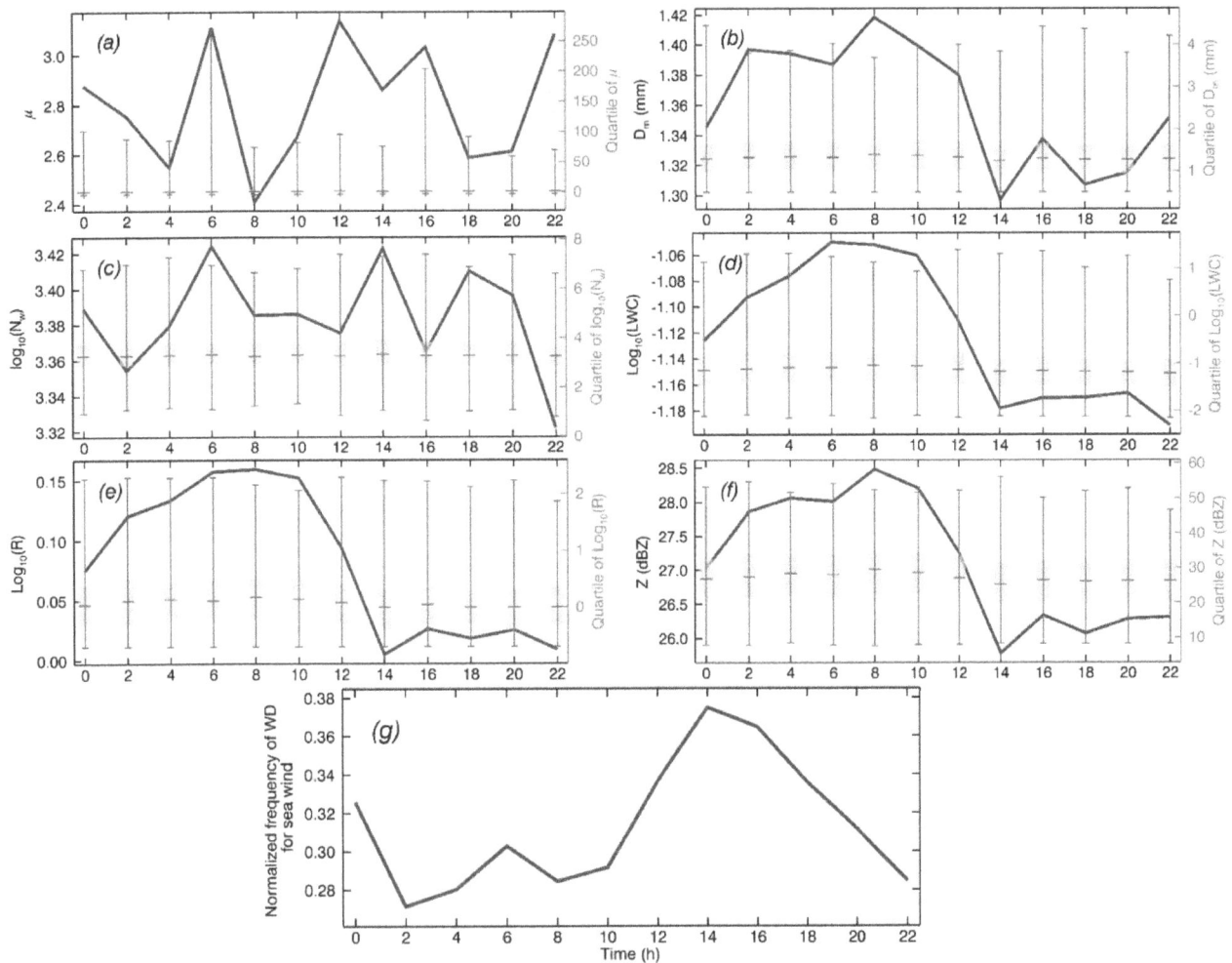

Figure 8. Two-hour interval time series of (a) μ, (b) D_m, (c) $\log_{10}(N_w)$, (d) $\log_{10}(R)$, (e) $\log_{10}(\text{LWC})$, (f) Z_h, and (g) normalized frequency of wind direction for sea wind (45 to 225°) with quartiles for the total period. Solid lines are quartiles for each time.

Table 5. Sum of the normalized wind direction frequencies between summer and winter.

Sum of the normalized wind direction frequencies				
Season	Summer		Winter	
Type	Sea wind	Land wind	Sea wind	Land wind
Frequency	0.4139	0.5861	0.3137	0.6863

Difference of the normalized wind direction frequency between DT and NT (DT–NT)				
Season	Summer		Winter	
Type	Sea wind	Land wind	Sea wind	Land wind
Frequency	0.0731	−0.0731	−0.0697	0.0697

Gadanki (southern India), Singapore, and Kototabang (West Sumatra) during the summer monsoon season. All regions displayed maximum R at approximately 16:00 LST, except for Gadanki. Also, Qian (2008) analyzed the diurnal variability of wind direction and R on Java during the summer

season using 30 years (from 1971 to 2000) of NCEP/NCAR reanalyzed data. They found that a land wind occurred from 01:00 to 10:00 LST and a sea wind from 13:00 to 22:00 LST (Fig. 7 of Qian, 2008). Normalized wind frequency for each direction is a similar pattern to the results of Qian (2008), but the pattern of R is different from that of Kozu et al. (2006). The diurnal variation of rain rate in the present study from 02:00 (12:00) to 10:00 (20:00) KST shows relatively smaller (larger) frequencies of sea wind. It is a different pattern from the result of Kozu et al. (2006). However, these patterns matched with the time series of D_m and $\log_{10}(N_w)$. A larger frequency of sea wind direction shows a counter-proportional (proportional) relationship to the smaller (larger) frequency of D_m ($\log_{10}(N_w)$).

Variability of the D_m time series for winter is the inverse of the summer time series (Fig. 11a). The mean value of D_m steadily increases from 00:00 to 16:00 KST and then decreases from 16:00 to 22:00 KST. The winter $\log_{10}(N_w)$ time series displays a clear inverse pattern compared with

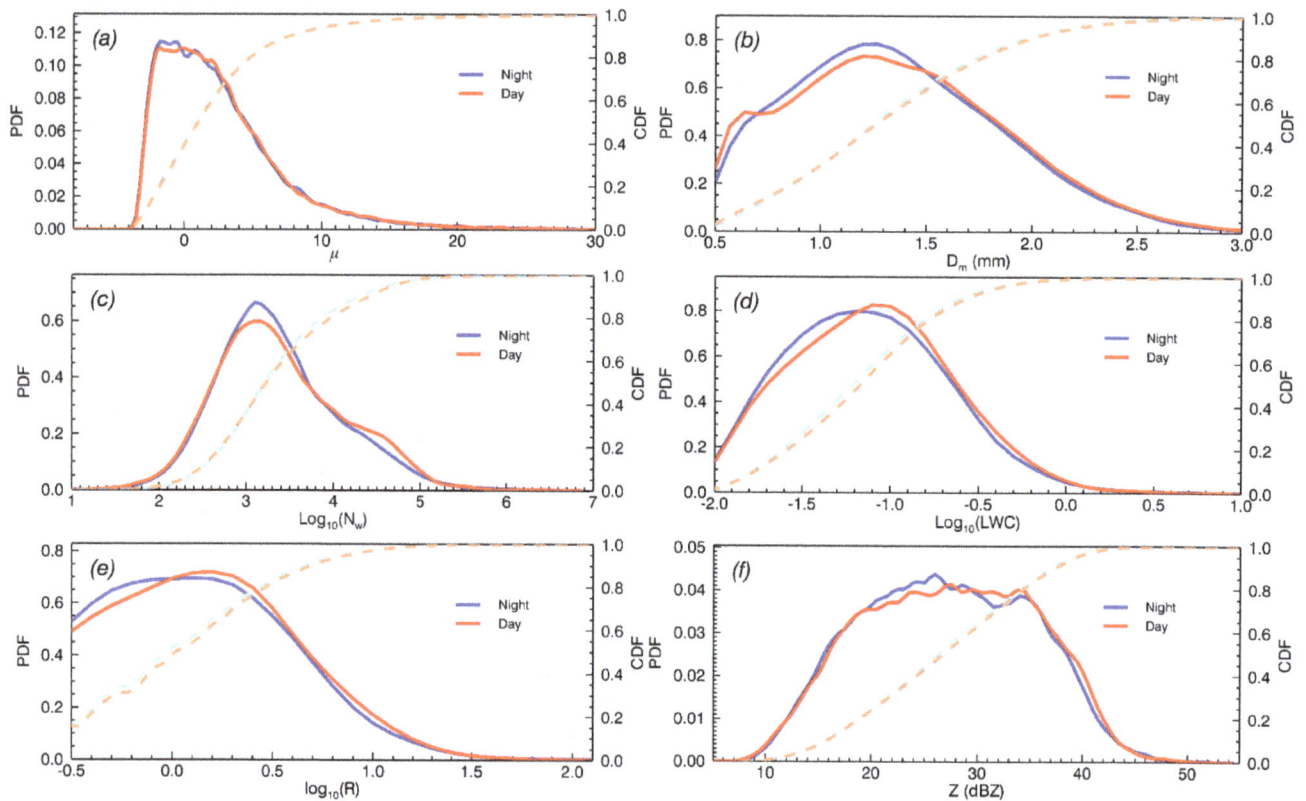

Figure 9. PDF and CDF curves for **(a)** μ, **(b)** D_m, **(c)** $\log_{10}(N_\mathrm{w})$, **(d)** $\log_{10}(R)$, **(e)** $\log_{10}(\mathrm{LWC})$, and **(f)** Z for DT and NT according to the entire period. The solid red and blue lines represent the PDF for DT and NT, respectively. The broken light red and blue lines represent the CDF for DT and NT, respectively.

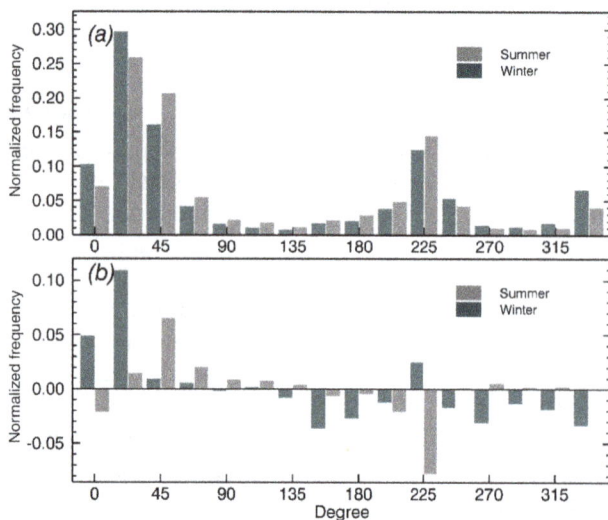

Figure 10. Histogram of normalized frequency for 16 wind directions in **(a)** the entire period and **(b)** differences of normalized frequency with respect to wind directions between DT and NT (DT–NT) for the summer (light grey) and winter (dark grey) seasons.

the D_m variation with time and increases from 16:00 KST to 04:00 KST and then steadily decreases from 04:00 to 16:00 KST (Fig. 11b). The peak of $\log_{10}(N_\mathrm{w})$ occurs at 04:00 KST. However, the time series of $\log_{10}(R)$ for the winter season shows a similar pattern to that of summer, unlike other parameters (Fig. 11c). Based on the diurnal variation of R, the variations of D_m and N_w would be independent of R.

Similar to D_m and $\log_{10}(N_\mathrm{w})$, normalized wind frequency of wind direction for the winter season shows an inverse relationship with that of the summer season (Fig. 11d). The value of the frequency generally decreases (increases) from 04:00 (14:00) to 14:00 (04:00) KST. Also, it shows a symmetry pattern with that of the summer season.

The PDF distribution of summer D_m displays a relatively large DT frequency compared with NT when $D_\mathrm{m} < 1.65$ mm, except for the range between 0.6 and 0.9 mm. However, in the range of $D_\mathrm{m} > 1.65$ mm, the NT PDF displays a larger frequency (Fig. 12a). The PDF of $\log_{10}(N_\mathrm{w})$ for DT (NT) has a larger frequency than NT (DT) when $\log_{10}(N_\mathrm{w}) > (<) \ 3.3$, but a smaller frequency when $\log_{10}(N_\mathrm{w}) < (>) \ 3.3$ (Fig. 12c).

The DT and NT PDFs of D_m and $\log_{10}(N_\mathrm{w})$ during winter display an inverse distribution to that of summer. For the PDF of D_m, there is a considerable frequency for NT (DT) when $D_\mathrm{m} < (>) \ 1.6$ mm (Fig. 12b). The PDF of $\log_{10}(N_\mathrm{w})$

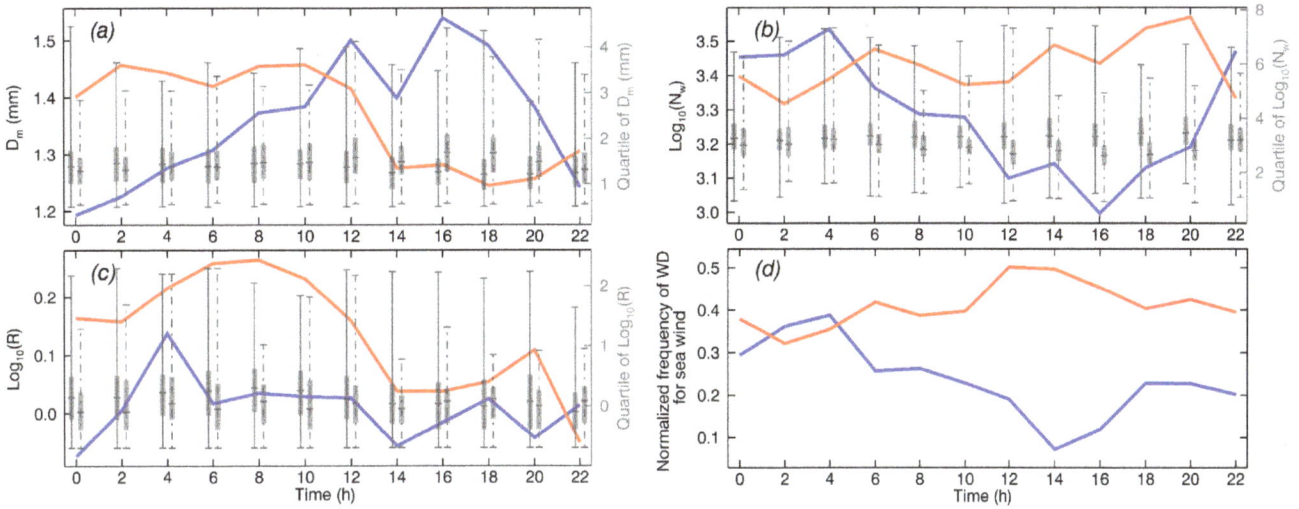

Figure 11. Two-hour interval time series and quartiles of (a) D_m, (b) $\log_{10}$, (N_w) (c) $\log_{10}(R)$, and (d) normalized frequency of wind direction for sea wind (45 to 225°) for the summer (red) and winter (blue) seasons. Solid (broken) lines are quartiles of summer (winter) for each time, respectively.

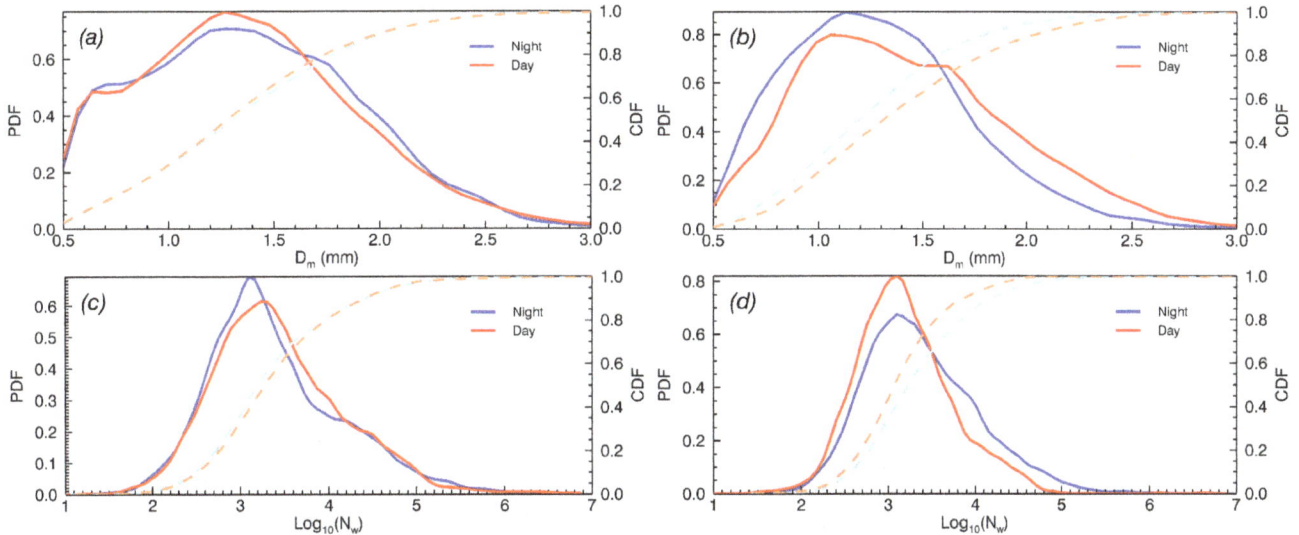

Figure 12. PDF and CDF of (a) D_m ((b) D_m) and (c) N_w ((d) N_w) for all rainfall types in the summer (winter) season. Red and blue solid lines represent the PDF of DT and NT, respectively. The light red and blue broken lines represent the CDF for each season.

of the summer season for NT (DT) is larger than that of DT when $\log_{10}(N_w) < (>)$ 3.5 (Fig. 12d). In the PDF analysis, relatively large (small) D_m and small (large) $\log_{10}(N_w)$ are displayed during NT (DT) when a land wind (sea wind) occurs.

Bringi et al. (2003) referred to the fact that the convective rainfall type is able to be classified as the continental and maritime-like precipitation using D_m and N_w. As the previous study result, we analyzed the PDF of DSDs for summer and winter with respect to convective rainfall type. These feature would be shown more clearly in the convective type. The convective rainfall type of PDFs of DT and

NT for summer show a similar shape of distribution to that of all rainfall types (Fig. 3a). For the PDF of D_m, there is a higher frequency for DT (NT) than NT (DT) when $D_m < (>)$ 2.0 mm, except for between 0.7 and 1.2 mm (Fig. 13a). The PDF of convective rainfall type $\log_{10}(N_w)$ for DT (NT) has a larger frequency than NT (DT) when $\log_{10}(N_w) > (<)$ 3.4, except for between 4.3 and 5.5 (Fig. 12c). PDF distributions for the winter season show a clearer pattern compared with those of the entire rainfall type. The values of PDF for D_m in DT (NT) are considerably larger than NT (DT) when $D_m > (<)$ 1.9 mm, especially between 2.15 and 2.3 mm (Fig. 13b). Also, those for $\log_{10}(N_w)$ in DT (NT)

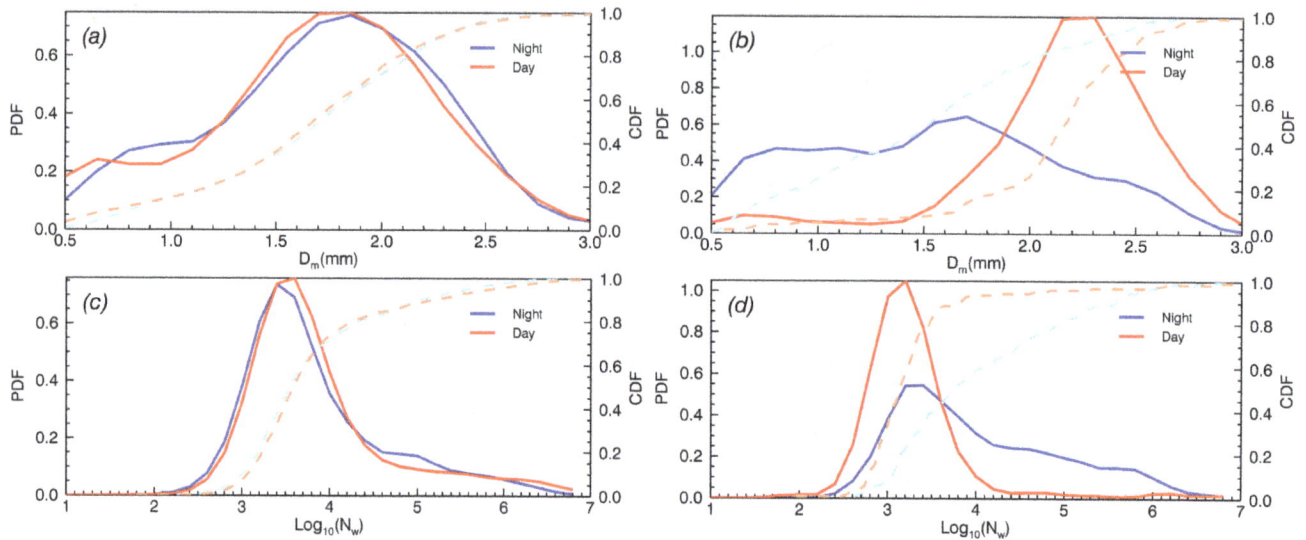

Figure 13. As in Fig. 12 but for convective rainfall type.

show dramatic values when $\log_{10}(N_w) < (>)$ 3.6. Furthermore, PDF values were significantly concentrated between $3 < \log_{10}(N_w) < 3.2$ (Fig. 13d). In short, considering the DSD parameters with wind directions, the maritime (continental)-like precipitation would depend on the sea (land) wind.

4   Summary and conclusion

Climatological characteristics of DSDs in Busan were analyzed using the DSD data observed by POSS over a 4-year period from 24 February 2001 to 24 December 2004. Observed DSDs were filtered to remove errors by performing several quality control measures, and an AWS rain gauge installed nearby was used to verify the rainfall amount recorded by the POSS. We analyzed DSD characteristics of convective and stratiform rainfall types, as defined by Bringi et al. (2003). The rainfall data set was thus divided into stratiform and convective rainfall and their contributions to the total rainfall were 62.93 and 6.11 %, respectively. Also, to find the climatological characteristics of DSD for the rainfall case, the entire rainfall data were classified as 10 rainfall categories including the entire period case.

According to the study by Bringi et al. (2003), the rainfall in Busan shows maritime climatological DSD characteristics. The mean values of D_m and N_w for stratiform rainfall are relatively small compared with the average line of stratiform rainfall produced by Bringi et al. (2003), except for heavy rainfall events. In the case of convective type, mean values of D_m and N_w are converged around the maritime cluster, except for the typhoon category. The convective rainfall associated with a typhoon has considerably smaller D_m and larger N_w values compared with the other rainfall categories. This is likely caused by an increased raindrop breakup mechanism as a result of strong wind effects. Further-

more, the distributions of mean D_m and N_w values for all rainfall categories associated with convective rainfall display a linear relationship including the typhoon category.

The analysis of diurnal variation in DSD yielded the following results: first, the frequency of μ is higher at NT than during DT in the negative value. The PDF of R is higher at NT than during DT when $\log_{10}(R) > 0.6$. The value of PDF for D_m during DT is larger than NT smaller than 0.65 mm. For N_w, which tends to be inversely related to D_m, its frequency is higher at NT than DT when $\log_{10}(N_w) > 3.8$. This feature is matched with the time series of normalized frequency of sea wind, which shows an inverse relationship with D_m. Smaller D_m corresponds to the larger sea wind frequency. In short, maritime (continental)-like precipitation is observed in DT (NT) more often than in NT (DT) according to the features of wind. The above-mentioned DSD characteristics are likely due to the land and sea wind caused by differences in specific heat between the land and ocean. These features are also apparent in the seasonal diurnal distribution. The PDF of DT and NT for convective rainfall type during the summer is similar to the PDF of the entire period; however, those of the winter display the significant inverse distribution compared to the summer because of obvious seasonal differences in wind direction.

Author contributions. C.-H. You designed the study. S.-H. Suh modified the original study theme and performed the study. C.-H. You and S.-H. Suh performed research, obtained the results and prepared the manuscript along with contributions from all of the co-authors. D.-I. Lee examined the results and checked the manuscript.

Acknowledgements. This work was funded by the Korea Meteorological Industry Promotion Agency under grant KMIPA 2015-1050. The authors acknowledge provision of the weather chart and AWS data for this work from the Korea Meteorological Administration, and codes for scattering simulation from V. N. Bringi at Colorado State University.

References

Andsager, K., Beard, K. V., and Laird, N. F.: Laboratory measurements of axis ratios for large raindrops, J. Atmos. Sci., 56, 2673–2683, 1999.

Atlas, D., Srivastava, R., and Sekhon, R. S.: Doppler radar characteristics of precipitation at vertical incidence, Rev. Geophys., 11, 1–35, 1973.

Atlas, D., Ulbrich, C. W., Marks, F. D., Amitai, E., and Williams, C. R.: Systematic variation of drop size and radar-rainfall relations, J. Geophys. Res.-Atmos., 104, 6155–6169, 1999.

Beard, K. V. and Chuang, C.: A new model for the equilibrium shape of raindrops, J. Atmos. Sci., 44, 1509–1524, 1987.

Brandes, E. A., Zhang, G., and Vivekanandan, J.: Experiments in rainfall estimation with a polarimetric radar in a subtropical environment, J. Appl. Meteorol., 41, 674–685, 2002.

Bringi, V., Chandrasekar, V., Hubbert, J., Gorgucci, E., Randeu, W., and Schoenhuber, M.: Raindrop size distribution in different climatic regimes from disdrometer and dual-polarized radar analysis, J. Atmos. Sci., 60, 354–365, 2003.

Chang, W.-Y., Wang, T.-C. C., and Lin, P.-L.: Characteristics of the raindrop size distribution and drop shape relation in Typhoon systems in the western Pacific from the 2D video disdrometer and NCU C-band polarimetric radar, J. Atmos. Ocean. Tech., 26, 1973–1993, 2009.

Dou, X., Testud, J., Amayenc, P., and Black, R.: The parameterization of rain for a weather radar, Comptes Rendus de l'Académie des Sciences-Series IIA-Earth and Planetary Science, 328, 577–582, 1999.

Feingold, G. and Levin, Z.: The lognormal fit to raindrop spectra from frontal convective clouds in Israel, J. Clim. Appl. Meteorol., 25, 1346–1363, 1986.

Gamache, J. F. and Houze Jr., R. A.: Mesoscale air motions associated with a tropical squall line, Mon. Weather Rev., 110, 118–135, 1982.

Göke, S., Ochs III, H. T., and Rauber, R. M.: Radar analysis of precipitation initiation in maritime versus continental clouds near the Florida coast: Inferences concerning the role of CCN and giant nuclei, J. Atmos. Sci., 64, 3695–3707, 2007.

Gunn, R. and Kinzer, G. D.: The terminal velocity of fall for water droplets in stagnant air, J. Meteorol., 6, 243–248, 1949.

Hu, Z. and Srivastava, R.: Evolution of raindrop size distribution by coalescence, breakup, and evaporation: Theory and observations, J. Atmos. Sci., 52, 1761–1783, 1995.

Johnson, R. H. and Hamilton, P. J.: The relationship of surface pressure features to the precipitation and airflow structure of an intense midlatitude squall line, Mon. Weather Rev., 116, 1444–1473, 1988.

Kozu, T., Reddy, K. K., Mori, S., Thurai, M., Ong, J. T., Rao, D. N., and Shimomai, T.: Seasonal and diurnal variations of raindrop size distribution in Asian monsoon region, J. Meteorol. Soc. Jpn., 84, 195–209, 2006.

Leinonen, J., Moisseev, D., Leskinen, M., and Petersen, W. A.: A climatology of disdrometer measurements of rainfall in Finland over five years with implications for global radar observations, J. Appl. Meteorol. Clim., 51, 392–404, 2012.

Levin, Z.: Charge separation by splashing of naturally falling raindrops, J. Atmos. Sci., 28, 543–548, 1971.

Mapes, B. E. and Houze Jr., R. A.: Cloud clusters and superclusters over the oceanic warm pool, Mon. Weather Rev., 121, 1398–1416, 1993.

Markowitz, A. H.: Raindrop size distribution expressions, J. Appl. Meteorol., 15, 1029–1031, 1976.

Marshall, J. S. and Palmer, W. M. K.: The distribution of raindrops with size, J. Meteorol., 5, 165–166, 1948.

Mueller, E. A. and Sims, A. L.: Radar cross sections from drop size spectra, Tech. Rep. ECOM-00032-F, Contract DA-28-043 AMC-00032(E), Illinois State Water Survey, Urbana, 110 pp., AD-645218, 1966.

Pruppacher, H. and Beard, K.: A wind tunnel investigation of the internal circulation and shape of water drops falling at terminal velocity in air, Q. J. Roy. Meteor. Soc., 96, 247–256, 1970.

Qian, J.-H.: Why precipitation is mostly concentrated over islands in the Maritime Continent, J. Atmos. Sci., 65, 1428–1441, 2008.

Ray, P. S.: Broadband complex refractive indices of ice and water, Appl. Optics, 11, 1836–1844, 1972.

Sauvageot, H. and Lacaux, J.-P.: The shape of averaged drop size distributions, J. Atmos. Sci., 52, 1070–1083, 1995.

Seliga, T. and Bringi, V.: Potential use of radar differential reflectivity measurements at orthogonal polarizations for measuring precipitation, J. Appl. Meteorol., 15, 69–76, 1976.

Sheppard, B. E.: Measurement of raindrop size distributions using a small Doppler radar, J. Atmos. Ocean. Tech., 7, 255–268, 1990.

Sheppard, B. E. and Joe, P. I.: Comparison of raindrop size distribution measurements by a Joss–Waldvogel disdrometer, a PMS 2DG spectrometer and a POSS Doppler radar, J. Atmos. Ocean. Tech., 11, 874–887, 1994.

Sheppard, B. and Joe, P.: Performance of the precipitation occurrence sensor system as a precipitation gauge, J. Atmos. Ocean. Tech., 25, 196–212, 2008.

Steiner, M., Houze Jr., R. A., and Yuter, S. E.: Climatological characterization of three-dimensional storm structure from operational radar and rain gauge data, J. Appl. Meteorol., 34, 1978–2007, 1995.

Testud, J., Oury, S., Black, R. A., Amayenc, P., and Dou, X.: The concept of "normalized" distribution to describe raindrop spectra: A tool for cloud physics and cloud remote sensing, J. Appl. Meteorol., 40, 1118–1140, 2001.

Tokay, A. and Short, D. A.: Evidence from tropical raindrop spectra of the origin of rain from stratiform versus convective clouds, J. Appl. Meteorol., 35, 355–371, 1996.

Ulbrich, C. W.: Natural variations in the analytical form of the raindrop size distribution, J. Clim. Appl. Meteorol., 22, 1764–1775, 1983.

Ulbrich, C. W. and Atlas, D.: Rainfall microphysics and radar properties: Analysis methods for drop size spectra, J. Appl. Meteorol., 37, 912–923, 1998.

Waterman, P. C.: Matrix formulation of electromagnetic scattering, Proc. IEEE, 53, 805–812, 1965.

Waterman, P. C.: Symmetry, unitarity, and geometry in electromagnetic scattering, Phys. Rev. D, 3, 825–839, 1971.

Willis, P. T.: Functional fits to some observed drop size distributions and parameterization of rain, J. Atmos. Sci., 41, 1648–1661, 1984.

You, C.-H., Lee, D.-I., and Kang, M.-Y.: Rainfall estimation using specific differential phase for the first operational polarimetric radar in Korea, Advances in Meteorology, 1–10, doi:10.1155/2014/413717, 2014.

6

Approximate analysis of three-dimensional groundwater flow toward a radial collector well in a finite-extent unconfined aquifer

C.-S. Huang, J.-J. Chen, and H.-D. Yeh

Institute of Environmental Engineering, National Chiao Tung University, Hsinchu, Taiwan

Correspondence to: H.-D. Yeh (hdyeh@mail.nctu.edu.tw)

Abstract. This study develops a three-dimensional (3-D) mathematical model for describing transient hydraulic head distributions due to pumping at a radial collector well (RCW) in a rectangular confined or unconfined aquifer bounded by two parallel streams and no-flow boundaries. The streams with low-permeability streambeds fully penetrate the aquifer. The governing equation with a point-sink term is employed. A first-order free surface equation delineating the water table decline induced by the well is considered. Robin boundary conditions are adopted to describe fluxes across the streambeds. The head solution for the point sink is derived by applying the methods of finite integral transform and Laplace transform. The head solution for a RCW is obtained by integrating the point-sink solution along the laterals of the RCW and then dividing the integration result by the sum of lateral lengths. On the basis of Darcy's law and head distributions along the streams, the solution for the stream depletion rate (SDR) can also be developed. With the aid of the head and SDR solutions, the sensitivity analysis can then be performed to explore the response of the hydraulic head to the change in a specific parameter such as the horizontal and vertical hydraulic conductivities, streambed permeability, specific storage, specific yield, lateral length, and well depth. Spatial head distributions subject to the anisotropy of aquifer hydraulic conductivities are analyzed. A quantitative criterion is provided to identify whether groundwater flow at a specific region is 3-D or 2-D without the vertical component. In addition, another criterion is also given to allow for the neglect of vertical flow effect on SDR. Conventional 2-D flow models can be used to provide accurate head and SDR predictions if satisfying these two criteria.

1 Introduction

The applications of a radial collector well (RCW) have received much attention in the aspects of water resource supply and groundwater remediation since rapid advances in drilling technology. An average yield for the well approximates $27\,000\,\mathrm{m^3\,day^{-1}}$ (Todd and Mays, 2005). As compared to vertical wells, RCWs require less operating cost, produce smaller drawdown, and have better efficiency of withdrawing water from thin aquifers. In addition, RCWs can extract water from an aquifer underlying obstacles such as buildings, but vertical wells cannot. Recently, Huang et al. (2012) reviewed semi-analytical and analytical solutions associated with RCWs. Since then, Yeh and Chang (2013) provided a valuable overview of articles associated with RCWs.

A variety of analytical models involving a horizontal well, a specific case of a RCW with a single lateral, in aquifers were developed (e.g., Park and Zhan, 2003; Hunt, 2005; Anderson, 2013). The flux along the well screen is commonly assumed to be uniform. The equation describing three-dimensional (3-D) flow is used. Kawecki (2000) developed analytical solutions of the hydraulic heads for the early linear flow perpendicular to a horizontal well and late pseudo-radial flow toward the middle of the well in confined aquifers. They also developed an approximate solution for unconfined aquifers on the basis of the head solution and an unconfined flow modification. The applicability of the approximate solution was later evaluated in comparison with a finite difference solution developed by Kawecki and Al-Subaikhy (2005). Zhan et al. (2001) presented an analytical solution for drawdown induced by a horizontal well in confined aquifers and compared the difference in the type curves based on the well and a vertical well. Zhan and Zlotnik (2002) developed a semi-analytical solution of drawdown due to pumping from a nonvertical well in an unconfined aquifer accounting for the effect of instantaneous drainage or delayed yield when the free surface declines. They discussed the influences of the length, depth, and inclination of the well on temporal drawdown distributions. Park and Zhan (2002) developed a semi-analytical drawdown solution considering the effects of a finite diameter, the wellbore storage, and a skin zone around a horizontal well in anisotropic leaky aquifers. They found that those effects cause significant change in drawdown at an early pumping period. Zhan and Park (2003) provided a general semi-analytical solution for pumping-induced drawdown in a confined aquifer, an unconfined aquifer on a leaky bottom, or a leaky aquifer below a water reservoir. Temporal drawdown distributions subject to the aquitard storage effect were compared with those without that effect. Sun and Zhan (2006) derived a semi-analytical solution of drawdown due to pumping at a horizontal well in a leaky aquifer. A transient 1-D flow equation describing the vertical flow across the aquitard was considered. The derived solution was used to evaluate the Zhan and Park (2003) solution, which assumed steady-state vertical flow in the aquitard.

Sophisticated numerical models involved in RCWs or horizontal wells were also reported. Steward (1999) applied the analytic element method to approximate 3-D steady-state flow induced by horizontal wells in contaminated aquifers. They discussed the relation between the pumping rate and the size of a polluted area. Chen et al. (2003) utilized the polygon finite difference method to deal with three kinds of seepage-pipe flows including laminar, turbulent, and transitional flows within a finite-diameter horizontal well. A sandbox experiment was also carried out to verify the prediction made by the method. Mohamad and Rushton (2006) used MODFLOW to predict flows inside an aquifer, from the aquifer to a horizontal well, and within the well. The predicted head distributions were compared with field data measured in Sarawak, Malaysia. Su et al. (2007) used software TOUGH2 based on the integrated finite difference method to handle irregular configurations of several laterals of two RCWs installed beside the Russian River, Forestville, California, and analyzed pumping-induced unsaturated regions beneath the river. Lee et al. (2012) developed a finite element solution with triangle elements to assess whether the operation of a RCW near Nakdong River in South Korea can induce riverbank filtration. They concluded that the well can be used for sustainable water supply at the study site. In addition, Rushton and Brassington (2013a) extended Mohamad and Rushton (2006) study by enhancing the Darcy–Weisbach formula to describe frictional head lose inside a horizontal well. The spatial distributions of predicted flux along the well revealed that the flux at the pumping end is 4 times the magnitude of that at the far end. Later, Rushton and Brassington (2013b) applied the same model to a field experiment at the Seton Coast, northwest England.

Well pumping in aquifers near streams may cause groundwater–surface water interactions (e.g., Rodriguez et al., 2013; Chen et al., 2013; Zhou et al., 2013; Exner-Kittridge et al., 2014; Flipo et al., 2014; Unland et al., 2014). The stream depletion rate (SDR), commonly used to quantify stream water filtration into the adjacent aquifer, is defined as the ratio of the filtration rate to a pumping rate. The SDR ranges from zero to a certain value, which could be equal to or less than unity (Zlotnik, 2004). Tsou et al. (2010) developed an analytical solution of the SDR for a slanted well in confined aquifers adjacent to a stream treated as a constant-head boundary. They indicated that a horizontal well parallel to the stream induces the steady-state SDR of unity more quickly than a slanted well. Huang et al. (2011) developed an analytical SDR solution for a horizontal well in unconfined aquifers near a stream regarded as a constant-head boundary. Huang et al. (2012) provided an analytical solution for SDR induced by a RCW in unconfined aquifers adjacent to a stream with a low-permeability streambed under the Robin condition. The influence of the configuration of the laterals on temporal SDR and spatial drawdown distributions was analyzed. Recently, Huang et al. (2014) gave an exhaustive review on analytical and semi-analytical SDR solutions and classified these solutions into two categories. One group involved 2-D flow toward a fully-penetrating vertical well according to aquifer types and stream treatments. The other group included the solutions involving 3-D and quasi-3-D flows according to aquifer types, well types, and stream treatments.

At present, existing analytical solutions associated with flow toward a RCW in unconfined aquifers have involved laborious calculation (Huang et al., 2012) and predicted approximate results (Hantush and Papadopoulos, 1962). The Huang et al. (2012) solution involves numerical integration of a triple integral in predicting the hydraulic head and a quintuple integral in predicting SDR. The integrand is expressed in terms of an infinite series expanded by roots of nonlinear equations. The integration variables are related to those roots. The application of their solution is therefore limited to those who are familiar with numerical methods. In addition, the accuracy of the Hantush and Papadopoulos (1962) solution is limited to some parts of a pumping period; that is, it gives accurate drawdown predictions at early and late times but divergent ones at middle times.

The objective of this study is to present new analytical solutions of the head and SDR, which overcome the above-mentioned limitations, for 3-D flow toward a RCW. A mathematical model is built to describe 3-D spatiotemporal hydraulic head distributions in a rectangular unconfined aquifer bounded by two parallel streams and by the no-flow stratums in the other two sides. The flux across the well screen is assumed to be uniform along each of the laterals. The assumption is valid for a short lateral within 150 m verified by agreement on drawdown observed in field experiments and predicted by existing analytical solutions (Huang et al.,

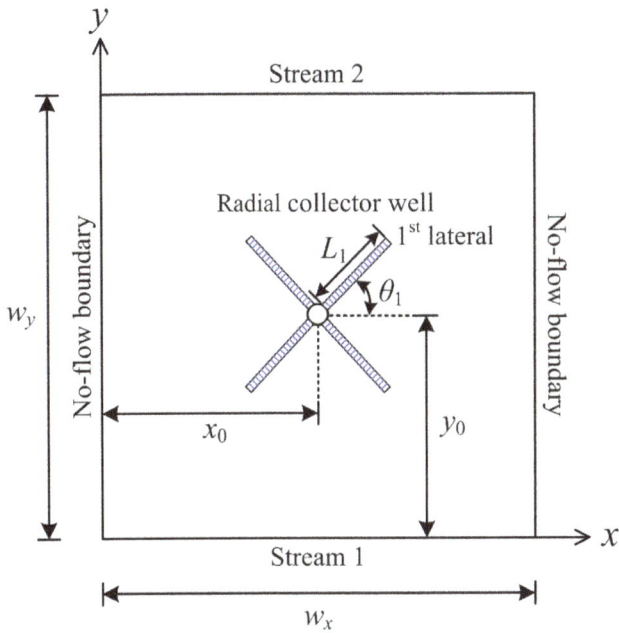

Figure 1. Schematic diagram of a radial collector well in a rectangular unconfined aquifer.

2011, 2012). The streams fully penetrate the aquifer and connect the aquifer with low-permeability streambeds. The model for the aquifer system with two parallel streams can be used to determine the fraction of water filtration from the streams and solve the associated water right problem (Sun and Zhan, 2007). The transient 3-D groundwater flow equation with a point-sink term is considered. The first-order free surface equation is used to describe water table decline due to pumping. Robin boundary conditions are adopted to describe fluxes across the streambeds. The head solution for a point sink is derived by the methods of Laplace transform and finite integral transform. The analytical head solution for a RCW is then obtained by integrating the point-sink solution along the well and dividing the integration result by the total lateral length. The RCW head solution is expressed in terms of a triple series expanded by eigenvalues, which can be obtained by a numerical algorithm such as Newton's method. On the basis of Darcy's law and the RCW head solution, the SDR solution can then be obtained in terms of a double series with fast convergence. With the aid of both solutions, the sensitivity analysis is performed to investigate the response of the hydraulic head to the change in each of aquifer's parameters. Spatial head distributions subject to the anisotropy of aquifer hydraulic conductivities are analyzed. The influences of the vertical flow and well depth on temporal SDR distributions are investigated. Moreover, temporal SDR distributions induced by a RCW and a fully penetrating vertical well in confined aquifers are also compared. A quantitative criterion is provided to identify whether groundwater flow at a specific region is 3-D or 2-D without the vertical compo-

nent. In addition, another criterion is also given to judge the suitability of neglecting the vertical flow effect on SDR.

2 Methodology

2.1 Mathematical model

Consider a RCW in a rectangular unconfined aquifer bounded by two parallel streams and no-flow stratums as illustrated in Fig. 1. The symbols for variables and parameters are defined in Table 1. The origin of the Cartesian coordinate is located at the lower left corner. The aquifer domain falls in the range of $0 \leq x \leq w_x$, $0 \leq y \leq w_y$, and $-H \leq z \leq 0$. The RCW consists of a caisson and several laterals, each of which extends with length L_k and counterclockwise with angle θ_k, where $k \in 1, 2, \ldots N$ and N is the number of laterals. The caisson is located at (x_0, y_0), and the surrounding laterals are at $z = -z_0$.

First, a mathematical model describing 3-D flow toward a point sink in the aquifer is proposed. The equation describing 3-D hydraulic head distribution $h(x, y, z, t)$ is expressed as

$$K_x \frac{\partial^2 h}{\partial x^2} + K_y \frac{\partial^2 h}{\partial y^2} + K_z \frac{\partial^2 h}{\partial z^2} = \qquad (1)$$

$$S_s \frac{\partial h}{\partial t} + Q\delta(x - x_0')\delta(y - y_0')\delta(z + z_0'),$$

where $\delta(\ )$ is the Dirac delta function, the second term on the right-hand side (RHS) indicates the point sink, and Q is positive for pumping and negative for injection. The first term on the RHS of Eq. (1) depicts aquifer storage release based on the concept of effective stress proposed by Terzaghi (see for example, Bear, 1979; Charbeneau, 2000), which is valid under the assumption of constant total stress. By choosing the water table as a reference datum where the elevation head is set to zero, the initial condition can therefore be denoted as

$$h = 0 \text{ at } t = 0. \qquad (2)$$

Note that Eq. (2) introduces a negative hydraulic head for pumping, and the absolute value of the head equals drawdown.

The aquifer boundaries at $x = 0$ and $x = w_x$ are considered to be impermeable and thus expressed as

$$\partial h / \partial x = 0 \text{ at } x = 0 \qquad (3)$$

and

$$\partial h / \partial x = 0 \text{ at } x = w_x. \qquad (4)$$

Streambed permeability is usually less than the adjacent aquifer formation. The fluxes across the streambeds can be described by Robin boundary conditions as

$$K_y \frac{\partial h}{\partial y} - \frac{K_1}{b_1} h = 0 \text{ at } y = 0 \qquad (5)$$

Table 1. Symbols used in the text and their definitions.

Symbol	Definition
a	Shortest horizontal distance between stream 1 and the far end of lateral
$\overline{a}$	a/y_0
b_1, b_2	Thicknesses of streambeds 1 and 2, respectively
d	Shortest horizontal distance between the far end of lateral and location of having only horizontal flow
$\overline{d}$	d/y_0
H	Aquifer thickness
h	Hydraulic head
$\overline{h}$	$(K_y H h)/Q$
K_x, K_y, K_z	Aquifer hydraulic conductivities in x, y, and z directions, respectively
(K_1, K_2)	Hydraulic conductivities of streambeds 1 and 2, respectively
L_k	Length of kth lateral where $k \in (1, 2, \ldots N)$
$\overline{L}_k$	L_k/y_0
N	The number of laterals
Q	Pumping rate of point sink or radial collector well
p	Laplace parameter
p_i	$-\kappa_z \lambda_i^2 - \kappa_x \alpha_m^2 - \beta_n^2$
p_i'	$-\kappa_z \lambda_i^2 - \beta_n^2$
p_0	$\kappa_z \lambda_0^2 - \kappa_x \alpha_m^2 - \beta_n^2$
p_0'	$\kappa_z \lambda_0^2 - \beta_n^2$
R	Shortest horizontal distance between the far end of lateral and aquifer lateral boundary
S_s, S_y	Specific storage and specific yield, respectively
t	Time since pumping
$\overline{t}$	$(K_y t)/\left(S_s y_0^2\right)$
w_x, w_y	Aquifer widths in x and y directions, respectively
$\overline{w}_x, \overline{w}_y$	$w_x/y_0, w_y/y_0$
X_n	Equaling $X_{m,n}$ defined in Eq. (39) with $\alpha_m = 0$
$X_{n,k}$	Defined in Eq. (45)
x, y, z	Cartesian coordinate system
$\overline{x}, \overline{y}, \overline{z}$	$x/y_0, y/y_0, z/H$
$\overline{x}_k$	Coordinate $\overline{x}$ of the far end of the kth lateral
x_0, y_0, z_0	Location of center of RCW
$\overline{x}_0, \overline{y}_0, \overline{z}_0$	$x_0/y_0, 1, z_0/H$
x_0', y_0', z_0'	Location of point sink
$\overline{x}_0', \overline{y}_0', \overline{z}_0'$	$x_0'/y_0, y_0'/y_0, z_0'/H$
α_m	$m\pi/\overline{w}_x$
β_n	Roots of Eq. (19)
ϕ_n	Equaling $\phi_{m,n}$ defined in Eq. (32) with $\alpha_m = 0$
γ	$S_y/(S_s H)$
κ_x, κ_z	$K_x/K_y, \left(K_z y_0^2\right)/(K_y H^2)$
κ_1, κ_2	$(K_1 y_0)/(K_y b_1), (K_2 y_0)/(K_y b_2)$
λ_0, λ_i	Roots of Eqs. (40) and (41), respectively
λ_s, λ_s'	$\sqrt{\left(\kappa_x \alpha_m^2 + \beta_n^2\right)/\kappa_z}, \beta_n/\sqrt{\kappa_z}$
$\mu_{n,0}$	Equaling $\mu_{m,n,0}$ defined in Eq. (36) with $\alpha_m = 0$
$\nu_{n,i}$	Equaling $\nu_{m,n,i}$ defined in Eq. (37) with $\alpha_m = 0$
θ_k	Counterclockwise angle from x axis to kth lateral where $k \in (1, 2, \ldots N)$
$\max \overline{x}_k, \min \overline{x}_k$	Maximum and minimum of $\overline{x}_k$, respectively, where $k \in (1, 2, \ldots N)$

and

$$K_y \frac{\partial h}{\partial y} + \frac{K_2}{b_2} h = 0 \text{ at } y = w_y. \tag{6}$$

The free surface equation describing the water table decline is written as

$$K_x \left(\frac{\partial h}{\partial x}\right)^2 + K_y \left(\frac{\partial h}{\partial y}\right)^2 + K_z \left(\frac{\partial h}{\partial z}\right)^2 - K_z \frac{\partial h}{\partial z} \tag{7}$$

$$= S_y \frac{\partial h}{\partial t} \text{ at } z = h.$$

Neuman (1972) indicated that the effect of the second-order terms in Eq. (7) can generally be ignored in developing analytical solutions. Equation (7) is thus linearized by neglecting the quadratic terms, and the position of the water table is fixed at the initial condition (i.e., $z = 0$). The result is written as

$$K_z \frac{\partial h}{\partial z} = -S_y \frac{\partial h}{\partial t} \text{ at } z = 0. \tag{8}$$

Notice that Eq. (8) is applicable when the conditions $|h|/H \leq 0.1$ and $|\partial h/\partial x| + |\partial h/\partial y| \leq 0.01$ are satisfied. These two conditions have been studied and verified by simulations in, for example, Nyholm et al. (2002), Goldscheider and Drew (2007) and Yeh et al. (2010). Nyholm et al. (2002) achieved agreement on drawdown measured in a field pumping test and predicted by MODFLOW, which models flow in the study site as confined behavior because of $|h|/H \leq 0.1$ in the pumping well. Goldscheider and Drew (2007) revealed that pumping drawdown predicted by the analytical solution by Neuman (1972) based on Eq. (8) agrees well with that obtained in a field pumping test. In addition, Yeh et al. (2010) also achieved agreement on the hydraulic head predicted by their analytical solution based on Eq. (8), their finite difference solution based on Eq. (7) with $\partial h/\partial y = 0$, and the Teo et al. (2003) solution derived by applying the perturbation technique to deal with Eq. (7) with $\partial h/\partial y = 0$ when $|h|/H = 0.1$ and $|\partial h/\partial x| = 0.01$ (i.e., $\alpha = 0.1$ and $\varphi |\partial/\partial x| = 0.01$ at $x = 0$ in Yeh et al., 2010, Fig. 5a). On the other hand, the bottom of the aquifer is considered as a no-flow boundary condition denoted as

$$\partial h/\partial z = 0 \text{ at } z = -H \tag{9}$$

Define dimensionless variables as $\bar{h} = (K_y H h)/Q$, $\bar{t} = (K_y t)/S_s y_0^2$, $\bar{x} = x/y_0$, $\bar{y} = y/y_0$, $\bar{z} = z/H$, $\bar{x}_0' = x_0'/y_0$, $\bar{y}_0' = y_0'/y_0$, $\bar{z}_0' = z_0'/H$, $\bar{w}_x = w_x/y_0$, and $\bar{w}_y = w_y/y_0$, where the overbar denotes a dimensionless symbol, H is the initial aquifer thickness, and y_0 is a distance between stream 1 and the center of the RCW. On the basis of the definitions, Eq. (1) can be written as

$$\kappa_x \frac{\partial^2 \bar{h}}{\partial \bar{x}^2} + \frac{\partial^2 \bar{h}}{\partial \bar{y}^2} + \kappa_z \frac{\partial^2 \bar{h}}{\partial \bar{z}^2} \tag{10}$$

$$= \frac{\partial \bar{h}}{\partial \bar{t}} + \delta(\bar{x} - \bar{x}_0')\delta(\bar{y}' - \bar{y}_0')\delta(\bar{z} + \bar{z}_0'),$$

where $\kappa_x = K_x/K_y$ and $\kappa_z = (K_z y_0^2)/(K_y H^2)$.

Similarly, the initial and boundary conditions are expressed as

$$\bar{h} = 0 \text{ at } \bar{t} = 0, \tag{11}$$

$$\partial \bar{h}/\partial \bar{x} = 0 \text{ at } \bar{x} = 0, \tag{12}$$

$$\partial \bar{h}/\partial \bar{x} = 0 \text{ at } \bar{x} = \bar{w}_x, \tag{13}$$

$$\partial \bar{h}/\partial \bar{y} - \kappa_1 \bar{h} = 0 \text{ at } \bar{y} = 0, \tag{14}$$

$$\partial \bar{h}/\partial \bar{y} + \kappa_2 \bar{h} = 0 \text{ at } \bar{y} = \bar{w}_y, \tag{15}$$

$$\frac{\partial \bar{h}}{\partial \bar{z}} = -\frac{\gamma}{\kappa_z} \frac{\partial \bar{h}}{\partial \bar{t}} \text{ at } \bar{z} = 0, \tag{16}$$

$$\partial \bar{h}/\partial \bar{z} = 0 \text{ at } \bar{z} = -1, \tag{17}$$

where $\kappa_1 = (K_1 y_0)/(K_y b_1)$, $\kappa_2 = (K_2 y_0)/(K_y b_2)$ and $\gamma = S_y (S_s H)$.

2.2 Head solution for point sink

The model, Eqs. (10)–(17), reduces to an ordinary differential equation (ODE) with two boundary conditions in terms of $\bar{z}$ after taking Laplace transform and finite integral transform. The former transform converts $\bar{h}(\bar{x}, \bar{y}, \bar{z}, \bar{t})$ into $\hat{h}(\bar{x}, \bar{y}, \bar{z}, p)$, $\delta(\bar{x} - \bar{x}_0')\delta(\bar{y} - \bar{y}_0')\delta(\bar{z} - \bar{z}_0')$ in Eq. (10) into $\delta(\bar{x} - \bar{x}_0')\delta(\bar{y} - \bar{y}_0')\delta(\bar{z} - \bar{z}_0')/p$, and $\partial \bar{h}/\partial \bar{t}$ in Eqs. (10) and (16) into $p\hat{h} - \bar{h}|_{\bar{t}=0}$, where p is the Laplace parameter, and the second term, the initial condition in Eq. (11), equals zero (Kreyszig, 1999). The transformed model becomes a boundary value problem written as

$$\kappa_x \frac{\partial^2 \hat{h}}{\partial \bar{x}^2} + \frac{\partial^2 \hat{h}}{\partial \bar{y}^2} + \kappa_z \frac{\partial^2 \hat{h}}{\partial \bar{z}^2} \tag{18}$$

$$= p\hat{h} + \delta(\bar{x} - \bar{x}_0')\delta(\bar{y}' - \bar{y}_0')\delta(\bar{z} + \bar{z}_0')/p$$

with boundary conditions $\partial \hat{h}/\partial \bar{x} = 0$ at $\bar{x} = 0$ and $\bar{x} = \bar{w}_x$, $\partial \hat{h}/\partial \bar{y} - \kappa_1 \hat{h} = 0$ at $\bar{y} = 0$, $\partial \hat{h}/\partial \bar{y} + \kappa_2 \hat{h} = 0$ at $\bar{y} = \bar{w}_y$, $\partial \hat{h}/\partial \bar{z} = -p\gamma \hat{h}/\kappa_z$ at $\bar{z} = 0$, and $\partial \bar{h}/\partial \bar{z} = 0$ at $\bar{z} = -1$. We then apply finite integral transform to the problem. One can refer to Appendix A for its detailed definition. The transform converts $\hat{h}(\bar{x}, \bar{y}, \bar{z}, p)$ in the problem into $\tilde{h}(\alpha_m \beta_n \bar{z} p)$, $\delta(\bar{x} - \bar{x}_0')\delta(\bar{y} - \bar{y}_0')$ in Eq. (18) into $\cos(\alpha_m \bar{x}_0')K(\bar{y}_0')$, and $\kappa_x \partial^2 \hat{h}/\partial \bar{x}^2 + \partial^2 \hat{h}/\partial \bar{y}^2$ in Eq. (18) into $-(\kappa_x \alpha_m^2 + \beta_n^2)\tilde{h}$, where $(m, n) \in 1, 2, 3, \ldots \infty$, $\alpha_m = m\pi/\bar{w}_x$, $K(\bar{y}_0')$ is defined in Eq. (A2) with $\bar{y} = \bar{y}_0'$, and β_n represents eigenvalues equaling the roots of the following equation as (Latinopoulos, 1985)

$$\tan(\beta_n \bar{w}_y) = \frac{\beta_n(\kappa_1 + \kappa_2)}{\beta_n^2 - \kappa_1 \kappa_2}. \tag{19}$$

The method to determine the roots is discussed in Sect. 2.3. In turn, Eq. (18) becomes a second-order ODE defined by

$$\kappa_z \frac{\partial^2 \tilde{h}}{\partial \bar{z}^2} - (\kappa_x \alpha_m^2 + \beta_n^2 + p)\tilde{h} \tag{20}$$
$$= \cos(\alpha_m \bar{x}_0') K(\bar{y}_0')\delta(\bar{z} + \bar{z}_0')/p$$

with two boundary conditions denoted as

$$\frac{\partial \tilde{h}}{\partial \bar{z}} = -\frac{p\,\gamma}{\kappa_z}\tilde{h} \text{ at } \bar{z} = 0 \tag{21}$$

and

$$\partial \tilde{h}/\partial \bar{z} = 0 \text{ at } \bar{z} = -1. \tag{22}$$

Equation (20) can be separated into two homogeneous ODEs as

$$\kappa_z \frac{\partial^2 \tilde{h}_a}{\partial \bar{z}^2} - (\kappa_x \alpha_m^2 + \beta_n^2 + p)\tilde{h}_a = 0 \text{ for } -\bar{z}_0' \le \bar{z} \le 0 \tag{23}$$

and

$$\kappa_z \frac{\partial^2 \tilde{h}_b}{\partial \bar{z}^2} - (\kappa_x \alpha_m^2 + \beta_n^2 + p)\tilde{h}_b = 0 \text{ for } -1 \le \bar{z} \le -\bar{z}_0', \tag{24}$$

where h_a and h_b, respectively, represent the heads above and below $\bar{z} = -\bar{z}_0'$, where the point sink is located. Two continuity requirements should be imposed at $\bar{z} = -\bar{z}_0'$. The first is the continuity of the hydraulic head denoted as

$$\tilde{h}_a = \tilde{h}_b \text{ at } \bar{z} = -\bar{z}_0'. \tag{25}$$

The second describes the discontinuity of the flux due to point pumping represented by the Dirac delta function in Eq. (20). It can be derived by integrating Eq. (20) from $\bar{z} = -\bar{z}_0'^{-}$ to $\bar{z} = -\bar{z}_0'^{+}$ as

$$\frac{\partial \tilde{h}_a}{\partial \bar{z}} - \frac{\partial \tilde{h}_b}{\partial \bar{z}} = \frac{\cos(\alpha_m \bar{x}_0') K(\bar{y}_0')}{p\kappa_z} \text{ at } \bar{z} = -\bar{z}_0'. \tag{26}$$

Solving Eqs. (23) and (24) simultaneously with Eqs. (21), (22), (25), and (26) yields the Laplace-domain head solution as

$$\tilde{h}_a(\alpha_m, \beta_n, \bar{z}, p) = \Omega(-\bar{z}_0', \bar{z}, 1) \text{ for } -\bar{z}_0' \le \bar{z} \le 0 \tag{27a}$$

and

$$\tilde{h}_b(\alpha_m, \beta_n, \bar{z}, p) = \Omega(\bar{z}, \bar{z}_0', -1) \text{ for } -1 \le \bar{z} \le -\bar{z}_0' \tag{27b}$$

with

$$\Omega(a, b, c) = \tag{28}$$
$$\frac{\cosh[(1+a)\lambda][-\kappa_z\lambda\cosh(b\lambda) + cp\gamma\sinh(b\lambda)]\cos(\alpha_m\bar{x}_0)K(\bar{y}_0)}{p\kappa_z\lambda(p\gamma\cosh\lambda + \kappa_z\lambda\sinh\lambda)},$$

$$\lambda = \sqrt{(\kappa_x\alpha_m^2 + \beta_n^2 + p)/\kappa_z}, \tag{29}$$

where a, b, and c are arguments. Taking the inverse Laplace transform and finite integral transform to Eq. (28) results in Eq. (31). One is referred to Appendix B for the detailed derivation. A time-domain head solution for a point sink is therefore written as

$$\bar{h}(\bar{x}, \bar{y}, \bar{z}, \bar{t}) = \begin{cases} \Phi(-\bar{z}_0', \bar{z}, 1) \text{ for } -\bar{z}_0' \le \bar{z} \le 0 \\ \Phi(\bar{z}, \bar{z}_0', -1) \text{ for } -1 \le \bar{z} \le -\bar{z}_0' \end{cases} \tag{30}$$

with

$$\Phi(a, b, c) = \tag{31}$$
$$\frac{2}{w_x}\left\{\sum_{n=1}^{\infty}\left[\phi_n X_n + 2\sum_{m=1}^{\infty}\phi_{m,n}X_{m,n}\cos(\alpha_m\bar{x})\right]Y_n\right\},$$

$$\phi_{m,n} = \psi_{m,n} + \psi_{m,n,0} + \sum_{i=1}^{\infty}\psi_{m,n,i}, \tag{32}$$

$$\psi_{m,n} = -\cosh[(1+a)\lambda_s]\cosh(b\lambda_s)/(\kappa_z\lambda_s\sinh\lambda_s), \tag{33}$$

$$\psi_{m,n,0} = \mu_{m,n,0}\cosh[(1+a)\lambda_0] \tag{34}$$
$$[-\kappa_z\lambda_0\cosh(b\lambda_0) + cp_0\gamma\sinh(b\lambda_0)],$$

$$\psi_{m,n,i} = v_{m,n,i}\cos[(1+a)\lambda_i] \tag{35}$$
$$[-\kappa_z\lambda_i\cos(b\lambda_i) + cp_i\gamma\sin(b\lambda_i)],$$

$$\mu_{m,n,0} = 2\exp(p_0\bar{t})/\{p_0[(1+2\gamma)\kappa_z\lambda_0 \tag{36}$$
$$\cosh\lambda_0 + (p_0\gamma + \kappa_z)\sinh\lambda_0]\},$$

$$v_{m,n,i} = 2\exp(p_i\bar{t})/\{p_i[(1+2\gamma)\kappa_z\lambda_i \tag{37}$$
$$\cos\lambda_i + (p_i\gamma + \kappa_z)\sin\lambda_i]\},$$

$$Y_n = \frac{\beta_n\cos(\beta_n\bar{y}) + \kappa_1\sin(\beta_n\bar{y})}{(\beta_n^2 + \kappa_1^2)[\bar{w}_y + \kappa_2/(\beta_n^2 + \kappa_2^2)] + \kappa_1}, \tag{38}$$

$$X_{m,n} = \cos(\alpha_m\bar{x}_0')[\beta_n\cos(\beta_n\bar{y}_0') + \kappa_1\sin(\beta_n\bar{y}_0')], \tag{39}$$

where $\lambda_s = \sqrt{(\kappa_x\alpha_m^2 + \beta_n^2)/\kappa_z}$, $p_0 = \kappa_z\lambda_0^2 - \kappa_x\alpha_m^2 - \beta_n^2$, $p_i = -\kappa_z\lambda_i^2 - \kappa_x\alpha_m^2 - \beta_n^2$, ϕ_n and X_n equal $\phi_{m,n}$ and $X_{m,n}$ with $\alpha_m = 0$, respectively, and the eigenvalues λ_0 and λ_i are, respectively, the roots of the following equations

$$e^{2\lambda_0} = \frac{-\gamma\kappa_z\lambda_0^2 + \kappa_z\lambda_0 + \gamma(\kappa_x\alpha_m^2 + \beta_n^2)}{\gamma\kappa_z\lambda_0^2 + \kappa_z\lambda_0 - \gamma(\kappa_x\alpha_m^2 + \beta_n^2)}, \tag{40}$$

$$\tan\lambda_i = \frac{-\gamma(\kappa_z\lambda_i^2 + \kappa_x\alpha_m^2 + \beta_n^2)}{\kappa_z\lambda_i}. \tag{41}$$

The determination for those eigenvalues is introduced in the next section. Notice that the solution consists of a simple series expanded in β_n, double series expanded in β_n and λ_i (or α_m and β_n), and triple series expanded in α_m, β_n, and λ_i.

2.3 Evaluations for β_n, λ_0 and λ_i

Application of Newton's method with proper initial guesses to determine the eigenvalues β_n, λ_0, and λ_i has been proposed by Huang et al. (2014) and is briefly introduced herein. The eigenvalues are situated at the intersection points of the left-hand side (LHS) and RHS functions of Eq. (19) for β_n, Eq. (40) for λ_0, and Eq. (41) for λ_i. Hence, the initial guesses for β_n are considered as $\beta_v - \delta$ if $\beta_v > (\kappa_1 \kappa_2)^{0.5}$ and as $\beta_v + \delta$ if $\beta_v < (\kappa_1 \kappa_2)^{0.5}$, where $\beta_v = (2n-1)\pi/(2\overline{w}_y)$ and δ is a chosen small value such as 10^{-8} for avoiding being right at the vertical asymptote. In addition, the guess for λ_0 can be formulated as

$$\lambda_{0\text{ initial}} = \delta + \left\{ -\kappa_z - \sqrt{\kappa_z[\kappa_z + 4\gamma^2(\kappa_x \alpha_m^2 + \beta_n^2)]} \right\} / (2\gamma \kappa_z), \tag{42}$$

where the RHS second term represents the location of the vertical asymptote derived by letting the denominator of the RHS function in Eq. (40) to be zero and solving λ_0 in the resultant equation. Moreover, the guessed value for λ_i is $(2i-1)\pi/2 + \delta$.

2.4 Head solution for radial collector well

The lateral of RCW is approximately represented by a line sink composed of a series of adjoining point sinks. The locations of these point sinks are expressed in terms of $(\overline{x}_0 + \overline{l}\cos\theta, \overline{y}_0 + \overline{l}\sin\theta, \overline{z}_0)$ where $(\overline{x}_0, \overline{y}_0, \overline{z}_0) = (x_0/y_0, 1, z_0/H)$ is the central of the lateral, and $\overline{l}$ is a variable to define different locations of the point sink. The solution of head $\overline{h}_w(\overline{x}, \overline{y}, \overline{z}, \overline{t})$ for a lateral can therefore be derived by substituting $\overline{x}_0' = \overline{x}_0 + \overline{l}\cos\theta$, $\overline{y}_0' = 1 + \overline{l}\sin\theta$, and $\overline{z}_0' = \overline{z}_0$ into the point-sink solution, Eq. (30), then by integrating the resultant solution to $\overline{l}$, and finally by dividing the integration result into the sum of lateral lengths. The derivation can be denoted as

$$\overline{h}_w(\overline{x}, \overline{y}, \overline{z}, \overline{t}) = \left(\sum_{k=1}^{N} \overline{L}_k\right)^{-1} \sum_{k=1}^{N} \int_0^{\overline{L}_k} \overline{h}(\overline{x}, \overline{y}, \overline{z}, \overline{t}) \, d\overline{l}, \tag{43}$$

where $\overline{L}_k = L_k/y_0$ is the kth dimensionless lateral length. Note that the integration variable $\overline{l}$ (i.e., $\overline{x}_0'$ and $\overline{y}_0'$) appears only in X_n and $X_{m,n}$ in Eq. (31). The integral in Eq. (43) can thus be done analytically by integrating X_n and $X_{m,n}$ with respect to $\overline{l}$. After the integration, Eq. (43) can be expressed as

$$\overline{h}_w(\overline{x}, \overline{y}, \overline{z}, \overline{t}) = \tag{44}$$

$$\left(\sum_{k=1}^{N} \overline{L}_k\right)^{-1} \sum_{k=1}^{N} \left\{ \begin{array}{l} \Phi(-\overline{z}_0, \overline{z}, 1) \text{ for } -\overline{z}_0 \le \overline{z} \le 0 \\ \Phi(\overline{z}, \overline{z}_0, -1) \text{ for } -1 \le \overline{z} \le -\overline{z}_0 \end{array} \right.,$$

where Φ is defined by Eqs. (31)–(38), and X_n and $X_{m,n}$ in Eq. (31) are replaced, respectively, by

$$X_{n,k} = -G_k/(\beta_n \sin\theta_k) \tag{45}$$

and

$$X_{m,n,k} = \frac{\alpha_m F_k \cos\theta_k + \beta_n G_k \sin\theta_k}{\alpha_m^2 \cos^2\theta_k - \beta_n^2 \sin^2\theta_k} \tag{46}$$

with

$$F_k = \sin(X\alpha_m)[\beta_n \cos(Y\beta_n) + \kappa_1 \sin(Y\beta_n)] \tag{47}$$
$$\quad - \sin(\overline{x}_0 \alpha_m)(\beta_n \cos\beta_n + \kappa_1 \sin\beta_n),$$

$$G_k = \cos(X\alpha_m)[\kappa_1 \cos(Y\beta_n) - \beta_n \sin(Y\beta_n)] \tag{48}$$
$$\quad - \cos(\overline{x}_0 \alpha_m)(\kappa_1 \cos\beta_n - \beta_n \sin\beta_n),$$

where $X = \overline{x}_0 + \overline{L}_k \cos\theta_k$ and $Y = 1 + \overline{L}_k \sin\theta_k$. Notice that Eq. (45) is obtained by substituting $\alpha_m = 0$ into Eq. (46). When $\theta_k = 0$ or π, Eq. (45) reduces to Eq. (49) by applying L'Hospital's rule.

$$X_{n,k} = \overline{L}_k(\beta_n \cos\beta_n + \kappa_1 \sin\beta_n) \tag{49}$$

2.5 SDR solution for radial collector well

On the basis of Darcy's law and the head solution for a RCW, the SDR from streams 1 and 2 can be defined, respectively, as

$$SDR_1(\overline{t}) = \tag{50}$$

$$-\int_{\overline{x}=0}^{\overline{x}=\overline{w}_x} \left(\int_{\overline{z}=-\overline{z}_0}^{\overline{z}=0} \frac{\partial \overline{h}_w}{\partial \overline{y}} d\overline{z} + \int_{\overline{z}=-1}^{\overline{z}=-\overline{z}_0} \frac{\partial \overline{h}_w}{\partial \overline{y}} d\overline{z} \right) d\overline{x} \text{ at } \overline{y}=0$$

and

$$SDR_2(\overline{t}) = \tag{51}$$

$$\int_{\overline{x}=0}^{\overline{x}=\overline{w}_x} \left(\int_{\overline{z}=-\overline{z}_0}^{\overline{z}=0} \frac{\partial \overline{h}_w}{\partial \overline{y}} d\overline{z} + \int_{\overline{z}=-1}^{\overline{z}=-\overline{z}_0} \frac{\partial \overline{h}_w}{\partial \overline{y}} d\overline{z} \right) d\overline{x} \text{ at } \overline{y}=\overline{w}_y.$$

Again, the double integrals in both equations can be done analytically. Notice that the series term of $2\sum_{m=1}^{\infty} \phi_{m,n} X_{m,n} \cos(\alpha_m \overline{x})$ in Eq. (31) disappears due to the consideration of Eqs. (3) and (4) and the integration with respect to $\overline{x}$ in Eqs. (50) and (51) when deriving the SDR solution. The SDR_1 and SDR_2 are therefore expressed in terms of double series and given below:

$$SDR_1(\overline{t}) = \tag{52}$$

$$-\frac{2}{\sum_{k=1}^{N} \overline{L}_k} \sum_{k=1}^{N} \sum_{n=1}^{\infty} \left(\psi_n' + \psi_{n,0}' + \sum_{i=1}^{\infty} \psi_{n,i}' \right) X_{n,k} Y_n'(0)$$

and

$$\text{SDR}_2(\overline{t}) = \tag{53}$$

$$\frac{2}{\sum_{k=1}^{N} \overline{L}_k} \sum_{k=1}^{N} \sum_{n=1}^{\infty} \left(\psi_n' + \psi_{n,0}' + \sum_{i=1}^{\infty} \psi_{n,i}' \right) X_{n,k} Y_n'(\overline{w}_y)$$

with

$$Y_n'(\overline{y}) = \frac{\kappa_1 \beta_n \cos(\beta_n \overline{y}) - \beta_n^2 \sin(\beta_n \overline{y})}{(\beta_n^2 + \kappa_1^2)[\overline{w}_y + \kappa_2/(\beta_n^2 + \kappa_2^2)] + \kappa_1}, \tag{54}$$

$$\psi_n' = -\left\{ \sinh(\overline{z}_0 \lambda_s') \cosh[(1 - \overline{z}_0)\lambda_s'] \right. \tag{55}$$
$$\left. + \sinh[(1 - \overline{z}_0)\lambda_s'] \cosh(\overline{z}_0 \lambda_s') \right\} / (\kappa_z {\lambda_s'}^2 \sinh \lambda_s'),$$

$$\psi_{n,0}' = -\mu_{n,0}(\theta_{n,0} + \vartheta_{n,0})/\lambda_0, \tag{56}$$

$$\theta_{n,0} = \cosh[(1 - \overline{z}_0)\lambda_0] \left\{ p_0' \gamma \left[-1 + \cosh(\overline{z}_0 \lambda_0) \right. \right. \tag{57}$$
$$\left. \left. + \kappa_z \lambda_0 \sinh(\overline{z}_0 \lambda_0) \right] \right\},$$

$$\vartheta_{n,0} = \sinh[(1 - \overline{z}_0)\lambda_0] \tag{58}$$
$$[\kappa_z \lambda_0 \cosh(\overline{z}_0 \lambda_0) + p_0' \gamma \sinh(\overline{z}_0 \lambda_0)],$$

$$\psi_{n,i}' = \nu_{n,i}(\sigma_{n,i} - \eta_{n,i})/\lambda_i, \tag{59}$$

$$\sigma_{n,i} = \cos[(1 - \overline{z}_0)\lambda_i] \left\{ p_i' \gamma \left[-1 + \cos(\overline{z}_0 \lambda_i) \right] \right. \tag{60}$$
$$\left. - \kappa_z \lambda_i \sin(\overline{z}_0 \lambda_i) \right\},$$

$$\eta_{n,i} = \sin[(1 - \overline{z}_0)\lambda_i] \left[\kappa_z \lambda_i \cos(\overline{z}_0 \lambda_i) + p_i' \gamma \sin(\overline{z}_0 \lambda_i) \right], \tag{61}$$

where $\lambda_s' = \beta_n/\sqrt{\kappa_z}$; $p_0' = \kappa_z \lambda_0^2 - \beta_n^2$; $p_i' = -\kappa_z \lambda_i^2 - \beta_n^2$; $\mu_{n,0} = \mu_{m,n,0}$ in Eq. (36) with $\alpha_m = 0$; $\nu_{n,i} = \nu_{m,n,i}$ in Eq. (37) with $\alpha_m = 0$; $X_{n,k}$ is defined in Eq. (45) for $\theta_k \neq 0$ or π and Eq. (49) for $\theta_k = 0$ or π; and λ_0 and λ_i are the roots of Eqs. (40) and (41) with $\alpha_m = 0$.

2.6 Special cases of the present solution

2.6.1 Confined aquifer of finite extent

If $\gamma = 0$ (i.e., $S_y = 0$ in Eq. 8), the top boundary is regarded as an impermeable stratum. The aquifer is then a confined system. Under this circumstance, Eq. (40) reduces to $e^{2\lambda_0} = 1$ having the root of $\lambda_0 = 0$, and Eq. (41) yields $\tan \lambda_i = 0$ having the roots of $\lambda_i = i\pi$, where $i \in 1, 2, 3, \ldots \infty$. With

$\gamma = 0$, $\lambda_0 = 0$ and $\lambda_i = i\pi$, the head solution for a confined aquifer can be expressed as Eq. (44) with Eqs. (31)–(38) and (45)–(49) where $\psi_{m,n,0}$ in Eq. (32) is replaced by

$$\psi_{m,n,0} = -\exp(p_0 \overline{t})/p_0. \tag{62}$$

Similarly, the SDR solution for a confined aquifer can be written as Eqs. (52) and (53), where the RHS function in Eq. (56) reduces to that in Eq. (62) by applying L'Hospital's rule with $\gamma = 0$ and $\lambda_0 = 0$.

2.6.2 Confined aquifer of infinite extent

The head solution introduced in Sect. 2.6.1 is applicable to spatiotemporal head distributions in confined aquifers of infinite extent before the lateral boundary effect comes. Wang and Yeh (2008) indicated that the time can be quantified, in our notation, as $t = R^2 S_s/(16 K_y)$ (i.e., $\overline{t} = R^2/(16 y_0^2)$ for dimensionless time), where R is the shortest distance between a RCW and aquifer lateral boundary. Prior to the time, the present head solution with $N = 1$ for a horizontal well in a confined aquifer gives very close results given in Zhan et al. (2001).

2.6.3 Unconfined aquifer of infinite extent

Prior to the beginning time mentioned in Sect. 2.6.2, the absolute value calculated by the present head solution, Eqs. (44) with $N = 1$, represents drawdown induced by a horizontal well in unconfined aquifers of infinite extent. The calculated drawdown should be close to that of the solution from Zhan and Zlotnik (2002) for the case of the instantaneous drainage from water table decline.

2.6.4 Unconfined aquifer of semi-infinite extent

When $\kappa_1 \to \infty$ (i.e., $b_1 = 0$), Eq. (14) reduces to the Dirichlet condition of $\overline{h} = 0$ for stream 1 in the absence from a low-permeability streambed, and Eq. (19) becomes $\tan(\beta_n \overline{w}_y) = -\beta_n/\kappa_2$. In addition, the boundary effect occurring at the other three sides of the aquifer can be neglected prior to the beginning time. Moreover, when $N = 1$ and $\theta_1 = 0$, a RCW can be regarded as a horizontal well parallel to stream 1. Under these three conditions, the present head and SDR predictions are close to those in Huang et al. (2011), the head solution of which agrees well with measured data from a field experiment executed by Mohamed and Rushton (2006). On the other hand, before the time when the boundary effect occurs at $\overline{x} = 0$, $\overline{x} = \overline{w}_x$, and $\overline{y} = \overline{w}_y$, the present head and SDR solutions for a RCW give close predictions to those in Huang et al. (2012), the head and SDR solutions of which agree well with observation data taken from two field experiments carried out by Schafer (2006) and Jasperse (2009), respectively.

2.7 Sensitivity analysis

The hydraulic parameters determined from field observed data are inevitably subject to measurement errors. Consequently, head predictions from the analytical model have uncertainty due to the propagation of measurement errors. Sensitivity analysis can be considered as a tool of exploring the response of the head to the change in a specific parameter (Zheng and Bennett, 2002). One may define the normalized sensitivity coefficient as

$$S_{i,t} = \frac{P_i}{H}\frac{\partial h}{\partial P_i}, \qquad (63)$$

where $S_{i,t}$ is the normalized sensitivity coefficient for the ith parameter at time t, and P_i represents the magnitude of the ith parameter. Equation (63) can be approximated as

$$S_{i,t} = \frac{h(P_i + \Delta P_i) - h(P_i)}{\Delta P_i} \times \frac{P_i}{H}, \qquad (64)$$

where ΔP_i is an increment chosen as $10^{-3}P_i$ (Yeh et al., 2008).

3 Results and discussion

This section demonstrates head and SDR predictions and explores some physical insights regarding flow behavior. In Sect. 3.1, equipotential lines are drawn to identify 3-D or 2-D flow without the vertical flow at a specific region. In Sect. 3.2, the influence of anisotropy on spatial head and temporal SDR distributions is studied. In Sect. 3.3, the sensitivity analysis is performed to investigate the response of the head to the change in each hydraulic parameter. In Sect. 3.4, the effects of the vertical flow and well depth on temporal SDR distributions for confined and unconfined aquifers are investigated. For conciseness, we consider a RCW with two laterals with $N = 2, \overline{L}_1 = \overline{L}_2 = 0.5, \theta_1 = 0$, and $\theta_2 = \pi$. The well can be viewed as a horizontal well parallel to streams 1 and 2. The default values for the other dimensionless parameters are $\overline{w}_x = \overline{w}_y = 2, \gamma = 100, \overline{x}_0 = 1, \overline{y}_0 = 1, \overline{z}_0 = 0.5, \kappa_x = \kappa_z = 1$, and $\kappa_1 = \kappa_2 = 20$.

3.1 Identification of 3-D or 2-D flow at observation point

Most existing models assume 2-D flow by neglecting the vertical flow for pumping at a horizontal well (e.g., Mohamed and Rushton, 2006; Haitjema et al., 2010). The head distributions predicted by those models are inaccurate if an observation point is close to the region where the vertical flow prevails. Figure 2 demonstrates the equipotential lines predicted by the present solution for a horizontal well in an unconfined aquifer for $\overline{x}_0 = 10, \overline{w}_x = \overline{w}_y = 20$, and $\kappa_z = 0.1$, 1, and 10. The well is located at $9.5 \le \overline{x} \le 10.5, \overline{y} = 1$, and $\overline{z} = 0.5$ as illustrated in the figure. The equipotential lines are

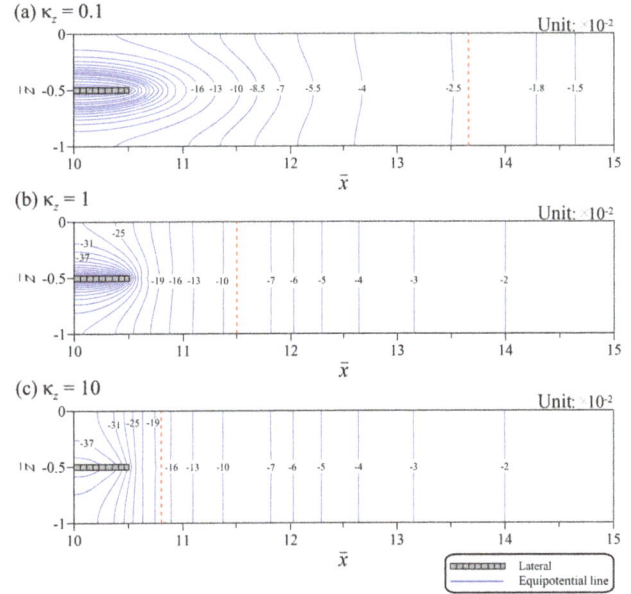

Figure 2. Equipotential lines predicted by the present solution for $\kappa_z =$ (**a**) 0.1, (**b**) 1, and (**c**) 10.

based on steady-state head distributions plotted by Eq. (44) with $\overline{y} = 1$ and $\overline{t} = 10^7$. When $\kappa_z = 0.1$, in the range of $10 \le \overline{x} \le 13.66$, the contours of the hydraulic head are in a curved path, and the flow toward the well is thus slanted. Moreover, the range decreases to $10 \le \overline{x} \le 11.5$ when $\kappa_z = 1$ and to $10 \le \overline{x} \le 10.82$ when $\kappa_z = 10$. Beyond these ranges, the head contours are nearly vertical, and the flow is essentially horizontal. Define $\overline{d} = d/y_0$ as a shortest dimensionless horizontal distance between the well and a nearest location of only horizontal flow. The $\overline{d}$ is therefore chosen as 3.16, 1, and 0.32 for the cases of $\kappa_z = 0.1$, 1, and 10, respectively. Substituting $(\kappa_z, \overline{d}) = (0.1, 3.16), (1, 1)$, and $(10, 0.32)$ into $\kappa_z \overline{d}^2$ leads to about unity. We may therefore conclude that the vertical flow at an observation point is negligible if its location is beyond the range of $\overline{d} < \sqrt{1/\kappa_z}$ (i.e., $d < H\sqrt{K_y/K_z}$) for thin aquifers, an observation point far from the well, and/or a small ratio of K_y/K_z.

3.2 Anisotropy analysis of hydraulic head and stream depletion rate

Previous articles have seldom analyzed flow behavior for anisotropic aquifers, i.e., $\kappa_x (K_x/K_y) \neq 1$. Head predictions based on the models, developed for isotropic aquifers, will be inaccurate if $\kappa_x \neq 1$. Consider $\overline{w}_x = \overline{w}_y = 2, \overline{t} = 10^7$ for steady-state head distributions, and a RCW with $\overline{L}_1 = \overline{L}_2 = 0.25, \theta_1 = 0, \theta_2 = \pi$, and $(\overline{x}_0\overline{y}_0\overline{z}_0) = (1, 1, -0.5)$ for symmetry. The contours of the dimensionless head at $\overline{z} = -0.5$ are shown in Fig. 3a–d for $\kappa_x = 1, 10$ and $50, 10^{-3}$, and 10^{-4}, respectively. The figure indicates that the anisotropy causes a significant effect on the head distributions in com-

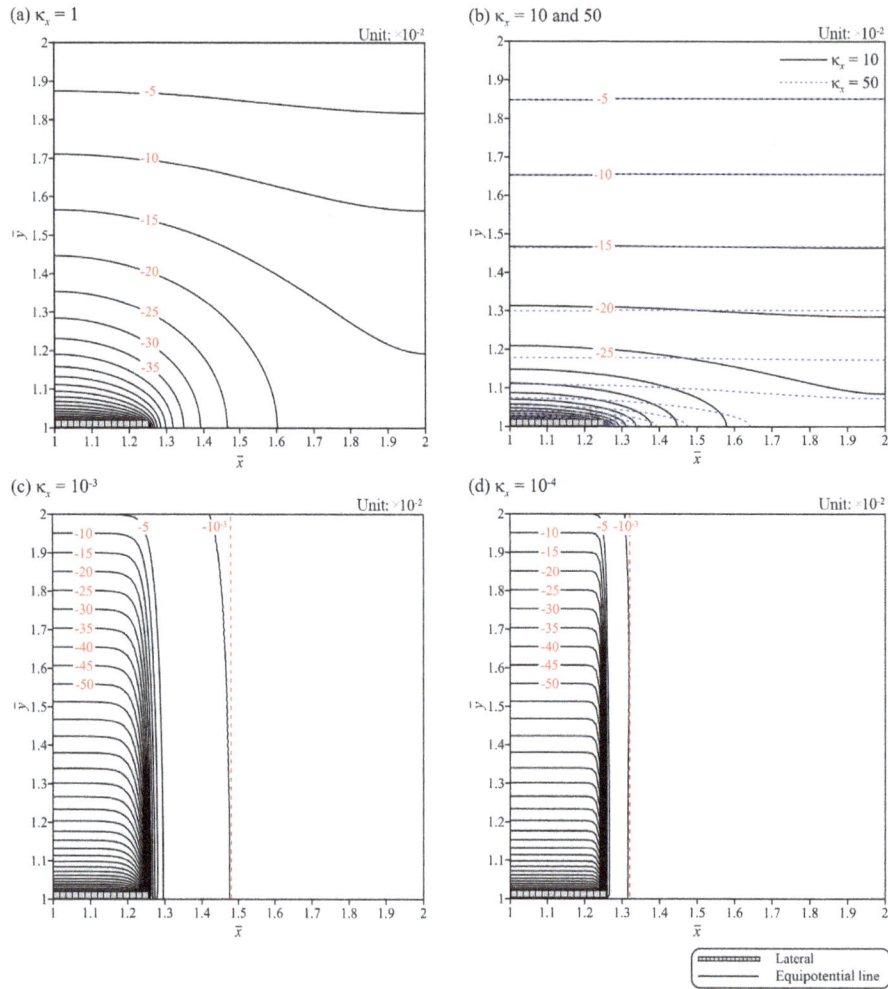

Figure 3. Spatial distributions of the dimensionless head predicted by the present head solution for $\kappa_x = $ **(a)** 1, **(b)** 10 and 50, **(c)** 10^{-3} and **(d)** 10^{-4}.

parison with the case of $\kappa_x = 1$. In Fig. 3b, the contours exhibit smooth curves in the strip regions of $1 \leq \bar{y} \leq 1.45$ for the case of $\kappa_x = 10$ and $1 \leq \bar{y} \leq 1.2$ for the case of $\kappa_x = 50$. For the region of $\bar{y} \geq 1.45$, the predicted heads for both cases agree well, and all the contour lines are parallel, indicating that the flow is essentially unidirectional. Substituting $(\kappa_x, \bar{y}) = (10, 1.45)$ and $(50, 1.2)$ into $\kappa_x(\bar{y} - 1)^2$ results in a value of about 2. Accordingly, we may draw the conclusion that plots from the inequality of $\kappa_x(\bar{y} - 1)^2 \leq 2$ indicate the strip region for κ_x being greater than 10. Some existing models assuming 2-D flow in a vertical plane with neglecting the flow component along a horizontal well give accurate head predictions beyond the region (e.g., Anderson, 2000, 2003; Kompani-Zare et al., 2005).

Aquifers with $K_y H \geq 10^3 \, \text{m}^2 \, \text{day}^{-1}$ can efficiently produce plenty of water from a well. RCWs usually operate with $Q \leq 10^5 \, \text{m}^3 \, \text{day}^{-1}$ for field experiments (e.g., Schafer, 2006; Jasperse, 2009). We therefore define significant dimensionless head drop as $|\bar{h}| > 10^{-5}$ (i.e., $|h| > 1$ mm). The

anisotropy of $\kappa_x < 1$ produces the drop in the strip areas of $1 \leq \bar{x} \leq 1.48$ for the case of $\kappa_x = 10^{-3}$ in Fig. 3c and $1 \leq \bar{x} \leq 1.32$ for the case of $\kappa_x = 10^{-4}$ in Fig. 3d. Substituting $(\kappa_x, \bar{x}) = (10^{-3}, 1.48)$ and $(10^{-4}, 1.32)$ into $(\bar{x} - \bar{x}_0 - \overline{L}_1)^2 / \kappa_x$ approximates 52.9. This result leads to the conclusion that the area can be determined by the inequalities of $(\bar{x} - \bar{x}_0 - \overline{L}_1)^2 \leq 52.9\kappa_x$ and $(\bar{x} - \bar{x}_0 + \overline{L}_2)^2 \leq 52.9\kappa_x$ for any value of κ_x in the range $\kappa_x < 1$. For a RCW with irregular lateral configurations, the inequalities become $(\bar{x} - \max \bar{x}_k)^2 \leq 52.9\kappa_x$ and $(\bar{x} - \min \bar{x}_k)^2 \leq 52.9\kappa_x$, where $\bar{x}_k$ is coordinate $\bar{x}$ of the far end of the kth lateral. The conclusion applies in principle to reduction in grid points for numerical solutions based on finite difference methods or finite element methods. On the other hand, we have found that Eq. (52) or (53) with various κ_x predicts the same temporal SDR distribution (not shown), indicating that the SDR is independent of κ_x.

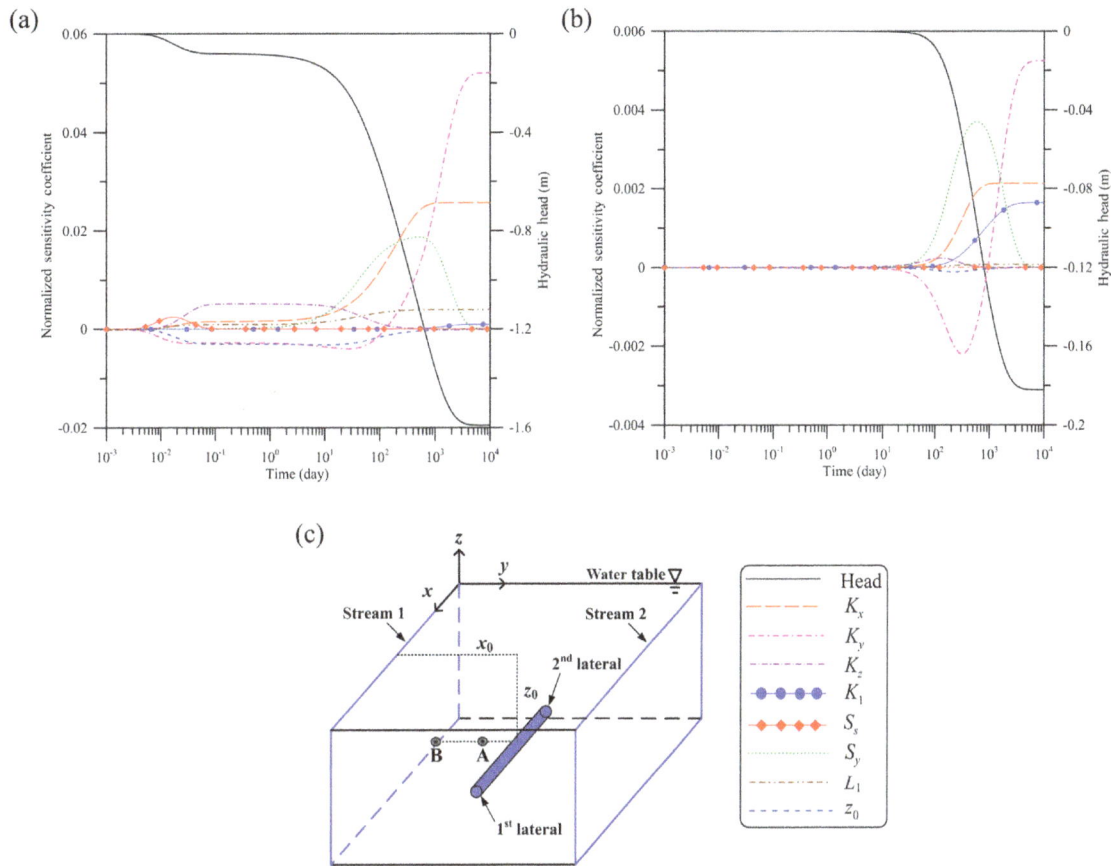

Figure 4. Temporal distribution curves of the normalized sensitivity coefficients for parameters K_x, K_y, K_z, S_s, S_y, K_1, L_1, and z_0 observed at piezometers (**a**) A of $(400, 340, -10\,\text{m})$ and (**b**) B of $(400, 80, -10\,\text{m})$.

3.3 Sensitivity analysis of hydraulic head

Consider an unconfined aquifer of $H = 20\,\text{m}$ and $w_x = w_y = 800\,\text{m}$ with a RCW having two laterals of $L_1 = L_2 = 50\,\text{m}$, $\theta_1 = 0$, and $\theta_2 = \pi$ and two piezometers installed at point A of $(400, 340, -10\,\text{m})$ and point B of $(400, 80, -10\,\text{m})$ illustrated in Fig. 4. As discussed in Sect. 3.1, the temporal head distribution at point A exhibits the unconfined behavior in Fig. 4a because of $\kappa_z \overline{d}^2 < 1$ while at point B displays the confined one in Fig. 4b due to $\kappa_z \overline{d}^2 > 1$. The sensitivity analysis is conducted with the aid of Eq. (64) to observe head responses at these two piezometers to the change in each of K_x, K_y, K_z, S_s, S_y, K_1, L_1 and z_0. The temporal distribution curves of the normalized sensitivity coefficients for those eight parameters are shown in Fig. 4a for point A and 4b for point B when $K_x = K_y = 1\,\text{m}\,\text{day}^{-1}$, $K_z = 0.1\,\text{m}\,\text{day}^{-1}$, $S_s = 10^{-5}\,\text{m}^{-1}$, $S_y = 0.2$, $K_1 = K_2 = 0.1\,\text{m}\,\text{day}^{-1}$, $b_1 = b_2 = 1\,\text{m}$, $Q = 100\,\text{m}^3\,\text{day}^{-1}$, $x_0 = y_0 = 400\,\text{m}$, and $z_0 = 10\,\text{m}$. The figure demonstrates that the hydraulic heads at both piezometers are most sensitive to the change in K_y, second-most sensitive to the change in K_x, and third-most sensitive to the change in S_y, indicating that

K_y, K_x, and S_y are the most crucial factors in designing a pumping system. This figure also shows that the heads at point A is sensitive to the change in S_s at the early period of 4×10^{-3} day $< t < 10^{-1}$ day but at point B is insensitive to the change over the entire period. In addition, the head at point A is sensitive to the changes in K_z and z_0 due to 3-D flow (i.e., $\kappa_z \overline{d}^2 < 1$) as discussed in Sect. 3.1. In contrast, the head at point B is insensitive to the changes in K_z and z_0 because the vertical flow diminishes (i.e., $\kappa_z \overline{d}^2 > 1$). Moreover, the head at point A is sensitive to the change in L_1 but the head at point B is not because its location is far away from the well. Furthermore, the normalized sensitivity coefficient of K_1 for point A away from stream 1 approaches zero but for point B in the vicinity of stream 1 increases with time and finally maintains a certain value at the steady state. Regarding the sensitivity analysis of SDR, Huang et al. (2014) has performed the sensitivity analysis of normalized coefficients of SDR_1 to the changes in K_y, K_1, and S_s for a confined aquifer and in K_y, K_z, K_1, S_s, and S_y for an unconfined aquifer.

(a)

(b)

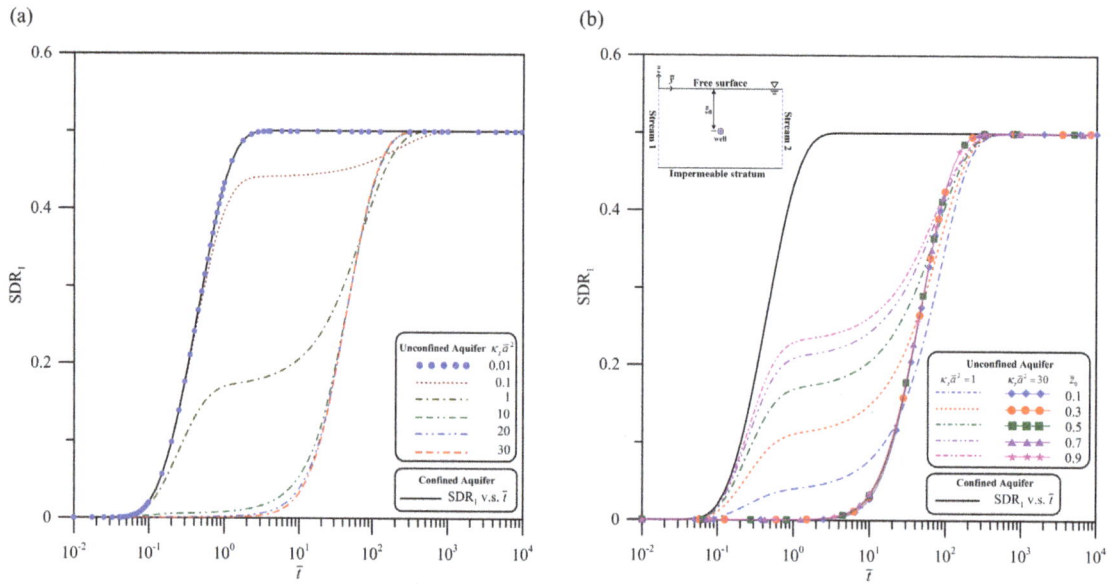

Figure 5. Temporal SDR_1 distributions predicted by Eq. (52) for stream 1 with various values of **(a)** $\kappa_z \bar{a}^2$ and **(b)** $\bar{z}_0$.

3.4 Effects of vertical flow and well depth on stream depletion rate

Huang et al. (2014) revealed that the effect of the vertical flow on SDR induced by a vertical well is dominated by the magnitude of the key factor κ_z (i.e., $K_z y_0^2/(K_y H^2)$), where y_0 herein is a distance between stream 1 and the vertical well. They concluded that the effect is negligible when $\kappa_z \geq 10$ for a leaky aquifer. The factor should be replaced by $\kappa_z \bar{a}^2$ (i.e., $K_z a^2/(K_y H^2)$) where a is a shortest distance measured from stream 1 to the end of a lateral of a RCW, and $\bar{a} = a/y_0 = 1$ in this study due to $N = 2$, $\theta_1 = 0$, and $\theta_2 = \pi$. We investigate SDR in response to various $\bar{z}_0$ and $\kappa_z \bar{a}$ for unconfined and confined aquifers. The temporal SDR_1 distributions predicted by Eq. (52) for stream 1 adjacent to an unconfined aquifer are shown in Fig. 5a for $\bar{z}_0 = 0.5$ and $\kappa_z \bar{a}^2 = 0.01$, 0.1, 1, 10, 20, and 30 and Fig. 5b for $\kappa_z \bar{a}^2 = 1$ and 30 when $\bar{z}_0 = 0.1, 0.3, 0.5, 0.7,$ and 0.9. The curves of SDR_1 versus $\bar{t}$ are plotted in both panels by the present SDR solution for a confined aquifer. In Fig. 5a, the present solution for an unconfined aquifer predicts a close SDR_1 to that for the confined aquifer when $\kappa_z \bar{a}^2 = 0.01$, indicating that the vertical flow in the unconfined aquifer is ignorable. The SDR_1 for the unconfined aquifer with $\kappa_z \bar{a}^2 = 30$ behaves like that for a confined one, indicating the vertical flow can also be ignored. The SDR_1 is therefore independent of well depths $\bar{z}_0$ when $\kappa_z \bar{a}^2 = 30$ as shown in Fig. 5b. We may therefore conclude that, under the condition of $\kappa_z \bar{a}^2 \leq 0.01$ or $\kappa_z \bar{a}^2 \geq 30$, a 2-D horizontal flow model can give good predictions in SDR_1 for unconfined aquifers. In contrast, SDR_1 increases with decreasing $\kappa_z \bar{a}^2$ when $0.01 < \kappa_z \bar{a}^2 < 30$ in Fig. 5a, indicating that the vertical flow component induced by pumping in unconfined aquifers significantly affects SDR_1. The

effect of well depth $\bar{z}_0$ on SDR_1 is also significant as shown in Fig. 5b when $\kappa_z \bar{a}^2 = 1$. Obviously, the vertical flow effect should be considered in a model when $0.01 < \kappa_z \bar{a}^2 < 30$ for unconfined aquifers.

It is interesting to note that the SDR_1 or SDR_2 induced by two laterals (i.e., $\theta_1 = 0$ and $\theta_2 = \pi$) parallel to the streams adjacent to a confined aquifer is independent of $\kappa_z \bar{a}^2$ and $\bar{z}_0$ but depends on the aquifer width of $\bar{w}_y$. The temporal SDR distribution curves based on Eqs. (52) and (53) with $\gamma = 0$ for a confined aquifer with $\bar{w}_y = 2, 4, 6, 10,$ and 20 are plotted in Fig. 6. The dimensionless distance between the well and stream 1 is set to unity (i.e., $\bar{y}_0 = 1$) for each case. The SDR_1 predicted by the solution by Hunt (1999) based on a vertical well in a confined aquifer extending infinitely is considered. The present solution for each $\bar{w}_y$ gives the same SDR_1 as the Hunt solution before the time when stream 2 contributes filtration water to the aquifer and influences the supply of SDR_1. It is interesting to note that the sum of steady-state SDR_1 and SDR_2 is always unity for a fixed $\bar{w}_y$. The former and latter can be estimated by $(\bar{w}_y - 1)/\bar{w}_y$ and $1/\bar{w}_y$, respectively. Such a result corresponds with that in Sun and Zhan (2007), which investigates the distribution of steady-state SDR_1 and SDR_2 induced by a vertical well.

4 Concluding remarks

This study develops a new analytical model describing 3-D flow induced by a RCW in a rectangular confined or unconfined aquifer bounded by two parallel streams and no-flow stratums in the other two sides. The flow equation in terms of the hydraulic head with a point sink term is employed. Both streams fully penetrate the aquifer and are un-

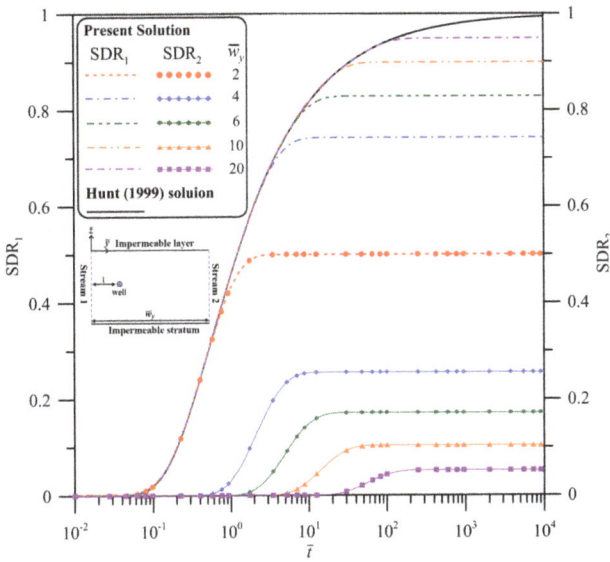

Figure 6. Temporal SDR distribution curves predicted by Eqs. (52) and (53) with $\gamma = 0$ for confined aquifers when $\overline{w}_y = 2, 4, 6, 10$, and 20.

der the Robin condition in the presence of low-permeability streambeds. A first-order free surface Eq. (8) describing the water table decline gives good predictions when the conditions $|h|/H \leq 0.1$ and $|\partial h/\partial x| + |\partial h/\partial y| \leq 0.01$ are satisfied. The flux across the well screen might be uniform on a lateral within 150 m. The head solution for the point sink is expressed in terms of a triple series derived by the methods of Laplace transform and finite integral transform. The head solution for a RCW is then obtained by integrating the point-sink solution along the laterals and dividing the integration result by the sum of lateral lengths. The integration can be done analytically due to the aquifer of finite extent with Eqs. (3)–(6). On the basis of Darcy's law and the head solution, the SDR solution for two streams can also be acquired. The double integrals of defining the SDR in Eqs. (50) and (51) can also be done analytically due to considerations of Eqs. (3)–(6). The sensitivity analysis is performed to explore the response of the head to the change in each of the hydraulic parameters and variables. New findings regarding the responses of flow and SDR to pumping at a RCW are summarized below.

Groundwater flow in a region based on $\overline{d} < \sqrt{1/\kappa_z}$ is 3-D, and temporal head distributions exhibit the unconfined behavior. A mathematical model should consider 3-D flow when predicting the hydraulic head in the region. Beyond this region, groundwater flow is horizontal, and temporal head distributions display the confined behavior. A 2-D flow model can predict accurate hydraulic head.

The aquifer anisotropy of $\kappa_x > 10$ causes unidirectional flow in the strip region determined based on $\kappa_x (\overline{y} - 1)^2 > 2$ for a horizontal well. Existing models assuming 2-D flow in a vertical plane with neglecting the flow component along the well give accurate head predictions in the region.

The aquifer anisotropy of $\kappa_x < 1$ produces significant change in the head (i.e., $|\overline{h}| > 10^{-5}$ or $|h| > 1\,\text{mm}$) in the strip area determined by $(\overline{x} - \max\overline{x}_k)^2 \leq 52.9\kappa_x$ and $(\overline{x} - \min\overline{x}_k)^2 \leq 52.9\kappa_x$ for a RCW with irregular lateral configurations.

The hydraulic head in the whole domain is most sensitive to the change in K_y, second-most sensitive to the change in K_x, and third-most sensitive to the change in S_y. They are thus the most crucial factors in designing a pumping system.

The hydraulic head is sensitive to changes in K_z, S_s, z_0, and L_k in the region of $\overline{d} < \sqrt{1/\kappa_z}$ and is insensitive to the changes of them beyond the region.

The hydraulic head at observation points near stream 1 is sensitive to the change in K_1 but away from the stream is not.

The effect of the vertical flow on SDR is ignorable when $\kappa_z \overline{a}^2 \leq 0.01$ or $\kappa_z \overline{a}^2 \geq 30$ for unconfined aquifers. In contrast, neglecting the effect will underestimate SDR when $0.01 < \kappa_z \overline{a}^2 < 30$.

For unconfined aquifers, SDR increases with dimensionless well depth $\overline{z}_0$ when $0.01 < \kappa_z < 30$ and is independent of $\overline{z}_0$ when $\kappa_z \leq 0.01$ or $\kappa_z \geq 30$. For confined aquifers, SDR is independent of $\overline{z}_0$ and κ_z. For both kinds of aquifers, the distribution curve of SDR versus $\overline{t}$ is independent of aquifer anisotropy κ_x.

Appendix A: Finite integral transform

Latinopoulos (1985) provided the finite integral transform for a rectangular aquifer domain where each side can be under either the Dirichlet, no-flow, or Robin condition. The transform associated with the boundary conditions, Eqs. (12)–(15), is defined as

$$\tilde{h}(\alpha_m, \beta_n) = \Im\{\bar{h}(\bar{x}, \bar{y})\} \tag{A1}$$

$$= \int_0^{\bar{w}_x} \int_0^{\bar{w}_y} \bar{h}(\bar{x}, \bar{y}) \cos(\alpha_m \bar{x}) K(\bar{y}) d\bar{y} d\bar{x}$$

with

$$K(\bar{y}) = \sqrt{2} \frac{\beta_n \cos(\beta_n \bar{y}) + \kappa_1 \sin(\beta_n \bar{y})}{\sqrt{(\beta_n^2 + \kappa_1^2)[\bar{w}_y + \kappa_2/(\beta_n^2 + \kappa_2^2)] + \kappa_1}}, \tag{A2}$$

where $\cos(\alpha_m \bar{x}) K(\bar{y})$ is the kernel function. According to Latinopoulos (1985, Eq. 9), the transform has the property of

$$\Im\left\{\kappa_x \frac{\partial^2 \bar{h}}{\partial \bar{x}^2} + \frac{\partial^2 \bar{h}}{\partial \bar{y}^2}\right\} = -(\kappa_x \alpha_m^2 + \beta_n^2) \tilde{h}(\alpha_m, \beta_n). \tag{A3}$$

The formula for the inverse finite integral transform can be written as (Latinopoulos, 1985, Eq. 14)

$$\bar{h}(\bar{x}, \bar{y}) = \Im^{-1}\{\tilde{h}(\alpha_m, \beta_n)\} = \tag{A4}$$

$$\frac{1}{\bar{w}_x}\left[\sum_{n=1}^{\infty} \tilde{h}(0, \beta_n) K(\bar{y}) + 2\sum_{m=1}^{\infty}\sum_{n=1}^{\infty} \tilde{h}(\alpha_m, \beta_n) \cos(\alpha_m \bar{x}) K(\bar{y})\right].$$

Appendix B: Derivation of Eq. (31)

The function of p in Eq. (28) is defined as

$$F(p) = \frac{\cosh[(1+a)\lambda][-\kappa_z \lambda \cosh(b\lambda) + cp\gamma \sinh(b\lambda)]}{p\kappa_z \lambda (p\gamma \cosh\lambda + \kappa_z \lambda \sinh\lambda)}. \tag{B1}$$

Notice that the term $\cos(\alpha_m \bar{x}_0) K(\bar{y}_0)$ in Eq. (28) is excluded because it is independent of p. $F(p)$ is a single-value function with respect to p. On the basis of the residue theorem, the inverse Laplace transform for $F(p)$ equals the summation of residues of poles in the complex plane. The residue of a simple pole can be derived according to the formula below:

$$\text{Res}|_{p=p_i} = F(p)\exp(p\bar{t})(p - p_i), \tag{B2}$$

where p_i is the location of the pole in the complex plane.

The locations of poles are the roots of the equation obtained by letting the denominator in Eq. (B1) to be zero, denoted as

$$p\kappa_z \lambda (p\gamma \cosh\lambda + \kappa_z \lambda \sinh\lambda) = 0, \tag{B3}$$

where λ is defined in Eq. (29). Notice that $p = -\kappa_x\alpha_m^2 - \beta_n^2$ obtained by $\lambda = 0$ is not a pole in spite of being a root. Apparently, one pole is at $p = 0$, and the residue based on Eq. (B2) with $p_i = 0$ is expressed as

$$\text{Res}|_{p=0} = \tag{B4}$$

$$\frac{\cosh[(1+a)\lambda][-\kappa_z \lambda \cosh(b\lambda) + cp\gamma \sinh(b\lambda)]}{\kappa_z \lambda (p\gamma \cosh\lambda + \kappa_z \lambda \sinh\lambda)}\exp(p\bar{t})$$

with $p = 0$ and $\lambda = \lambda_s$ reduces to $\psi_{m,n}$ in Eq. (33). Other poles are determined by the equation of

$$p\gamma \cosh\lambda + \kappa_z \lambda \sinh\lambda = 0, \tag{B5}$$

which comes from Eq. (B3). One pole is at $p = p_0$ between $p = 0$ and $p = -\kappa_x\alpha_m^2 - \beta_n^2$ in the negative part of the real axis. Newton's method can be used to obtain the value of p_0. In order to have a proper initial guess for Newton's method, we let $\lambda = \lambda_0$ and then have $p = \kappa_z\lambda_0^2 - \kappa_x\alpha_m^2 - \beta_n^2$ based on Eq. (29). Substituting $\lambda = \lambda_0$, $p = \kappa_z\lambda_0^2 - \kappa_x\alpha_m^2 - \beta_n^2$, $\cosh\lambda_0 = (e^{\lambda_0} + e^{-\lambda_0})/2$, and $\sinh\lambda_0 = (e^{\lambda_0} - e^{-\lambda_0})/2$ into Eq. (B5) and rearranging the result leads to Eq. (40). The initial guess for finding root λ_0 of Eq. (40) is discussed in Sect. 2.3. With a known value of λ_0, one can obtain $p_0 = \kappa_z\lambda_0^2 - \kappa_x\alpha_m^2 - \beta_n^2$. According to Eq. (B2), the residue of the simple pole at $p = p_0$ is written as

$$\text{Res}|_{p=p_0} = \tag{B6}$$

$$\frac{\cosh[(1+a)\lambda][-\kappa_z \lambda \cosh(b\lambda) + cp\gamma \sinh(b\lambda)]}{p\kappa_z \lambda (p\gamma \cosh\lambda + \kappa_z \lambda \sinh\lambda)}\exp(p\bar{t})(p - p_0),$$

where both the denominator and nominator equal zero when $p = p_0$. Applying L'Hospital's rule to Eq. (B6) results in

$$\text{Res}|_{p=p_0} = \tag{B7}$$

$$\frac{2\cosh[(1+a)\lambda][-\kappa_z \lambda \cosh(b\lambda) + cp\gamma \sinh(b\lambda)]}{p[(1+2\gamma)\kappa_z \lambda \cosh\lambda + (\gamma p + \kappa_z)\sinh\lambda]}\exp(p\bar{t})$$

with $p = p_0$ and $\lambda = \lambda_0$ reduces to $\psi_{m,n,0}$ in Eq. (34).

On the other hand, infinite poles are at $p = p_i$ behind $p = -\kappa_x\alpha_m^2 - \beta_n^2$. Similar to the derivation of Eq. (40), we let $\lambda = \sqrt{-1}\lambda_i$ and then have $p = -\kappa_z\lambda_i^2 - \kappa_x\alpha_m^2 - \beta_n^2$ based on Eq. (29). Substituting $\lambda = \sqrt{-1}\lambda_i$, $p = -\kappa_z\lambda_i^2 - \kappa_x\alpha_m^2 - \beta_n^2$, $\cosh\lambda = \cos\lambda_i$, and $\sinh\lambda = \sqrt{-1}\sin\lambda_i$ into Eq. (B3) and rearranging the result yields Eq. (41). The determination of λ_i is discussed in Sect. 2.3. With known value λ_i, one can have $p_i = -\kappa_z\lambda_i^2 - \kappa_x\alpha_m^2 - \beta_n^2$. The residues of those simple poles at $p = p_i$ can be expressed as $\psi_{m,n,i}$ in Eq. (35) by substituting $p_0 = p_i$, $p = p_i$, $\lambda = \sqrt{-1}\lambda_i$, $\cosh\lambda = \cos\lambda_i$, and $\sinh\lambda = \sqrt{-1}\sin\lambda_i$ into Eq. (B7). Eventually, the inverse Laplace transform for $F(p)$ equals the sum of those residues (i.e., $\phi_{m,n} = \psi_{m,n} + \psi_{m,n,0} + \sum_{i=1}^{\infty}\psi_{m,n,i}$). The time-domain result of $\Omega(a, b, c)$ in Eq. (28) is then obtained as $\phi_{m,n}\cos(\alpha_m\bar{x}_0)K(\bar{y}_0)$. By substituting $\tilde{h}(\alpha_m, \beta_n) = \phi_{m,n}\cos(\alpha_m\bar{x}_0)K(\bar{y}_0)$ and $\tilde{h}(0, \beta_n) =$

$\phi_n K(\overline{y}_0)$ into Eq. (A4) and letting $\overline{h}(\overline{x}, \overline{y})$ be $\Phi(a, b, c)$, the inverse finite integral transform for the result can be derived as

$$\Phi(a, b, c) = \tag{B8}$$

$$\frac{1}{\overline{w}_x} \left[\sum_{n=1}^{\infty} (\phi_n K(\overline{y}_0) K(\overline{y}) + 2 \sum_{m=1}^{\infty} \phi_{m,n} \cos(\alpha_m \overline{x}_0) \right.$$

$$\left. K(\overline{y}_0) \cos(\alpha_m \overline{x}) K(\overline{y})) \right].$$

Moreover, Eq. (B8) reduces to Eq. (31) when letting the terms of $K(\overline{y}_0) K(\overline{y})$ and $\cos(\alpha_m \overline{x}_0) K(\overline{y}_0) K(\overline{y})$ to be $2X_n Y_n$ and $2X_{m,n} Y_n$, respectively.

Acknowledgements. Research leading to this paper has been partially supported by the grants from the Taiwan Ministry of Science and Technology under the contract NSC 102-2221-E-009-072-MY2, MOST 103-2221-E-009-156, and MOST 104-2221-E-009-148-MY2.

References

Anderson, E. I.: The method of images for leaky boundaries, Adv. Water Resour., 23, 461–474, doi:10.1016/S0309-1708(99)00044-5, 2000.

Anderson, E. I.: An analytical solution representing groundwater-surface water interaction, Water Resour. Res., 39, 1071, doi:10.1029/2002WR001536, 2003.

Anderson, E. I.: Stable pumping rates for horizontal wells in bank filtration systems, Adv. Water Resour., 54, 57–66, doi:10.1016/j.advwatres.2012.12.012, 2013.

Bear, J.: Hydraulics of Groundwater, McGraw-Hill, New York, 84 pp., 1979.

Charbeneau, R. J.: Groundwater Hydraulics and Pollutant Transport, Prentice-Hall, NJ, 57 pp., 2000.

Chen, C. X., Wan, J. W., and Zhan, H. B.: Theoretical and experimental studies of coupled seepage-pipe flow to a horizontal well, J. Hydrol., 281, 159–171, doi:10.1016/S0022-1694(03)00207-5, 2003.

Chen, X., Dong, W., Ou, G., Wang, Z., and Liu, C.: Gaining and losing stream reaches have opposite hydraulic conductivity distribution patterns, Hydrol. Earth Syst. Sci., 17, 2569–2579, doi:10.5194/hess-17-2569-2013, 2013.

Exner-Kittridge, M., Salinas, J. L., and Zessner, M.: An evaluation of analytical stream to groundwater exchange models: a comparison of gross exchanges based on different spatial flow distribution assumptions, Hydrol. Earth Syst. Sci., 18, 2715–2734, doi:10.5194/hess-18-2715-2014, 2014.

Flipo, N., Mouhri, A., Labarthe, B., Biancamaria, S., Rivière, A., and Weill, P.: Continental hydrosystem modelling: the concept of nested stream–aquifer interfaces, Hydrol. Earth Syst. Sci., 18, 3121–3149, doi:10.5194/hess-18-3121-2014, 2014.

Goldscheider, N. and Drew, D.: Methods in karst hydrology, Taylor and Francis Group, London, UK, 88 pp., 2007.

Haitjema, H., Kuzin, S., Kelson, V., and Abrams, D.: Modeling flow into horizontal wells in a Dupuit-Forchheimer model, Ground Water, 48, 878–883, doi:10.1111/j.1745-6584.2010.00694.x, 2010.

Hantush, M. S. and Papadopoulos, I. S.: Flow of groundwater to collector wells, J. Hydr. Eng. Div., 88, 221–244, 1962.

Huang, C. S., Chen, Y. L., and Yeh, H. D.: A general analytical solution for flow to a single horizontal well by Fourier and Laplace transforms, Adv. Water Resour., 34, 640–648, doi:10.1016/j.advwatres.2011.02.015, 2011.

Huang, C. S., Tsou, P. R., and Yeh, H. D.: An analytical solution for a radial collector well near a stream with a low-permeability streambed, J. Hydrol., 446, 48–58, doi:10.1016/j.jhydrol.2012.04.028, 2012.

Huang, C. S., Lin, W. S., and Yeh, H. D.: Stream filtration induced by pumping in a confined, unconfined or leaky aquifer bounded by two parallel streams or by a stream and an impervious stratum, J. Hydrol., 513, 28–44, doi:10.1016/j.jhydrol.2014.03.039, 2014.

Hunt, B.: Unsteady stream depletion from ground water pumping, Ground Water, 37, 98–102, doi:10.1111/j.1745-6584.1999.tb00962.x, 1999.

Hunt, B.: Flow to vertical and nonvertical wells in leaky aquifers, J. Hydrol. Eng., 10, 477–484, doi:10.1061/(ASCE)1084-0699(2005)10:6(477), 2005.

Jasperse, J.: Planning, design and operations of collector 6, Sonoma County Water Agency, NATO Sci. Peace Secur., 169–202, doi:10.1007/978-94-007-0026-0_11, 2009.

Kawecki, M. W.: Transient flow to a horizontal water well, Ground Water, 38, 842–850, doi:10.1111/j.1745-6584.2000.tb00682.x, 2000.

Kawecki, M. W. and Al-Subaikhy, H. N.: Unconfined linear flow to a horizontal well, Ground Water, 43, 606–610, doi:10.1111/j.1745-6584.2005.0059.x, 2005.

Kompani-Zare, M., Zhan, H., and Samani, N.: Analytical study of capture zone of a horizontal well in a confined aquifer, J. Hydrol., 307, 48–59, doi:10.1016/j.jhydrol.2004.09.021, 2005.

Kreyszig, E.: Advanced engineering mathematics, John Wiley and Sons, New York, 258 pp., 1999.

Latinopoulos, P.: Analytical solutions for periodic well recharge in rectangular aquifers with third-kind boundary conditions, J. Hydrol., 77, 293–306, 1985.

Lee, E., Hyun, Y., Lee, K. K., and Shin, J.: Hydraulic analysis of a radial collector well for riverbank filtration near Nakdong River, South Korea, Hydrogeol. J., 20, 575–589, doi:10.1007/s10040-011-0821-3, 2012.

Mohamed, A. and Rushton, K.: Horizontal wells in shallow aquifers: Field experiment and numerical model, J. Hydrol., 329, 98–109, doi:10.1016/j.jhydrol.2006.02.006, 2006.

Neuman, S. P.: Theory of flow in unconfined aquifers considering delayed response of the water table, Water Resour. Res., 8, 1031–1045, 1972.

Nyholm, T., Christensen, S., and Rasmussen, K. R.: Flow depletion in a small stream caused by ground water abstraction from wells, Ground Water, 40, 425–437, 2002.

Park, E. and Zhan, H. B.: Hydraulics of a finite-diameter horizontal well with wellbore storage and skin effect, Adv. Water Resour., 25, 389–400, doi:10.1016/S0309-1708(02)00011-8, 2002.

Park, E. and Zhan, H. B.: Hydraulics of horizontal wells in fractured shallow aquifer systems, J. Hydrol., 281, 147–158, doi:10.1016/S0022-1694(03)00206-3, 2003.

Rodríguez, L., Vives, L., and Gomez, A.: Conceptual and numerical modeling approach of the Guarani Aquifer System, Hydrol. Earth Syst. Sci., 17, 295–314, doi:10.5194/hess-17-295-2013, 2013.

Rushton, K. R. and Brassington, F. C.: Significance of hydraulic head gradients within horizontal wells in unconfined aquifers of limited saturated thickness, J. Hydrol., 492, 281–289, doi:10.1016/j.jhydrol.2013.04.006, 2013a.

Rushton, K. R. and Brassington, F. C.: Hydraulic behavior and regional impact of a horizontal well in a shallow aquifer: example from the Sefton Coast, northwest England (UK), Hydrogeol. J., 21, 1117–1128, doi:10.1007/s10040-013-0985-0, 2013b.

Schafer, D. C.: Use of aquifer testing and groundwater modeling to evaluate aquifer/river hydraulics at Louisville Water Company,

Louisville, Kentucky, USA, NATO Sci. Ser. IV Earth Enviro. Sci., 60, 179–198, doi:10.1007/978-1-4020-3938-6_8, 2006.

Steward, D. R.: Threedimensional analysis of the capture of contaminated leachate by fully penetrating, partially penetrating, and horizontal wells, Water Resour. Res., 35, 461–468, doi:10.1029/1998WR900022, 1999.

Su, G. W., Jasperse, J., Seymour, D., Constantz, J., and Zhou, Q.: Analysis of pumping-induced unsaturated regions beneath a perennial river, Water Resour. Res., 43, W08421, doi:10.1029/2006WR005389, 2007.

Sun, D. M. and Zhan, H. B.: Flow to a horizontal well in an aquitard-aquifer system, J. Hydrol., 321, 364–376, doi:10.1016/j.jhydrol.2005.08.008, 2006.

Sun, D. M. and Zhan, H. B.: Pumping induced depletion from two streams, Adv. Water Resour., 30, 1016–1026, doi:10.1016/j.advwatres.2006.09.001, 2007.

Todd, D. K. and Mays, L. W.: Groundwater hydrology, John Wiley & Sons, Inc., New Jersey, USA, 240, 2005.

Tsou, P.-R., Feng, Z.-Y., Yeh, H.-D., and Huang, C.-S.: Stream depletion rate with horizontal or slanted wells in confined aquifers near a stream, Hydrol. Earth Syst. Sci., 14, 1477–1485, doi:10.5194/hess-14-1477-2010, 2010.

Unland, N. P., Cartwright, I., Cendón, D. I., and Chisari, R.: Residence times and mixing of water in river banks: implications for recharge and groundwater-surface water exchange, Hydrol. Earth Syst. Sci., 18, 5109–5124, doi:10.5194/hess-18-5109-2014, 2014.

Wang, C. T. and Yeh, H. D.: Obtaining the steady-state drawdown solutions of constant-head and constant-flux tests, Hydrol. Process., 22, 3456–3461, doi:10.1002/hyp.6950, 2008.

Yeh, H. D. and Chang, Y. C.: Recent advances in modeling of well hydraulics, Adv. Water Resour., 51, 27–51, doi:10.1016/j.advwatres.2012.03.006, 2013.

Yeh, H. D., Chang, Y. C., and Zlotnik, V. A.: Stream depletion rate and volume from groundwater pumping in wedge-shaped aquifers, J. Hydrol., 349, 501–511, doi:10.1016/j.jhydrol.2007.11.025, 2008.

Yeh, H. D., Huang, C. S., Chang, Y. C., and Jeng, D. S.: An analytical solution for tidal fluctuations in unconfined aquifers with a vertical beach, Water Resour. Res., 46, W10535, doi:10.1029/2009WR008746, 2010.

Zhan, H. B. and Zlotnik, V. A.: Groundwater flow to a horizontal or slanted well in an unconfined aquifer, Water Resour. Res., 38, doi:10.1029/2001WR000401, 2002.

Zhan, H. B. and Park, E.: Horizontal well hydraulics in leaky aquifers, J. Hydrol., 281, 129–146, doi::10.1016/S0022-1694(03)00205-1, 2003.

Zhan, H. B., Wang, L. V., and Park, E.: On the horizontal-well pumping tests in anisotropic confined aquifers, J. Hydrol., 252, 37–50, doi:10.1016/S0022-1694(01)00453-X, 2001.

Zheng, C. and Bennett, G. D.: Applied contaminant transport modeling, 2nd ed., Wiley-Interscience, N.Y., 287, 2002.

Zhou, Y., Wenninger, J., Yang, Z., Yin, L., Huang, J., Hou, L., Wang, X., Zhang, D., and Uhlenbrook, S.: Groundwater-surface water interactions, vegetation dependencies and implications for water resources management in the semi-arid Hailiutu River catchment, China – a synthesis, Hydrol. Earth Syst. Sci., 17, 2435–2447, doi:10.5194/hess-17-2435-2013, 2013.

Zlotnik, V. A.: A concept of maximum stream depletion rate for leaky aquifers in alluvial valleys, Water Resour. Res., 40, W06507, doi:10.1029/2003WR002932, 2004.

7

Aggregation in environmental systems – Part 1: Seasonal tracer cycles quantify young water fractions, but not mean transit times, in spatially heterogeneous catchments

J. W. Kirchner[1,2]

[1]ETH Zürich, Zurich, Switzerland
[2]Swiss Federal Research Institute WSL, Birmensdorf, Switzerland

Correspondence to: J. W. Kirchner (kirchner@ethz.ch)

Abstract. Environmental heterogeneity is ubiquitous, but environmental systems are often analyzed as if they were homogeneous instead, resulting in aggregation errors that are rarely explored and almost never quantified. Here I use simple benchmark tests to explore this general problem in one specific context: the use of seasonal cycles in chemical or isotopic tracers (such as Cl^-, $\delta^{18}O$, or δ^2H) to estimate timescales of storage in catchments. Timescales of catchment storage are typically quantified by the mean transit time, meaning the average time that elapses between parcels of water entering as precipitation and leaving again as streamflow. Longer mean transit times imply greater damping of seasonal tracer cycles. Thus, the amplitudes of tracer cycles in precipitation and streamflow are commonly used to calculate catchment mean transit times. Here I show that these calculations will typically be wrong by several hundred percent, when applied to catchments with realistic degrees of spatial heterogeneity. This aggregation bias arises from the strong nonlinearity in the relationship between tracer cycle amplitude and mean travel time. I propose an alternative storage metric, the young water fraction in streamflow, defined as the fraction of runoff with transit times of less than roughly 0.2 years. I show that this young water fraction (not to be confused with event-based "new water" in hydrograph separations) is accurately predicted by seasonal tracer cycles within a precision of a few percent, across the entire range of mean transit times from almost zero to almost infinity. Importantly, this relationship is also virtually free from aggregation error. That is, seasonal tracer cycles also accurately predict the young water fraction in runoff from highly heterogeneous mixtures of

subcatchments with strongly contrasting transit-time distributions. Thus, although tracer cycle amplitudes yield biased and unreliable estimates of catchment mean travel times in heterogeneous catchments, they can be used to reliably estimate the fraction of young water in runoff.

1 Introduction

Environmental systems are characteristically complex and heterogeneous. Their processes and properties are often difficult to quantify at small scales and difficult to extrapolate to larger scales. Thus, translating process inferences across scales and aggregating across heterogeneity are fundamental challenges for environmental scientists. These ubiquitous aggregation problems have been a focus of research in some environmental fields, such as ecological modeling (e.g., Rastetter et al., 1992), but have received surprisingly little attention elsewhere. In the catchment hydrology literature, for example, spatial heterogeneity has been widely recognized as a fundamental problem but has rarely been the subject of rigorous analysis.

Instead, it is often tacitly assumed (although *hoped* might be a better word) that any problems introduced by spatial heterogeneity will be solved or masked by model parameter calibration. This is an intuitively appealing notion. After all, we are often not particularly interested in understanding or predicting point-scale processes within the system, but rather in predicting the resulting ensemble behavior at the whole-catchment scale, such as streamflow, stream chemistry, evapotranspiration losses, ecosystem carbon uptake, and so forth. Furthermore, we rarely have point-scale information from the system under study, and when we do, we have no clear way to translate it to larger scales. Instead, often our most reliable and readily available measurements are at the whole-catchment scale: streamflow, stream chemistry, weather variables, etc. Would it not be nice if these whole-catchment measurements could be used to estimate spatially aggregated model parameters that somehow subsume the spatial heterogeneity of the system, at least well enough to generate reliable predictions of whole-catchment behavior?

This is a testable proposition, and the answer will depend partly on the nature of the underlying model. All models obscure a system's spatial heterogeneity to some degree, and many conceptual models obscure it completely, by treating spatially heterogeneous catchments as if they were spatially homogeneous instead. Doing so is not automatically disqualifying, but neither is it obviously valid. Rather, this spatial aggregation is a modeling choice, whose consequences should be explicitly analyzed and quantified. What do I mean by "explicitly analyzed and quantified?". As an example, consider the Kirchner et al. (1993) analysis of how spatial heterogeneity affected a particular geochemical model for estimating catchment buffering of acid deposition. The authors began by noting that spatial heterogeneities will not "average out" in nonlinear model equations and by showing that the resulting aggregation bias will be proportional to the nonlinearity in the model equations (which can be directly estimated) and proportional to the variance in the heterogeneous real-world parameter values (which is typically unknown but may at least be given a plausible upper bound). They then showed that their geochemical model's governing equations were sufficiently linear that the effects of spatial heterogeneity were likely to be small. They then confirmed this theoretical result by mixing measured runoff chemistry time series from random pairs of geochemically diverse catchments (which do not flow together in the real world). They showed that the geochemical model correctly predicted the buffering behavior of these spatially heterogeneous pseudo-catchments, without knowing that those catchments were heterogeneous and without knowing anything about the nature of their heterogeneities.

Here I use similar thought experiments to explore the consequences of spatial heterogeneity for catchment mean transit-time estimates derived from seasonal tracer cycles in precipitation and streamflow. Catchment *transit time* or, equivalently, *travel time* – the time that it takes for rainfall to travel through a catchment and emerge as streamflow – is a fundamental hydraulic parameter that controls the retention and release of contaminants and thus the downstream consequences of pollution episodes (Kirchner et al., 2000; McDonnell et al., 2010). In many geological settings, catchment transit times also control chemical weathering rates, geochemical solute production, and the long-term carbon cy-

cle (Burns et al., 2003; Godsey et al., 2009; Maher, 2010; Maher and Chamberlain, 2014).

A catchment is characterized by its travel-time distribution (TTD), which reflects the diversity of flowpaths (and their velocities) connecting each point on the landscape with the stream. Because these flowpaths and velocities change with hydrologic forcing, the TTD is nonstationary (Kirchner et al., 2001; Tetzlaff et al., 2007; Botter et al., 2010; Hrachowitz et al., 2010a; Van der Velde et al., 2010; Birkel et al., 2012; Heidbüchel et al., 2012; Peters et al., 2014); but time-varying TTDs are difficult to estimate in practice, so most catchment studies have focused on estimating time-averaged TTDs instead. Both the shape of the TTD and its corresponding mean travel time (MTT) reflect storage and mixing processes in the catchment (Kirchner et al., 2000, 2001; Godsey et al., 2010; Hrachowitz et al., 2010a). However, due to the difficulty in reliably estimating the shape of the TTD, and the volumes of data required to do so, many catchment studies have simply assumed that the TTD has a given shape, and have estimated only its MTT. As a result, and also because of its obvious physical interpretation as the ratio between the storage volume and the average water flux (in steady state), the MTT is by far the most universally reported parameter in catchment travel-time studies. Estimates of MTTs have been correlated with a wide range of catchment characteristics, including drainage density, aspect, hillslope gradient, depth to groundwater, hydraulic conductivity, and the prevalence of hydrologically responsive soils (e.g., McGuire et al., 2005; Soulsby et al., 2006; Tetzlaff et al., 2009; Broxton et al., 2009; Hrachowitz et al., 2009, 2010b; Asano and Uchida, 2012; Heidbüchel et al., 2013).

Travel-time distributions and mean travel times cannot be measured directly, and they differ – often by orders of magnitude – from the hydrologic response timescale, because the former is determined by the velocity of water flow, and the latter is determined by the celerity of hydraulic potentials (Horton and Hawkins, 1965; Hewlett and Hibbert, 1967; Beven, 1982; Kirchner et al., 2000; McDonnell and Beven, 2014). Nor can travel-time characteristics be reliably determined a priori from theory. Instead, they must be determined from chemical or isotopic tracers, such as Cl^-, ^{18}O, and ^2H, in precipitation and streamflow. These passive tracers "follow the water"; thus, their temporal fluctuations reflect the transport, storage, and mixing of rainfall as it is transformed into runoff. (Groundwaters can also be dated using dissolved gases such as CFCs and $^3H/^3He$, but these tracers are not conserved in surface waters or in the vadose zone, so they are not well suited to estimating whole-catchment travel times.)

As reviewed by McGuire and McDonnell (2006), three methods are commonly used to infer catchment travel times from conservative tracer time series: (1) time-domain convolution of the input time series to simulate the output time series, with parameters of the convolution kernel (the travel-time distribution) fitted by iterative search techniques; (2) Fourier transform spectral analysis of the input and output time series; and (3) sine-wave fitting to the seasonal tracer variation in the input and output. In all three methods, the greater the damping of the input signal in the output, the longer the inferred mean travel time. Sine-wave fitting can be viewed as the simplest possible version of both spectral analysis (examining the Fourier transform at just the annual frequency) and time-domain convolution (approximating the input and output as sinusoids, for which the convolution relationship is particularly easy to calculate). Whereas time-domain convolution methods require continuous, unbroken precipitation isotopic records spanning at least several times the MTT (McGuire and McDonnell, 2006; Hrachowitz et al., 2011), and spectral methods require time series spanning a wide range of timescales (Feng et al., 2004), sine-wave fitting can be performed on sparse, irregularly sampled data sets. Because sine-wave fitting is mathematically straightforward, and because its data requirements are modest compared to the other two methods, it is arguably the best candidate for comparison studies based on large multi-site data sets of isotopic measurements in precipitation and river flow. For that reason – and because it presents an interesting test case of the general aggregation issues alluded to above, in which some key results can be derived analytically – the sinusoidal fitting method will be the focus of my analysis.

The isotopic composition of precipitation varies seasonally as shifts in meridional circulation alter atmospheric vapor transport pathways (Feng et al., 2009) and as shifts in temperature and storm intensity alter the degree of rainout-driven fractionation that air masses undergo (Bowen, 2008). The resulting seasonal cycles in precipitation (e.g., Fig. 1a) are damped and phase-shifted as they are transmitted through catchments (e.g., Fig. 1b), by amounts that depend on – and thus can be used to infer properties of – the travel-time distribution. Figure 1 shows an example of sinusoidal fits to seasonal δ^{18}O cycles in precipitation and baseflow at one particular field site. The visually obvious damping of the isotopic cycle in baseflow relative to precipitation implies, in this case, an estimated MTT of 1.4 years (DeWalle et al., 1997) under the assumption that the TTD is exponential.

That particular estimate of mean transit time, like practically all such estimates in the literature, was made by methods that assume that the catchment is homogeneous and therefore that the shape of its TTD can be straightforwardly characterized. Typical catchments violate this assumption, but the consequences for estimating MTTs have not been systematically investigated, either for sine-wave fitting or for any other methods that infer travel times from tracer data. Are any of these estimation methods reliable under realistic degrees of spatial heterogeneity? Are they biased, and by how much? We simply do not know, because they have not been tested. Instead, we have been directly applying theoretical results, derived for idealized hypothetical cases, to complex real-world situations that do not share those idealized characteristics. Methods for estimating catchment travel

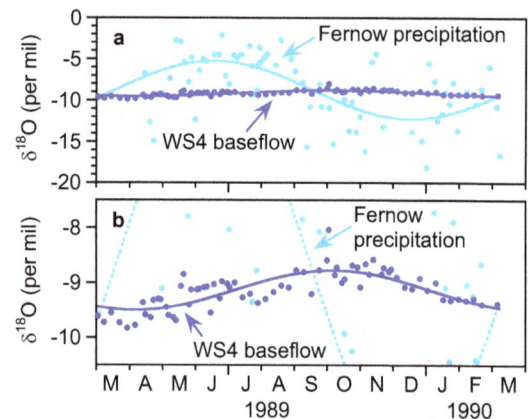

Figure 1. Seasonal cycles in δ^{18}O in precipitation and baseflow at catchment WS4, Fernow Experimental Forest, West Virginia, USA (DeWalle et al., 1997). Both panels show the same data; the axes of (**b**) are expanded to more clearly show the seasonal cycle in baseflow. Sinusoidal cycles are fitted by iteratively reweighted least squares regression (IRLS), a robust fitting technique that limits the influence of outliers.

times urgently need benchmark testing. The work presented below is intended as one small step toward filling that gap.

2   Mathematical preliminaries: tracer cycles in homogeneous catchments

Any method for inferring transit-time distributions (or their parameters, such as mean transit time) must make simplifying assumptions about the system under study. Most such methods assume that conservative tracers in streamflow can be modeled as the convolution of the catchment's transit time distribution with the tracer time series in precipitation (Maloszewski et al., 1983; Maloszewski and Zuber, 1993; Barnes and Bonell, 1996; Kirchner et al., 2000).

$$c_S(t) = \int_0^\infty h(\tau)c_P(t-\tau)\mathrm{d}\tau, \qquad (1)$$

where $c_S(t)$ is the concentration in the stream at time t, $c_P(t-\tau)$ is the concentration in precipitation at any previous time $t-\tau$, and $h(\tau)$ is the distribution of transit times τ separating the arrival of tracer molecules in precipitation and their delivery in streamflow. The concentrations $c_S(t)$ and $c_P(t-\tau)$ can also represent ratios of stable isotopes in the familiar δ notation (e.g., δ^{18}O or δ^2H); the mathematics are the same in either case.

The transit-time distribution $h(\tau)$ expresses the fractional contribution of past inputs to present runoff. Equation (1) implicitly assumes that the catchment is a linear time-invariant system and, thus, that the convolution kernel $h(\tau)$ is stationary (i.e., constant through time). This is never strictly true, most obviously because if no precipitation falls on a particular day, it cannot contribute any tracer to the stream τ days later, and because higher precipitation rates will increase the rate at which water and tracers are flushed through the catchment. Thus, real-world TTDs vary through time, depending on the history of prior precipitation (Kirchner et al., 2001; Tetzlaff et al., 2007; Botter et al., 2010; Hrachowitz et al., 2010a; Van der Velde et al., 2010; Birkel et al., 2012; Heidbüchel et al., 2012; Peters et al., 2014). However, in applications using real-world data, $h(\tau)$ is conventionally interpreted as a time-invariant ensemble average, taken over an ensemble of precipitation histories, which obviously will differ from one another in detail. Mathematically, the ensemble averaging embodied in Eq. (1) is equivalent to the simplifying assumption that water fluxes in precipitation and streamflow are constant over time. (One can relax this assumption somewhat by integrating over the cumulative water flux rather than time, as proposed by Niemi (1977). If the rates of transport and mixing vary proportionally to the flow rate through the catchment, this yields a stationary distribution in flow-equivalent time.) A further simplification inherent in Eq. (1) is that evapotranspiration and its effects on tracer signatures are ignored.

2.1 A class of transit-time distributions

In much of the analysis that follows, I will assume that the transit-time distribution $h(\tau)$ belongs to the family of gamma distributions:

$$h(\tau) = \frac{\tau^{\alpha-1}}{\beta^\alpha \Gamma(\alpha)} e^{-\tau/\beta} = \frac{\tau^{\alpha-1}}{(\overline{\tau}/\alpha)^\alpha \Gamma(\alpha)} e^{-\alpha\tau/\overline{\tau}}, \quad (2)$$

where α and β are a shape factor and scale factor, respectively, τ is the transit time, and $\overline{\tau} = \alpha\beta$ is the mean transit time. I make this assumption mostly so that some key results can be calculated exactly, but as I show below, the key results extend beyond this (already broad) class of distributions.

Figure 2 shows gamma distributions spanning a range of shape factors α. For the special case of $\alpha = 1$, the gamma distribution becomes the exponential distribution. Exponential distributions describe the behavior of continuously mixed reservoirs of constant volume, and they have been widely used to model catchment storage and mixing. The gamma distribution expresses the TTD of a Nash cascade (Nash, 1957) of α identical linear reservoirs connected in series, and the analogy to a Nash cascade holds even for noninteger α, through the use of fractional integration. For $\alpha > 1$, the gamma distribution rises to a peak and then falls off, similarly to a typical storm hydrograph, which is why Nash cascades have often been used to model rainfall–runoff relationships. For $\alpha < 1$, however, the gamma distribution has a completely different shape, having maximum weight at lags near zero and a relatively long tail. These characteristics represent problematic contaminant behavior, with rapid release of an intense contaminant spike followed by persistent lower-level contamination far into the future. Tracer time series from

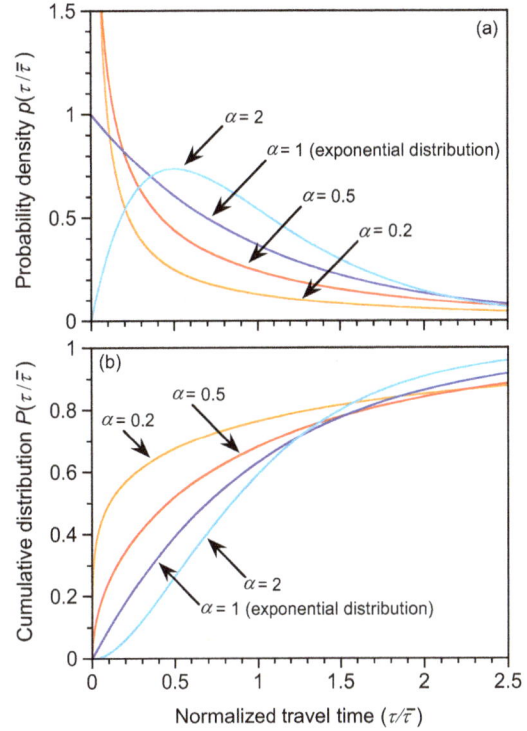

Figure 2. Gamma distributions for the range of shape factors $\alpha = 0.2$–2 considered in this analysis. Horizontal axes are normalized by the mean transit time $\overline{\tau}$ and thus are dimensionless.

many catchments have been shown to exhibit fractal $1/f$ scaling, which is consistent with gamma TTDs with $\alpha \approx 0.5$ (Kirchner et al., 2000, 2001; Godsey et al., 2010; Kirchner and Neal, 2013; Aubert et al., 2014).

For present purposes, it is sufficient to note that the family of gamma distributions encompasses a wide range of shapes which approximate many plausible TTDs (Fig. 2). The moments of the gamma distribution vary systematically with the shape factor α (Walck, 2007):

$$\text{mean}(\tau) = \beta\alpha = \overline{\tau}, \quad (3a)$$

$$\text{SD}(\tau) = \beta\sqrt{\alpha} = \overline{\tau}/\sqrt{\alpha}, \quad (3b)$$

$$\text{skewness}(\tau) = 2/\sqrt{\alpha}, \text{ and} \quad (3c)$$

$$\text{kurtosis}(\tau) = 6/\alpha. \quad (3d)$$

As α increases above 1, the standard deviation (SD) declines in relation to the mean, and the shape of the distribution becomes more normal. But as α decreases below 1, the SD grows in relation to the mean, implying greater variability in transit times for the same average (in other words: more short transit times, more long transit times, and fewer close to the mean). Likewise the skewness and kurtosis grow with decreasing α, reflecting greater dominance by the tails of the distribution.

Studies that have used tracers to constrain the shape of catchment TTDs have generally found shape factors α ranging from 0.3 to 0.7, corresponding to spectral slopes of the transfer function between roughly 0.6 and 1.4 (Kirchner et al., 2000, 2001; Godsey et al., 2010; Hrachowitz et al., 2010a; Kirchner and Neal, 2013; Aubert et al., 2014). Other studies – including those that have used annual tracer cycles to estimate mean transit times – have assumed that the catchment is a well-mixed reservoir and thus that $\alpha = 1$. Here I will assume that α falls in the range of 0.5–1 for typical catchment transit-time distributions, but I will also show some key results for the somewhat wider range of $\alpha = 0.2$–2, for illustrative purposes. The results reported here will not necessarily apply to TTDs that rise to a peak after a long delay, such as the gamma distribution with $\alpha \gg 2$. However, one would not expect such a distribution to characterize whole-catchment TTDs in the first place because, except in very unusual catchments, a substantial amount of precipitation can fall close to the stream and enter it relatively quickly, thus producing a strong peak at a short lag (Kirchner et al., 2001).

2.2   Estimating mean transit time from tracer cycles

Because convolutions (Eq. 1) are linear operators, they transform any sinusoidal cycle in the precipitation time series $c_P(t)$ into a sinusoidal cycle of the same frequency, but a different amplitude and/or phase, in the streamflow time series $c_S(t)$. Real-world transit-time distributions $h(\tau)$ are causal (i.e., $h(\tau) = 0$ for $t < 0$) and mass-conserving (i.e., $\int h(\tau) = 1$), implying that $c_S(t)$ will be damped and phase-shifted relative to $c_P(t)$ and also implying that one can use the relative amplitudes and phases of cycles in $c_S(t)$ and $c_P(t)$ to infer characteristics of $h(\tau)$. This mathematical property forms the basis for sine-wave fitting, and also for the spectral methods of Kirchner et al. (2000, 2001), which can be viewed as sine-wave fitting across many different timescales.

The amplitudes A and phases φ of seasonal cycles in precipitation and streamflow can be estimated by nonlinear fitting,

$$c_P(t) = A_P \sin(2\pi f t - \varphi_P) + k_P,$$
$$c_S(t) = A_S \sin(2\pi f t - \varphi_S) + k_S, \tag{4}$$

or by determining the cosine and sine coefficients a and b via multiple linear regression,

$$c_P(t) = a_P \cos(2\pi f t) + b_P \sin(2\pi f t) + k_P,$$
$$c_S(t) = a_S \cos(2\pi f t) + b_S \sin(2\pi f t) + k_S, \tag{5}$$

and then calculating the amplitudes and phases using the conventional identities:

$$A_P = \sqrt{a_P^2 + b_P^2}, \; A_S = \sqrt{a_S^2 + b_S^2}, \; \varphi_P = \arctan(b_P/a_P)$$
$$\text{and } \varphi_S = \arctan(b_S/a_S). \tag{6}$$

In Eqs. (4)–(6) above, t is time, f is the frequency of the cycle ($f = 1 \text{ year}^{-1}$ for a seasonal cycle), and the subscripts P

and S refer to precipitation and streamflow. In fitting sinusoidal cycles to real-world data, robust estimation techniques such as iteratively reweighted least squares (IRLS) regression can help in limiting the influence of outliers. Also, because precipitation and streamflow rates vary through time, it may be useful to weight each tracer sample by its associated volume, for example to reduce the influence of small rainfall events (for more on the implications of volume-weighting, see Kirchner, 2016). An R script for performing volume-weighted IRLS is available from the author.

The key to calculating the amplitude damping and phase shift that will result from convolving a sinusoidal input with a gamma-distributed $h(\tau)$ is the gamma distribution's Fourier transform, also called, in this context, its "characteristic function" (Walck, 2007):

$$H(f) = (1 - i2\pi f \beta)^{-\alpha} = (1 - i2\pi f \bar{\tau}/\alpha)^{-\alpha}. \tag{7}$$

From Eq. (7), one can derive how the shape factor α and the mean transit time $\bar{\tau}$ affect the amplitude ratio A_S/A_P between the streamflow and precipitation cycles,

$$\frac{A_S}{A_P} = \left(1 + (2\pi f \beta)^2\right)^{-\alpha/2}, \tag{8}$$

and also the phase shift between them,

$$\varphi_S - \varphi_P = \alpha \arctan(2\pi f \beta), \tag{9}$$

where $\beta = \bar{\tau}/\alpha$. Figure 3a and b show the expected amplitude ratios and phase shifts for a range of shape factors and mean transit times.

If the shape factor α is known (or can be assumed), the mean transit time can be calculated directly from the amplitude ratio A_S/A_P by inverting Eq. (8):

$$\bar{\tau} = \alpha\beta, \; \beta = \frac{1}{2\pi f}\sqrt{(A_S/A_P)^{-2/\alpha} - 1}. \tag{10}$$

Equation (10), with $\alpha = 1$, is the standard tool for estimating MTTs from seasonal tracer cycles in precipitation and streamflow. Alternatively, as Fig. 3c shows, both the shape factor α and the mean transit time $\bar{\tau}$ can be jointly determined from the phase shift $\varphi_S - \varphi_P$ and the amplitude ratio A_S/A_P, if these can both be quantified with sufficient accuracy. Mathematically, this joint solution can be achieved by substituting Eq. (10) in Eq. (9), yielding the following implicit expression for α:

$$\varphi_S - \varphi_P = \alpha \arctan\left(\sqrt{(A_S/A_P)^{-2/\alpha} - 1}\right), \tag{11}$$

which can be solved using nonlinear search techniques such as Newton's method. Once α has been determined, the mean transit time $\bar{\tau}$ can be calculated straightforwardly using Eq. (10). However, when precipitation is episodic, the phase shift $\varphi_S - \varphi_P$ may be difficult to estimate accurately, which

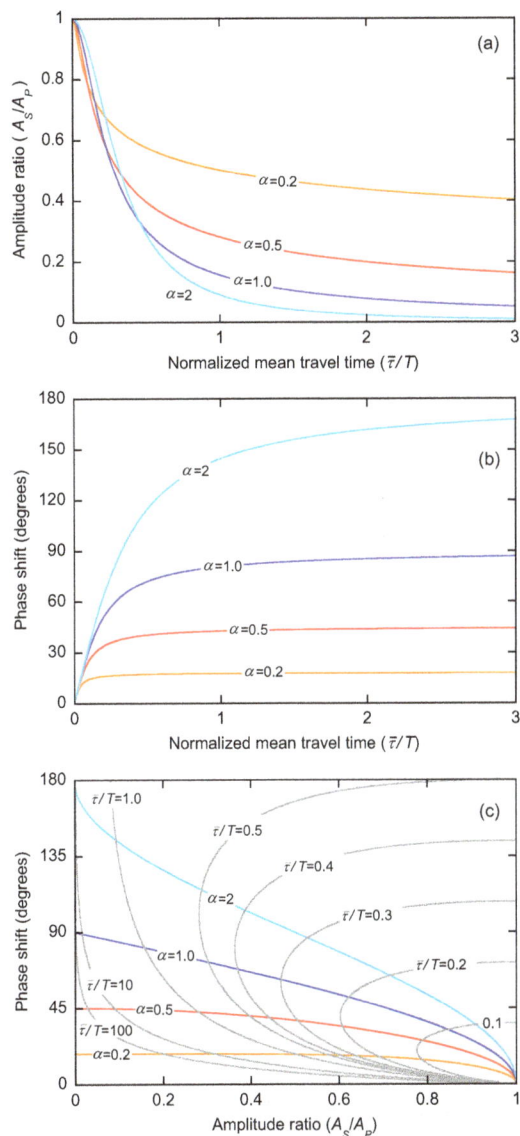

Figure 3. Amplitude ratio and phase shift between seasonal cycles in precipitation and streamflow, for gamma-distributed catchment transit-time distributions with a range of shape factors α (colored lines). **(a)** Ratio of seasonal cycle amplitudes in streamflow and precipitation (A_S/A_P) as a function of mean transit time ($\bar{\tau}$) normalized by the period ($T = 1/f$) of the tracer cycle. **(b)** Phase lag between streamflow and precipitation cycles, as a function of mean transit time normalized by the tracer cycle period ($\bar{\tau}/T$). **(c)** Relationship between phase lag and amplitude ratio, with contours of shape factor (α) ranging from 0.2 to 8 (colored lines), and contours of mean transit time normalized by tracer cycle period $\bar{\tau}/T$ (gray lines). For seasonal tracer cycles, $T = 1/f = 1$ year and normalized transit time equals time in years.

can result in large errors in α and thus $\bar{\tau}$, particularly if the phase shift is near zero. Perhaps for this reason or because (to the best of my knowledge) the relevant math has not previ-

ously been presented, tracer cycle phase information has not typically been used in estimating α and MTT.

3   Transit times and tracer cycles in heterogeneous catchments: a thought experiment

The methods outlined above can be applied straightforwardly in a homogeneous catchment characterized by a single transit-time distribution. Real-world catchments, however, are generally heterogeneous; they combine different landscapes with different characteristics and thus different TTDs. The implications of this heterogeneity can be demonstrated with a simple thought experiment. What if, instead of a single homogeneous catchment, we have two subcatchments with different MTTs and therefore different tracer cycles, which then flow together, as shown in Fig. 4? If we observed only the tracer cycle in the combined runoff (the solid blue line in Fig. 4), and not the tracer cycles in the individual subcatchments (the red and orange lines in Fig. 4), would we correctly infer the whole-catchment MTT? Note that although I refer to the different runoff sources as "subcatchments", they could equally well represent alternate slopes draining to the same stream channel or even independent flowpaths down the same hillslope; nothing in this thought experiment specifies the scale of the analysis. And, of course, real-world catchments are much more complex than the simple thought experiment diagrammed in Fig. 4, but this two-component model is sufficient to illustrate the key issues at hand.

From assumed MTTs $\bar{\tau}$ and shape factors α for each of the subcatchments, one can calculate the amplitude ratios A_S/A_P and phase shifts $\varphi_S - \varphi_P$ of their tracer cycles using Eqs. (8) and (9), and then average these cycles together using the conventional trigonometric identities. (Equivalently, one can estimate the cosine and sine coefficients of the individual subcatchments' tracer cycles from the real and imaginary parts of Eq. (7) and algebraically average them together.) The shares of the two subcatchments in the average will depend on their relative drainage areas and/or water yields. For simplicity, I combine the runoff from the two subcatchments in a 1 : 1 ratio; this also guarantees that the combined runoff will be as different as possible from each of the two sources. I then ask the question: from the tracer behavior in the combined runoff (the solid blue line in Fig. 4), would I correctly estimate the mean transit time for the whole catchment? That is, would I infer a MTT that is close to the average of the MTTs of the two subcatchments?

One can immediately see that this situation is highly prone to aggregation bias, following the Kirchner et al. (1993) rule of thumb that the degree of aggregation bias is proportional to the nonlinearity in the governing equations and the variance in the heterogeneous parameters. The amplitude ratios A_S/A_P and phase shifts $\varphi_S - \varphi_P$ of seasonal tracer cycles are strongly nonlinear functions of the MTT (see Eqs. 8 and 10),

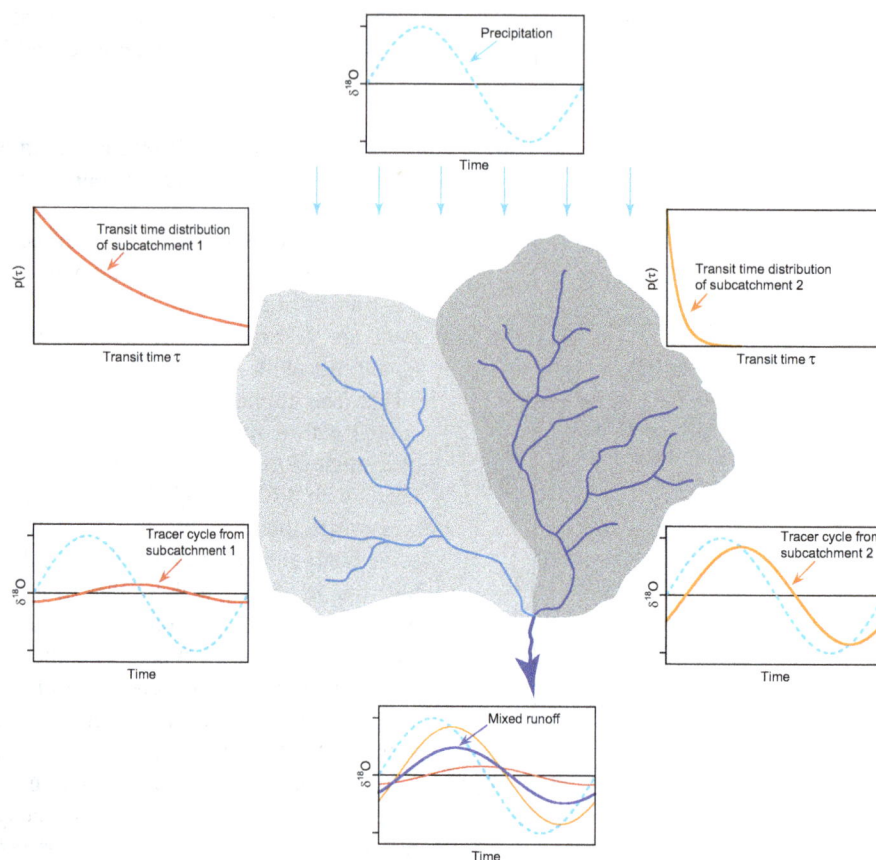

Figure 4. Conceptual diagram illustrating the mixture of seasonal tracer cycles in runoff from a heterogeneous catchment, comprising two subcatchments with strongly contrasting MTTs, and which thus damp the tracer cycle in precipitation (light blue dashed line) by different amounts. The tracer cycle in the combined runoff from the two subcatchments (dark blue solid line) will average together the highly damped cycle from subcatchment 1, with long MTT (solid red line), and the less damped cycle from subcatchment 2, with short MTT (solid orange line).

as illustrated in Fig. 3a and b. And, importantly, the likely range of variation in subcatchment MTTs (from, say, fractions of a year to perhaps several years) straddles the nonlinearity in the governing equations. Thus, we should expect to see significant aggregation bias in estimates of MTT.

Figure 5 illustrates the crux of the problem. The plotted curve shows the relationship between A_S/A_P and MTT for exponential transit-time distributions ($\alpha = 1$); other realistic transit-time distributions will give somewhat different relationships, but they will all be curved. Seasonal cycles from the two subcatchments (the red and orange squares) will mix along the dashed gray line (which is nearly straight but not exactly so, owing to phase differences between the two cycles). A 50:50 mixture of tracer cycles from the two subcatchments will plot as the solid blue square, with an amplitude ratio A_S/A_P of 0.43 and a MTT of just over 2 years in this particular example. But the crux of the problem is that if we use this amplitude ratio to infer the corresponding MTT, we will do so where the amplitude ratio intersects with the black curve (Eq. 10), yielding an inferred MTT of

only 0.33 years (the open square), which underestimates the true MTT of the mixed runoff by more than a factor of six. Bethke and Johnson (2008) pointed out that nonlinear averaging can lead to bias in groundwater dating by radioactive tracers; Fig. 5 illustrates how a similar bias can also arise in age determinations based on fluctuation damping in passive tracers.

Combining flows from two subcatchments with different mean transit times will result in a combined TTD that differs in shape, not just in scale, from the TTDs of either of the subcatchments. For example, combining two exponential distributions with different mean transit times does not result in another exponential distribution but rather a hyperexponential distribution, as shown in Fig. 6. The characteristic function of the hyperexponential distribution (Walck, 2007) yields the following expression for the amplitude ratio of tracer cycles in precipitation and streamflow,

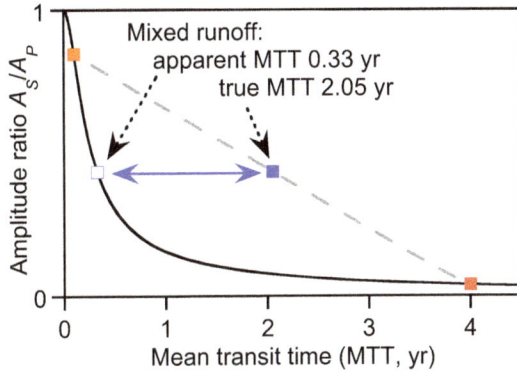

Figure 5. Illustration of the aggregation error that arises when mean transit time is inferred from seasonal tracer cycles in mixed runoff from two landscapes with contrasting transit-time distributions (e.g., Fig. 4). The relationship between MTT and the amplitude ratio (A_S/A_P) of annual cycles in streamflow and precipitation is strongly nonlinear (black curve). Seasonal cycles from subcatchments with MTT of 0.1 years ($A_S/A_P = 0.85$, orange square) and 4 years ($A_S/A_P = 0.04$, red square) will mix along the dashed gray line. A 50 : 50 mixture of the two sources will have a MTT of $(4+0.1)/2 = 2.05$ years and an amplitude ratio A_S/A_P of 0.43 (blue square). But if this amplitude ratio is interpreted as coming from a single catchment (Eq. 10), it implies a MTT of only 0.33 years (open square), 6 times shorter than the true MTT of the mixed runoff.

$$\frac{A_S}{A_P} = \left(\left(\frac{p}{1+(2\pi f \overline{\tau}_1)^2} + \frac{q}{1+(2\pi f \overline{\tau}_2)^2} \right)^2 \right.$$

$$\left. + \left(\frac{p2\pi f \overline{\tau}_1}{1+(2\pi f \overline{\tau}_1)^2} + \frac{q2\pi f \overline{\tau}_2}{1+(2\pi f \overline{\tau}_2)^2} \right)^2 \right)^{1/2}, \qquad (12)$$

where $\overline{\tau}_1$ and $\overline{\tau}_2$ are the mean transit times of the two exponential distributions, and p and $q = 1 - p$ are their proportions in the mixed runoff. Equation (12) describes the dashed gray line in Fig. 5, and one can see by inspection that in a 1 : 1 mixture ($p = q$) the amplitude ratio A_S/A_P will be determined primarily by the shorter of the two mean transit times. As Fig. 5 shows, the amplitude ratio implied by Eq. (12) is greater – often much greater – than Eq. (8) would predict for an exponential distribution with an equivalent mean transit time $\overline{\tau} = p\overline{\tau}_1 + q\overline{\tau}_2$. In other words, when amplitude ratios are interpreted as if they were generated by individual uniform catchments (i.e., Eq. 8) rather than a heterogeneous collection of subcatchments (i.e., Eq. 12), the inferred mean transit time will be underestimated, potentially by large factors.

To test the generality of this result, I repeated the thought experiment outlined above for 1000 hypothetical pairs of subcatchments, each with individual MTTs randomly chosen from a uniform distribution of logarithms spanning the interval between 0.1 and 20 years (Fig. 7). Pairs with MTTs that differed by less than a factor of 2 were excluded, so that the entire sample consisted of truly heterogeneous catchments. I

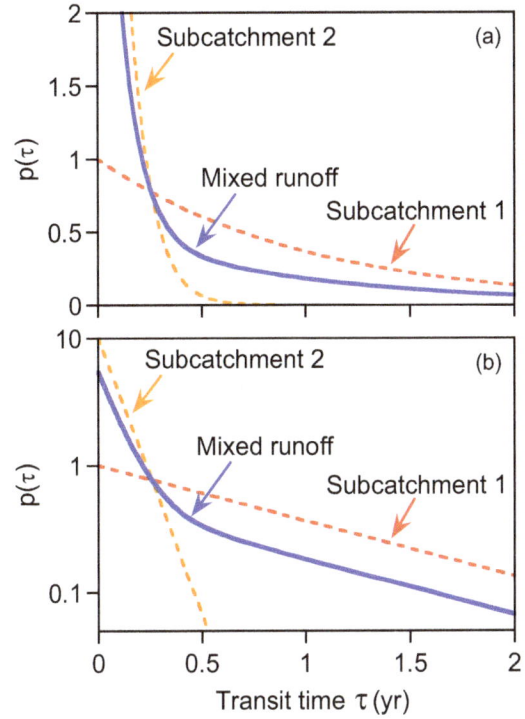

Figure 6. Exponential transit-time distributions for subcatchments 1 and 2 in Fig. 4 (with mean transit times of 1 and 0.1 years, shown by the orange and red dashed lines, respectively), and the hyperexponential distribution formed by merging them in equal proportions (solid blue line). (**a**) and (**b**) show linear and logarithmic axes.

then applied Eq. (10) to calculate the apparent MTT from the inferred runoff. As Fig. 7 shows, apparent MTTs calculated from the combined runoff of the two subcatchments can underestimate true whole-catchment MTTs by an order of magnitude or more, and this strong underestimation bias persists across a wide range of shape factors α. MTTs are reliably estimated (with values close to the 1 : 1 line in Fig. 7) only when both subcatchments have MTTs of much less than 1 year.

In most real-world cases, unlike these hypothetical thought experiments, one will only have measurements or samples from the whole catchment's runoff. The properties of the individual subcatchments and thus the degree of heterogeneity in the system will generally be unknown. And even if data were available for the subcatchments, those subcatchments would be composed of sub-subcatchments, which would themselves be heterogeneous to some unknown degree, and so on. Thus, it will generally be difficult or impossible to characterize the system's heterogeneity, but that is no justification for pretending that this heterogeneity does not exist. Nonetheless, in such situations it will be tempting to treat the whole system as if it were homogeneous, perhaps using terms like "apparent age" or "model age" to preserve a sense of rigor. But whatever the semantics, as Fig. 7 shows, assum-

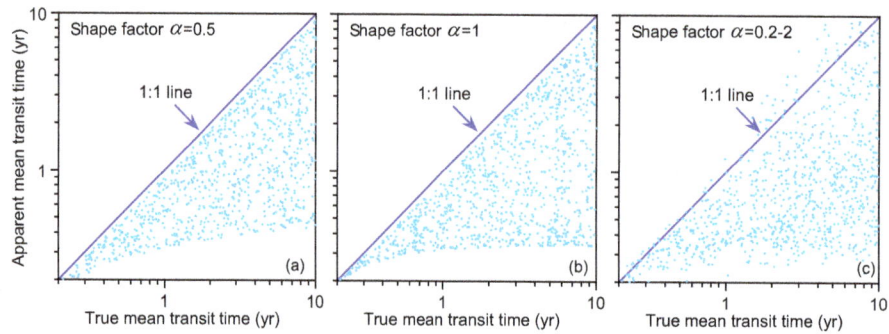

Figure 7. Apparent MTT inferred from seasonal tracer cycles, showing order-of-magnitude deviations from true MTT for 1000 synthetic catchments. Each synthetic catchment comprises two subcatchments with individual MTTs randomly chosen from a uniform distribution of logarithms spanning the interval between 0.1 and 20 years, with each pair differing by at least a factor of 2. In (a) and (b), both subcatchments have shape factors α of 0.5 and 1, respectively; in (c), the subcatchments' shape factors are independently chosen from the range of 0.2–2. Apparent MTTs were inferred from the amplitude ratio A_S/A_P of the combined runoff using Eq. (10), with an assumed value of $\alpha = 0.5$ for (a), $\alpha = 1$ for (b), and also $\alpha = 1$ for (c), both because $\alpha = 1$ is close to the average of the randomized α values and because $\alpha = 1$ is typically assumed whenever Eq. (10) is applied to real catchment data.

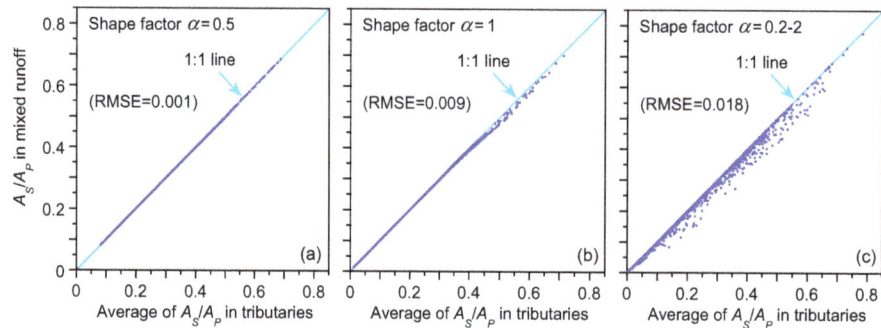

Figure 8. Amplitude ratio (A_S/A_P) of tracer cycles in precipitation and mixed runoff from the same 1000 synthetic catchments shown in Fig. 7 (vertical axes), compared to the average of the tracer cycle amplitude ratios in the two tributaries (horizontal axes). As in Fig. 7, each synthetic catchment comprises two subcatchments with individual MTTs randomly chosen from a uniform distribution of logarithms spanning the interval between 0.1 and 20 years, and with each pair of MTTs differing by at least a factor of 2. In (a) and (b), all subcatchments have the same shape factor α. In (c), shape factors for each subcatchment are randomly chosen from a uniform distribution between $\alpha = 0.2$ and $\alpha = 2$. The close fits to the 1 : 1 lines, and the small root-mean-square error (RMSE) values, show that the tracer cycle amplitudes from the tributaries are averaged almost exactly in the mixed runoff.

ing homogeneity in heterogeneous catchments will result in strongly biased estimates of whole-catchment mean transit times.

4   Quantifying the young water component of streamflow

The analysis in Sect. 3 demonstrates what can be termed an "aggregation error": in heterogeneous systems, mean transit times estimated from seasonal tracer cycles yield inconsistent results at different levels of aggregation. The aggregation bias demonstrated in Figs. 5 and 7 implies that seasonal cycles of conservative tracers are unreliable estimators of catchment mean transit times. This observation raises the obvious question: is there anything *else* that can be estimated from

seasonal tracer cycles and that is relatively free from the aggregation bias that afflicts estimates of mean transit times?

One hint is provided by the observation that when two tributaries are mixed, the tracer cycle amplitude in the mixture will almost exactly equal the average of the tracer cycle amplitudes in the two tributaries (Fig. 8). This is not intuitively obvious, because the tributary cycles will generally be somewhat out of phase with each other, so their amplitudes will not average exactly linearly. But when the tributary cycles are far out of phase (because the subcatchments have markedly different mean transit times or shape factors), the two amplitudes will also generally be very different and thus the phase angle between the tributary cycles will have little effect on the amplitude of the mixed cycle.

Because tracer cycle amplitudes will average almost linearly when two streams merge and thus are virtually free

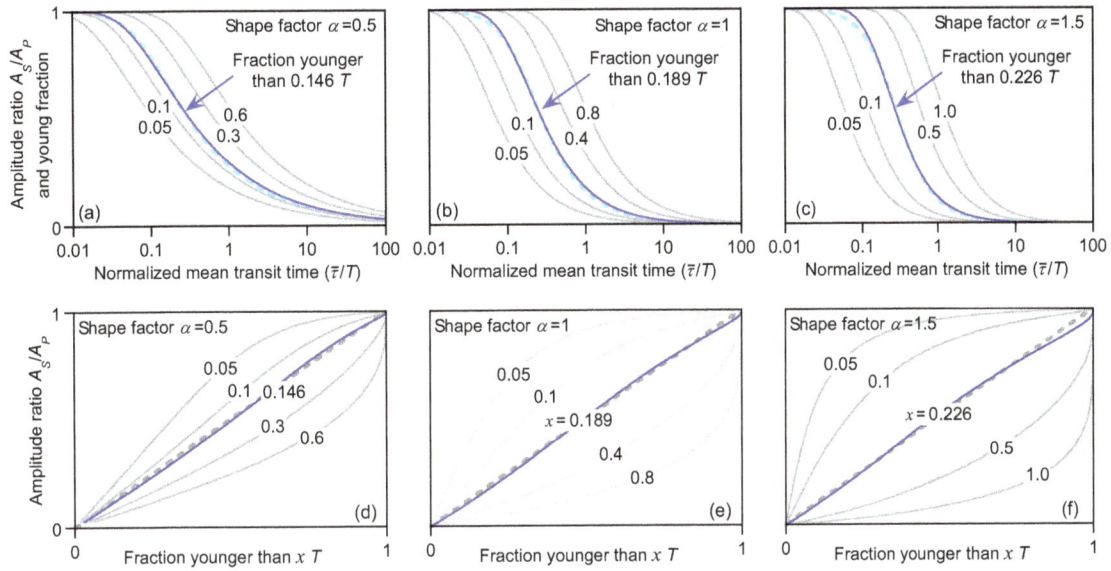

Figure 9. (a)–(c) show the amplitude ratios A_S/A_P in precipitation and streamflow tracer cycles (light blue dashed line) as function of mean transit time $\bar{\tau}$, compared to the fraction of water younger than several threshold ages (gray lines), and the best-fit age threshold (dark blue line). **(d)–(f)** show the relationship between amplitude ratio and the fraction of water younger than several age thresholds (gray lines) and the best-fit age threshold (dark blue line), with the 1 : 1 line (dashed gray) for comparison. Panels show results for three different gamma distributions, with shape factors $\alpha = 0.5$, $\alpha = 1$, and $\alpha = 1.5$. Root-mean-squared errors (RMSEs) for amplitude ratios A_S/A_P as predictors of the best-fit young water fractions are 0.012, 0.011, and 0.015 for **(d)–(f)**, respectively. In all panels, threshold age and mean transit time are normalized by T, the period of the tracer cycle. For seasonal tracer cycles, $T = 1$ year and thus threshold age and mean transit time are in years.

from aggregation bias (Fig. 8), anything that is proportional to tracer cycle amplitude will also be virtually free from aggregation bias. So, what is proportional to tracer cycle amplitude? One hint is provided by the observation that in Fig. 5, for example, the tracer cycle amplitude in the mixture is highly sensitive to transit times that are much shorter than the period of the tracer cycle (for a seasonal cycle, this period is $T = 1$ year) but highly insensitive to transit times that are much longer than the period of the tracer cycle. As a thought experiment, one can imagine a catchment in which some fraction of precipitation bypasses storage entirely (and thus transmits the precipitation tracer cycle directly to the stream), while the remainder is stored and mixed over very long timescales (and thus its tracer cycles are completely obliterated by mixing). In this idealized catchment, the amplitude ratio A_S/A_P between the tracer cycles in the stream and precipitation will be proportional to (indeed it will be exactly *equal to*) the fraction of precipitation that bypasses storage (and thus has a near-zero transit time).

4.1 Young water

These lines of reasoning lead to the conjecture that for many realistic transit-time distributions, the amplitude ratio A_S/A_P may be a good estimator of the fraction of streamflow that is younger than some threshold age. This young water threshold should be expected to vary somewhat with the shape of

the TTD. It should also be proportional to the tracer cycle period T because, as dimensional scaling arguments require and as Eq. (8) shows for the specific case of gamma distributions, convolving the tracer cycle with the TTD will yield amplitude ratios A_S/A_P that are functions of $f\bar{\tau} = \bar{\tau}/T$.

Numerical experiments verify these conjectures for gamma distributions spanning a wide range of shape factors (see Fig. 9). I define the young water fraction F_{yw} as the proportion of the transit-time distribution younger than a threshold age τ_{yw} and calculate this proportion via the regularized lower incomplete gamma function:

$$F_{yw} = P\left(\tau < \tau_{yw}\right) = \Gamma\left(\tau_{yw}, \alpha, \beta\right) = \int_{\tau=0}^{\tau_{yw}} \frac{\tau^{\alpha-1}}{\beta^{\alpha}\Gamma(\alpha)}e^{-\tau/\beta}d\tau, \quad (13)$$

where, as before, $\beta = \bar{\tau}/\alpha$. I then numerically search for the threshold age for which (for a given shape factor α) the amplitude ratio A_S/A_P closely approximates F_{yw} across a wide range of scale factors β (or equivalently, a wide range of mean transit times $\bar{\tau}$). As Fig. 9 shows, this young water fraction nearly equals the amplitude ratio A_S/A_P, with the threshold for "young" water varying from 1.7 to 2.7 months as the shape factor α ranges from 0.5 to 1.5. The amplitude ratio A_S/A_P and the young water fraction F_{yw} are both dimensionless and they both range from 0 to 1, so they can be directly compared without further calibration, beyond the determination of the threshold age τ_{yw}. As Fig. 10 shows, the best-fit threshold age varies modestly as a function of the shape factor α:

$$\tau_{yw}/T \approx 0.0949 + 0.1065\alpha - 0.0126\alpha^2. \tag{14}$$

Across the entire range of $\alpha = 0.2$ to $\alpha = 2$ shown in Fig. 10, and across the entire range of amplitude ratios from 0 to 1 (and thus mean transit times from zero to near-infinity), the amplitude ratio A_S/A_P estimates the young water fraction with a root mean square error of less than 0.023 or 2.3 %.

The young water fraction F_{yw}, as defined here, has the inevitable drawback that, because the shape factors of individual tributaries will usually be unknown, the threshold age τ_{yw} will necessarily be somewhat imprecise. However, F_{yw} has the considerable advantage that it is virtually immune to aggregation bias in heterogeneous catchments because it is nearly equal to the amplitude ratio A_S/A_P (Fig. 9), which itself aggregates with very little bias and also with very little random error (Fig. 8). This observation leads to the important implication that A_S/A_P should reliably estimate F_{yw}, not only in individual subcatchments but also in the combined runoff from heterogeneous landscapes. To test this proposition, I calculated the young water fractions F_{yw} for 1000 heterogeneous pairs of synthetic subcatchments (with the same MTTs and shape factors shown in Fig. 7) using Eqs. (13) and (14), and compared each pair's average F_{yw} to the amplitude ratio A_S/A_P in the merged runoff. Figure 11 shows that, as hypothesized, A_S/A_P estimates the young water fraction in the merged runoff with very little scatter or bias. The root-mean-square error in Fig. 11 is roughly 2 % or less, in marked contrast to errors of several hundred percent shown in Fig. 7 for estimates of mean transit time from the same synthetic catchments.

4.2 Sensitivity to assumed TTD shape and threshold age

The analysis presented in Sect. 4.1 shows that the amplitude ratio A_S/A_P accurately estimates the fraction of streamflow younger than a threshold age. But this threshold age depends on the shape factor α of the subcatchment TTDs, which will generally be uncertain. Consider, for example, a hypothetical case where we measure an amplitude ratio of $A_S/A_P = 0.2$ in the seasonal tracer cycles in a particular catchment, but we do not know whether its subcatchments are characterized by $\alpha = 1$, $\alpha = 0.5$, or a mixture of distributions between these shape factors. How much does this uncertainty in α, and thus in the threshold age, affect the inferences we can draw from A_S/A_P? We can approach this question from two different perspectives.

We can interpret the uncertainty in α as creating ambiguity in either the threshold age τ_{yw} (which defines *young* in the young water fraction) or in the proportion of water younger than any fixed threshold age (the "fraction" in the young water fraction).

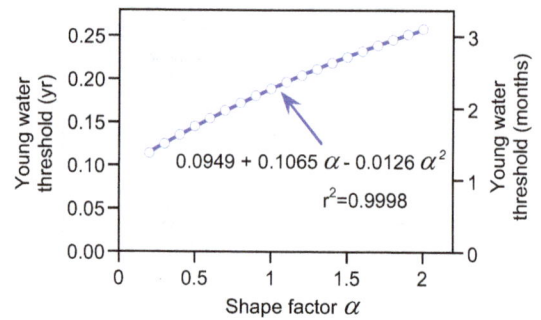

Figure 10. Best-fit young water thresholds for gamma transit-time distributions, as a function of shape factors α ranging from 0.2 to 2.0. The young water threshold τ_{yw} is defined such that the fraction of the distribution with ages less than τ_{yw} approximately equals the amplitude ratio (A_S/A_P) of annual cycles in streamflow and precipitation (see Fig. 9).

First, from Fig. 10 we can estimate how uncertainty in α affects the threshold age τ_{yw} that defines what counts as "young" streamflow. One can see that across the plausible range of shape factors, the young water threshold (that is, the threshold defining whatever young water fraction will aggregate correctly) varies from about $\tau_{yw} = 1.75$ months for $\alpha = 0.5$ to $\tau_{yw} = 2.27$ months for $\alpha = 1$. Thus, the ambiguity in α translates into an ambiguity of 0.52 months (or about two weeks) in the threshold that defines "young" water. If some subcatchments are characterized by $\alpha = 0.5$, others by $\alpha = 1$, and still others by values in between, then the effective threshold age for the ensemble will lie somewhere between 1.75 and 2.27 months. If the range of uncertainty in α is wider, then the range of uncertainty in τ_{yw} will be wider as well, spanning over a factor of 2 (1.37–3.10 months) for values of α spanning the full order-of-magnitude range shown in Fig. 2 ($\alpha = 0.2$–2).

Alternatively, we can treat the uncertainty in α as creating, for any fixed threshold age, an ambiguity in the fraction of streamflow that is younger than that age. Consider the hypothetical case outlined above, in which $A_S/A_P = 0.2$. If we assume that the subcatchments are characterized by $\alpha = 1$ (and thus $\tau_{yw} = 2.27$ months), then we would infer that roughly 20 % of streamflow is younger than 2.27 months (the exact young water fraction, using Eqs. (10) and (13), is 0.215). But if the subcatchments are characterized by $\alpha = 0.5$ instead, then according to Eqs. (10) and (13) the fraction younger than 2.27 months will be 0.242 instead of 0.215. Thus, the uncertainty in α corresponds to an uncertainty in the young water fraction of 3 % (of the range of a priori uncertainty in F_{yw}, which is between 0 and 1) or 13 % (of the original estimate for $\alpha = 1$).

For comparison, we can contrast this uncertainty with the corresponding uncertainty in the mean transit time $\bar{\tau}$ calculated from Eq. (10). A seasonal tracer cycle amplitude ratio of $A_S/A_P = 0.2$ implies a mean transit time of $\bar{\tau} = 0.80$ years

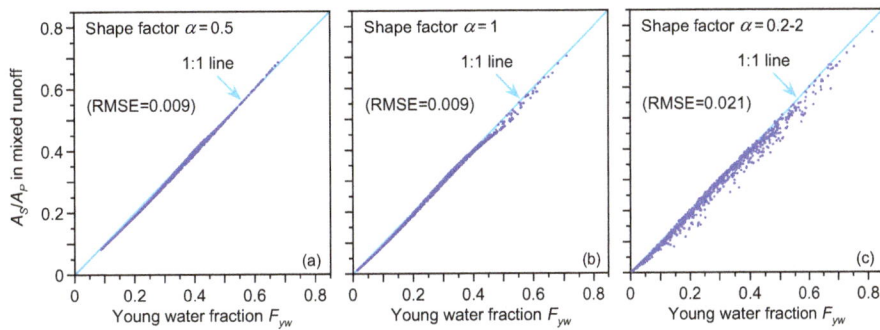

Figure 11. True and apparent young water fractions for the same 1000 synthetic catchments shown in Fig. 7. The tracer cycle amplitude ratio in the combined runoff of the two subcatchments (vertical axes) corresponds closely to the average young water fraction in the combined runoff (horizontal axes). As in Fig. 7, each synthetic catchment comprises two subcatchments with individual MTTs randomly chosen from a uniform distribution of logarithms spanning the interval between 0.1 and 20 years, and with each pair of MTTs differing by at least a factor of 2. In (**a**) and (**b**), all subcatchments have the same shape factor α. In (**c**), shape factors for each subcatchment are randomly chosen from a uniform distribution between $\alpha = 0.2$ and $\alpha = 2$.

if $\alpha = 1$, but $\bar{\tau} = 1.99$ years if $\alpha = 0.5$. Thus, the uncertainty in the mean transit time is a factor of 2.5, compared to a few percent for the young water fraction.

We can extend these sample calculations over a range of shape factors α and amplitude ratios A_S/A_P (see Fig. 12). As Fig. 12 shows, when the shape factor is uncertain in the range of $0.5 < \alpha < 1$, the corresponding uncertainty in the young water fraction F_{yw} is typically several percent, but the corresponding uncertainty in the MTT is typically a factor of 2 or more. For a factor of 10 uncertainty in the shape factor ($0.2 < \alpha < 2$), the uncertainty in the young water fraction is consistently less than a factor of 2, whereas the uncertainty in the MTT can exceed a factor of 100.

Similar sensitivity of mean transit time to model assumptions was also observed by Kirchner et al. (2010) in two Scottish streams and by Seeger and Weiler (2014) in their study calibrating three different transit-time models to monthly $\delta^{18}O$ time series from 24 mesoscale Swiss catchments. The three transit-time models of Seeger and Weiler yielded MTT estimates that were often inconsistent by orders of magnitude but yielded much more consistent estimates of the fraction of water younger than 3 months, foreshadowing the sensitivity analysis presented here.

4.3 Young water estimation with nongamma distributions

Because both the young water fraction F_{yw} and the tracer cycle amplitude ratio A_S/A_P aggregate nearly linearly, the results shown in Fig. 11 will also approximately hold at higher levels of aggregation. That is, we can merge each catchment in Fig. 11, which has two tributaries, with another two-tributary catchment to form a four-tributary catchment, which we can merge with another four-tributary catchment to form an eight-tributary catchment, and so on. Figure 13 shows the outcome of this thought experiment. One can see that just like in the two-tributary case, the tracer cycle am-

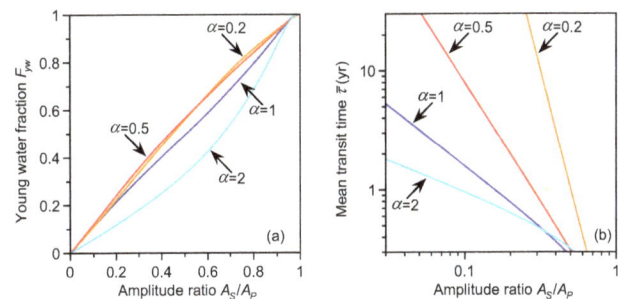

Figure 12. Sensitivity analysis showing how variations in shape factor α affect young water fractions F_{yw} (**a**) and mean transit times $\bar{\tau}$ (**b**) inferred from the amplitude ratio A_S/A_P of seasonal tracer cycles in precipitation and streamflow. Curves are shown for the four shape factors shown in Figs. 2 and 3. For a plausible range of uncertainty in the shape factor ($0.5 < \alpha < 1$; see Sect. 2.1), estimated young water fractions vary by a few percent (**a**), whereas estimated mean transit times vary by large multiples (note the logarithmic axes in **b**). (**a**) shows the fractions of water younger than $\tau_{yw} = 2.27$ months, which are closely approximated by A_S/A_P if $\alpha = 1$ (the dark blue curve). In (**b**), the axis scales are chosen to span transit times ranging from several months to several years, as is commonly observed in transit-time studies (McGuire and McDonnell, 2006).

plitude ratio A_S/A_P in the merged runoff predicts the average young water fraction F_{yw} with relatively little scatter. There is a slight underestimation bias, which is more visible in Fig. 13 than for the two-tributary case in Fig. 11. In contrast to the minimal estimation bias in F_{yw}, MTT is underestimated by large factors in both the two-tributary case and the eight-tributary case.

It is important to recognize that the two-tributary catchments that were merged in Fig. 13 are not characterized by gamma transit-time distributions (although their tributaries are), because mixing two gamma distributions does not cre-

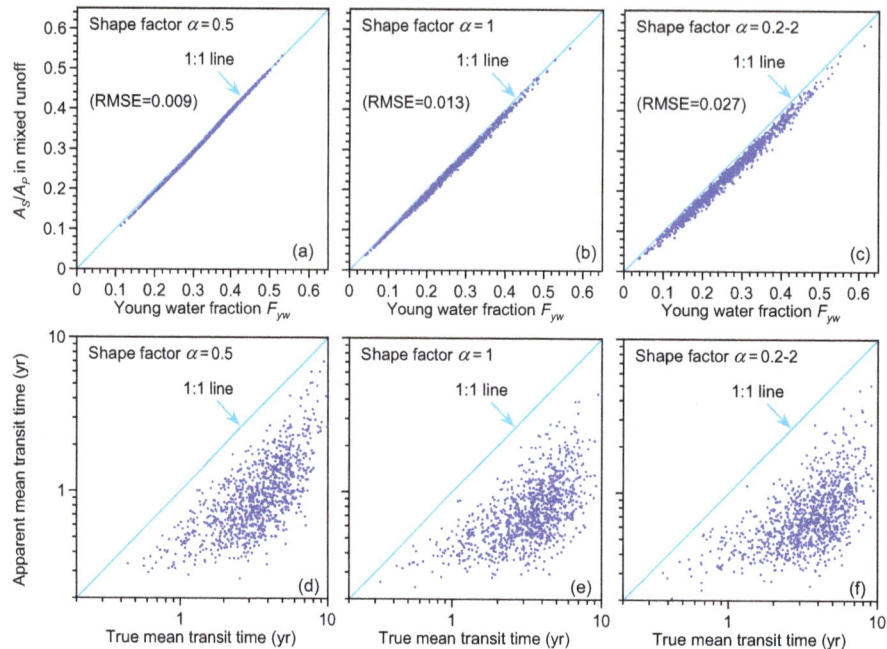

Figure 13. True and apparent young water fractions F_{yw} for 1000 synthetic catchments, each consisting of eight subcatchments with randomly chosen mean transit times between 0.1 and 20 years (top panels) and true and apparent mean transit times for the same catchments (bottom panels). The tracer cycle amplitude ratio in the combined runoff predicts the true young water fraction with a slight underestimation bias (top panels). Mean transit times inferred from tracer cycle amplitude ratios show severe underestimation bias (bottom panels). In **(a)**, **(b)**, **(d)**, and **(e)**, all subcatchments have the same shape factor α. In **(c)** and **(f)**, shape factors for each subcatchment are randomly chosen from a uniform distribution between $\alpha = 0.2$ and $\alpha = 2$.

ate another gamma distribution. Thus, Fig. 13 demonstrates the important result that although the analysis presented here was based on gamma distributions for mathematical convenience, the general principles developed here – namely, that the amplitude ratio A_S/A_P estimates the young water fraction F_{yw}, and that estimates of F_{yw} are relatively immune to aggregation bias in heterogeneous catchments – are not limited to distributions within the gamma family.

For example, as Fig. 6 showed, mixing two exponential distributions will not create another exponential distribution, nor any other member of the gamma family but rather a hyperexponential distribution. Thus, Fig. 13b implies that A_S/A_P also estimates F_{yw} accurately for mixtures of exponentials, that is, for any distribution of the form

$$h(\tau) = \frac{1}{\sum\limits_{i=1}^{n} k_i} \sum_{i=1}^{n} \frac{k_i}{\overline{\tau}_i} e^{-\tau/\overline{\tau}_i}, \tag{15}$$

where the weights k_i and mean transit times $\overline{\tau}_i$ can take on any positive real values. Likewise, Fig. 13c implies that A_S/A_P estimates F_{yw} reasonably accurately for mixtures of gamma distributions, that is, for any distribution of the form

$$h(\tau) = \frac{1}{\sum\limits_{i=1}^{n} k_i} \sum_{i=1}^{n} \frac{k_i \tau^{\alpha_i - 1}}{(\overline{\tau}_i/\alpha_i)^{\alpha_i} \Gamma(\alpha_i)} e^{-\alpha_i \tau/\overline{\tau}_i}, \tag{16}$$

where, as before, the weights k_i and mean transit times $\overline{\tau}_i$ can take on any positive real values, and the shape factors α_i can take on any values between 0.2 and 2. In the continuum limit, n could potentially be infinite in Eq. (15) or (16), whereupon the summations become integrals. Equations (15) and (16) describe very broad classes of distributions, suggesting that the results reported here also apply to a very wide range of catchment transit-time distributions, well beyond the (already broad) family of gamma distributions with shape factors $\alpha < 2$.

4.4 Incorporating phase information in estimating young water fractions and mean transit times

One interpretation of the strong aggregation bias in mean transit-time estimates, as documented in Figs. 7 and 13, is that when the transit-time distributions of the individual tributaries are averaged together, the result has a different shape (i.e., averages of exponentials are not exponentials and averages of gamma distributions are not gamma-distributed). Thus, it is unsurprising that a formula for estimating mean travel times based on exponential distributions (for example) will be inaccurate when applied to nonexponential distributions. The practical issue in the real world, of course, is that the shape of the transit-time distribution will usually be un-

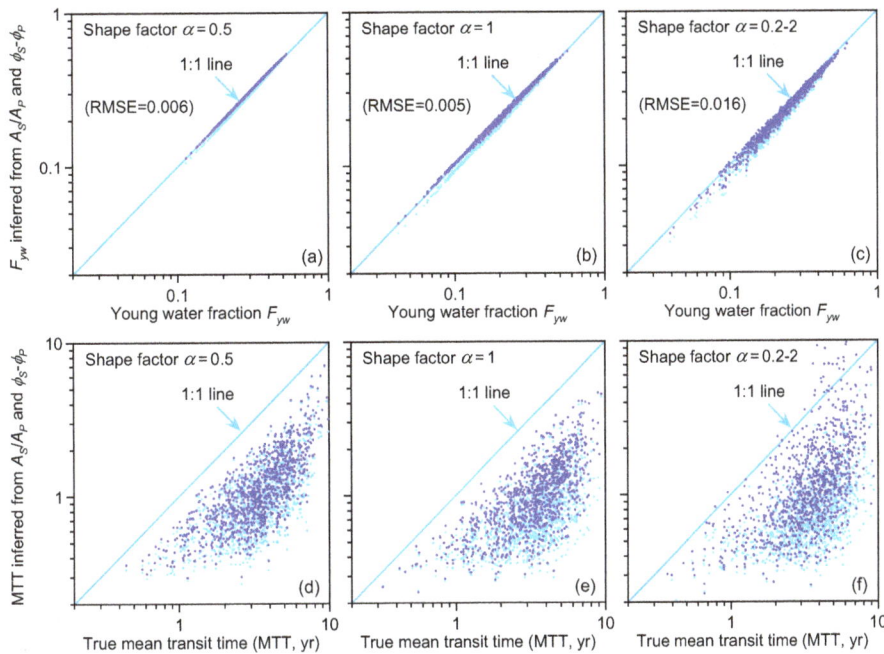

Figure 14. Effect of including phase information in estimates of young water fraction (F_{yw}) and MTT. Light symbols show F_{yw} and MTT estimates derived from tracer cycle amplitude ratios (A_S/A_P) alone; dark symbols show the same estimates derived from amplitude ratios and phase shifts ($\varphi_S - \varphi_P$). Data points come from the same 1000 synthetic catchments shown in Fig. 13, each consisting of eight subcatchments with randomly chosen mean transit times between 0.1 and 20 years. Adding phase shift information eliminates much of the (already small) bias in F_{yw} estimates, particularly when F_{yw} is small. Adding phase information reduces the bias in MTT estimates as well, but a severe underestimation bias remains.

known, so the problem of fitting the "wrong" distribution will be difficult to solve.

In the specific case of fitting seasonal sinusoidal patterns, the only information one has for estimating the transit-time distribution is the amplitude ratio and the phase shift of streamflow relative to precipitation. The phase shift has heretofore been ignored as a source of additional information. Could it be helpful?

As described in Sect. 2.2, one can use the amplitude ratio and phase shift to jointly estimate the shape factor α by iteratively solving Eq. 11 and then estimating the scale factor β via Eq. (10). The mean transit time can then be estimated as $\alpha\beta$ (Eq. 3a). From the fitted value of α, one can also use Eq. (14) to estimate the threshold age τ_{yw} for young water fractions that should aggregate nearly linearly and then, finally, estimate the young water fraction as $F_{yw} = \Gamma(\tau_{yw}, \alpha, \beta)$ (Eq. 13). The lower incomplete gamma function $\Gamma(\tau_{yw}, \alpha, \beta)$ is readily available in many software packages (for example, the igamma function in R or the GAMMA.DIST function in Microsoft Excel).

This approach assumes that the catchment's transit times are gamma-distributed. To test whether it can nonetheless improve estimates of the mean transit time or the young water fraction, even in catchments whose transit times are not gamma-distributed, I applied this method to the eight-tributary synthetic catchments shown in Fig. 13. As pointed out in Sect. 4.3, the TTDs of these catchments (and even their two-subcatchment tributaries) will be sums of gammas and thus not gamma-distributed themselves. Figure 14 shows the new estimates based on amplitude ratios and phase shifts (in dark blue), superimposed on the previous estimates from Fig. 13, based on amplitude ratios alone, as reference (in light blue). Mean transit-time estimates based on both phase and amplitude information are somewhat more accurate than those based on amplitude ratios alone (Fig. 14d–f), but they still exhibit very large aggregation bias. Incorporating phase information in estimates of F_{yw} (Fig. 14a–c) eliminates much of the (already small) bias in F_{yw} estimates obtained from amplitude ratios alone. (The logarithmic axes of Fig. 14a–c make this bias more visible than it is on the linear axes of Fig. 13a–c.) The top and bottom rows of Fig. 14 are plotted on consistent axes (both are logarithmic scales spanning a factor of 50), so they provide a direct visual comparison of the reliability of estimates of F_{yw} and MTT.

5 Implications

Two main results emerge from the analysis presented above. First, MTTs estimated from seasonal tracer cycles exhibit severe aggregation bias in heterogeneous catchments, underestimating the true MTT by large factors. Second, seasonal tracer cycle amplitudes accurately reflect the fraction of young water in streamflow and exhibit very little aggregation bias. Both of these results have important implications for catchment hydrology.

5.1 Biases in mean transit times

Figures 7, 13, and 14 indicate that in spatially heterogeneous catchments (which is to say, all real-world catchments), MTTs estimated from seasonal tracer cycles are fundamentally unreliable. The relationship between true and inferred MTTs shown in these figures is not only strongly biased, but also wildly scattered – so much so, that it can only be visualized on logarithmic axes. The huge scatter in the relationship means that there is little point in trying to correct the bias with a calibration curve, because most of the resulting estimates would still be wrong by large factors. This scatter also implies that one should be careful about drawing inferences from site-to-site comparisons of MTT values derived from seasonal cycles, since a large part of their variability may be aggregation noise.

The underestimation bias in MTT estimates arises because, as Figs. 3a and 5 show, travel times significantly shorter than 1 year have a much bigger effect on seasonal tracer cycles than travel times of roughly 1 year and longer. DeWalle et al. (1997) calculated that an exponential TTD with a MTT of 5 years would result in such a small isotopic cycle in streamflow that it would approach the analytical detection limit of isotope measurements. But while this may be the hypothetical upper limit to MTTs determined from seasonal isotope cycles, my results show that even MTTs far below that limit cannot be reliably estimated in heterogeneous landscapes. Indeed, Fig. 7 shows that MTTs can only be reliably estimated (that is, they will fall close to the 1 : 1 line) in heterogeneous systems where the MTT is roughly 0.2 years or so – in other words, only when most of the streamflow is "young" water.

It is becoming widely recognized that stable isotopes are effectively blind to the long tails of travel-time distributions (Stewart et al., 2010, 2012; Seeger and Weiler, 2014). The results presented here reinforce this point, showing how in heterogeneous catchments any stable isotope cycles from long-MTT subcatchments (or flowpaths) will be overwhelmed by much larger cycles from short-MTT subcatchments (or flowpaths). Furthermore, the nonlinearities in the governing equations (Figs. 3, 5) imply that the shorter-MTT components will dominate MTT estimates, which will thus be biased low. This underestimation bias may help to explain the discrepancy between MTT estimates derived from stable isotopes and those derived from other tracers, such as tritium (Stewart et al., 2010, 2012). However, one should note that, like any radioactive tracer, tritium ages should themselves be vulnerable to underestimation bias in heterogeneous systems (Bethke and Johnson, 2008). Until tritium ages are subjected to benchmark tests like those I have presented here for sta-

ble isotopes, one cannot estimate how much they, too, are distorted by aggregation bias.

5.2 Other methods for estimating MTTs from tracers

Sine-wave fitting to seasonal tracer cycles is just one of several methods for estimating MTTs from tracer data. I have focused on this method because the relevant calculations are easily posed, and several key results can be obtained analytically. My results show that MTT estimates from sine-wave fitting are subject to severe aggregation bias, but they do not show whether other methods are better or worse in this regard. This is unknown at present and needs to be tested. But, until this is done, there is little basis for optimism that other methods will be immune to the biases identified here. One would expect that the results presented here should translate straightforwardly to spectral methods for estimating MTTs, as these methods essentially perform sine-wave fitting across a range of timescales. Thus, one should expect aggregation bias at each timescale. The upper limit of reliable MTT estimates should be expected to be a fraction of the longest observable cycles in the data (as it is for the annual cycles measured here). Thus, this upper limit will depend on the lengths of the tracer time series and also on whether they contain significant input and output variability on long wavelengths (longer records will not help, unless the tracer concentrations are actually variable on those longer timescales). The same principles are likely to apply to convolution modeling of tracer time series, due to the formal equivalence of the time and frequency domains under Fourier's theorem. Furthermore, to the extent that seasonal cycles are the dominant features of many natural tracer time series, convolution modeling of tracer time series may effectively be an elaborate form of sine-wave fitting, with all the attendant biases outlined here. Until these conjectures are tested, however, they will remain speculative. Given the severe aggregation bias identified here, there is an urgent need for benchmark testing of the other common methods for MTT estimation.

It should also be noted that methods for estimating MTTs assume not only homogeneity but also stationarity, and real-world catchments violate both of these assumptions. The results presented here suggest that nonstationarity (which is, very loosely speaking, heterogeneity in time) is likely to create its own aggregation bias, in addition to the spatial aggregation bias identified here. This aggregation bias can also be characterized using benchmark tests, as I show in a companion paper (Kirchner, 2016).

5.3 Implications for mechanistic interpretations of MTTs

The analysis presented here implies that many literature values of MTT are likely to be underestimated by large factors or, in other words, that typical catchment travel times are probably several times longer than we previously thought they were. This result sharpens the "rapid mobilization of old water" paradox: how do catchments store water for weeks or months, and then release it within minutes or hours in response to precipitation events (Kirchner, 2003)? This result also sharpens an even more basic puzzle: where can catchments store so much water, that it can be so old, on average?

Many studies have sought to link MTTs to catchment characteristics, often with inconsistent results. For example, McGuire et al. (2005) reported that MTT was positively correlated with the ratio of flowpath distance to average hillslope gradient at experimental catchments in Oregon, but Tetzlaff et al. (2009) reported that MTT was *negatively* correlated with the same ratio and positively correlated with the extent of hydrologically responsive soils at several Scottish catchments. Hrachowitz et al. (2009) reported that MTT was related to precipitation intensity, soil characteristics, drainage density, and topographic wetness index across a larger network of Scottish catchments, whereas Asano and Uchida (2012) reported that subsurface flowpath depth was the main control on baseflow MTT at their Japanese field sites. Heidbüchel et al. (2013) reported that MTT was correlated with soil depth, hydraulic conductivity, or planform curvature, with different characteristics becoming more important under different rainfall regimes. And, most recently, Seeger and Weiler (2014) reported that most of the observed correlations between MTT and terrain characteristics across 24 Swiss catchments became nonsignificant when the variation in mean annual discharge was taken into account. My analysis casts much of this literature in a different light. Given that a large component of MTT estimates in the literature may be aggregation noise (Figs. 7, 13, 14), one should not be surprised if MTT estimates exhibit weak and inconsistent correlations with catchment characteristics, even if those characteristics are important controls on real-world MTTs.

5.4 The young water fraction F_{yw} as an alternative travel-time metric

More generally, though, my analysis implies that the young water fraction F_{yw} is a more useful metric of catchment travel time than MTT is, for the simple reason that F_{yw} can be reliably determined in heterogeneous catchments but MTT cannot. Of course, if we know the young water fraction in runoff, we obviously also know the fraction of "old" water as well (meaning water older than the "young water" threshold). But we do not know – and my analysis implies that we generally *cannot* know – how old this "old" water is, at least from analyses of seasonal tracer cycles.

Of course, because F_{yw} is nearly equal to the amplitude ratio and because MTT can also be expressed as a function of the amplitude ratio for TTDs of any known shape, one might conclude that MTT and F_{yw} are just transforms of one another. But that conclusion presumes that the shape of the TTD is known, and my analysis shows that in heterogeneous catchments the shape of the TTD will be unpredictable. Because the MTT is sensitive to the shape of the TTD – and in particular to the long-time tail, which is particularly poorly constrained – it cannot be reliably estimated. By contrast, my analysis shows that despite the uncertainty in the shape of the TTD in heterogeneous catchments, the F_{yw} can be reliably estimated from the amplitude ratio of seasonal tracer cycles in precipitation and runoff. The fact that this is possible is neither a miracle nor a fortuitous accident; instead, F_{yw} has been defined with exactly this result in mind. The F_{yw} entails an unavoidable ambiguity in what, exactly, the threshold age of young water is (because this depends on the shape of the TTD, which is usually unknown), but this uncertainty is small (Fig. 10) compared to the very large uncertainty in the MTT.

It should be kept in mind that in real-world data, unlike the thought experiments analyzed here, the tracer measurements themselves will be somewhat uncertain, and this uncertainty will also flow through to estimates of either MTT or F_{yw}. In particular, although my analysis has focused on the effects of spatial heterogeneity in catchment properties (as reflected in the TTDs of the individual tributary subcatchments), it has ignored any spatial heterogeneity in the atmospheric inputs themselves. Furthermore, estimates of MTT or F_{yw} typically assume that any patterns in stream tracer concentrations arise only from the convolution of varying input concentrations and not, for example, from seasonal evapoconcentration effects (for chemical tracers) or evaporative fractionation (for isotopes). If this assumption is violated, the resulting structural errors are potentially much more consequential than random errors in tracer measurements.

5.5 Potential applications for young water fractions

Since young water fractions are estimated from amplitude ratios and phase shifts of seasonal tracer cycles, one could ask whether they add any new information or whether we could characterize catchments equally well by their amplitude ratios and phase shifts instead. One obvious answer is that amplitude ratios and phase shifts, by themselves, are purely phenomenological descriptions of input–output behavior. Young water fractions, by contrast, offer a mechanistic explanation for how that behavior arises, showing how it is linked to the fraction of precipitation that reaches the stream in much less than 1 year. Not only is this potentially useful for understanding the transport of contaminants and nutrients, it also directly quantifies the importance of relatively fast flowpaths in the catchment. These fast flowpaths are likely to be shallow (since permeability typically decreases rapidly with depth: Brooks et al., 2004; Bishop et al., 2011) and to originate relatively close to flowing channels. One would expect F_{yw} to increase under wetter conditions, as the water table rises into more permeable near-surface zones and as the flowing channel network extends to more finely dissect the landscape (Godsey and Kirchner, 2014), thus shortening the path length of subsurface flows as well as multiplying the wetted catchment area in riparian zones. In a companion paper (Kirchner, 2016), I show that young water fractions can be estimated separately for individual flow regimes, allowing one to infer how shifts in hydraulic forcing alter the fraction of streamflow that is generated via fast flowpaths. I further demonstrate how one can estimate the chemistry of "young water" and "old water" end-members, based on comparisons of F_{yw} and solute concentrations across different flow regimes.

Because one can estimate F_{yw} from irregularly and sparsely sampled tracer time series, it can be used to facilitate intercomparisons among many catchments that lack more detailed tracer data. For example, Jasechko et al. (2016) have recently used the approach outlined here to calculate young water fractions for hundreds of catchments around the globe, ranging from small research watersheds to continental-scale river basins, and to examine how they respond to variations in catchment characteristics.

One final note: it has not escaped my notice that because the young water threshold is defined as a fraction of the period of the fitted sinusoid (here, an annual cycle), and because spectral analysis is equivalent to fitting sinusoids across a range of timescales, the input and output spectra of conservative tracers can be re-expressed as a series of young water fractions for a series of young water thresholds. In principle, then, this cascade of young water fractions (and their associated threshold ages) should directly express the catchment's cumulative distribution of travel times, thus solving the long-standing problem of measuring the shape of the transit-time distribution. A proof-of-concept study of this direct approach to deconvolution is currently underway.

6 Summary and conclusions

I used benchmark tests with data from simple synthetic catchments (Fig. 4) to test how catchment heterogeneity affects estimates of mean transit times (MTTs) derived from seasonal tracer cycles in precipitation and streamflow (e.g., Fig. 1). The relationship between tracer cycle amplitude and MTT is strongly nonlinear (Fig. 3), with the result that tracer cycles from heterogeneous catchments will underestimate their average MTTs (Fig. 5). In heterogeneous catchments, furthermore, the shape of the transit-time distribution (TTD) in the mixed runoff will differ from that of the tributaries; e.g., mixtures of exponential distributions are not exponentials (Fig. 6) and mixtures of gamma distributions are not gamma-distributed. These two effects combine to make seasonal tracer cycles highly unreliable as estimators of MTTs, with large scatter and strong underestimation bias in heterogeneous catchments (Figs. 7, 13). These results imply that many literature values of MTT are likely to be underestimated by large factors and thus that typical catchment travel times are much longer than previously thought.

However, seasonal tracer cycles can be used to reliably estimate the *young water fraction* (F_{yw}) in runoff, defined as the fraction younger than approximately 0.15–0.25 years (i.e., ~ 2–3 months), depending on the shape of the underlying travel-time distribution (Figs. 9, 10). The amplitude ratio of seasonal tracer cycles in precipitation and runoff predicts F_{yw} with an accuracy of roughly 2 % or better, across the entire range of plausible TTD shape factors from $\alpha = 0.2$ to $\alpha = 2$ and across the entire range of mean transit times from nearly zero to near-infinity (Fig. 9). Most importantly, this relationship is virtually immune to aggregation bias, so the amplitude ratio reliably predicts the young water fraction in the combined runoff from heterogeneous landscapes, with little bias or scatter (Figs. 11, 13). Incorporating phase as well as amplitude information virtually eliminates the (already small) bias in F_{yw} estimates obtained from amplitude information alone (Fig. 14). Thus, my analysis not only reveals large aggregation errors in MTT, which have been widely used to characterize catchment transit time, it also proposes an alternative metric, F_{yw}, which should be reliable in heterogeneous catchments.

More generally, these results vividly illustrate how the pervasive heterogeneity of environmental systems can confound the simple conceptual models that are often used to analyze them. But not all properties of environmental systems are equally susceptible to aggregation error. Although environmental heterogeneity makes some measures (like MTT) highly unreliable, it has little effect on others (like F_{yw}). Benchmark tests are essential for determining which measures are highly susceptible to aggregation error and which are relatively immune. Thus, these results highlight the broader need for benchmark testing to diagnose aggregation errors in environmental measurements and models, beyond the specific illustrative case analyzed here.

Acknowledgements. This analysis was motivated by intensive discussions with Scott Jasechko and Jeff McDonnell; I thank them for their encouragement, and for many insightful comments. I also appreciate the comments by Markus Weiler and two anonymous reviewers, which spurred improvements in the final version of the manuscript.

References

Asano, Y. and Uchida, T.: Flow path depth is the main controller of mean base flow transit times in a mountainous catchment, Water Resour. Res., 48, W03512, doi:10.1029/2011wr010906, 2012.

Aubert, A. H., Kirchner, J. W., Gascuel-Odoux, C., Facheux, M., Gruau, G., and Merot, P.: Fractal water quality fluctuations spanning the periodic table in an intensively farmed watershed, Environ. Sci. Technol., 48, 930–937, doi:10.1021/es403723r, 2014.

Barnes, C. J. and Bonell, M.: Application of unit hydrograph techniques to solute transport in catchments, Hydrol. Process., 10, 793–802, 1996.

Bethke, C. M. and Johnson, T. M.: Groundwater age and groundwater age dating, Annu. Rev. Earth Planet. Sci., 36, 121–152, doi:10.1146/annurev.earth.36.031207.124210, 2008.

Beven, K.: On subsurface stormflow: predictions with simple kinematic theory for saturated and unsaturated flows, Water Resour. Res., 18, 1627–1633, 1982.

Birkel, C., Soulsby, C., Tetzlaff, D., Dunn, S., and Spezia, L.: High-frequency storm event isotope sampling reveals time-variant transit time distributions and influence of diurnal cycles, Hydrol. Process., 26, 308–316, doi:10.1002/hyp.8210, 2012.

Bishop, K., Seibert, J., Nyberg, L., and Rodhe, A.: Water storage in a till catchment II: Implications of transmissivity feedback for flow paths and turnover times, Hydrol. Process. 25, 3950–3959, doi:10.1002/hyp.8355, 2011.

Botter, G., Bertuzzo, E., and Rinaldo, A.: Transport in the hydrological response: Travel time distributions, soil moisture dynamics, and the old water paradox, Water Resour. Res., 46, W03514, doi:10.1029/2009WR008371, 2010.

Bowen, G. J.: Spatial analysis of the intra-annual variation of precipitation isotope ratios and its climatological corollaries, J. Geophys. Res.-Atmos., 113, D05113, doi:10.1029/2007jd009295, 2008.

Brooks, E. S., Boll, J., and McDaniel, P. A.: A hillslope-scale experiment to measure lateral saturated hydraulic conductivity, Water Resour. Res., 40, W04208, doi:10.1029/2003WR002858, 2004.

Broxton, P. D., Troch, P. A., and Lyon, S. W.: On the role of aspect to quantify water transit times in small mountainous catchments, Water Resour. Res., 45, W08427, doi:10.1029/2008wr007438, 2009.

Burns, D. A., Plummer, L. N., McDonnell, J. J., Busenberg, E., Casile, G. C., Kendall, C., Hooper, R. P., Freer, J. E., Peters, N. E., Beven, K. J., and Schlosser, P.: The geochemical evolution of riparian ground water in a forested piedmont catchment, Ground Water, 41, 913–925, 2003.

DeWalle, D. R., Edwards, P. J., Swistock, B. R., Aravena, R., and Drimmie, R. J.: Seasonal isotope hydrology of three Appalachian forest catchments, Hydrol. Process., 11, 1895–1906, 1997.

Feng, X. H., Kirchner, J. W., and Neal, C.: Spectral analysis of chemical time series from long-term catchment monitoring studies: Hydrochemical insights and data requirements, Water Air Soil Poll. Focus, 4, 221–235, 2004.

Feng, X. H., Faiia, A. M., and Posmentier, E. S.: Seasonality of isotopes in precipitation: A global perspective, J. Geophys. Res.-Atmos., 114, D08116, doi:10.1029/2008jd011279, 2009.

Godsey, S. E., Kirchner, J. W., and Clow, D. W.: Concentration-discharge relationships reflect chemostatic characteristics of US catchments, Hydrol. Process., 23, 1844–1864, doi:10.1002/hyp.7315, 2009.

Godsey, S. E., Aas, W., Clair, T. A., de Wit, H. A., Fernandez, I. J., Kahl, J. S., Malcolm, I. A., Neal, C., Neal, M., Nelson, S. J., Norton, S. A., Palucis, M. C., Skjelkvåle, B. L., Soulsby, C., Tetzlaff, D., and Kirchner, J. W.: Generality of fractal $1/f$ scaling in catchment tracer time series, and its implications for catchment travel time distributions, Hydrol. Process., 24, 1660–1671, doi:10.1002/hyp.7677, 2010.

Godsey, S. E. and Kirchner, J. W.: Dynamic, discontinuous stream networks: hydrologically driven variations in active drainage density, flowing channels, and stream order, Hydrol. Process., 28, 5791–5803, doi:10.1002/hyp.10310, 2014.

Heidbüchel, I., Troch, P. A., Lyon, S. W., and Weiler, M.: The master transit time distribution of variable flow systems, Water Resour. Res., 48, W06520, doi:10.1029/2011WR011293, 2012.

Heidbüchel, I., Troch, P. A., and Lyon, S. W.: Separating physical and meteorological controls of variable transit times in zero-order catchments, Water Resour. Res., 49, 7644–7657, doi:10.1002/2012wr013149, 2013.

Hewlett, J. D. and Hibbert, A. R.: Factors affecting the response of small watersheds to precipitation in humid regions, in: Forest Hydrology, edited by: Sopper, W. E. and Lull, H. W., Pergamon Press, Oxford, 275–290, 1967.

Horton, J. H. and Hawkins, R. H.: the path of rain from the soil surface to the water table, Soil Science, 100, 377–383, 1965.

Hrachowitz, M., Soulsby, C., Tetzlaff, D., Dawson, J. J. C., and Malcolm, I. A.: Regionalization of transit time estimates in montane catchments by integrating landscape controls, Water Resour. Res., 45, W05421, doi:10.1029/2008wr007496, 2009.

Hrachowitz, M., Soulsby, C., Tetzlaff, D., Malcolm, I. A., and Schoups, G.: Gamma distribution models for transit time estimation in catchments: Physical interpretation of parameters and implications for time-variant transit time assessment, Water Resour. Res., 46, W10536, doi:10.1029/2010wr009148, 2010a.

Hrachowitz, M., Soulsby, C., Tetzlaff, D., and Speed, M.: Catchment transit times and landscape controls-does scale matter?, Hydrol. Process., 24, 117–125, doi:10.1002/hyp.7510, 2010b.

Hrachowitz, M., Soulsby, C., Tetzlaff, D., and Malcolm, I. A.: Sensitivity of mean transit time estimates to model conditioning and data availability, Hydrol. Process., 25, 980–990, doi:10.1002/hyp.7922, 2011.

Jasechko, S., Kirchner, J. W., Welker, J. M., and McDonnell, J. J.: Substantial proportion of global streamflow less than three months old, Nat. Geosci., in press, 2016.

Kirchner, J. W.: A double paradox in catchment hydrology and geochemistry, Hydrol. Process., 17, 871–874, 2003.

Kirchner, J. W.: Aggregation in environmental systems – Part 2: Catchment mean transit times and young water fractions under hydrologic nonstationarity, Hydrol. Earth Syst. Sci., 20, 299–328, doi:10.5194/hess-20-299-2016, 2016.

Kirchner, J. W. and Neal, C.: Universal fractal scaling in stream chemistry and its implications for solute transport and water quality trend detection, P. Natl. Acad. Sci. USA, 110, 12213–12218, doi:10.1073/pnas.1304328110, 2013.

Kirchner, J. W., Dillon, P. J., and LaZerte, B. D.: Predictability of geochemical buffering and runoff acidification in spatially heterogeneous catchments, Water Resour. Res., 29, 3891–3901, 1993.

Kirchner, J. W., Feng, X., and Neal, C.: Fractal stream chemistry and its implications for contaminant transport in catchments, Nature, 403, 524–527, 2000.

Kirchner, J. W., Feng, X., and Neal, C.: Catchment-scale advection and dispersion as a mechanism for fractal scaling in stream tracer concentrations, J. Hydrol., 254, 81–100, 2001.

Kirchner, J. W., Tetzlaff, D., and Soulsby, C.: Comparing chloride and water isotopes as hydrological tracers in two Scottish catchments, Hydrol. Process., 24, 1631–1645, doi:10.1002/hyp.7676, 2010.

Maher, K.: The dependence of chemical weathering rates on fluid residence time, Earth. Planet. Sc. Lett., 294, 101–110, 2010.

Maher, K. and Chamberlain, C. P.: Hydrologic regulation of chemical weathering and the geologic carbon cycle, Science, 343, 1502–1504, 2014.

Maloszewski, P. and Zuber, A.: Principles and practice of calibration and validation of mathematical models for the interpretation of environmental tracer data in aquifers, Adv. Water Resour., 16, 173–190, 1993.

Maloszewski, P., Rauert, W., Stichler, W., and Herrmann, A.: Application of flow models in an alpine catchment area using tritium and deuterium data, J. Hydrol., 66, 319–330, 1983.

McDonnell, J. J. and Beven, K.: Debates-The future of hydrological sciences: A (common) path forward? A call to action aimed at understanding velocities, celerities and residence time distributions of the headwater hydrograph, Water Resour. Res., 50, 5342–5350, doi:10.1002/2013wr015141, 2014.

McDonnell, J. J., McGuire, K., Aggarwal, P., Beven, K. J., Biondi, D., Destouni, G., Dunn, S., James, A., Kirchner, J., Kraft, P., Lyon, S., Maloszewski, P., Newman, B., Pfister, L., Rinaldo, A., Rodhe, A., Sayama, T., Seibert, J., Solomon, K., Soulsby, C., Stewart, M., Tetzlaff, D., Tobin, C., Troch, P., Weiler, M., Western, A., Worman, A., and Wrede, S.: How old is streamwater? Open questions in catchment transit time conceptualization, modelling and analysis, Hydrol. Process., 24, 1745–1754, doi:10.1002/hyp.7796, 2010.

McGuire, K. J. and McDonnell, J. J.: A review and evaluation of catchment transit time modeling, J. Hydrol., 330, 543–563, 2006.

McGuire, K. J., McDonnell, J. J., Weiler, M., Kendall, C., McGlynn, B. L., Welker, J. M., and Seibert, J.: The role of topography on catchment-scale water residence time, Water Resour. Res., 41, W05002, doi:10.1029/2004WR003657, 2005.

Nash, J. E.: The form of the instantaneous unit hydrograph, Comptes Rendus et Rapports, IASH General Assembly Toronto 1957, Publ. No. 45, Int. Assoc. Sci. Hydrol. (Gentbrugge), 3, 114–121, 1957.

Niemi, A. J.: Residence time distributions of variable flow processes, Int. J. Appl. Radiat. Isotop., 28, 855–860, 1977.

Peters, N. E., Burns, D. A., and Aulenbach, B. T.: Evaluation of high-frequency mean streamwater transit-time estimates using groundwater age and dissolved silica concentrations in a small forested watershed, Aquat. Geochem., 20, 183–202, 2014.

Rastetter, E. B., King, A. W., Cosby, B. J., Hornberger, G. M., O'Neill, R. V., and Hobbie, J. E.: Aggregating fine-scale ecological knowledge to model coarser-scale attributes of ecosystems, Ecol. Appl., 2, 55–70, 1992.

Seeger, S. and Weiler, M.: Reevaluation of transit time distributions, mean transit times and their relation to catchment topography, Hydrol. Earth Syst. Sci., 18, 4751–4771, doi:10.5194/hess-18-4751-2014, 2014.

Soulsby, C., Tetzlaff, D., Rodgers, P., Dunn, S., and Waldron, A.: Runoff processes, stream water residence times and controlling landscape characteristics in a mesoscale catchment: an initial evaluation, J. Hydrol., 325, 197–221, 2006.

Stewart, M. K., Morgenstern, U., and McDonnell, J. J.: Truncation of stream residence time: how the use of stable isotopes has skewed our concept of streamwater age and origin, Hydrol. Process., 24, 1646–1659, doi:10.1002/hyp.7576, 2010.

Stewart, M. K., Morgenstern, U., McDonnell, J. J., and Pfister, L.: The 'hidden streamflow' challenge in catchment hydrology: a call to action for stream water transit time analysis, Hydrol. Process., 26, 2061–2066, doi:10.1002/hyp.9262, 2012.

Tetzlaff, D., Malcolm, I. A., and Soulsby, C.: Influence of forestry, environmental change and climatic variability on the hydrology, hydrochemistry and residence times of upland catchments, J. Hydrol., 346, 93–111, 2007.

Tetzlaff, D., Seibert, J., and Soulsby, C.: Inter-catchment comparison to assess the influence of topography and soils on catchment transit times in a geomorphic province; the Cairngorm mountains, Scotland, Hydrol. Process., 23, 1874–1886, doi:10.1002/hyp.7318, 2009.

Van der Velde, Y., De Rooij, G. H., Rozemeijer, J. C., van Geer, F. C., and Broers, H. P.: The nitrate response of a lowland catchment: on the relation between stream concentration and travel time distribution dynamics, Water Resour. Res., 46, W11534, doi:10.1029/2010WR009105, 2010.

Walck, C.: Handbook on statistical distributions for experimentalists, Particle Physics Group, University of Stockholm, Stockholm, 202 pp., 2007.

8

Aggregation in environmental systems – Part 2: Catchment mean transit times and young water fractions under hydrologic nonstationarity

J. W. Kirchner[1,2]

[1]ETH Zürich, Zurich, Switzerland
[2]Swiss Federal Research Institute WSL, Birmensdorf, Switzerland

Correspondence to: J. W. Kirchner (kirchner@ethz.ch)

Abstract. Methods for estimating mean transit times from chemical or isotopic tracers (such as Cl^-, $\delta^{18}O$, or δ^2H) commonly assume that catchments are stationary (i.e., time-invariant) and homogeneous. Real catchments are neither. In a companion paper, I showed that catchment mean transit times estimated from seasonal tracer cycles are highly vulnerable to aggregation error, exhibiting strong bias and large scatter in spatially heterogeneous catchments. I proposed the young water fraction, which is virtually immune to aggregation error under spatial heterogeneity, as a better measure of transit times. Here I extend this analysis by exploring how nonstationarity affects mean transit times and young water fractions estimated from seasonal tracer cycles, using benchmark tests based on a simple two-box model. The model exhibits complex nonstationary behavior, with striking volatility in tracer concentrations, young water fractions, and mean transit times, driven by rapid shifts in the mixing ratios of fluxes from the upper and lower boxes. The transit-time distribution in streamflow becomes increasingly skewed at higher discharges, with marked increases in the young water fraction and decreases in the mean water age, reflecting the increased dominance of the upper box at higher flows. This simple two-box model exhibits strong equifinality, which can be partly resolved by simple parameter transformations. However, transit times are primarily determined by residual storage, which cannot be constrained through hydrograph calibration and must instead be estimated by tracer behavior.

Seasonal tracer cycles in the two-box model are very poor predictors of mean transit times, with typical errors of several hundred percent. However, the same tracer cycles predict time-averaged young water fractions (F_{yw}) within a few percent, even in model catchments that are both nonstationary and spatially heterogeneous (although they may be biased by roughly 0.1–0.2 at sites where strong precipitation seasonality is correlated with precipitation tracer concentrations). Flow-weighted fits to the seasonal tracer cycles accurately predict the flow-weighted average F_{yw} in streamflow, while unweighted fits to the seasonal tracer cycles accurately predict the unweighted average F_{yw}. Young water fractions can also be estimated separately for individual flow regimes, again with a precision of a few percent, allowing direct determination of how shifts in a catchment's hydraulic regime alter the fraction of water reaching the stream by fast flow-paths. One can also estimate the chemical composition of idealized "young water" and "old water" end-members, using relationships between young water fractions and solute concentrations across different flow regimes. These results demonstrate that mean transit times cannot be estimated reliably from seasonal tracer cycles and that, by contrast, the young water fraction is a robust and useful metric of transit times, even in catchments that exhibit strong nonstationarity and heterogeneity.

1 Introduction

In a companion paper (Kirchner, 2016, hereafter referred to as Paper 1), I pointed out that although catchments are pervasively heterogeneous, we often model them, and interpret measurements from them, as if they were homogeneous. This makes our measurements and models vulnerable to so-called "aggregation error", meaning that they yield inconsistent results at different levels of aggregation. I illustrated this general problem with the specific example of mean transit times (MTTs) estimated from seasonal tracer cycles in precipitation and discharge. Using simple numerical experiments with synthetic data, I showed that these MTT estimates will typically exhibit strong bias and large scatter when they are derived from spatially heterogeneous catchments. Given that spatial heterogeneity is ubiquitous in real-world catchments, these findings pose a fundamental challenge to the use of MTTs to characterize catchment behavior.

In Paper 1 I also showed that seasonal tracer cycles in precipitation and streamflow can be used to estimate the young water fraction F_{yw}, defined as the fraction of discharge that is younger than a threshold age of approximately 2–3 months. I further showed that F_{yw} estimates, unlike MTT estimates, are robust against extreme spatial heterogeneity. Thus, Paper 1 demonstrates the feasibility of determining the proportions of "young" and "old" water (F_{yw} and $1 - F_{yw}$, respectively) in spatially heterogeneous catchments.

But real-world catchments are not only heterogeneous. They are also nonstationary: their travel-time distributions shift with changes in their flow regimes, due to shifts in the relative water fluxes and flow speeds of different flowpaths (e.g., Kirchner et al., 2001; Tetzlaff et al., 2007; Hrachowitz et al., 2010; Botter et al., 2010; Van der Velde et al., 2010; Birkel et al., 2012; Heidbüchel et al., 2012; Peters et al., 2014). This nonstationarity is more than simply a time-domain analogue to the heterogeneity problem explored in Paper 1, because variations in flow regime may alter both the transit-time distributions of individual flowpaths and the mixing ratios between them. Intuition suggests that catchment nonstationarity could play havoc with estimates of MTTs, and perhaps also with estimates of the young water fraction.

This paper explores three central questions. First, does nonstationarity lead to aggregation errors in MTT and thus to bias or scatter in MTT estimates derived from seasonal tracer cycles? Second, is the young water fraction F_{yw} also vulnerable to aggregation errors under nonstationarity or is it relatively immune, like it is to aggregation errors arising from spatial heterogeneity? Third, can either MTT or F_{yw} be estimated reliably from seasonal tracer cycles, in catchments that are both nonstationary and heterogeneous, as real catchments are?

In keeping with the spirit of the approach developed in Paper 1, here I explore the consequences of catchment nonstationarity through simple thought experiments. These thought experiments are based on a simple two-compartment conceptual model (Fig. 1). This model greatly simplifies the complexities of real-world catchments, but it is sufficient to illustrate the key issues at hand. It is not intended to simulate the behavior of a specific real-world catchment, and thus its

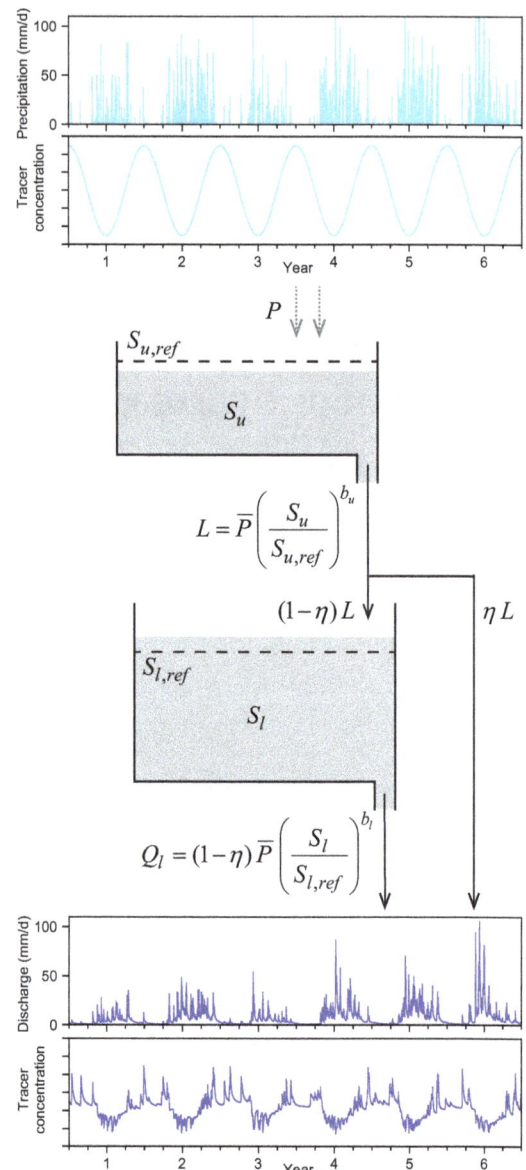

Figure 1. Schematic diagram of conceptual model. Drainage from the upper and lower boxes is determined by power functions of the storage volumes S_u and S_l (depicted by gray, shaded regions) as ratios of the reference storage levels $S_{u,ref}$ and $S_{l,ref}$ (depicted by dashed lines). The partition coefficient splits the upper box drainage L into direct discharge and infiltration to the lower box.

"goodness of fit" to any particular catchment time series is unimportant. Instead, its purpose is to simulate how nonstationary dynamics may influence tracer concentrations across wide ranges of catchment behavior and thus to serve as a numerical "test bed" for exploring how catchment nonstationarity affects our ability to infer catchment transit times from tracer concentrations. One can of course construct more complicated and (perhaps) realistic models, but that is not the point here. The point here is to explore the consequences of catchment nonstationarity, in the context of one of the simplest possible models which nonetheless exhibits a wide range of nonstationary behaviors.

2 A simple conceptual model for exploring nonstationarity

2.1 Structure and basic equations

The model catchment consists of two compartments, an upper box and a lower box (Fig. 1). In typical conceptual models the upper box might represent soil water storage and the lower box might represent groundwater, but for the present purposes it is unnecessary to assign the two boxes to specific domains in the catchment. The upper box storage S_u is filled by precipitation P, and drains at a leakage rate L that is a power function of storage; for simplicity, evapotranspiration is ignored. Thus, storage in the upper box evolves according to

$$\frac{dS_u}{dt} = P - L = P - k_u S_u^{b_u}, \tag{1}$$

where the coefficient k_u and the exponent b_u are parameters. A third parameter $0 < \eta < 1$ partitions the leakage L from the upper box into an amount ηL that flows directly to discharge and an amount $(1 - \eta)L$ that recharges the lower box. The lower box storage S_l is recharged by leakage from the upper box and drains to streamflow at a discharge rate Q_l that is another power function of storage:

$$\frac{dS_l}{dt} = (1 - \eta)L - Q_l = (1 - \eta)L - k_l S_l^{b_l}, \tag{2}$$

where the coefficient k_l and the exponent b_l are the final two parameters. The stream discharge is the sum of the contributions from the upper and lower boxes, or

$$Q_S = \eta L + Q_l. \tag{3}$$

All storages are in millimeters of water equivalent depth, and all fluxes are in millimeters per day. The age distribution in each box is explicitly tracked at daily resolution for the youngest 90 days and by accounting for the aggregate "age mass" (Bethke and Johnson, 2008) of each box's water that is older than 90 days. The young water fraction F_{yw} is calculated as the fraction of water in each box that is up to (and including) 69 days old; this threshold age equals 0.189 years, which was shown in Paper 1 to be the theoretical young-water threshold age for seasonal cycles in systems with exponential transit-time distributions.

Discharge from both boxes is assumed to be non-age-selective, meaning that discharge is taken proportionally from each part of the age distribution; thus, the flow from each box will have the same tracer concentration, the same young water fraction F_{yw}, and the same mean age as the averages of those quantities in that box (at that moment in time).

Tracer concentrations and mean ages are tracked under the assumption that both boxes are each well-mixed but also separate from one another, so their tracer concentrations and water ages will differ. The tracer concentrations, young water fractions, and mean water ages in streamflow are the flux-weighted averages of the contributions from the two boxes.

The model is solved on a daily time step, using a weighted combination of the partly implicit trapezoidal method (for greater accuracy) and the fully implicit backward Euler method (for guaranteed stability). Details of the solution scheme are outlined in Appendix A.

2.2 Parameters and initialization

The drainage coefficients k_u and k_l are problematic as model parameters, because their values and dimensions are strongly dependent on the exponents b_u and b_l. Therefore, I instead parameterize the model drainage functions by the (dimensionless) exponents b_u and b_l and by the (dimensional) "reference" storage values $S_{u,ref}$ and $S_{l,ref}$. These reference values represent the storage levels at which the drainage rates of each box will equal their long-term average input rates. That is, $S_{u,ref}$ is the level of upper-box storage at which the leakage rate L equals the long-term average input rate $\overline{P}$. Likewise, $S_{l,ref}$ is the level of lower-box storage at which the discharge rate Q_l equals the average rate of recharge $(1 - \eta)\overline{L}$ (which, due to conservation of mass in the upper box, also equals $(1 - \eta)\overline{P}$). The drainage function coefficients are calculated from the reference storage values as follows:

$$\begin{aligned} k_u S_{u,ref}^{b_u} &= \overline{P}, & k_u &= \overline{P} S_{u,ref}^{-b_u}, \\ k_l S_{l,ref}^{b_l} &= (1 - \eta)\overline{P}, & k_l &= (1 - \eta)\overline{P} S_{l,ref}^{-b_l}. \end{aligned} \tag{4}$$

Expressing k_u and k_l in this way is equivalent to writing the drainage equations for the two boxes in dimensionless form, with the drainage rate expressed with reference to the long-term input rate as follows:

$$\frac{L}{\overline{P}} = \left(\frac{S_u}{S_{u,ref}}\right)^{b_u}, \tag{5}$$

$$\frac{Q_l}{(1 - \eta)\overline{P}} = \left(\frac{S_l}{S_{l,ref}}\right)^{b_l}. \tag{6}$$

One advantage of this approach is that, whereas the drainage coefficients k_u and k_l have no clear meaning and their numerical values and dimensions can vary wildly, the reference storage values are measured in millimeters of water equivalent depth, and their interpretation is straightforward. A further advantage of this approach is that it provides for varying degrees of residual storage without requiring any additional parameters to do so. Because $S_{u,ref}$ and $S_{l,ref}$ are the storage levels at which long-term mass balance is achieved, they represent the equilibria around which S_u and S_l will tend to fluctuate, with the range of those fluctuations largely determined by the variability in precipitation rates and by the stiffness of

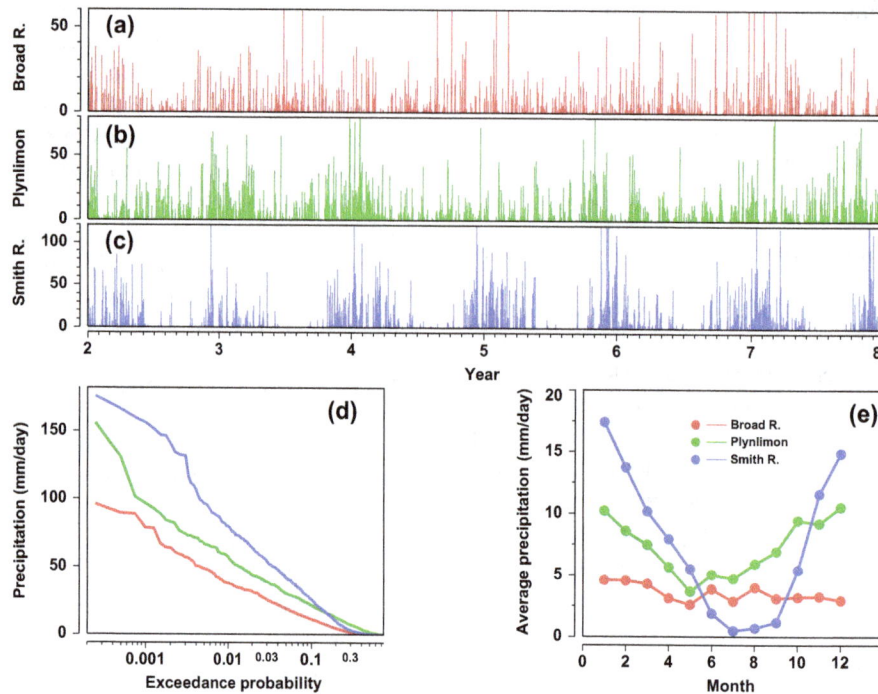

Figure 2. Excerpts of daily precipitation records used to drive the model: (**a**) Broad River, Georgia, USA (humid temperate climate; Köppen climate zone Cfa) in red, (**b**) Plynlimon, Wales (humid maritime climate; Köppen climate zone Cfb) in green, and (**c**) Smith River, California, USA (Mediterranean climate; Köppen climate zone Csb) in blue. Axes are expanded to make typical storms visible; thus, the largest storms, some of which extend to roughly twice the axis limits, are cut off. Exceedance probability (**d**) shows a steeper magnitude–frequency relationship for Smith River than for the other two records. Monthly precipitation averages (**e**) show clear differences in seasonality among the three sites.

the drainage functions, as specified by the exponents b_u and b_l (see Sect. 3.2).

The storages are initialized at the reference values $S_{u,ref}$ and $S_{l,ref}$. The tracer concentrations are initialized at equilibrium (that is, at the volume-weighted mean of the precipitation tracer concentration). Likewise, the mean ages in each box are initialized at their steady-state equilibrium values: $S_{u,ref}/\overline{P}$ in the upper box and $S_{u,ref}/\overline{P} + S_{l,ref}/[\overline{P}(1-\eta)]$ in the lower box. After a 1-year spin-up period, I run the model for 10 more years; the results for those 10 years are reported here.

2.3 Parameter ranges and precipitation drivers

Here I drive the model with three different real-world rainfall time series, representing a range of climatic regimes: a humid maritime climate with frequent rainfall and moderate seasonality (Plynlimon, Wales; Köppen climate zone Cfb), a Mediterranean climate marked by wet winters and very dry summers (Smith River, California, USA; Köppen climate zone Csb), and a humid temperate climate with very little seasonal variation in average rainfall (Broad River, Georgia, USA; Köppen climate zone Cfa). Figure 2 shows the contrasting frequency distributions and seasonalities of the three rainfall records. The Plynlimon rain gauge

data were provided by the Centre for Ecology and Hydrology (UK), and the Smith River and Broad River precipitation data are reanalysis products from the MOPEX (Model Parameter Estimation Experiment) project (Duan et al., 2006; ftp://hydrology.nws.noaa.gov/pub/gcip/mopex/US_Data/). The use of these real-world precipitation time series obviates the need to generate statistically realistic synthetic precipitation to drive the model.

The model used here shares a similar overall structure with many other conceptual models (e.g., Benettin et al., 2013), with several simplifications. However, although the model used here is typical in many respects, I will use it in an unusual way. Typically, one calibrates a model to reproduce the behavior of a real-world catchment and then draws inferences about that catchment from the parameters and behavior of the calibrated model. Here, however, the model is not intended to represent any particular real-world system. Instead, the model itself is the system under study, across wide ranges of parameter values, because the goal is to gain insight into how nonstationarity affects general patterns of tracer behavior. Thus, the fidelity of the model in representing any particular catchment is not a central issue.

For the simulations shown here, the drainage exponents b_u and b_l are randomly chosen from uniform distributions spanning the ranges of 1–20 and 1–50, respectively, the partitioning coefficient η is randomly chosen from a uniform distribution ranging from 0.1 to 0.9, and the reference storage levels $S_{u,ref}$ and $S_{l,ref}$ are randomly chosen from a uniform distribution of logarithms spanning the ranges of 20–500 and 500–10 000 mm, respectively. These parameter distributions are designed to encompass a wide range of possible behaviors, including both strong and damped response to rainfall inputs and small and large residual storage. To illustrate the behavior of the model for one concrete case, I use a "reference" parameter set with values taken from roughly the middle of each of these parameter distributions ($b_u = 10$, $b_l = 20$, $\eta = 0.5$, $S_{u,ref} = 100$ mm, and $S_{l,ref} = 2000$ mm). These parameter values are not "better" than any others in any particular sense; they are simply a point of reference (hence the name) for discussing the model's behavior.

3   Results and discussion

3.1   Nonstationarity in the two-box model

My main purpose is to use the simple two-box model to explore how catchment nonstationarity affects our ability to infer water ages from tracer time series. I will take up that issue beginning in Sect. 3.3. As background for that analysis, however, it is helpful to first characterize the nonstationary behavior of the simple model system.

Figure 3 shows excerpts from the time series generated by the model with the Smith River (Mediterranean climate) precipitation time series and the reference parameter set. One can immediately see that the upper and lower boxes have markedly different mean ages (Fig. 3e), young water fractions (Fig. 3d), and tracer concentrations (Fig. 3c), which also vary differently through time. Tracer concentrations in the upper box (the orange line in Fig. 3c) show a blocky, irregular pattern, remaining almost constant during periods of little rainfall, and then changing rapidly when the box is episodically flushed by large precipitation events. The lower box's tracer concentrations (the red line in Fig. 3c) are much more stable than the upper box's, because its mean residence time is roughly 40 times longer ($S_{l,ref}$ is 20 times $S_{u,ref}$, and with $\eta = 0.5$ the flux through the lower box is only half of the flux through the upper box). Because much more rain falls during the winters than the summers, the mean tracer concentration in the lower box is closer to the winter concentrations than the summer concentrations. During the wet winter season, rapid flushing keeps the young water fraction near 100 % in the upper box (the orange line in Fig. 3d) and can raise the young water fraction to 30–40 % in the lower box (the red line in Fig. 3d). Conversely, during the late summer the young water fraction in the upper box temporarily dips to 50 % or less, and the young water fraction in the lower box declines to nearly zero. The small volume in the upper box means that its water age (the orange line in Fig. 3e) is only a small fraction of a year. The mean water age in the lower box

(the red line in Fig. 3e) is much older and exhibits both seasonal variation and inter-annual drift, reflecting year-to-year variations in total precipitation. Thus, the two components of this simple system have strongly contrasting characteristics and behavior. These internal states of any real-world system would not be observable, except as they are reflected in the volume and composition of streamflow.

In this regard, the most striking feature of Fig. 3 is the volatility of the tracer concentrations, young water fractions, and mean transit times in discharge (the dark blue lines in Fig. 3c–e), as the mixing ratio between the two boxes (Fig. 3b) shifts in response to precipitation events. This mixing ratio is not a simple function of discharge (Fig. 4c); instead it is both hysteretic and nonstationary, varying in response both to precipitation forcing and to the antecedent moisture status of the two boxes (and thus to the prior history of precipitation). This dependence on prior precipitation reflects the fact that the boxes typically retain their water age and tracer signatures over timescales much longer than the timescale of hydraulic response, because their residual storage is large compared to their dynamic storage (see Sect. 3.2). As a result, both the young water fraction and mean age of discharge and storage are widely scattered functions of discharge (Fig. 4a, b). Likewise, there is no simple relationship between either the young water fraction or mean age in storage and the corresponding quantities in discharge (Fig. 4d), although there is a strong overall bias toward water in discharge being much younger than the average water in storage.

Even though drainage from each box is non-age-selective (that is, the young water fraction and mean age in drainage from each box are identical to those in storage), this is emphatically not true at the level of the two-box system, because the two boxes account for different proportions of discharge than of storage. Furthermore, because the fractional contributions to streamflow from the (younger, smaller) upper box and the (older, larger) lower box are highly variable, the water age and young water fraction in discharge are not only strongly biased, but also highly scattered, indicators of the same quantities in storage (Fig. 4d).

The aggregate long-term implications of these dynamics are evident in the marginal (time-averaged) age distributions of storage and discharge (Fig. 5). From Fig. 5 it is immediately obvious that the age distributions in discharge are strongly skewed toward young ages, compared to the age distributions in storage, both for each box individually and for the catchment as a whole. This skew toward young ages arises for two main reasons. First, although drainage from each box is not age-selective, more outflow occurs during periods of stronger precipitation forcing and thus shorter residence times. Thus, the average ages of the outflow and the storage can differ greatly. Second, under high-flow conditions a larger proportion of discharge is derived from the upper box (which has a relatively short transit time), and at base flow more discharge is derived from the lower box (which

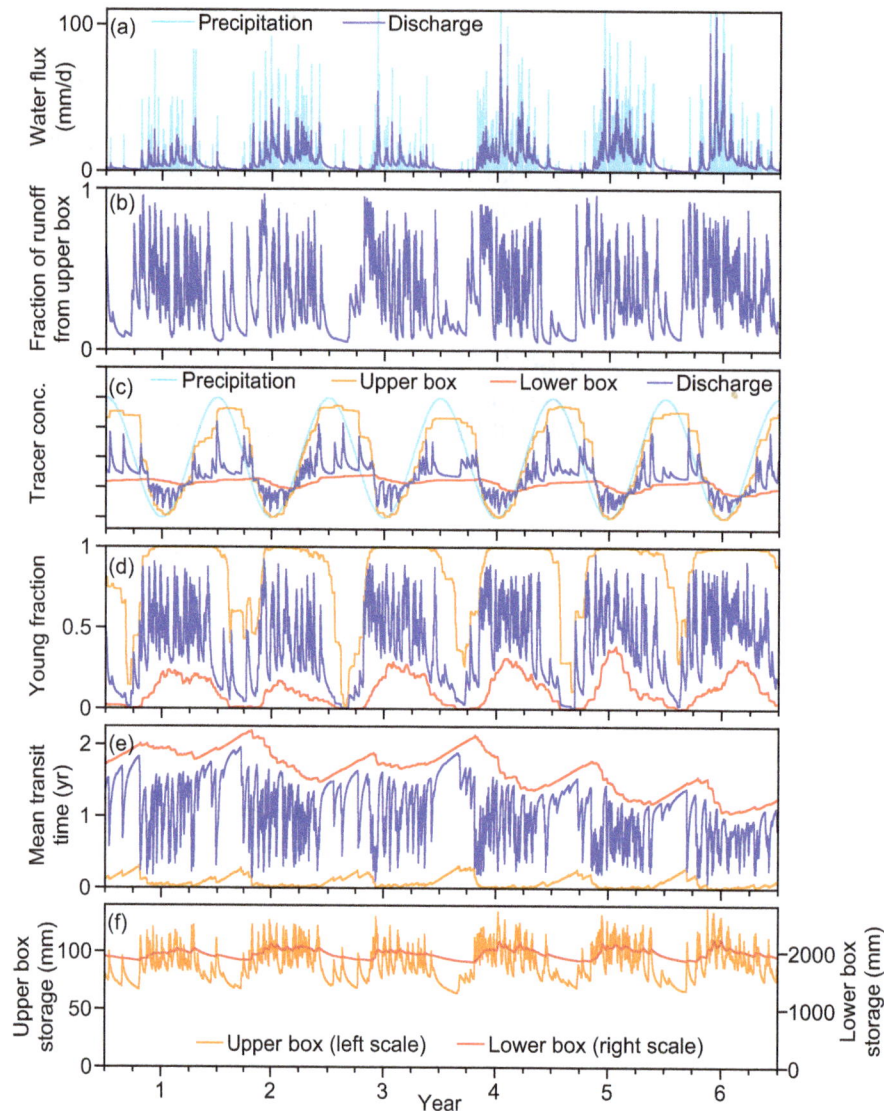

Figure 3. Illustrative time series from the two-box model, using the reference parameter set and the Smith River (Mediterranean climate) precipitation time series. Responses to precipitation events (**a**) entail rapid shifts in the proportions of discharge coming from the upper and lower boxes (**b**). The smaller, upper box, shown in orange, has a larger young water fraction (**d**) and a younger mean age (**e**) than the larger, lower box, shown in red, and thus its tracer concentration (**c**) is less lagged and damped relative to the hypothetical precipitation concentration, shown by the cosine wave in (**c**). Mean ages increase (**e**) and young water fractions decrease (**d**), in both boxes, throughout the dry summer periods. The proportions of streamflow originating from the upper and lower boxes shift dramatically in response to transient precipitation inputs; thus, the tracer concentrations, young water fractions, and mean ages in discharge (dark blue, **c**–**e**) vary widely between the time-varying end-members represented by the upper and lower boxes. Storage volumes fluctuate in a relatively narrow range (**f**) while discharge varies by orders of magnitude, because the drainage rates from both boxes are strongly nonlinear functions of storage. Thus, both boxes have sizeable residual storage, which is not drained even under extreme low-flow conditions.

has a larger volume and a relatively long transit time). Thus, the short-transit-time components of the system dominate the discharge, while the long-transit-time components of the system dominate the storage. As a result, the mean age in discharge will generally be much younger than the mean age in whole-catchment storage, and likewise the young water fraction in discharge will be much larger than the young water fraction in storage. Note that this is the opposite of what

one would expect from conceptual models like those of Botter (2012), in which the mean water age in discharge either equals the mean age in storage (for well-mixed systems) or is older than the mean age in storage (for piston-flow systems).

More generally, and more importantly, these results imply that estimates of water age in streamflow cannot be translated straightforwardly into estimates of water age in storage. Instead, they may underestimate the age of water in

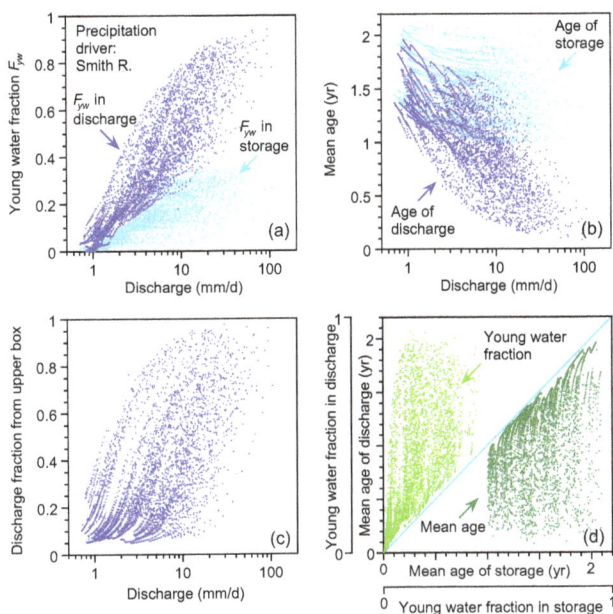

Figure 4. Daily values of young water fractions F_{yw} (**a**) and mean water ages (**b**) in storage (light blue) and discharge (dark blue) in the two-box model with reference parameter values and Smith River (Mediterranean climate) precipitation. The young water fraction and mean age are both highly scattered functions of discharge (**a, b**), as is the fractional contribution from the upper box to streamflow (**c**), reflecting the effects of variations in antecedent rainfall. The average age and F_{yw} of water in discharge are strongly biased, and highly scattered, measures of the same quantities in storage (**d**).

the steady-state approximation will be a misleading guide to the non-steady-state behavior of the system, *even on average*. That is, even over timescales where inputs equal outputs and the long-term average fluxes are essentially constant – and thus the steady-state approximation, on average, holds – the average behavior of the non-steady-state system can differ significantly from the average behavior of an equivalent steady-state system.

One can further explore these issues by examining the marginal (time-averaged) age distributions for separate ranges of discharge (Fig. 6). Figure 6 shows that at higher discharges, age distributions in streamflow are much more strongly skewed toward younger ages, reflecting the increased dominance of the upper box at higher flows. For the upper half of all discharges, the age distributions are more skewed than exponential; that is, they plot as upward-curving lines in Fig. 6b. For the top 25 % of discharges, water ages follow approximate power-law distributions, plotting as nearly straight lines in Fig. 6c. The slopes of these lines are steeper than 1, however, implying that the distributions must deviate from this trend at very short ages; otherwise their integrals (i.e., their cumulative distributions) would become infinite. It is important to note the mean ages quoted in Fig. 6a imply that the tails of the distributions all extend far beyond the plot axes, which are truncated at 90 days. Note also that the distributions shown in Fig. 6 have different shapes in different flow regimes, suggesting that the model's high-flow behavior is not simply a re-scaled transform of its low-flow behavior.

3.2 Residual storage and the disconnect between transit time and hydraulic response timescales

The model's complex, nonstationary water age and tracer dynamics arise from the disconnect between the timescales of hydraulic response and catchment storage in each box, and from the divergence in both these timescales between the two boxes. These contrasting timescales can be estimated through simple scaling and perturbation analyses, as outlined in this section.

Total catchment storage consists of two components: the dynamic storage that is linked to discharge fluctuations through storage–discharge relationships like Eqs. (6)–(7), plus the residual or "passive" storage that remains when discharge has declined to very slow rates. The range of dynamic storage exerts an important control on timescales of catchment hydrologic response, while the much larger residual (or "passive") storage has little effect on water fluxes but is an essential control on residence times (Kirchner, 2009; Birkel et al., 2011).

In real-world catchments, sharply nonlinear storage–discharge relationships (Kirchner, 2009) guarantee that dynamic storage will be small compared to residual storage. This behavior is mirrored in the model, where if Eqs. (6) and (7) are strongly nonlinear (i.e., if the drainage exponents b_u

storage by large factors, although in the particular example shown in Fig. 5, the difference is only about a factor of 2. Three closely related theoretical functions have recently been proposed to quantify the long-recognized (Kreft and Zuber, 1978) disconnect between the age distributions in storage and in discharge. These include the time-dependent StorAge Selection (SAS) function ω_Q of Botter et al. (2011), the Storage Outflow Probability (STOP) functions of Van der Velde et al. (2012), and the rank StorAge Selection (rSAS) function of Harman (2015). While these functions are all grounded in elaborate theoretical frameworks, it remains to be seen whether they can be reliably estimated in practice using real-world data.

A further implication of the analysis above is that the marginal age distributions are not exponential, even for individual boxes, and even though drainage from each box is not age-selective. In steady state, non-age-selective drainage (i.e., the well-mixed assumption) would yield an exponential distribution of ages in the upper box and in the short-time age distribution in streamflow. However, when the system is not in steady state and we aggregate its behavior over time, we are combining different age distributions from different moments in time with different precipitation forcing. This creates an aggregation error in the time domain, in the sense that

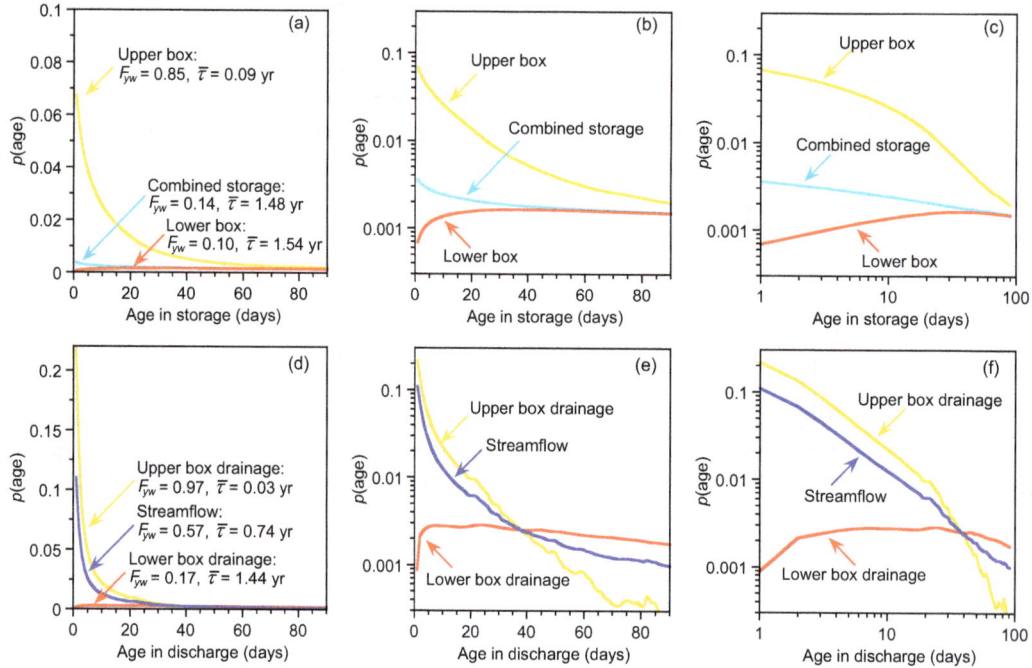

Figure 5. Marginal (time-averaged) age distributions in storage (**a–c**) and drainage (**d–f**) in the reference case simulation (Fig. 3), shown on linear (**a, d**), log-linear (**b, e**), and double-log (**c, f**) axes. Distributions in drainage (lower panels) are skewed toward younger ages than the storage distributions that they come from (upper panels). This arises, even though drainage is not age-selective, because storage is flushed more quickly (and thus is younger) during periods of higher discharge. Age distributions in the upper box, combined storage, and streamflow are more skewed than exponentials (i.e., they are upward-curving in the middle panels). The age distributions in the combined storage and streamflow (blue lines) are approximate power laws; i.e., they are nearly straight in the right-hand panels, with markedly different power-law slopes. The light blue line in the upper panels shows the age distribution of the combined upper and lower boxes, which resembles the age distribution of the lower box because the reference parameter values imply that the lower box comprises about 95 % of total storage. However, direct drainage from the upper box comprises 50 % of streamflow; thus, the streamflow age distribution (shown by the dark blue line in lower panels) reflects the strong skew of the upper box age distribution. Although both boxes are well mixed and have nearly constant volumes, the age distribution of discharge clearly differs from the distribution that would be expected in steady state, which would be exponential in the short-time limit.

and b_l are much greater than 1), the volumes in the upper and lower boxes will vary by only a small fraction of their reference storage values $S_{u,ref}$ and $S_{l,ref}$ (e.g., Fig. 3f). They will remain relatively constant because, when the drainage exponents b_u and b_l are large, the storage volumes cannot become much smaller than $S_{u,ref}$ and $S_{l,ref}$ without drainage rates falling to near zero (thus stopping further decreases in storage) and, conversely, the storage volumes also cannot become much larger than $S_{u,ref}$ and $S_{l,ref}$ without drainage rates becoming very high (thus stopping further increases in storage). Thus, $S_{u,ref}$ and $S_{l,ref}$ will be good approximations to the residual storage volume, whenever the drainage exponents are much greater than 1.

One can express this concept more quantitatively (though only approximately) using a simple perturbation analysis. A first-order Taylor expansion of Eqs. (6) and (7) shows directly that the fractional variability in drainage rates and storage are related by the drainage exponents in the two boxes:

$$\frac{\Delta L}{\overline{P}} \approx b_u \frac{\Delta S_u}{S_{u,ref}}, \tag{7}$$

$$\frac{\Delta Q_l}{(1-\eta)\overline{P}} \approx b_l \frac{\Delta S_l}{S_{l,ref}}. \tag{8}$$

The variability in drainage rates from the upper and lower boxes, denoted as ΔL and ΔQ_l, will be controlled by the temporal variability in precipitation; thus, for a given precipitation climatology, the dynamic variability in storage (denoted as ΔS_u and ΔS_l) will scale according to the ratios $S_{u,ref}/b_u$ and $S_{l,ref}/b_l$. For example, when the model is driven by Smith River precipitation and uses the reference parameters (Fig. 3), the variability in discharge from the lower box, as measured by its standard deviation, is $3.7 \, \text{mm day}^{-1}$, nearly equal to the average lower box discharge of $3.8 \, \text{mm day}^{-1}$. Because the reference value of b_l is 20, Eq. (9) implies that the standard deviation of lower box storage should be approximately 1/20th of the reference storage $S_{l,ref}$, or roughly 100 mm. Consistent with this estimate, the actual standard deviation of S_l is 84 mm or about

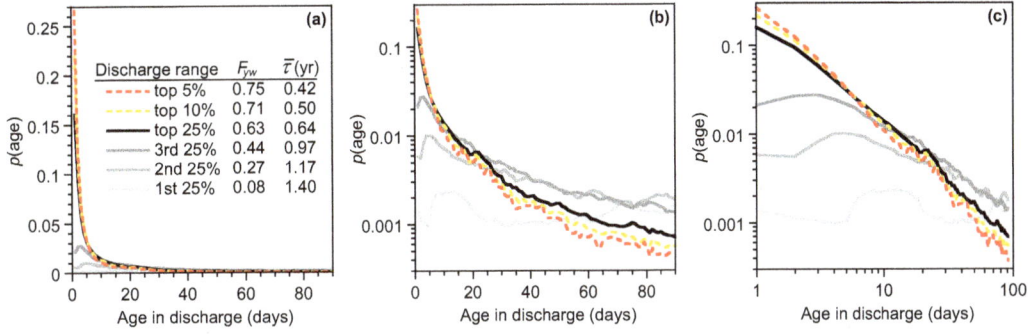

Figure 6. Marginal (time-averaged) transit-time distributions (TTDs) for selected ranges of daily discharges in the two-box model, with the reference parameter set and Smith River (Mediterranean climate) precipitation forcing, on linear **(a)**, log-linear **(b)**, and double-log **(c)** axes. The TTD becomes increasingly skewed at higher discharges (a), with a marked increase in the young water fraction F_{yw} and decrease in the mean water age $\overline{\tau}$. For the upper half of all discharges, the age distribution is upward-curving on log-linear axes **(b)**, implying that it is more skewed than exponential. Discharges in the top 25 % and above have approximately power-law age distributions, plotting as nearly straight lines on double-log axes **(c)**.

4 % of the total. Figure 3f shows that at least 90 % of $S_{l,ref}$ is residual storage that never drains during the 10-year simulation, roughly consistent with the perturbation analysis.

The perturbation analysis also yields estimates for the timescale of hydraulic response (which controls how "flashy" the discharge will be), through a rearrangement of Eqs. (8) and (9) as follows:

$$\frac{\Delta S_u}{\Delta L} \approx \frac{S_{u,ref}}{b_u \overline{P}} \quad \text{(hydraulic response timescale, upper box),} \quad (9)$$

$$\frac{\Delta S_l}{\Delta Q_l} \approx \frac{S_{l,ref}}{b_l (1-\eta)\overline{P}} \quad \text{(hydraulic response timescale, lower box).} \quad (10)$$

Again, using the reference parameter values and Smith River precipitation (for which $\overline{P}$ is roughly 7.6 mm day^{-1}), Eqs. (10) and (11) imply a hydraulic response time of roughly 1.3 days (for $b_u = 10$) in the upper box and of roughly 26 days (for $b_l = 20$) in the lower box. These timescales are factors b_u and b_l smaller than the steady-state mean transit times, which are determined by the ratios between the volumes and water fluxes,

$$\frac{S_{u,ref}}{\overline{P}} \quad \text{(steady-state mean transit time, upper box),} \quad (11)$$

$$\frac{S_{l,ref}}{(1-\eta)\overline{P}} \quad \text{(steady-state mean transit time, lower box).} \quad (12)$$

From Eqs. (12) and (13) one can also directly estimate the steady-state mean travel time in the combined discharge, as the weighted average of streamflow derived directly from the upper box, and water that flows through the upper and lower boxes in series,

$$\eta \frac{S_{u,ref}}{\overline{P}} + (1-\eta)\left(\frac{S_{u,ref}}{\overline{P}} + \frac{S_{l,ref}}{(1-\eta)\overline{P}}\right) = \frac{S_{u,ref} + S_{l,ref}}{\overline{P}}, \quad (13)$$

which is the expected result for any system at steady state: regardless of its internal configuration, the mean transit time

in any steady-state system will equal the ratio between its storage volume and its throughput rate. For the reference parameter set and Smith River precipitation, Eq. (14) becomes (100 mm + 2000 mm)/7.6 mm day^{-1}, or roughly 0.76 years, in good agreement with the whole-catchment mean transit time of 0.74 years determined from age tracking (see Fig. 5d). Note, however, that the *distribution* of these transit times will be markedly different from the exponential distribution that would be expected in steady state. This makes estimating mean transit times from tracer fluctuations difficult, as shown in Sect. 3.3.

Equations (12) and (13) imply that the mean transit times in the upper and lower boxes should be roughly 13 days (or 0.036 years) and 529 days (or 1.45 years), respectively, in good agreement with the mean transit times of 0.03 and 1.44 years determined from age tracking (Fig. 5d). However, Eqs. (10) and (11) imply that these transit times will differ by factors of 10 and 20 (the values of b_u and b_l, respectively) from the hydraulic response timescales that regulate catchment runoff response. The disconnect between hydraulic response times and mean transit times is the counterpart, in lumped conceptual models, to the disconnect between the velocity of water transport and the celerity of hydraulic head propagation in more realistic, physically extended systems (Beven, 1982; Kirchner et al., 2000; McDonnell and Beven, 2014). This contrast between hydraulic response times and mean transit times (or dynamic and total storage, or celerity and velocity) is a simple explanation for the apparent paradox of prompt discharge of old water during storm events (Kirchner, 2003).

3.3 Inferring MTT and F_{yw} from seasonal tracer cycles in nonstationary catchments

The analysis above shows that the simple two-box model gives hydrograph and tracer behavior that is complex and nonstationary (Figs. 3–6). Furthermore, even this simple five-parameter model exhibits strong equifinality (Appendix B). Much of this equifinality can be alleviated (compare Figs. B1 and B2) through parameter transformations based on the perturbation analysis outlined above. However, because the timescales of catchment storage and hydraulic response are controlled by different combinations of parameters, parameter calibration to the hydrograph cannot constrain the storage volumes or streamwater age (Figs. B2, B3). These model results demonstrate general principles that have been recognized for years: (a) the hydrograph responds to and, thus, can help to constrain dynamic storage but not passive storage; and (b) because passive storage is often large, timescales of hydrologic response and catchment water storage are decoupled from one another, such that water ages cannot be inferred from hydrograph dynamics. Thus, for understanding how catchments store and mix water, tracer data are essential.

But how should these tracer data be used? One approach is to explicitly include tracers in a catchment model and calibrate that model against both the hydrograph and the tracer chemograph (e.g., Birkel et al., 2011; Benettin et al., 2013; Hrachowitz et al., 2013). The usefulness of that approach depends on whether the model parameters can be constrained and, more importantly, whether the model structure adequately characterizes the system under study (which is usually unknown, and possibly unknowable). Except in multi-model studies, it will be unclear how much the conclusions depend on the particular model that was used and on the particular way that it was fitted to the data. Furthermore, adequate tracer data for calibrating such models are rare, particularly because dynamic models require input data with no gaps. The mismatch between model complexity and data availability means that, in some cases, all the data are used for calibration and validation must be skipped, leaving the reproducibility of the model results unclear (e.g., Benettin et al., 2015).

For all of these reasons, there will be an ongoing need for methods of inferring water ages that have modest data requirements and that are not dependent on specific model structures and parameters. Sine-wave fitting of seasonal tracer cycles, for example, is not based on a particular mechanistic model but, instead, is based on a broader conceptual framework in which stream output is some convolution of previous precipitation inputs. That premise is of course open to question but, nevertheless, seasonal tracer cycles (of, e.g., ^{18}O, ^2H, and Cl^-) have been widely used to estimate mean catchment transit times (see McGuire and McDonnell (2006) and references therein), largely because this particular method has modest data requirements. In particular, it does not need unbroken records of either precipitation inputs or streamflow outputs.

As detailed more fully in Paper 1, the seasonal tracer cycle method is based on the principle that when one convolves a sinusoidal tracer input with a TTD, one obtains a sinusoidal output that is damped and phase-lagged by an amount that depends on the shape of the TTD and also on its scale, as expressed, for example, by its MTT. Conventionally one assumes an exponential TTD, which is the steady-state solution for a well-mixed reservoir. More generally, one might assume that transit times are gamma-distributed, recognizing that the exponential distribution is a special case of the gamma distribution (with the shape factor α equal to 1). A sinusoidal tracer cycle that has been convolved with a gamma TTD will be damped and phase-lagged as described in Eqs. (8) and (9) of Paper 1. These equations can then be inverted to infer the shape and scale of the TTD from the seasonal tracer cycles in precipitation and streamflow.

The procedure is as follows. One first measures the amplitudes and phases of the seasonal tracer cycles in precipitation and streamflow using Eqs. (4)–(6) of Paper 1. If one assumes an exponential TTD, one can estimate the MTT directly from the amplitude ratio A_S/A_P in streamflow and precipitation using Eq. (10) of Paper 1 with $\alpha = 1$. Where I plot results from this procedure (i.e., Fig. 7) the corresponding axis will say "MTT inferred from A_S/A_P". This is the approach that is conventionally used in the literature. Alternatively, as I showed in Sect. 4.4 of Paper 1, one can use the tracer cycle amplitude ratio A_S/A_P and phase shift $\varphi_S - \varphi_P$ to jointly estimate the shape factor α and the MTT (assuming the TTD is gamma-distributed, which is less restrictive than assuming that it is exponential). To do this one estimates the shape factor α from A_S/A_P and $\varphi_S - \varphi_P$, using Eq. (11) from Paper 1, and then estimates the scale factor β using Eq. (10) from Paper 1; the MTT is α times β. MTTs estimated by this procedure are shown in Figs. 10–12 as "MTT inferred from A_S/A_P and $\varphi_S - \varphi_P$".

Paper 1 shows that both of these MTT measures are extremely vulnerable to aggregation bias in spatially heterogeneous catchments. Therefore, Paper 1 proposes an alternative measure of travel times: the young water fraction F_{yw}, which is designed to be much less sensitive than MTT to aggregation artifacts. F_{yw} is the fraction of streamflow that is younger than a specified threshold age. For a seasonal cycle (i.e., with a period of 1 year) and reasonable range of TTD shapes, the threshold age varies between about 0.15 and 0.25 years or, equivalently, $\sim$2–3 months (see Eq. 14 and Fig. 10 in Paper 1). As described in Sect. 2, in the model simulations the "true" F_{yw} is defined by a threshold age of 0.189 years (69 days), which equals the threshold age for seasonal cycles convolved with an exponential TTD.

One can use seasonal tracer cycles to infer the young water fraction following either of two strategies. As shown in Sect. 4.1 of Paper 1, in many situations F_{yw} is approximately equal to the amplitude ratio A_S/A_P itself (indeed, it was designed to have this property). In figures where the amplitude ratio A_S/A_P is used as an estimate of F_{yw} (e.g., Fig. 7), the axis says simply "F_{yw} inferred from A_S/A_P". Alternatively, one can use both the amplitude ratio A_S/A_P and phase shift $\varphi_S - \varphi_P$ to estimate F_{yw}, as explained in Sect. 4.4 of Pa-

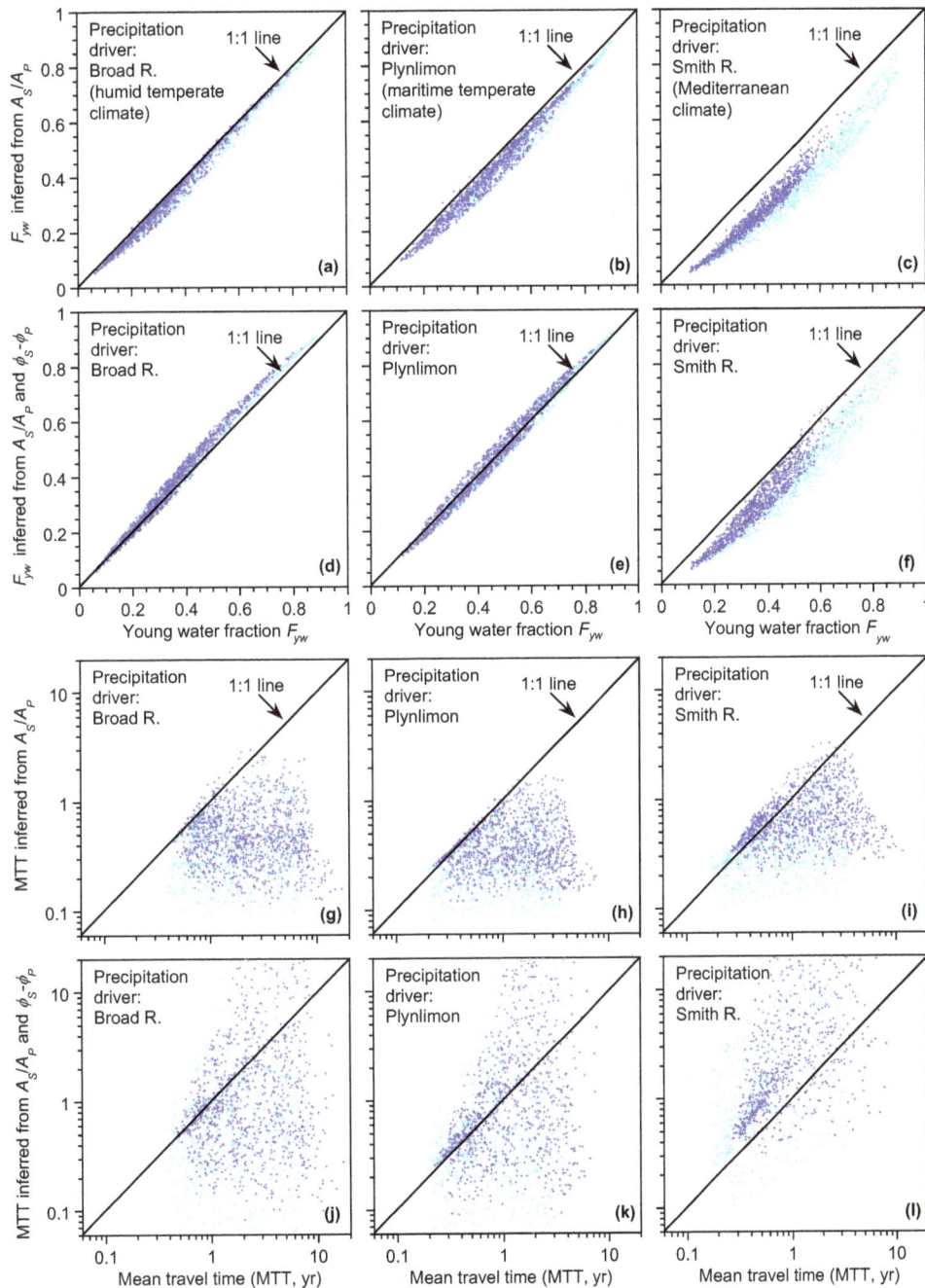

Figure 7. Young water fractions (F_{yw}, top panels) and mean transit times (MTT, bottom panels – note log scale) in streamflow from the two-box model. Upper panels compare the average F_{yw} in discharge, determined by age tracking within the model (on the horizontal axes) with the seasonal tracer cycle amplitude ratio A_S/A_P (**a–c**), and with F_{yw} inferred from the tracer cycle amplitude ratio A_S/A_P and phase shift $\varphi_S - \varphi_P$ (**d–f**). Lower panels compare the average MTT in discharge (again from age tracking) with MTT inferred from the tracer amplitude ratio (**g–i**) and from amplitude ratio and phase shift (**j–l**). Light blue points show flow-weighted average F_{yw} values and MTTs for each simulation, compared to estimates from flow-weighted fits to seasonal tracer cycles. Dark blue points show unweighted average F_{yw} values and MTTs, compared to estimates from unweighted fits to seasonal tracer cycles. Panels show results from 1000 random parameter sets and three contrasting precipitation drivers: Broad River (humid, temperate, with very little seasonality), Plynlimon (wet maritime climate with slight seasonality), and Smith River (Mediterranean climate with pronounced winter-wet, summer-dry seasonality). Seasonal tracer cycle amplitudes generally predict the average young water fraction, although they exhibit some systematic bias under strongly seasonal precipitation regimes like Smith River, where seasonal cycles in precipitation volume are correlated with seasonal cycles in tracer concentration. By contrast, mean transit-time estimates from seasonal tracer cycles are highly unreliable in all precipitation regimes.

per 1. First, one estimates the shape factor α from A_S/A_P and $\varphi_S - \varphi_P$ using Paper 1's Eq. (11). One then determines the threshold age τ_{yw} from α using Paper 1's Eq. (14), and the scale factor β from α and A_S/A_P using Paper 1's Eq. (10). Lastly, one estimates F_{yw} as lower incomplete gamma function $\Gamma(\tau_{yw}, \alpha, \beta)$ (Eq. 13 in Paper 1). Where I have followed this more complex procedure (e.g., Figs. 9–12), the figure axes say "F_{yw} inferred from A_S/A_P and $\varphi_S - \varphi_P$". All of these F_{yw} and MTTs are intended as temporal averages, reflecting whatever conditions (e.g., precipitation climatologies or flow regimes) have shaped the seasonal cycles that are used to estimate them.

These methods for inferring the young water fraction F_{yw} are derived from the properties of gamma TTDs. However, as I showed in Sects. 4.2–4.3 of Paper 1, these methods reliably estimate F_{yw} for very wide ranges of catchment TTDs (beyond the already broad family of gamma distributions), at least in catchments that are spatially heterogeneous but time-invariant. Here I explore whether these methods are also reliable in nonstationary catchments (and, in Sect. 3.5, in catchments that are both nonstationary and spatially heterogeneous).

Figure 7 shows the true young water fractions F_{yw} and MTTs in discharge from the two-box model, compared to estimates of F_{yw} and MTT inferred from the model's seasonal tracer cycles. As Fig. 7a–c show, the amplitude ratios A_S/A_P of seasonal tracer cycles reliably estimate the true young water fractions in the model streamflow, across 1000 random parameter sets encompassing a very wide range of nonstationary catchment behavior. The slight underestimation bias in Fig. 7a–c is reduced when both amplitude and phase information are used to estimate F_{yw} (Fig. 7d–f). Under strongly seasonal precipitation forcing (Smith River; right panels in Fig. 7), the seasonal tracer cycles underestimate F_{yw} by roughly 0.1–0.2, although the predicted and observed values of F_{yw} remain strongly correlated. For the other two precipitation drivers (Broad River and Plynlimon), the predicted and observed values of F_{yw} correspond almost exactly. Thus, Fig. 7 shows that the young water fraction is relatively insensitive to aggregation error under nonstationarity, mirroring its robustness against spatial heterogeneity (as shown in Paper 1). By contrast, estimates of MTT are strongly biased and widely scattered, even on logarithmic axes (lower panels, Fig. 7).

One additional complication in nonstationary situations, compared to the time-invariant examples explored in Paper 1, is that the young water fraction F_{yw} and MTT can be expressed either as simple averages over time (representing the F_{yw} or MTT of an average *day* of streamflow) or as flow-weighted averages (representing the F_{yw} or MTT of an average *liter* of streamflow). These quantities will not be equivalent, since higher flows will typically have higher F_{yw} and shorter MTTs (Figs. 3, 4). Likewise one can expect that amplitudes of flow-weighted and unweighted fits to the seasonal tracer cycles will be different. As the light blue points in

Fig. 7 show, amplitude ratios of flow-weighted fits to the seasonal tracer cycles accurately predict the flow-weighted F_{yw} in streamflow; likewise, as the dark blue points show, the amplitude ratios of unweighted fits accurately predict the unweighted F_{yw} in streamflow. The flow-weighted fits to the seasonal tracer cycles were calculated by weighted least squares, with weights proportional to streamflow or precipitation volume. (In real-world applications, a robust fitting technique like iteratively reweighted least squares (IRLS) can be used to limit the influence of outliers. An R script for performing volume-weighted IRLS is available from the author.)

The underestimation bias in F_{yw} observed under the Smith River precipitation forcing may arise because the assumed tracer cycle is correlated with the strong seasonality in precipitation, such that tracer concentrations peak during the summer, when almost no rain falls. Thus, the effective variability of tracer inputs to the catchment is less than one would infer from a sinusoidal fit to the precipitation tracer concentrations (and volume-weighting the fit does not help because in these synthetic precipitation data the fit is exact, so there are no residuals on which the weighting can have any effect). Because the tracer concentration amplitude overestimates the effective variability in tracer concentrations reaching the catchment, the tracer damping in the catchment is overestimated and thus the F_{yw} is underestimated. This underestimation bias disappears if one shifts the phase of the assumed precipitation tracer concentrations so that they peak in the spring or fall, and thus are uncorrelated with the seasonality in precipitation volumes. I have not done so here, however, because stable isotope ratios in precipitation typically peak in mid-summer at latitudes poleward of $\sim 35°$ (Feng et al., 2009), where most catchment studies have been conducted. Thus, Fig. 7 suggests the potential for bias in F_{yw} estimates at sites where isotope cycles are correlated with very strong precipitation seasonality. However, even under the strongly seasonal Smith River precipitation forcing, the bias in inferred F_{yw} values is small compared to the a priori uncertainty in F_{yw} (which is on the order of 1), and small compared to the bias in inferred MTTs (which is large even on logarithmic axes).

Panels g–i of Fig. 7 compare the MTT in streamflow with estimates of MTT as they are conventionally calculated, that is, from the seasonal tracer cycle amplitude assuming an exponential TTD. These plots show that these conventional estimates are subject to a strong underestimation bias, which can exceed an order of magnitude. Some of the MTT estimates do fall close to the 1 : 1 line, but these are mostly cases in which the partition coefficient η is very small, such that nearly all drainage from the upper box is routed through the lower box, thus transforming the two-box, nonstationary model into a nearly one-box, nearly stationary model. The strong aggregation bias in MTT under catchment nonstationarity shown in Fig. 7g–i mirrors the similarly strong bias under spatial heterogeneity that was demonstrated in Paper 1.

The implication of Fig. 7g–i (and of Paper 1) is that many of the MTT values in the literature are likely to be underestimated by large factors and, thus, that real-world catchment MTTs are likely to be much longer than we thought. This observation raises the question: where is all that water being stored? In steady state, the storage volume must equal the discharge multiplied by the MTT (see Sect. 3.2). Thus, if we have been underestimating MTTs by large factors, then we have also been underestimating catchment storage volumes by similar multiples. Where is the storage volume that can accommodate all this water?

One possible answer is that in a non-steady-state system, the MTT decreases with increasing discharge (e.g., Fig. 4b), and the storage volume equals the discharge multiplied by the *volume-weighted* MTT rather than the *time-averaged* MTT. Because the volume-weighted MTT is less (potentially much less) than the time-averaged MTT (see also Peters et al., 2014), the implied storage volume is correspondingly smaller. Furthermore, many MTT studies in the literature have been based on tracer sampling that excludes high flows, such that they infer the mean age of baseflow rather than of the average discharge (McGuire and McDonnell, 2006). To the extent that mean baseflow discharges are lower than mean total discharges, the stored volume of baseflow water will be less than what one might overestimate by multiplying the mean *total* discharge by the mean *baseflow* age. Beyond these general considerations, however, it makes little sense to draw precise inferences based on MTT estimates that are likely to be strongly biased and widely scattered (as shown here and also in Paper 1).

It is important to recognize that the predicted F_{yw} values are really predictions, unlike many "predictions" from calibrated models. The horizontal axes in Fig. 7 are calculated solely from the age-tracking within the model, with no information about the tracer concentrations. Likewise, the vertical axes in Fig. 7 are calculated from the modeled tracer cycles alone, without any information about the model that generated them and in particular without any information about the modeled age of streamflow. Thus, Fig. 7 gives some basis for confidence that estimates of F_{yw} will also be reliable in real-world catchments, where the true "model" can never be known.

3.4 Young water fractions in discrete flow regimes

Figures 3 and 4 show that high-flow periods are characterized by shorter mean transit times and higher young water fractions, reflecting the increased dominance of drainage from the upper box with its younger water ages. Although instantaneous transit-time distributions (TTDs) can be highly variable and, thus, instantaneous mean transit times and young water fractions can exhibit scattered relationships with discharge (Fig. 4), the marginal (time-averaged) TTDs in Fig. 6 clearly show a systematically stronger skew toward younger water ages in higher ranges of streamflow. Thus, as Fig. 6

shows, the TTD varies in shape, not just in scale, between different flow regimes.

This observation leads naturally to the question of whether these variations in TTDs are also reflected in streamflow tracer concentrations and whether those tracer signatures can be used to draw inferences about the TTDs that characterize individual flow regimes. Figure 3 shows that high-flow periods typically exhibit wider variations in tracer concentrations, reflecting greater contributions from the upper box, which has shorter residence times and thus more labile tracer concentrations than the lower box does. To test how systematic these variations in concentrations are, I ran the model with the reference parameter set and Plynlimon (temperate maritime) precipitation forcing and separated the resulting time series into six discharge ranges. Figure 8 shows these six discharge ranges and the corresponding tracer concentrations in dark blue, superimposed on the entire discharge and concentration time series in light gray. As Fig. 8 shows, seasonal tracer cycles at higher flows are systematically less damped and phase-shifted (relative to the tracer cycle in precipitation, shown by the dotted gray line), implying shorter MTTs and larger young water fractions.

To test whether these changes in the seasonal tracer cycles are quantitatively consistent with the shifts in water age across the six flow regimes, I fitted sinusoids separately to the tracer concentrations in each individual discharge range (Fig. 8). I compared these with a single sinusoid fitted to the entire precipitation tracer time series (because it is not possible to assign discrete precipitation events to individual discharge ranges). From the resulting amplitude ratios and phase shifts for each discharge range, I then estimated F_{yw} values nd MTT using the methods outlined in Sect. 3.3. Figure 9 presents the results of this thought experiment, showing that the time-averaged (but flow-specific) young water fraction F_{yw} in each discharge range is accurately predicted by the damping and phase shift of the corresponding seasonal tracer cycle.

To test whether this result is general, I repeated this thought experiment for 200 random parameter sets and all three precipitation drivers. The results are shown in Fig. 10, with each discharge range plotted in a different color. The colors overlap because the discharge ranges, F_{yw} values, and MTTs all vary substantially from one parameter set to the next. The amplitudes and phase shifts of the seasonal tracer cycles predict the time-averaged young water fractions F_{yw} in each discharge range with reasonable accuracy (upper panels, Fig. 10). Somewhat surprisingly, the F_{yw} underestimation bias seen in Fig. 7c and f under the highly seasonal Smith River precipitation forcing does not arise in the predicted F_{yw} values for the separate discharge ranges (Fig. 10c). In contrast to the generally close correspondence between the predicted and observed F_{yw} values, predicted MTTs are very widely scattered for all discharge ranges and all precipitation forcings (lower panels, Fig. 10).

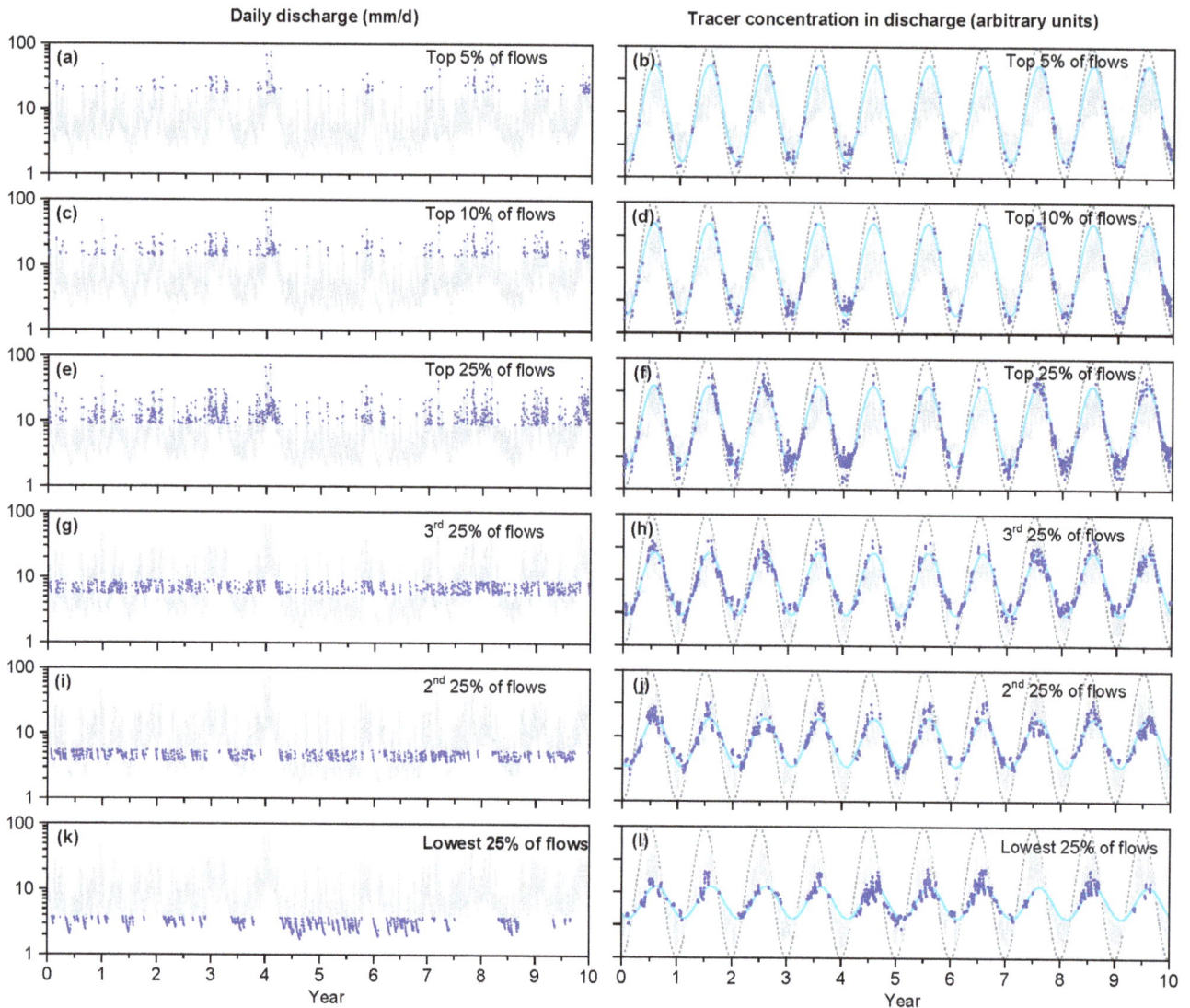

Figure 8. Daily discharges (left panels) and tracer concentrations (right panels) in streamflow from the two-box model with reference parameter values and Plynlimon precipitation forcing. Individual discharge ranges and corresponding tracer concentrations are highlighted in dark blue. In the right-hand panels, precipitation tracer concentrations are shown by dashed gray lines and sinusoidal fits to streamflow tracer concentrations are shown in light blue. At higher discharges, tracer cycles are less damped and less phase-shifted, indicating greater fractions of young water in streamflow.

3.5 Combined effects of nonstationarity and spatial heterogeneity

Paper 1 explored whether mean travel times and young water fractions can be reliably inferred from tracer dynamics in spatially heterogeneous (but stationary) catchments, composed of diverse subcatchments with different (but time-invariant) TTDs. The sections above have presented a similar analysis for nonstationary (but spatially homogeneous) catchments. However, real-world catchments are not *either* heterogeneous *or* nonstationary; instead they are *both* heterogeneous *and* nonstationary. That is, their subcatchments each exhibit nonstationary dynamics that may vary greatly

from one to the next. To explore the combined effects of nonstationarity and spatial heterogeneity, I merged the approach developed in Paper 1 with the model developed in Sect. 2.

As illustrated in Fig. 11, I ran eight copies of the nonstationary model developed in Sect. 2, representing eight different tributaries, each with a different, randomly chosen parameter set. I chose the number eight to provide a reasonable degree of complexity and heterogeneity while preserving a reasonable degree of computational efficiency. I supplied the same precipitation forcing (Fig. 11a) to all eight models (Fig. 11b) to simulate the behavior of the eight hypothetical tributary streams (Fig. 11c). I then simulated the merging of these streams by averaging their discharges, and

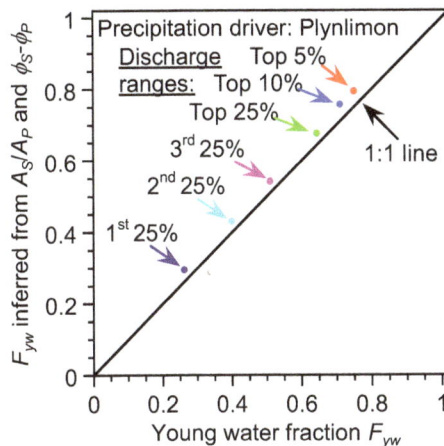

Figure 9. Time-averaged, flow-specific young water fractions F_{yw} for the six discharge ranges shown in Fig. 8, measured by age tracking in the model (with Plynlimon precipitation forcing and the reference parameter set), compared to F_{yw} values estimated from the amplitude ratios A_S/A_P and phase shifts $\varphi_S - \varphi_P$ of the tracer cycles shown in Fig. 8.

taking volume-weighted averages of their tracer concentrations, young water fractions, and water ages (Fig. 11d). Because the instantaneous flows from the eight tributaries vary differently through time, their mixing ratios also fluctuate. The individual random parameter sets create a wide range of model structures at the whole-catchment level, since the eight parallel subcatchments in Fig. 11 jointly comprise a 16-box, 40-parameter model incorporating wide ranges of large and small reservoirs with varying degrees of nonlinearity.

In any spatially heterogeneous catchment (which is to say, any real-world catchment), one will typically only have observations from the merged whole-catchment streamflow (i.e., the blue time series in Fig. 11d). One will typically have no information about the behavior of the individual tributaries (i.e., the colored time series in Fig. 11c), and if one did, then those tributaries would themselves have their own spatially heterogeneous tributary streams or flowpaths, and so on. Thus, the heterogeneity of any real-world catchment will remain poorly quantified (and possibly even unrecognized), and rigorously reductionist attempts to fully characterize such complex multiscale heterogeneity would be impractical.

Thus, we face the problem: how much can we infer from the behavior of the merged whole-catchment streamflow, given that it originates from processes that are heterogeneous and nonstationary (to a degree that is unknown and unknowable)? Figure 12 explores this general question in the specific context of young water fractions and mean travel times, presenting results from 200 iterations of the heterogeneous nonstationary model shown in Fig. 11 with all three precipitation drivers. In Fig. 12 the merged streamflow is separated into discrete flow regimes, following the approach outlined

in Sect. 3.4. As Fig. 12 shows, F_{yw} values inferred from the tracer cycles in each discharge range accurately predict the true fraction of young water in that discharge range, as determined from age tracking.

Figure 12 is analogous to Fig. 10, with the difference that Fig. 10 shows model runs for individual random parameter sets, whereas Fig. 12 shows results from eight runs merged together. Merging the model outputs will tend to average out the idiosyncrasies of the individual parameter sets, which is why the clusters of points in Fig. 12 are more compact than the corresponding point clouds in Fig. 10. As a result, the individual discharge ranges overlap less in Fig. 12 than in Fig. 10. The compact scatterplots shown in Fig. 12 show only small deviations from the 1 : 1 line for estimates of the young water fraction F_{yw}. By contrast, estimates of mean transit times in Fig. 12 exhibit substantial bias and scatter (note the logarithmic axes in Fig. 12d–f).

3.6 Hydrological and hydrochemical implications of young water fractions

The results reported above, together with the results reported in Paper 1, show that unlike mean transit times, young water fractions can be estimated reliably from seasonal tracer cycles in catchments that are spatially heterogeneous, nonstationary, or both. These findings then raise the obvious question: we can measure young water fractions reliably, but what are they good for? One answer is that young water fractions can be considered as a catchment characteristic, analogous (but far from equivalent) to MTT. In theory MTT should be particularly useful as a catchment descriptor, because the MTT times the mean annual discharge yields the total catchment storage. But because estimates of MTT will often be substantially in error, estimates of catchment storage derived from MTT are likely to be equally unreliable. If the shape of the TTD were known, of course, there would be a clear functional relationship between MTT and F_{yw}, and one could be calculated from the other. But if the shapes of the TTD were known, estimating the MTT itself would also be easy; the problem in estimating the MTT is the fact that the TTD's shape – particularly the length of its tail – is poorly constrained by tracer data. This is why F_{yw} can be estimated much more reliably than MTT. F_{yw}, like the amplitude of the seasonal tracer cycle, depends on the relative proportions of younger and older water, but is insensitive to how old the "older" water is. MTT depends critically on the age of the older water, which cannot be reliably determined because it has almost no effect on the seasonal tracer cycle (or on more elaborate convolution analyses; see Seeger and Weiler, 2014).

Because the young water fraction is indifferent to the age of the older water, it cannot be used to estimate residual storage. What F_{yw} estimates, instead, is the fraction of water reaching the stream by relatively fast (less than $\sim$ 2–3 month) flowpaths. In the context of the present model, this is re-

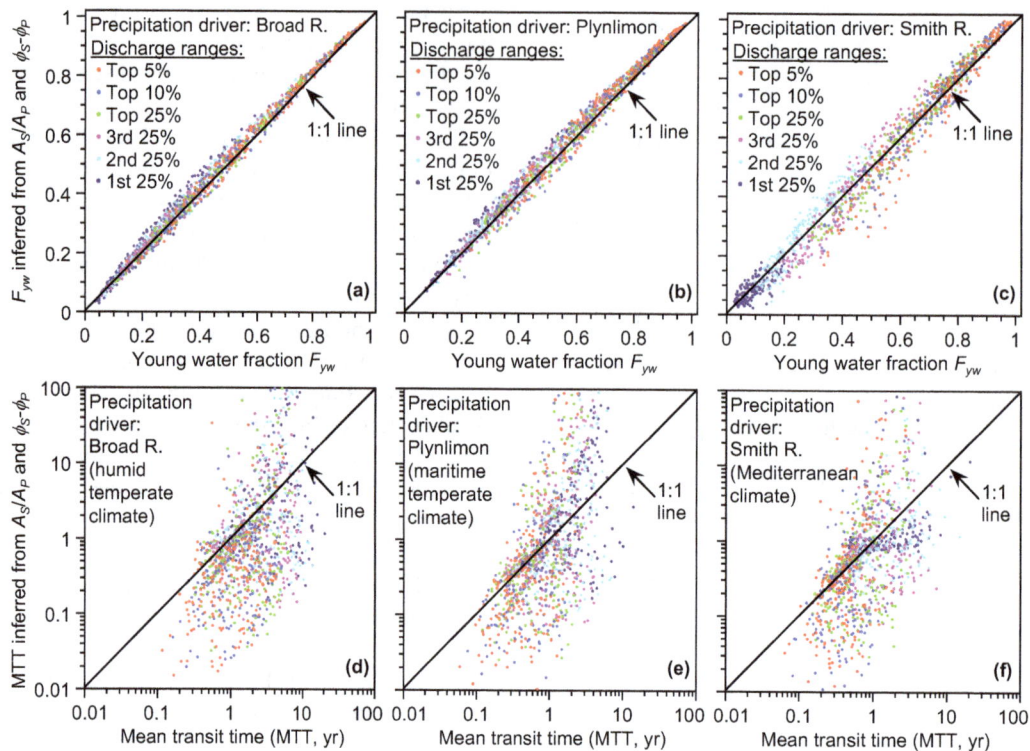

Figure 10. Young water fractions (F_{yw}) and MTTs in separate discharge ranges in streamflow from the two-box model. Upper panels compare the time-averaged, flow-specific F_{yw} for each discharge range (measured by age tracking in the model) with F_{yw} values estimated from the amplitude ratios A_S/A_P and phase shifts $\varphi_S - \varphi_P$ of the best-fit tracer cycle sinusoids in those discharge ranges (analogously to Fig. 8) using Eqs. (10), (11), (13) and (14) of Paper 1. Similar results (not shown) are also obtained for flow-weighted F_{yw} and flow-weighted tracer cycle sinusoids. Results obtained from tracer cycle amplitude alone (without phase information) are also similar, except in some cases where the amplitude ratio is small (particularly with Smith River precipitation forcing). Lower panels compare the MTT, determined by age tracking, with the MTT inferred from tracer amplitude ratios and phase shifts using Eqs. (10) and (11) from Paper 1. Each panel shows results from 200 random parameter sets and three contrasting precipitation drivers: Broad River (humid, temperate, with very little seasonality), Plynlimon (wet maritime climate with slight seasonality), and Smith River (Mediterranean climate with pronounced winter-wet, summer-dry seasonality). Tracer cycle amplitudes and phases generally predict the young water fractions in each discharge range, although with some modest scatter. Mean transit-time estimates, by contrast, are highly unreliable, exhibiting large scatter (note log scales).

flected in the correlation between F_{yw} and the partitioning parameter η (Fig. B2). This correlation is not exact, because F_{yw} will depend not only on how much streamflow comes from the upper box, but also on how much of the upper box is young water. That, in turn, will depend on precipitation climatology and the size of the upper box.

One can use F_{yw} not only to make comparisons across catchments but also, in an individual catchment, to compare how the proportions of flow traveling by fast flowpaths change across different flow regimes, as shown in Figs. 8–10 and 12. In turn it may be possible to draw inferences about how catchment processes change with flow regime. In this model, variations in F_{yw} across different flow regimes are strongly correlated with the fractional contributions of the upper box to streamflow (Fig. 13). The slopes and intercepts of the relationships vary among parameter sets, principally reflecting variations in the partitioning parameter η and the sizes of the upper and lower boxes. The strong correlations

shown in Fig. 13 are typical. Repeating the analysis shown in Fig. 13 for 200 random model "catchments" (i.e., different random parameter sets) yields an average correlation of over 0.99 (again, with different linear relationships for different parameter values). Of course these results – and, more generally, the interpretation of F_{yw} in terms of upper-box flow – are model-dependent. They are meant to demonstrate only that process inferences can be drawn from F_{yw}, not that these particular inferences should be applied literally to real-world catchments. Indeed one must remember that in the real world there is no "upper box"; it, like all model abstractions, should not be confused with reality.

The young water fraction F_{yw} may also be helpful in inferring chemical processes from streamflow concentrations of reactive chemical species. Many reactive species exhibit clear concentration–discharge relationships. Because one can determine how F_{yw} varies, on average, across different ranges of discharge (as demonstrated in Figs. 8–10

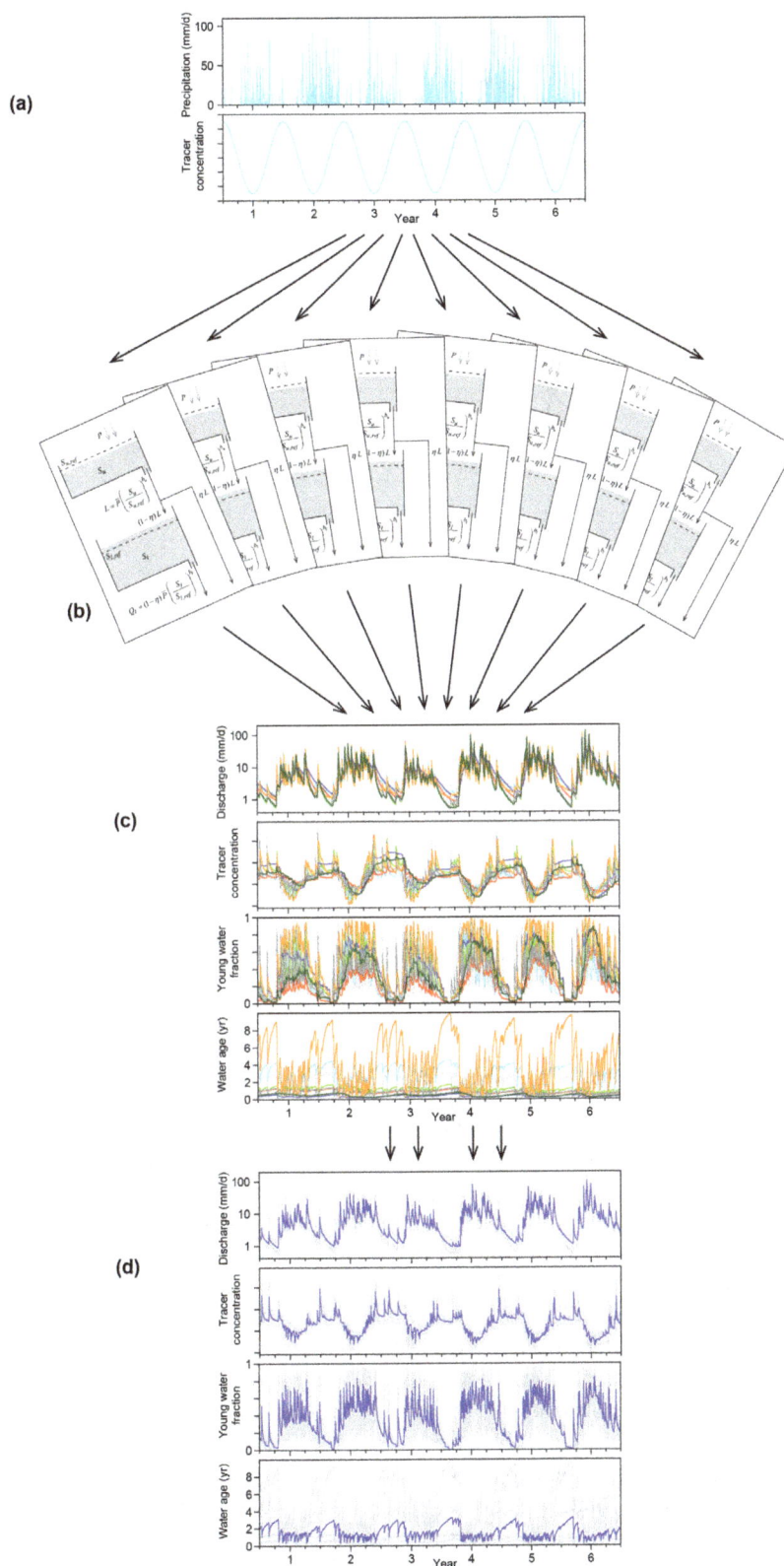

Figure 11. Scheme for simulating spatially heterogeneous catchments with nonstationary tributary subcatchments. A single precipitation time series (**a**) is used to drive eight copies of the model representing eight tributary streams (**b**), each with a different set of random parameter values. Streamflows, tracer concentrations, young water fractions, and water ages from these eight nonstationary tributaries (**c**, with each color representing a separate tributary stream) are mass-averaged to determine the time series that would be observed in the merged streamflow (**d**, with blue lines showing the merged streamflow and gray lines showing the tributaries).

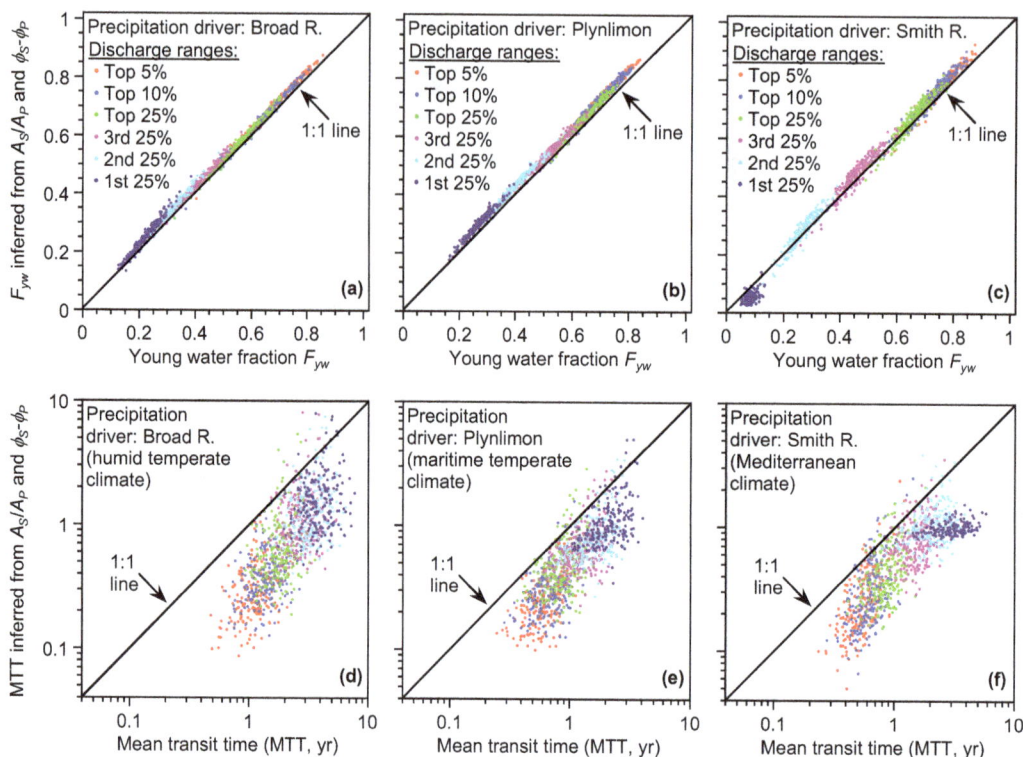

Figure 12. Actual and inferred young water fractions (F_{yw}, top panels) and MTTs (bottom panels) in separate discharge ranges, under combined effects of nonstationarity and spatial heterogeneity. Panels show results for 200 synthetic catchments, each consisting of eight copies of the two-box model with independent random parameter sets (Fig. 11). Upper panels compare average F_{yw} values with F_{yw} values predicted from amplitudes and phases of best-fit tracer cycle sinusoids for each discharge range (e.g., Fig. 8) using Eqs. (10), (11) and (13), (14) of Paper 1. Similar results (not shown) are also obtained for flow-weighted F_{yw} values and flow-weighted tracer cycle sinusoids. Results obtained from tracer cycle amplitude alone (without phase information) are also similar but exhibit slightly greater bias. Lower panels compare MTT with MTT predicted from tracer amplitude ratios and phase shifts using Eqs. (10) and (11) from Paper 1. Seasonal tracer cycle amplitudes and phases accurately predict young water fractions in separate flow regimes; the corresponding estimates of mean transit times exhibit substantial bias and scatter.

and 12), one can potentially construct mixing relationships between F_{yw} and the concentrations of reactive species. If the measurable range of F_{yw} is wide enough, one may even be able to estimate the end-member concentrations corresponding to idealized "young water" ($F_{yw} = 1$) and "old water" ($F_{yw} = 0$).

Figure 14 illustrates a preliminary proof of concept for this approach, based on 20–28 years of weekly precipitation and streamflow samples from three catchments at Plynlimon, Wales (Neal et al., 2011) with contrasting geochemical behavior. I separated the streamflow samples into five discharge ranges (lowest 20 %, next 20 %, and so on), then fitted the seasonal chloride concentration cycles in each discharge range and calculated the corresponding young water fractions using the approach outlined in Sect. 3.4. I then examined the relationships between these young water fractions and the mean streamwater concentrations of reactive chemical species in each discharge range. Figure 14 shows three different views of how reactive tracer chemistry varies with discharge across the three catchments. The left-hand panels

show the average concentrations in each discharge range, as functions of the logarithm of discharge. The middle panels show the same concentrations as functions of the inferred F_{yw}, with the vertical axis at $F_{yw} = 0$ indicating the hypothetical old water end-member. The right-hand panels show the concentrations plotted against the reciprocal of F_{yw}; here, the vertical axis at $1/F_{yw} = 1$ indicates the hypothetical young water end-member. The gray lines are fitted by hand to indicate general trends, and to suggest potential end-member concentrations.

The three catchments are characterized by contrasts in soil hydrology, with the abundance of impermeable gley soils and boulder clay tills increasing in the rank order Hafren < Hore < Tanllwyth. The same rank order is observed in the calculated young water fractions at high flows, reflecting the greater high-flow variability in chloride concentrations at sites with more impermeable soils. The three sites also exhibit contrasting concentration–discharge relationships for nitrate and aluminum (Fig. 14a, d), two solutes that are relatively abundant in near-surface soil solu-

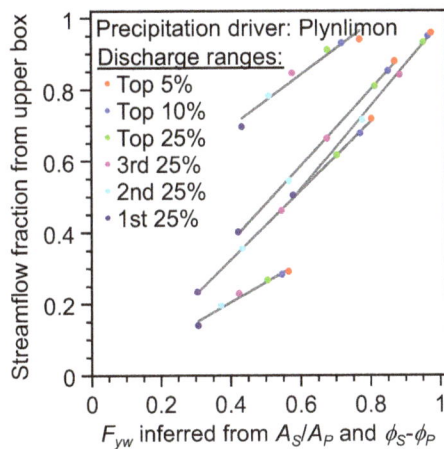

Figure 13. Correlations between flow-weighted young water fractions F_{yw} and fractional contributions of the upper box to streamflow across different discharge ranges, for five parameter sets illustrating the diversity of relationships that can arise in the model. The upper box contribution is strongly correlated with F_{yw} in all cases, although the slopes and the intercepts vary among parameter sets.

tions. When plotted against the young water fraction, however, these catchment-specific concentration–discharge relationships collapse to single concentration–F_{yw} relationships (Fig. 14b, e) in which the three sites are generally indistinguishable within error. These relationships can be extrapolated to reasonably well-constrained old water end-member concentrations of $\sim 0.1\,\mathrm{mg\,L^{-1}}\,NO_3$–N and $\sim 50\,\mu\mathrm{g\,L^{-1}}$ Al, and to comparably well-constrained young water end-member concentrations of $\sim 0.45\,\mathrm{mg\,L^{-1}}\,NO_3$–N and $\sim 600\,\mu\mathrm{g\,L^{-1}}$ Al (Fig. 14c, f). In the case of calcium, the three catchments have markedly different concentration–discharge relationships (Fig. 14g), reflecting differences in the abundance of calcite in their bedrock. As a result, the three catchments have different old water end-member calcium concentrations, ranging from ~ 1 to $\sim 4\,\mathrm{mg\,L^{-1}}$ (Fig. 14h). However, all three streams converge to similar concentrations of $\sim 0.5\,\mathrm{mg\,L^{-1}}$ Ca in the young water end-member (Fig. 14i).

It is tempting to interpret the concentration differences between the young and old end-members as reflecting chemical kinetics, but this should be approached with caution. A kinetic interpretation makes sense if the young and old end-members differ only in age (albeit by an unspecified amount since we cannot know how old the "old" end-member is), but not if they differ in other respects as well. At Plynlimon, for example, porewaters in the acidic soil layers have relatively high concentrations of aluminum and transition metals, and relatively low concentrations of base cations and silica, whereas waters infiltrating deep into the fractured bedrock react with calcite and layer lattice silicates and thus become enriched in base cations and silica, and depleted in aluminum and transition metals (Neal et al., 1997). Thus,

one must also consider the alternative hypothesis that the young end-member represents mostly soil water, that the old end-member represents mostly deeper groundwater, and that the two end-members exhibit different chemistry because of their sources rather than their ages. In this case, the end-member compositions identified through plots like Fig. 14 may help in characterizing the chemistries, and thus localizing the physical sources, of the young and old waters. In this proof-of-concept example, all three catchments appear to have geochemically similar young water end-members, with a composition suggesting a shallow soil source, but each has a different old water end-member, suggesting deeper groundwater sources with differing amounts of carbonate minerals. This is consistent with independent geochemical evidence at Plynlimon (Neal et al., 1997).

It is also important to note that if the ideal end-member mixing assumptions hold (i.e., the young and old end-members are invariant, and the mixture undergoes no further chemical reactions), then the mixing relationships in the middle plots of Fig. 14 should be straight lines, and they should extrapolate to physically realistic (non-negative) concentrations at both $F_{yw} = 0$ and $F_{yw} = 1$. To the extent that the mixing relationships are not straight, or imply unrealistic end-members, they indicate that these assumptions are not met.

3.7   General observations and caveats

It is important to recognize that the inferred young water fractions F_{yw} plotted in Figs. 7–12 are not in any way calibrated to the true values determined by age tracking. Nor do they make use of any information about the models that transform precipitation into streamflow (neither their structure, nor their parameter values). Thus, there is nothing artifactual about the close correspondence between predicted and observed values of F_{yw} in Figs. 7–12. Instead, these thought experiments provide strong evidence that seasonal tracer cycles can be used to reliably partition streamflow into young and old fractions (F_{yw} and $1 - F_{yw}$, respectively), even in catchments that are both nonstationary and spatially heterogeneous and whose real-world "models" (i.e., whose underlying processes) are poorly understood.

When these results are applied in practice, however, one must keep in mind that in contrast to typical field studies, these thought experiments are based on synthetic data sets that are dense (daily measurements for 10 years) and error-free. Furthermore, these thought experiments use a sinusoidal precipitation tracer signal that varies only seasonally, with no confounding variation on shorter or longer timescales. Further benchmark testing will be needed to test the accuracy of F_{yw} estimates derived from shorter, sparser, and messier data sets.

One can of course also question the realism of the particular model that I have used for these thought experiments. This model can be calibrated to reproduce the stream discharge with a Nash–Sutcliffe efficiency (NSE) of better than 0.85

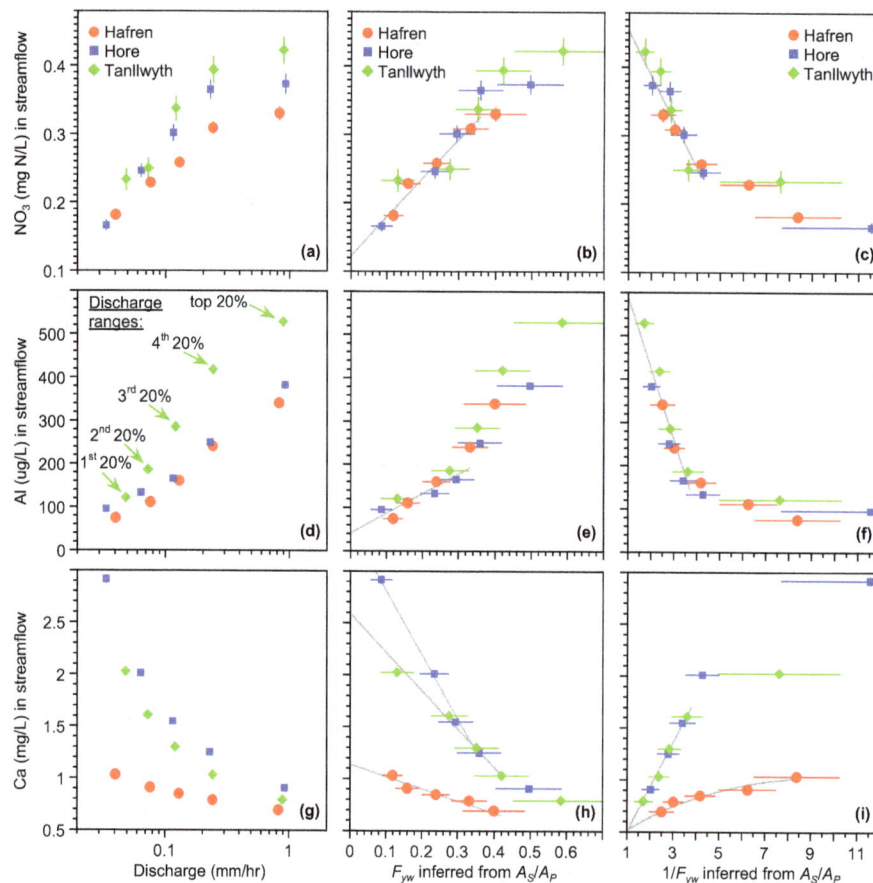

Figure 14. Concentrations of reactive chemical species as functions of discharge (left panels), young water fractions (middle panels), and reciprocal young water fractions (right panels) for streams draining three contrasting catchments at Plynlimon, Wales. Symbols show means for 20 % intervals of each catchment's discharge distribution, and error bars indicate ± 1 SE (standard error). Gray lines are drawn by hand to indicate general trends. Concentration–discharge relationships in nitrate and aluminum differ among the three catchments (**a, d**) but collapse to single concentration–F_{yw} relationships (**b–c, e–f**). These concentration–F_{yw} relationships extrapolate to broadly consistent old water end-members ($F_{yw} = 0$, **b** and **e**) and young water end-members ($F_{yw} = 1$, **d** and **f**). Calcium follows different concentration–F_{yw} relationships in the three streams, which extrapolate to three different old water end-members (**h**) but roughly the same young water end-members (**i**).

at two of the three sites, but there is no guarantee that it is getting the right answer for the right reasons. All models – whether lumped conceptual models or "physically based" spatially explicit models – necessarily involve approximations and simplifications. In plain language: any model, including this one, incorporates assumptions that are false and are known to be false. One obvious idealization (a less euphemistic word would be *fiction*) is the use of well-mixed boxes as the core of most lumped conceptual models, including the model presented here. Assuming that everything in each box is completely mixed or, equivalently, that it is randomly sampled in the outflow – regardless of where it is physically located in the landscape – clearly strains credibility, but this is what typical conceptual models must assume for mathematical convenience. The model presented here is no different.

What is different, however, is that here the model is used for purposes that make its literal realism unnecessary. Typi-

cal modeling studies draw conclusions about real-world systems from model behavior; thus, those conclusions depend critically on the realism of the model. Here, the primary goal is not to test how catchments work but instead to test specific methods for inferring water ages from complex, nonstationary time series of tracer concentrations. All the model must do is generate outputs with reasonable degrees of complexity and nonstationarity; it is not essential that the model generates these time series by the same mechanisms that real-world catchments do. The only inductive leap is the inference that if a method correctly infers water ages from tracer patterns in these complex, nonstationary time series, it will also correctly infer water ages in complex, nonstationary time series generated by real-world catchments.

It is important to highlight an essential difference between the approach developed here and typical studies that infer water ages or transit-time distributions from calibrated models (e.g., Birkel et al., 2011; Van der Velde et al., 2012; Heidbüchel et al., 2012; Hrachowitz et al., 2013; Benettin et al., 2013, 2015). When one draws inferences from a model, their validity depends on whether that model is structurally adequate and whether its parameter values are realistic, both of which are usually in doubt. Here, by contrast, I have developed an inferential method (for estimating the young water fraction F_{yw} from seasonal tracer cycles) that is not drawn from – and thus does not depend on – the model's structure or its parameter values. The model is used only to create synthetic data to test the inferential method.

The results reported here, together with those in Paper 1, show that MTTs cannot be estimated reliably by fitting sine waves to seasonal tracer cycles from nonstationary or spatially heterogeneous catchments. These results do not imply that other methods for estimating MTTs are any better; instead, they imply only that sine wave fitting has been subjected to rigorous benchmark testing and has failed. The other methods have not yet been similarly tested, and it is unclear whether they too will fail. Efforts to fill this knowledge gap are underway. But in the meantime, ignorance is not bliss; one should not simply assume that these other methods work as intended, just because they have not yet been rigorously tested. In that regard, the most general contribution of this analysis is not that it reveals specific problems with MTT estimation from seasonal tracer cycles, or that it demonstrates the reliability of F_{yw} as an alternative metric of catchment transit times, but rather that it illustrates the clarifying power of well-designed benchmark tests.

4   Summary and conclusions

The age of streamflow – i.e., the time that has elapsed since it fell as precipitation – is an essential descriptor of catchment functioning with broad implications for runoff generation, contaminant transport, and biogeochemical cycling (Kirchner et al., 2000; McGuire and McDonnell, 2006). The age of streamflow is commonly measured by its MTT, which in turn has often been estimated from the damping of seasonal cycles of chemical and isotopic tracers (such as Cl^-, $\delta^{18}O$, or δ^2H). In a companion paper (Paper 1: Kirchner, 2016), I demonstrated that MTT cannot be reliably estimated from seasonal tracer cycles in spatially heterogeneous catchments, and I proposed an alternative water age metric, the young water fraction F_{yw}, which is relatively immune to the errors and biases that afflict the MTT.

Here I have explored how catchment nonstationarity affects estimates of MTT and F_{yw}, using simple thought experiments based on a simple two-box conceptual model (Fig. 1) driven by three precipitation time series representing a range of precipitation climatologies (Fig. 2). The model exhibits complex nonstationary behavior (Fig. 3), with striking volatility in tracer concentrations, young water fractions, and mean transit times as the mixing ratio between the upper and lower boxes shifts in response to precipitation events. This

mixing ratio is both hysteretic and nonstationary, varying in response both to precipitation forcing and to the antecedent moisture status of the two boxes (Fig. 4).

Marginal (time-averaged) age distributions in drainage are skewed toward younger ages than the storage distributions they come from, because storage is flushed more quickly (and thus is younger) during periods of higher discharge (Fig. 5). The age distributions in whole-catchment storage and discharge are approximate power laws, with markedly different slopes (Fig. 5). The age distribution in streamflow becomes increasingly skewed at higher discharges, with a marked increase in the young water fraction and decrease in the mean water age (Fig. 6), reflecting the increased dominance of the upper box at higher flows. Flow-weighted average MTTs are typically close to the steady-state MTT, estimated as the ratio of the total storage to the throughput rate. However, the marginal age distributions are markedly different from the distributions that would be expected in steady state, demonstrating that steady-state approximations are misleading guides to the non-steady-state behavior of the system, *even on average*.

Even this simple two-box model exhibits strong equifinality (Fig. B1), with four of its five parameters having virtually no identifiability through hydrograph calibration. However, scaling arguments based on simple perturbation analyses (Sect. 3.2) reveal ratios of parameters that can be constrained through hydrograph calibration (Fig. B2), greatly reducing the equifinality in the parameter space. Unfortunately, water age is primarily controlled by residual storage, which cannot be constrained through hydrograph calibration (Fig. B2). Thus, parameter sets that yield virtually identical hydrographs imply widely differing young water fractions and mean water ages (Fig. B3).

The simple two-box model was used to simulate discharge, water ages, and the propagation of seasonal tracer cycles through the catchment, across wide ranges of random parameter sets. MTTs inferred from the damping and phase shift of the seasonal tracer cycles exhibited strong underestimation bias and large scatter (Fig. 7). This result implies that many literature MTT values (and thus also residual storage volumes) may have been underestimated by large factors. By contrast, the seasonal tracer cycles accurately predicted the actual F_{yw} in streamflow, as determined by age tracking within the model (Fig. 7).

Flow-weighted fits to the seasonal tracer cycles accurately predicted the flow-weighted average F_{yw} in streamflow, while unweighted fits to the seasonal tracer cycles accurately predicted the unweighted average F_{yw}. The streamflow time series can be separated into distinct flow regimes with their own seasonal tracer cycles (Fig. 8), which accurately reflect the F_{yw} in each flow regime (Figs. 9, 10). Seasonal tracer cycles also accurately predicted the F_{yw} in the merged streamflow from spatially heterogeneous assemblages of nonstationary model catchments (Fig. 12). Importantly, all of these F_{yw} predictions were really predictions; they were not calibrated in any way.

The relationship between F_{yw} and the flow regime reflects how the fluxes from short-term storages vary with hydrologic forcing (Fig. 13). In a preliminary proof of concept (Fig. 14), I showed that one can construct mixing relationships between solute concentrations and F_{yw} values for discrete flow regimes. From these mixing relationships one can estimate the chemical composition of idealized "young water" and "old water" end-members (Fig. 14).

These findings extend the results of Paper 1 by showing that estimates of MTT from seasonal tracer cycles are unreliable under nonstationarity as well as spatial heterogeneity. These findings also extend the results of Paper 1 by showing that F_{yw} can be reliably estimated in nonstationary catchments as well as spatially heterogeneous ones, and it can also be reliably estimated for discrete flow regimes. These results further demonstrate that F_{yw} can be reliably estimated for discrete flow regimes and can provide helpful insights into the hydrological and hydrochemical functioning of catchments. Most generally, these results, along with those of Paper 1, illustrate how well-posed benchmark tests can be essential in clarifying what is knowable – and, conversely, unknowable – in environmental research.

Appendix A: Solution scheme

For simplicity and efficiency, the hydrological model is solved on a fixed daily time step. This requires some care with the numerics, given the clear (though often overlooked) dangers in naive forward-stepping simulations of nonlinear equations (Clark and Kavetski, 2010; Kavetski and Clark, 2010, 2011). Here I use a weighted combination of the trapezoidal method (which is partly implicit, for enhanced accuracy) and the backward Euler method (which is fully implicit, for guaranteed stability). The hydrological solution scheme is illustrated here for the upper box; the lower box is handled analogously. The storage in the upper box is updated using the following equation:

$$S_u(t_{i+1}) - S_u(t_i) = \Delta t \left(P - \rho k_u S_u(t_{i+1})^{b_u} \right.$$
$$\left. - (1-\rho) k_u S_u(t_i)^{b_u} \right), \tag{A1}$$

where $S_u(t_i)$ is the storage in the upper box at the beginning of the ith time interval (with length Δt), $S_u(t_{i+1})$ is the storage at the end of that interval (and thus the beginning of the next), and P is the average precipitation rate over the interval. Equation (A1) is implicit and nonlinear; there is no closed-form solution for the future storage $S_u(t_{i+1})$, which instead is found using Newton's method. The relative dominance of the trapezoidal and backward Euler solutions is determined by the weighting factor ρ, which takes on values between $\rho = 0.5$ (trapezoidal method) and $\rho = 1$ (backward Euler method). The value of ρ in Eq. (A1) is determined for each time step using the simple stability criterion:

$$\rho = \min \left(0.5 + 0.5 \frac{\left(P - k_u S_u(t_i)^{b_u} \right) \Delta t}{(P/k_u)^{1/b_u} - S_u(t_i)}, \ 1 \right), \tag{A2}$$

where the numerator represents the amount that S_u would change during one time step if the instantaneous drainage rate L in Eq. (1) were projected forward in time, and the denominator represents the difference between S_u's current value and its equilibrium value at the precipitation rate P. Equation (A2) says that if the trapezoidal method would move S_u by only a small fraction of the distance to its equilibrium value (at the precipitation rate P), then the stability advantages of the backward Euler method are unnecessary and the more accurate trapezoidal method should dominate the solution instead ($\rho \approx 0.5$). On the other hand, if the trapezoidal method would overshoot the equilibrium value, then $\rho = 1$ and the fully implicit backward Euler method is used to solve Eq. (A1). The closer the trapezoidal method would come to overshooting the equilibrium, the larger the value of ρ and the greater the weight that is given to the backward Euler solution. The guaranteed stability of the backward Euler method is important when b_u or b_l is large, because the underlying equations can become quite stiff. After the final value of S_u is determined by Eq. (A1), the drainage from S_u between t_i and t_{i+1} is determined by mass balance:

$$L = P + (S_u(t_i) - S_u(t_{i+1}))/\Delta t, \tag{A3}$$

where L is the average drainage rate over the interval Δt between t_i and t_{i+1}.

The tracer concentrations are determined under the assumption that each box is well mixed, implying that individual water parcels within each box do not need to be tracked, and also that the concentration draining from each box equals the average concentration within the box. I make the simplifying assumption that each box's inflow and outflow rates (and also inflow concentrations) are constant over each day. Again taking the upper box as an example, these assumptions imply that starting from $t = t_i$ the tracer concentration will evolve as

$$\frac{dC_u}{dt} = \frac{P(C_P - C_u)}{S_u(t_i) + (P - L)(t - t_i)}, \tag{A4}$$

where C_P and C_u are the concentrations in precipitation and the upper box, respectively, and the denominator expresses how the volume in the box changes with time from its initial value of $S_u(t_i)$. Integrating Eq. (A4) over an interval Δt yields the concentration updating formula:

$$C_u(t_{i+1}) = C_P + (C_u(t_i) - C_P) \left(\frac{S_u(t_i)}{S_u(t_{i+1})} \right)^{(P/(P-L))}, \tag{A5}$$

where any quantities that are not shown as functions of time are constant at their average values over the interval. Equation (A5) could potentially become difficult to compute when P and L are nearly equal (differing by, say, less than 1 part in 1000), and the power function approaches its exponential limit. In such cases the change in volume in Eq. (A4) becomes trivially small, and one can replace Eq. (A5) with the more familiar exponential formula for a well-mixed box of constant volume:

$$C_u(t_{i+1}) = C_P + (C_u(t_i) - C_P) \exp(-P\Delta t/S_u). \tag{A6}$$

After the tracer concentrations are updated, the average concentrations in drainage are calculated by mass balance, as follows:

$$C_L = [C_P(t_i) P + C_u(t_i) S_u(t_i) - C_u(t_{i+1}) S_u(t_{i+1})]/L, \tag{A7}$$

where C_L is the average concentration in drainage over the time interval between t_i and t_{i+1}.

The mean age within each box is modeled analogously to the tracer concentrations, following the "age mass" concept widely used in groundwater hydrology. Here I will illustrate the approach using the example of the lower box, since it is the more complex case (for the upper box, the input age in precipitation is zero, but this is not true for the upper-box drainage that recharges the lower box). Assuming that the inflow and outflow rates $L(1-\eta)$ and Q_1 are constant over a day, as is the average age $\overline{\tau}_L$ of the inflow from the upper box, the mean age in the lower box should evolve according to

$$\frac{d\overline{\tau}_1}{dt} = \frac{L(1-\eta)(\overline{\tau}_L - \overline{\tau}_1)}{S_1(t_i) + (L(1-\eta) - Q_1)(t - t_i)} + 1, \quad (A8)$$

which is directly analogous to Eq. (A4), except for the additional term of $+1$, which accounts for the continual aging of the water in the box. The solution to Eq. (A8) is

$$\overline{\tau}_1(t_{i+1}) = \overline{\tau}_L + \frac{S_1(t_{i+1})}{2L(1-\eta) - Q_1} + \left(\overline{\tau}_1(t_i)\right.$$
$$\left. -\overline{\tau}_L - \frac{S_1(t_i)}{2L(1-\eta) - Q_1}\right)\left(\frac{S_1(t_i)}{S_1(t_{i+1})}\right)^{\left(\frac{L(1-\eta)}{L(1-\eta)-Q_1}\right)}, \quad (A9)$$

where $\overline{\tau}_1(t_i)$ and $\overline{\tau}_1(t_{i+1})$ are the mean age of the water in the lower box at the beginning and end of the time interval. Analogously to tracer concentrations, one can calculate the mean age of the drainage from the box based on the inputs and the change in mean age inside the box, using conservation of "age mass":

$$\overline{\tau}_{Q_1} = [\overline{\tau}_L(t_i)(1-\eta) + \overline{\tau}_1(t_i)S_1(t_i)$$
$$- (\overline{\tau}_1(t_{i+1}) - \Delta t)S_1(t_{i+1})]/Q_1, \quad (A10)$$

where the factor of $-\Delta t$ accounts for the aging of the contents of the box.

The approach used here for concentrations and water ages requires the assumption that input fluxes to each box are constant within each time interval (but constant at their average values, not their initial values). This is a reasonable approximation, particularly when we have no sub-daily precipitation data. And in exchange for this simplifying assumption, Eqs. (A5), (A6), and (A9) provide something important, namely, the exact analytical solution for the evolution of concentration and age during each time interval. Thus, these equations directly solve for the correct result even if, for example, an individual day's rainfall is much greater than the total volume of the upper box. The equations above will correctly calculate the consequences of the (potentially many-fold) flushing that occurs in such cases. The approach outlined above also guarantees exact consistency between stocks and fluxes (but note that this is not done in the usual way by updating stocks with fluxes, but rather by calculating output fluxes from inputs and changes in stocks). Readers should keep in mind that all stocks and properties of stocks (i.e., storage volumes, concentrations, and ages) are expressed as the instantaneous values at the beginning of each time interval, and that fluxes and properties of fluxes (i.e., water fluxes and their concentrations and ages) are expressed as averages over each time interval. Otherwise it could be difficult to make sense of the equations above.

Appendix B: Equifinality in hydraulic behavior and divergence in travel times

The analysis outlined in Sect. 3.2 implies that approximate equifinality is inevitable, even in such a simple model, because variations in the exponents b_u and b_l and the reference storage levels $S_{u,ref}$ and $S_{l,ref}$ will have nearly offsetting effects on the model's runoff response. Equations (10) and (11) show that, for a given average precipitation forcing, any parameter values for which the partitioning coefficient η and the ratios $S_{u,ref}/b_u$ and $S_{l,ref}/[(1-\eta)b_l]$ are invariant would give nearly equivalent hydrograph predictions, because the hydraulic response timescales of the upper and lower boxes, and their relative contributions to discharge, would be invariant. These conditions can be achieved for widely varying values of the individual parameters b_u, b_l, $S_{u,ref}$, and $S_{l,ref}$.

This equifinality problem can be readily visualized by plots like Fig. B1. To generate Fig. B1, I ran the model with Smith River precipitation forcing and the reference parameter set (shown by the red squares in Fig. B1) and used the resulting daily hydrograph (after the spin-up period) as virtual "ground truth" for model calibration. I then ran the model with 1000 random parameter sets and used the NSE of the logarithms of discharge to measure how well their hydrographs matched the reference hydrograph (thus the reference hydrograph has a NSE of 1 by definition). The 50 best-fitting parameter sets, all with NSE ≥ 0.98, are shown as dark blue points in Fig. B1. The bottom row of scatterplots shows the conventional "dotty plots". Their flat tops are the hallmark of equifinality, i.e., wide ranges of parameter values give equally good hydrograph predictions (Beven, 2006). Only the partition coefficient η, which performs well across half its range, can be even modestly constrained by calibration. (The other precipitation drivers yield results similar to those shown in Fig. B1.)

The other panels of the scatterplot matrix also give important clues to the origins of the observed equifinality. In particular, the best-fitting parameter sets show strong correlations between $S_{u,ref}$ and b_u, and between $S_{l,ref}$ and b_l, as expected from the perturbation analysis presented in Sect. 3.2. Thus, good model performance can be obtained across almost the entire range of these parameters but only for specific parameter combinations. These parameter combinations correspond to "valleys" in the model's response surface, a longstanding problem in model calibration (e.g., Ibbitt and O'Donnell, 1974). The interdependence of the parameters is visually obvious in the scatterplot matrix but is invisible in the conventional "dotty plots".

This information can be exploited to design parameter spaces that are more identifiable through calibration (e.g., Ibbitt and O'Donnell, 1974). An ideal parameter space would be one in which (1) all parameters are highly identifiable, meaning the goodness-of-fit surface is strongly curved along each parameter axis, and (2), in the best-fitting parameter sets, no parameters are strongly correlated with one another.

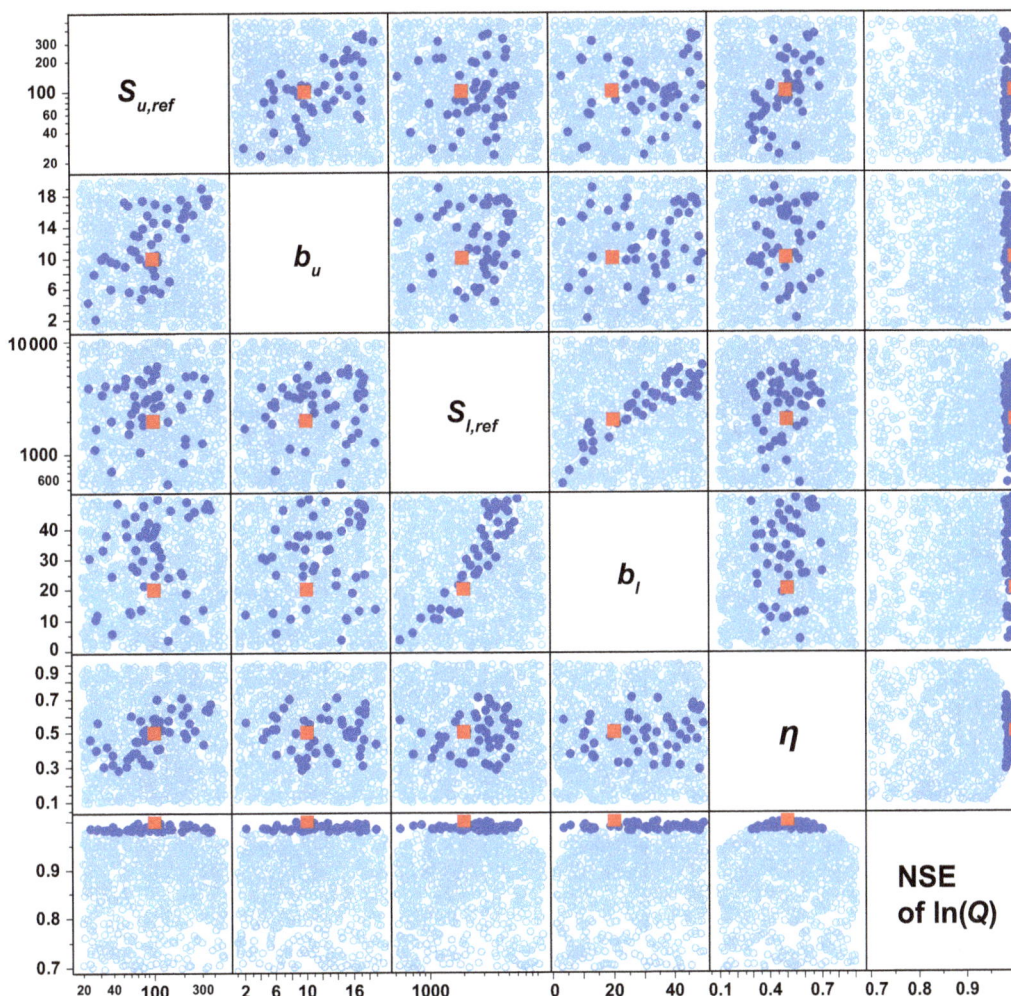

Figure B1. Equifinality in discharge predictions. The scatterplot matrix shows relationships among 1000 random parameter sets and the Nash–Sutcliffe efficiency (NSE) of discharge time series driven by Smith River (Mediterranean climate) precipitation forcing. The red square indicates the "reference" parameter set that was used to generate the discharge time series that the other parameter sets were tested against; these reference parameters thus correspond to NSE = 1.00 by definition. The dark blue dots show the best-fitting 50 (or 5 %) of the parameter sets, all with NSE ≥ 0.98. Excellent discharge predictions can be obtained across almost the full range of all five model parameters, except the partition coefficient η, which performs well across only about half its range. The dark blue dots show clear correlations between the reference storage levels in each box ($S_{u,ref}$, $S_{l,ref}$) and the corresponding drainage function exponents (b_u, b_l); these correlations delimit regions with nearly constant hydraulic response timescales, as defined by Eqs. (10) and (11).

The second of these criteria is necessary (although not sufficient) for the first, as Fig. B1 illustrates. A third criterion is that all parameters that are needed for simulating any quantities of interest must be determined somehow within the parameter space, either individually or through combinations of other parameters. Thus, for example, although the volumes of the boxes ($S_{u,ref}$ and $S_{l,ref}$) are strongly correlated with their exponents (b_u and b_l), the parameter space must allow them to be individually determined, because as Eqs. (12)–(14) suggest, the mean transit times will be controlled primarily by the volumes alone (not in combination with the exponents), whereas the runoff response will be controlled primarily by the ratios of volumes to exponents (Eqs. 10, 11). These crite-

ria, plus some trial and error, lead to a more identifiable parameter space, whose five axes are $S_{u,ref}$, $S_{l,ref}$, $S_{u,ref}/(\eta \cdot b_u)$, $S_{l,ref}/b_l$, and η.

Figure B2 shows that this parameter space exhibits much less equifinality than the parameter space shown in Fig. B1, although the underlying parameter sets and model simulations are exactly the same. All that has been done is to reproject the parameter space onto a different set of coordinate axes in which the curvature of the goodness-of-fit surface is more clearly visible. Thus, much of the apparent equifinality in the parameter space has been eliminated by simple transformations of variables. These transformations can be designed by eye in this case, because the dimensionality of

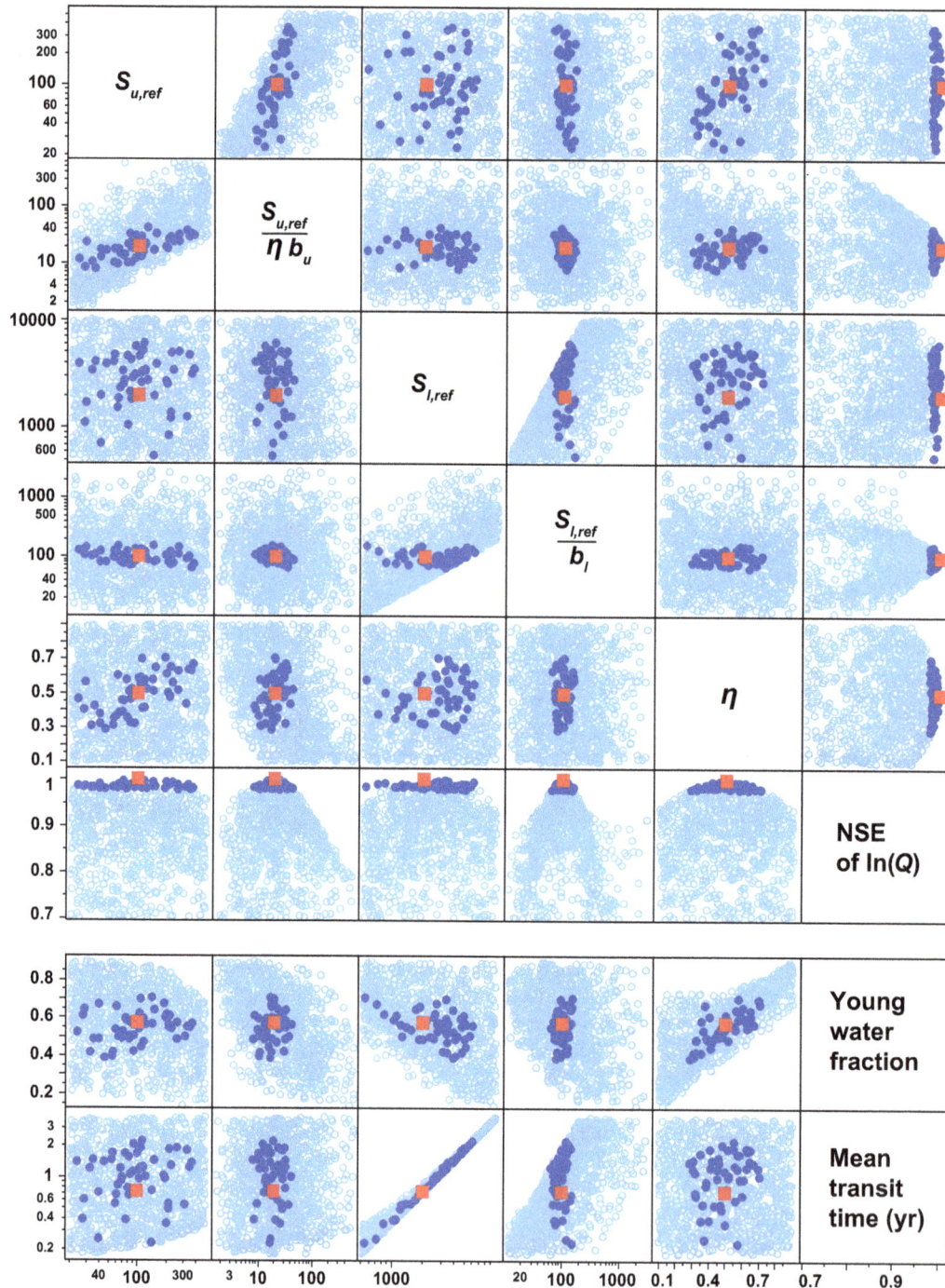

Figure B2. Equifinality partly cured by parameter transformations. The scatterplot matrix shows relationships among 1000 random parameter sets and the NSE of discharge time series driven by Smith River (Mediterranean climate) precipitation forcing, along with two key model outputs, the young water fraction and mean transit time in discharge (bottom two rows). As in Fig. B1, the red square indicates the "reference" parameter set that was used to generate the discharge time series that the other parameter sets were tested against; these reference parameters thus correspond to NSE = 1.00 by definition. The dark blue dots show the best-fitting 50 (or 5 %) of the parameter sets, all with NSE ≥ 0.98. In contrast to Fig. B1, three of the five parameters can be constrained by calibration against discharge (as shown by the clear peaks in NSE), and none of the parameters are strongly correlated with one another. However, the two reference storage volumes $S_{u,ref}$ and $S_{l,ref}$ remain poorly constrained. The mean transit time is determined almost entirely by $S_{l,ref}$, so it cannot be constrained by parameter calibration against the streamflow hydrograph.

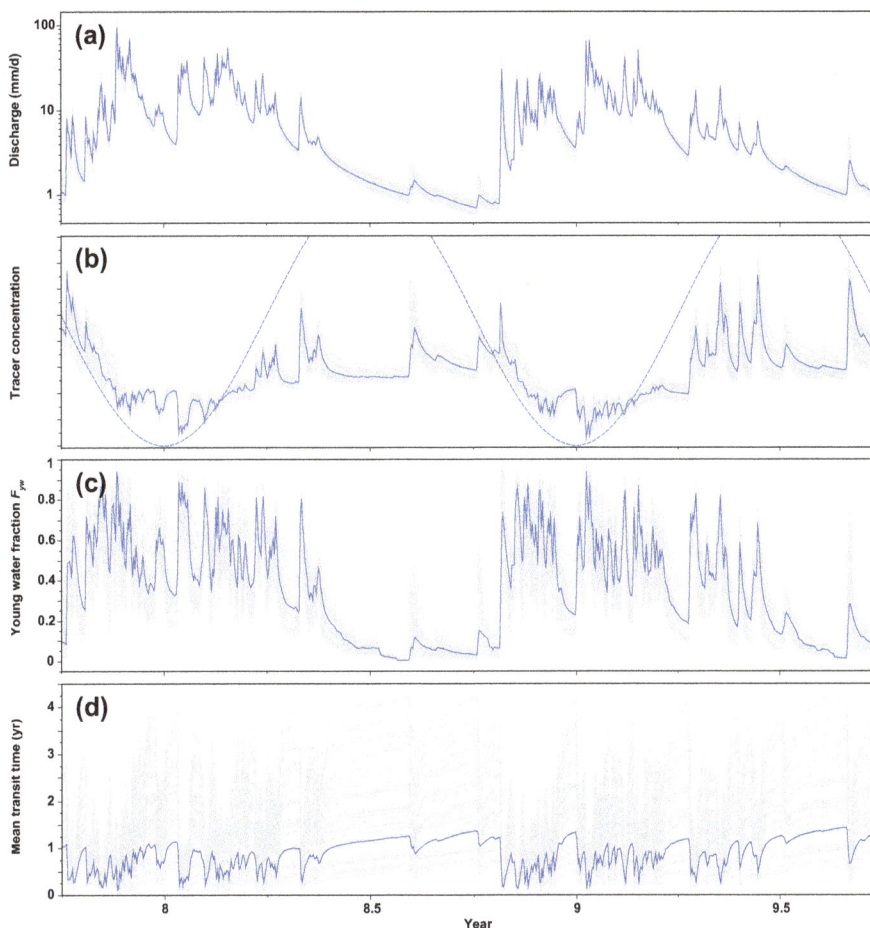

Figure B3. Excerpts from time series of discharge, tracer concentrations, young water fractions, and mean travel times in the two-box model with Smith River (Mediterranean climate) precipitation forcing and the reference parameter set (the dark lines, for the parameter values shown by the red squares in Figs. B1 and B2) and the 50 parameter sets that come closest to matching the reference discharge time series (the light gray lines, for the parameter sets shown by the solid blue dots in Figs. B1 and B2). The 50 gray hydrographs (**a**) cluster closely around the blue hydrograph (which is unsurprising because they have been selected to do so). The 50 gray tracer concentration curves (**b**) also generally follow the blue curve (the precipitation tracer sinusoid is shown for comparison by the dashed line). By contrast, the young water fraction F_{yw} (**c**) and mean transit time (**d**) are much more variable; the gray curves vary by an average range of 0.3 in F_{yw} and a factor of 9.5 in mean transit time.

the original parameter space is low. In higher-dimension parameter spaces, multivariate techniques such as factor analysis may be helpful. Nonetheless, given the obvious utility of this simple correlation analysis and the perturbation analysis of Sect. 3.2, it is surprising that they are not more widely used in hydrological modeling.

Despite the improved identifiability of the parameter space, however, it is still not possible to constrain the mean transit time by calibration to the hydrograph. As the bottom row of scatterplots in Fig. B2 shows, the MTT is almost entirely determined by the lower box's reference volume $S_{1,ref}$, as one would expect from Eq. (14). However, as predicted by the perturbation analysis in Sect. 3.2, and as shown by Fig. B2, the runoff response of the model system is essentially independent of $S_{1,ref}$ and therefore cannot be used to

constrain it. The runoff response does depend on the ratio of $S_{1,ref}$ to b_1, and thus can be used to constrain that ratio, but it cannot constrain $S_{1,ref}$ by itself, and thus it cannot constrain the MTT. For the young water fraction F_{yw} the outlook is not quite as bleak, because F_{yw} is correlated with the partition coefficient η, which can be constrained somewhat by calibration. As a result, it appears that F_{yw} could potentially be constrained within roughly 1/3 of its full range by parameter calibration to the hydrograph.

Figure B3 provides a different visualization of the same equifinality problem. Figure B3 shows a 2-year excerpt from the simulated time series of streamflows, tracer concentrations, young water fractions, and mean transit times for the reference parameter set (the blue curves), along with the 50 parameter sets that gave the best fit to the reference hydrograph (the gray curves). Because these 50 parameter sets were those that matched the reference hydrograph best, it is unsurprising that the 50 gray hydrographs generally follow the blue reference hydrograph in Fig. B3a. The 50 gray tracer concentration time series also follow the blue reference time series (Fig. B3b), but with somewhat greater variability than the hydrographs, indicating that the parameter values affect the chemographs and the hydrographs in somewhat different ways. But the most striking feature of Fig. B3 is the much greater variability among the young water fractions F_{yw} and (especially) the MTTs for these same parameter sets (Fig. B3c, d). Although all the parameter sets fit the reference hydrograph nearly perfectly, they vary over a range of 0.3 in F_{yw} (out of a total possible range of 1.0) and over a factor of 9.5 in MTT, on average, for the whole time period. Thus, these time series demonstrate, consistent with Fig. B2, that there are wide ranges of variability in F_{yw} and especially MTT that cannot be constrained by calibration to the hydrograph.

Acknowledgements. I thank Scott Jasechko and Jeff McDonnell for the intensive discussions that motivated this analysis, and Markus Weiler and an anonymous reviewer for their comments. I thank the Centre for Ecology and Hydrology for making the Plynlimon data available.

References

Benettin, P., van der Velde, Y., van der Zee, S., Rinaldo, A., and Botter, G.: Chloride circulation in a lowland catchment and the formulation of transport by travel time distributions, Water Resour. Res., 49, 4619–4632, doi:10.1002/wrcr.20309, 2013.

Benettin, P., Kirchner, J., Rinaldo, A., and Botter, G.: Modeling chloride transport using travel-time distributions at Plynlimon, Wales, Water Resour. Res., 51, 3259–3276, doi:10.1002/2014WR016600, 2015.

Bethke, C. M., and Johnson, T. M.: Groundwater age and groundwater age dating, Annu. Rev. Earth Planet. Sci., 36, 121–152, doi:10.1146/annurev.earth.36.031207.124210, 2008.

Beven, K.: On subsurface stormflow: predictions with simple kinematic theory for saturated and unsaturated flows, Water Resour. Res., 18, 1627–1633, 1982.

Beven, K.: A manifesto for the equifinality thesis, J. Hydrol., 320, 18–36, doi:10.1016/j.jhydrol.2005.07.007, 2006.

Birkel, C., Soulsby, C., and Tetzlaff, D.: Modelling catchment-scale water storage dynamics: reconciling dynamic storage with tracer-inferred passive storage, Hydrol. Process., 25, 3924–3936, 2011.

Birkel, C., Soulsby, C., Tetzlaff, D., Dunn, S., and Spezia, L.: High-frequency storm event isotope sampling reveals time-variant transit time distributions and influence of diurnal cycles, Hydrol. Process., 26, 308–316, doi:10.1002/hyp.8210, 2012.

Botter, G.: Catchment mixing processes and travel time distributions, Water Resour. Res., 48, 15, W05545, doi:10.1029/2011wr011160, 2012.

Botter, G., Bertuzzo, E., and Rinaldo, A.: Transport in the hydrological response: Travel time distributions, soil moisture dynamics, and the old water paradox, Water Resour. Res., 46, W03514, doi:10.1029/2009WR008371, 2010.

Botter, G., Bertuzzo, E., and Rinaldo, A.: Catchment residence and travel time distributions: The master equation, Geophys. Res. Lett., 38, L11403, doi:10.1029/2011GL047666, 2011.

Clark, M. P. and Kavetski, D.: Ancient numerical daemons of conceptual hydrological modeling: 1. Fidelity and efficiency of time stepping schemes, Water Resour. Res., 46, W10510, doi:10.1029/2009wr008894, 2010.

Duan, Q., Schaake, J., Andreassian, V., Franks, S., Goteti, G., Gupta, H. V., Gusev, Y. M., Habets, F., Hall, A., Hay, L., Hogue, T., Huang, M., Leavesley, G., Liang, X., Nasonova, O. N., Noilhan, J., Oudin, L., Sorooshian, S., Wagener, T., and Wood, E. F.: Model Parameter Estimation Experiment (MOPEX): An overview of science strategy and major results from the second and third workshops, J. Hydrol., 320, 3–17, doi:10.1016/j.jhydrol.2005.07.031, 2006.

Feng, X. H., Faiia, A. M., and Posmentier, E. S.: Seasonality of isotopes in precipitation: A global perspective, J. Geophys. Res.-Atmos., 114, D08116, doi:10.1029/2008jd011279, 2009.

Harman, C. J.: Time-variable transit time distributions and transport: Theory and application to storage-dependent transport of chloride in a watershed, Water Resour. Res., 51, 1–30, doi:10.1002/2014WR015707, 2015.

Heidbüchel, I., Troch, P. A., Lyon, S. W., and Weiler, M.: The master transit time distribution of variable flow systems, Water Resour. Res., 48, W06520, doi:10.1029/2011WR011293, 2012.

Hrachowitz, M., Soulsby, C., Tetzlaff, D., Malcolm, I. A., and Schoups, G.: Gamma distribution models for transit time estimation in catchments: Physical interpretation of parameters and implications for time-variant transit time assessment, Water Resour. Res., 46, W10536, doi:10.1029/2010wr009148, 2010.

Hrachowitz, M., Savenije, H., Bogaard, T. A., Tetzlaff, D., and Soulsby, C.: What can flux tracking teach us about water age distribution patterns and their temporal dynamics?, Hydrol. Earth Syst. Sci., 17, 533–564, doi:10.5194/hess-17-533-2013, 2013.

Ibbitt, R. P. and O'Donnell, T.: Designing conceptual catchment models for automatic fitting methods, in: Mathematical Models in Hydrology, International Association of Hydrological Sciences Publication, Wallingford, UK, 461–475, 1974.

Kavetski, D. and Clark, M. P.: Ancient numerical daemons of conceptual hydrological modeling: 2. Impact of time stepping schemes on model analysis and prediction, Water Resour. Res., 46, W10511, doi:10.1029/2009wr008896, 2010.

Kavetski, D. and Clark, M. P.: Numerical troubles in conceptual hydrology: Approximations, absurdities and impact on hypothesis testing, Hydrol. Process., 25, 661–670, doi:10.1002/hyp.7899, 2011.

Kirchner, J. W.: A double paradox in catchment hydrology and geochemistry, Hydrol. Process., 17, 871–874, 2003.

Kirchner, J. W.: Catchments as simple dynamical systems: catchment characterization, rainfall-runoff modeling, and doing hydrology backward, Water Resour. Res., 45, W02429, doi:10.1029/2008WR006912, 2009.

Kirchner, J. W.: Aggregation in environmental systems – Part 1: Seasonal tracer cycles quantify young water fractions, but not mean transit times, in spatially heterogeneous catchments, Hydrol. Earth Syst. Sci., 20, 279–297, doi:10.5194/hess-20-279-2016, 2016.

Kirchner, J. W., Feng, X., and Neal, C.: Fractal stream chemistry and its implications for contaminant transport in catchments, Nature, 403, 524–527, 2000.

Kirchner, J. W., Feng, X., and Neal, C.: Catchment-scale advection and dispersion as a mechanism for fractal scaling in stream tracer concentrations, J. Hydrol., 254, 81–100, 2001.

Kreft, A. and Zuber, A.: On the physical meaning of the dispersion equation and its solutions for different initial and boundary conditions, Chem. Eng. Sci., 33, 1471–1480, 1978.

McDonnell, J. J. and Beven, K.: Debates-The future of hydrological sciences: A (common) path forward? A call to action aimed at understanding velocities, celerities and residence time distributions of the headwater hydrograph, Water Resour. Res., 50, 5342–5350, doi:10.1002/2013wr015141, 2014.

McGuire, K. J. and McDonnell, J. J.: A review and evaluation of catchment transit time modeling, J. Hydrol., 330, 543–563, 2006.

Neal, C., Wilkinson, J., Neal, M., Harrow, M., Wickham, H., Hill, L., and Morfitt, C.: The hydrochemistry of the headwaters of the River Severn, Plynlimon, Hydrol. Earth Syst. Sci., 1, 583–617, doi:10.5194/hess-1-583-1997, 1997.

Neal, C., Reynolds, B., Norris, D., Kirchner, J. W., Neal, M., Rowland, P., Wickham, H., Harman, S., Armstrong, L., Sleep, D., Lawlor, A., Woods, C., Williams, B., Fry, M., Newton, G., and Wright, D.: Three decades of water quality measurements from the Upper Severn experimental catchments at Plynlimon, Wales: an openly accessible data resource for research, modelling, environmental management and education, Hydrol. Process., 25, 3818–3830, doi:10.1002/hyp.8191, 2011.

Peters, N. E., Burns, D. A., and Aulenbach, B. T.: Evaluation of high-frequency mean streamwater transit-time estimates using groundwater age and dissolved silica concentrations in a small forested watershed, Aquat. Geochem., 20, 183–202, 2014.

Seeger, S. and Weiler, M.: Reevaluation of transit time distributions, mean transit times and their relation to catchment topography, Hydrol. Earth Syst. Sci., 18, 4751–4771, doi:10.5194/hess-18-4751-2014, 2014.

Tetzlaff, D., Malcolm, I. A., and Soulsby, C.: Influence of forestry, environmental change and climatic variability on the hydrology, hydrochemistry and residence times of upland catchments, J. Hydrol., 346, 93–111, 2007.

Van der Velde, Y., De Rooij, G. H., Rozemeijer, J. C., van Geer, F. C., and Broers, H. P.: The nitrate response of a lowland catchment: on the relation between stream concentration and travel time distribution dynamics, Water Resour. Res., 46, W11534, doi:10.1029/2010WR009105, 2010.

Van der Velde, Y., Torfs, P. J. J. F., van der Zee, S. E. A. T. M., and Uijlenhoet, R.: Quantifying catchment-scale mixing and its effect on time-varying travel time distributions, Water Resour. Res., 48, W06536, doi:10.1029/2011WR011310, 2012.

9

Assessing various drought indicators in representing summer drought in boreal forests in Finland

Y. Gao[1], T. Markkanen[1], T. Thum[1], M. Aurela[1], A. Lohila[1], I. Mammarella[2], M. Kämäräinen[1], S. Hagemann[3], and T. Aalto[1]

[1]Finnish Meteorological Institute, P.O. Box 503, 00101 Helsinki, Finland
[2]University of Helsinki, Department of Physics, P.O. Box 48, 00014 Helsinki, Finland
[3]Max Planck Institute for Meteorology, Bundesstr. 53, 20146 Hamburg, Germany

Correspondence to: Y. Gao (yao.gao@fmi.fi)

Abstract. Droughts can have an impact on forest functioning and production, and even lead to tree mortality. However, drought is an elusive phenomenon that is difficult to quantify and define universally. In this study, we assessed the performance of a set of indicators that have been used to describe drought conditions in the summer months (June, July, August) over a 30-year period (1981–2010) in Finland. Those indicators include the Standardized Precipitation Index (SPI), the Standardized Precipitation–Evapotranspiration Index (SPEI), the Soil Moisture Index (SMI), and the Soil Moisture Anomaly (SMA). Herein, regional soil moisture was produced by the land surface model JSBACH of the Max Planck Institute for Meteorology Earth System Model (MPI-ESM). Results show that the buffering effect of soil moisture and the associated soil moisture memory can impact on the onset and duration of drought as indicated by the SMI and SMA, while the SPI and SPEI are directly controlled by meteorological conditions.

In particular, we investigated whether the SMI, SMA and SPEI are able to indicate the Extreme Drought affecting Forest health (EDF), which we defined according to the extreme drought that caused severe forest damages in Finland in 2006. The EDF thresholds for the aforementioned indicators are suggested, based on the reported statistics of forest damages in Finland in 2006. SMI was found to be the best indicator in capturing the spatial extent of forest damage induced by the extreme drought in 2006. In addition, through the application of the EDF thresholds over the summer months of the 30-year study period, the SPEI and SMA tended to show more frequent EDF events and a higher fraction of influenced area than SMI. This is because the SPEI and SMA are standardized indicators that show the degree of anomalies from statistical means over the aggregation period of climate conditions and soil moisture, respectively. However, in boreal forests in Finland, the high initial soil moisture or existence of peat often prevent the EDFs indicated by the SPEI and SMA to produce very low soil moisture that could be indicated as EDFs by the SMI. Therefore, we consider SMI is more appropriate for indicating EDFs in boreal forests. The selected EDF thresholds for those indicators could be calibrated when there are more forest health observation data available. Furthermore, in the context of future climate scenarios, assessments of EDF risks in northern areas should, in addition to climate data, rely on a land surface model capable of reliable prediction of soil moisture.

1 Introduction

Drought can be essentially defined as a prolonged and abnormal moisture deficiency (World Meteorological Organization, 2012). However, the cumulative nature of drought, the temporal and spatial variance during drought development, and the diverse systems that drought could have an impact on make drought difficult to quantify and define universally (Heim, 2002). The American Meteorological Society (1997) classifies drought into four categories: meteorological or climatological drought, agriculture or soil moisture drought, hydrological drought, and socio-economic drought. Drought is principally induced by a lack of precipitation. Furthermore, high atmospheric water demand, due to warm temperatures, low relative humidity, and changes in other environmental variables, often coincides with the absence of precipitation (Hirschi et al., 2011). Through land–atmosphere interactions, prolonged meteorological drought can further exacerbate soil moisture drought, or even hydrological drought (Mishra and Singh, 2010; Tallaksen and Van Lanen, 2004).

A number of drought indicators have been developed in the past in order to quantify the characteristics of the different drought types and their potential impacts on diverse ecosystems and societies (Heim, 2002). The most prominent and widely used drought indicator is the Standardized Precipitation Index (SPI), which has been recommended as a standard drought indicator by the World Meteorological Organization (WMO) due to its flexibility for various timescales, simplicity in input parameters and calculation, as well as effectiveness in decision making (Sheffield and Wood, 2011; Hayes et al., 2011). The SPI was developed to provide a spatially and temporally invariant comparison of drought determined by precipitation at different timescales (McKee et al., 1993, 1995). The Standardized Precipitation–Evapotranspiration Index (SPEI) is developed based on the SPI, and, in addition to precipitation, also accounts for temperature impacts on drought (Vicente-Serrano et al., 2010). Soil moisture status has been explored through the Soil Moisture Anomaly (SMA) and Soil Moisture Index (SMI). The SMA has been adopted in the Coupled Model Intercomparison Project (CMIP) in order to study soil moisture drought in present and future projections in Global Circulation Models (GCMs) (Orlowsky and Seneviratne, 2013). The SMI (also referred to as Relative Extractable Water – REW) is often used to investigate soil water related plant physiology issues, as it can represent the relative plant available water in the root zone (Lagergren and Lindroth, 2002; Granier et al., 1999). Those drought indicators are globally applicable. However, only few studies have examined drought indicators against drought impact data in regional level (Blauhut et al., 2015). Drought studies in northern Europe are quite rare due to the low occurrence of drought. Nevertheless, a soil moisture index calculated with simulated soil moisture have been tested with the forest health observation data in Finland in Muukkonen et al. (2015).

Boreal forests have been recognized as a *tipping element* of the Earth system as they are highly sensitive to climate warming (Lenton et al., 2008). Bi et al. (2013) reported that satellite observations show substantial greening of Eurasia with climate warming; however, browning in the boreal region of Eurasia has also been found despite a much larger fraction of browning in the boreal region of North America due to less precipitation. Forest damage induced by drought is a cumulative effect and is closely linked to soil moisture (Granier et al., 2007). Reduction of tree transpiration at the stand level induced by the low soil moisture condition has been broadly observed in most tree species (Irvine et al., 1998; Bréda et al., 1993; Clenciala et al., 1998). In

recent years, micrometeorological flux networks with intensive ancillary data have greatly supported the investigation of the relationship between drought and carbon fluxes over diverse ecosystems (Grossiord et al., 2014; Krishnan et al., 2006; Welp et al., 2007; Law et al., 2002; Grünzweig et al., 2003). In general, those studies observed a growth reduction in forests as a consequence of drought. In addition to a reduction in forest productivity (Ciais et al., 2005; Granier et al., 2007), severe drought can lead to tree mortality in boreal forests (Allen et al., 2010; Peng et al., 2011). In spite of the frequency, duration and severity of droughts, drought-induced forest damages may be connected to specific soil and plant characteristics, such as soil texture and depth, exposure, species and their composition, and life stage (Muukkonen et al., 2015; Grossiord et al., 2014; Dale et al., 2001; Gimbel et al., 2015). Disturbance of boreal forests could give rise to further feedbacks to the global climate system, due to the complex interactions between boreal forests and the climate system via the control on energy, water, and carbon cycles (Bonan, 2008; Ma et al., 2012).

Soil moisture strongly regulates transpiration and photosynthesis for most terrestrial plants, consequently modulating water and energy cycles of the landscape, as well as biogeochemical cycles of the plants (Seneviratne et al., 2010; Bréda et al., 2006). Nevertheless, ground observed soil moisture is limited in time and space (Seneviratne et al., 2010). Regional analysis is necessary to fully capture the spatial heterogeneity of the impacts of drought on ecosystem functioning (Aalto et al., 2015). In recent years, a multi-decadal global soil moisture record that incorporates passive and active microwave satellite retrievals has become available (Liu et al., 2012). However, microwave remote sensing can only provide surface soil moisture in the upper centimetres of the soil. Land surface models (LSMs) are valuable tools to derive spatial maps of soil moisture in deeper soil layers, for instance, the root-zone soil moisture, which is of particular importance in many climate studies (Hain et al., 2011; Rebel et al., 2012; Seneviratne et al., 2010).

This study aims to improve our understanding of the properties of different drought indicators (including SPI, SPEI, SMA, and SMI), and assess their ability to indicate the Extreme Drought that affects Forest health (EDF) in boreal forests in Finland. The EDF is defined in this study according to the extreme drought in Finland in 2006, which caused visible impacts on forest appearance compared to normal years (Muukkonen et al., 2015). For the soil moisture drought indicators (SMA, SMI), regional soil moisture was simulated by the JSBACH LSM of the Max Planck Institute for Meteorology Earth System Model (MPI-ESM) with its five layer soil hydrology scheme. Thus, this study also aims to gain insights into the capability of the five layer soil hydrology scheme with its parameters in the JSBACH LSM to simulate soil moisture dynamics across Finland. The outcome of this study provides suggestions on the selection and interpreta-

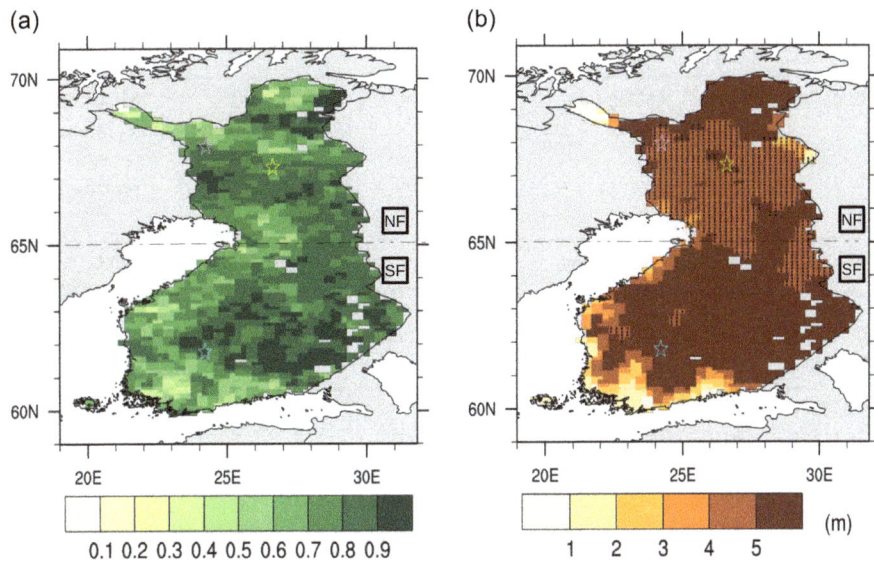

Figure 1. (a) The forest cover fraction over Finland in JSBACH derived according to Corine land cover 2006 data; **(b)** soil depth and the soil type (categorized as peatland and mineral soil) distributions in JSBACH over Finland (peatland area – dotted area; mineral soil area – area without dots). Northern (NF) and southern Finland (SF) are divided at the 65° N latitude. The location of the three ecosystem sites used in this study are marked as stars on the map (Blue–Hyytiälä; Yellow–Sodankylä; Pink–Kenttärova). The uncovered grid boxes (grey cells) in Finland represent inland lakes.

tion of drought indicators for estimating EDF risks in boreal forests in future climate scenarios.

2   Study area and observation-based data sets

2.1   Study area

Our study area is focused on Finland (Fig. 1). Finland is a northern European country, situated between 60 and 70° N in the north-western part of the Eurasian continent, close to the North Atlantic Ocean.

The temperature of Finland is generally moderate, compared to many other places at the same latitudes (Tikkanen, 2005). This is because the westerly winds bring warm air masses from the North Atlantic Ocean in winter, while in summer they bring clouds that decrease the amount of incoming solar radiation. However, the continental high pressure system located over the Eurasian continent occasionally influences the climate causing warm and cold spells in summer and winter, respectively. The precipitation in Finland is influenced by the Scandinavian mountain range, which blocks large amounts of moisture that are transported from west to east. Both temperature and precipitation show spatial variations along a south to north gradient. The annual mean surface temperature is about 5–6 °C in the south of Finland and extends below −2 °C in the coldest area located in northern Lapland. Annual precipitation, averaged over the 1971–2000 period, is more than 700 mm in the south and less than 400 mm in the north (Aalto et al., 2013; Drebs et al., 2002).

In addition to mineral soils, a high areal fraction of peatland is typical for Finland, especially in the north of the country. Shallow soil areas accompanied with bare rocks are mostly located around the coastline in southern Finland and are also found in north-west Finland, which is a part of the Scandinavian mountain range.

Coniferous forest, including Scots pine and Norway spruce, is the dominant forest type in Finnish boreal forests (Finnish Statistical Yearbook of Forestry 2012, 2012). Broadleaved forest accounts for less than 10 % of the forest area. In total, 75 % of the total forest land area is located on mineral soils. In the past, large areas of unproductive peatlands have been drained to grow forests in Finland, as a result of the originally high proportion of pristine peatlands and timber production requirements (Päivänen and Hånell, 2012).

2.2   Meteorological and soil moisture data

The gridded meteorological data compiled by the Finnish Meteorological Institute (FMI gridded observational data) are interpolated products from stand meteorological observations in Finland (Aalto et al., 2013). In this study, daily FMI gridded observational data were used on a 0.2° longitude × 0.1° latitude grid for the period 1981–2010. These data comprise daily mean, minimum, and maximum temperatures, precipitation, relative humidity, and incoming shortwave radiation. In addition, the 10 m wind speed of ECWMF ERA-Interim reanalysis data (Simmons et al., 2007) was used to calculate the reference evapotranspiration (ET_0) for

Table 1. Characteristics of the three micrometeorological sites. The plant functional types and the soil types in the JSBACH site simulations corresponding to observed tree species and soil types at the three sites are shown in brackets.

Site	Location	Period	Main tree specie	Soil type	Analysed measurement depth of soil moisture (cm)	Measurement technique for soil moisture	Reference
Hyytiälä	61°51′ N, 24°18′ E	1999–2009	Scots Pine (Conifers)	Haplicpodzol (Mineral)	−5 to −23; −23 to −60	TDR	Vesala et al. (2005)
Sodankylä	67°21′ N, 26°38′ E	2001–2008	Scots Pine (Conifers)	Sandy Podzol (Mineral)	−10, −20, −30 (averaged)	ThetaProbe	Thum et al. (2008)
Kenttärova	67°59′ N, 24°15′ E	2008–2010	Norway Spruce (Conifers)	Podzol (Mineral)	−10	ThetaProbe	Aurela et al. (2015)

SPEI from the Penman–Monteith equation (Allen et al., 1994).

In addition, meteorological and soil moisture data at three micrometeorological sites were used as meteorological forcing for site level simulations and for a comparison of modelled and observed soil moisture, respectively (Fig. 1; Table 1). Soil parameters derived from observations are only available for the Hyytiälä site (water content at saturation (θ_{SAT}) = 0.50 m^3 m^{-3}, water content at field capacity (θ_{FC}) = 0.30 m^3 m^{-3}, water content at wilting point (θ_{WILT}) = 0.08 m^3 m^{-3}). As explained in more detail below, we used the second layer of simulated soil moisture in the JSBACH soil profile (layer 2; 6.5–30 cm). Therefore, the observed soil moisture data were taken from existing measurement depths, which are consistent with the JSBACH layer 2 soil depth. For the Sodankylä site, an average of the measurements at soil depths −10, −20, and −30 cm was employed and for the Kenttärova site, the measurement at −10 cm was used. The two levels in the Hyytiälä soil moisture measurement, −5 to −23 and −23 to −60 cm, were both used.

2.3 Forest health observation data

We adopted the yearly forest drought damage percentage in Finland from Muukkonen et al. (2015), who based their analysis on the forest health observation data from a pan-European monitoring programme ICP Forests (the International Co-operative Programme on the Assessment and Monitoring of Air Pollution Effects on Forests). The visual forest damage symptom inspections have been carried out by 10–12 trained observers during July–August since 2005, following internationally standardized methods (Eichorn et al., 2010) and national field guidelines (e.g. Lindgren et al., 2005). When a single sample tree in a site showed drought symptoms, it was recognized as a drought damage site. Therefore, uncertainties can rise from different personal interpretations and inappropriate time point of the visual inspections.

A 4-year (2005–2008) period of forest health observation data were analysed in Muukkonen et al. (2015). The summer of 2006 was extremely dry, and 24.4 % of the 603 forest health observation sites over entire Finland were affected, in comparison to 2–4 % damaged sites in a normal year. In southern Finland, 30 % of the observational sites showed drought symptoms.

3 Methods

3.1 JSBACH land surface modelling

JSBACH (Raddatz et al., 2007; Reick et al., 2013) is the LSM of the Max Planck Institute for Meteorology Earth System Model (MPI-ESM) (Stevens et al., 2013; Roeckner et al., 1996). It simulates energy, hydrology, and carbon fluxes within the soil–vegetation continuum and between the land surface system and the atmosphere. Diversity of vegetation is represented by plant functional types (PFTs). A set of properties are attributed to PFTs with respect to the various processes JSBACH is accounting for. For soil hydrology, a bucket scheme was originally used, in which the maximum water that can be stored in the soil moisture reservoir (W_{cap}) corresponds to the root-zone water content (Hagemann, 2002). The bucket can be supplied from precipitation and snowmelt, but depleted through evapotranspiration (evaporation from the upper 10 cm of soil and plant transpiration from below), and lateral drainage. These processes are related to the amount of soil moisture in the bucket and are regulated by the Arno scheme, which separates rainfall and snowmelt into surface run-off and infiltration, and considers soil heterogeneity (Dümenil and Todini, 1992).

In order to more adequately simulate the soil hydrology, a five layer soil hydrology scheme has been newly introduced in JSBACH (Hagemann and Stacke, 2015). The five layer structure is defined with increasing layer thickness (0.065, 0.254, 0.913, 2.902, and 5.7 m) and reaches almost 10 m depth below the surface. However, the soil depth to the bed rock, determines the active soil layers. Therefore, in the five layer soil hydrology scheme, the root zone is differentiated into several layers, and there could be soil layers below the root zone, which transport water upwards for transpiration when the root zone has dried out. Moreover, evaporation from bare soil can occur when the uppermost layer is wet, while the whole soil moisture bucket must be largely saturated in the bucket scheme. For a more detailed description of the five layer soil hydrology scheme in JSBACH and how it affects soil moisture memory, see Hagemann and Stacke (2015).

In this work, the regional JSBACH simulation was driven by the prescribed meteorological data (1980–2011) simulated by the regional climate model REMO (Jacob, 2001; Jacob and Podzun, 1997), whose temperature and precipitation biases were corrected with the FMI gridded observational data (Aalto et al., 2013). A quantile–quantile type bias correction algorithm was applied to daily mean temperature (Räisänen and Räty, 2013), while daily cumulative precipitation was corrected using parametric quantile mapping (Räty et al., 2014). The ECWMF ERA-Interim reanalysis (Dee et al., 2011) was used as lateral boundary data for the climate variables and as the initial values of surface climate variables for the REMO simulation. Both the regional JSBACH and REMO simulations were conducted in the Fennoscandian domain centred on Finland with a spatial resolution of 0.167° (15–20 km). The land cover distribution for REMO and the corresponding PDF distribution for JSBACH over this domain were derived from the more up-to-date and more precise Corine land cover 2006 data (European Environment Agency, 2007) rather than the standard GLCCD (US Geological Survey, 2001), which is important for simulating land–atmosphere interactions (Gao et al., 2015; Törmä et al., 2015). Finland is a country predominantly covered by forests. The forest cover fraction over Finland in JSBACH derived according to the Corine land cover 2006 data is shown in Fig. 1a. Also, an improved FAO (Food and Agriculture Organization of the United Nations) soil type distribution is adopted in the JSBACH LSM (FAO/UNESCO (1971–1981); see Hagemann and Stacke (2015), for details), while the soil depth distribution is derived from the soil type data set and FAO soil profile data (Dunne and Willmott, 1996) (Fig. 1b).

In addition, simulations were carried out for the three measurement sites with the observed local meteorological forcing. The characteristics of the sites together with the corresponding model settings are described in Table 1.

Prior to the actual regional and site level JSBACH simulations, long-term spin-ups were conducted to obtain equilibrium for the soil water and soil heat.

3.2 Drought indicators

A set of hydro-meteorological indicators were analysed. The SPI, SPEI, and SMA are standardized indicators that show the degree of anomalies to long-term means over the aggregation period, while SMI describes the instantaneous soil moisture status normalized with total soil moisture storage available to plants. In this study, daily SMI was used. The SPI, SPEI, and SMA were calculated with a 4-week (28 days) aggregation time frame, but they were updated every day with running inputs over the 30-year period. Both the 4-week aggregation time frame and 30-year study period are considered to be of sufficient duration climatologically under WMO guidelines (World Meteorological Organization, 2012). The SPI and SPEI were calculated using both the FMI gridded observational data set and the regional JSBACH forcing data described in the previous chapter, while SMA and SMI were computed with the layer 2 soil moisture from the regional JSBACH simulation. In addition, the SMIs were derived from site soil moisture observations, as well as from site JSBACH simulations. The layer 2 soil moisture from the JSBACH simulations was used, because the soil moisture in the shallower layer (layer 1) is highly sensitive to small changes in climatic variables, and the soil moisture dynamics in the deeper layers are excessively suppressed. Furthermore, the layer 2 is representative of the root zone in forest soils.

3.2.1 Soil moisture index

The SMI is a measure of plant available soil water content relative to the maximum plant available water in the soil (Betts, 2004; Granier et al., 2007; Seneviratne et al., 2010). The soil water above field capacity cannot be retained, and produces gravitational drainage and usually flows laterally away. The soil water below the wilting point is strongly held by the soil matrix to such an extent that the plants are unable to overcome this suction to access the water (Hillel, 1998).

The SMI is calculated as follows:

$$\mathrm{SMI} = (\theta - \theta_{\mathrm{WILT}}) / (\theta_{\mathrm{FC}} - \theta_{\mathrm{WILT}}),$$

where θ is the volumetric soil moisture [$\mathrm{m}^3_{\mathrm{H_2O}} \, \mathrm{m}^{-3}_{\mathrm{soil}}$], θ_{FC} is the field capacity, θ_{WILT} is the permanent wilting point.

Note that soil water content can exceed θ_{FC} and reach water-holding capacity (i.e. saturation ratio) under certain circumstances. For those cases, the SMI is set to 1, indicating maximum plant available water. θ_{FC} and θ_{WILT} depend on soil types in this study, although θ_{WILT} is also related to PFTs in some other studies. At Hyytiälä, θ_{SAT} (saturation ratio) was used instead of θ_{FC} to be consistent with the JSBACH soil hydrology where θ_{FC} acts as a proxy for θ_{SAT} on the large ESM grid scale (Hagemann and Stacke, 2015).

3.2.2 Soil moisture anomaly

The SMA is an index relevant to plant functioning (Burke and Brown, 2008). The SMA depicts the deviation of the soil moisture status in a certain period of a year to the soil moisture climatology over this period. It can be normalized by the standard deviation of the soil moisture in this respective period over all years, for direct comparison with the other standardized drought indicators, e.g. SPI, SPEI.

The SMA in this study is calculated following the method of Orlowsky and Seneviratne (2013):

$$\mathrm{SMA} = \left(\bar{\theta} - \bar{\mu}\right)/\bar{\sigma},$$

where $\bar{\theta}$ denotes the averaged volumetric soil moisture over a certain period in a year, while $\bar{\mu}$ and $\bar{\sigma}$ denote the mean and standard deviation of the volumetric soil moisture of this period over all the studied years.

3.2.3 Standardized precipitation index

The SPI inspects the amplitudes of precipitation anomalies over a desired period with respect to the long-term normal. The homogenized precipitation series is fitted into a normal distribution to define the relationship of probability to precipitation (Edwards and McKee, 1997). In this work, a Pearson type III distribution is adopted because it is more flexible and universal with its three parameters in fitting the sample data than the two parameter Gamma distribution (Guttman, 1994, 1999). The parameters of the Pearson type III distribution are fitted by the unbiased probability-weighted moments method. Typically, the timescales of SPI range from 1–24 months. The reduced precipitation under various durations can illustrate the impacts of drought on different water resources (Sivakumar et al., 2011). A time frame of less than 1 month is not recommended as the strong variability in weekly precipitation may lead to erratic behaviour in the SPI (Wu et al., 2007). However, the "moving window" of a minimum of 4 weeks with daily updating is acceptable (World Meteorological Organization, 2012). Furthermore, attention should be paid when interpreting the 1 month SPI to prevent misunderstanding. Large values in the 1 month SPI can be caused by relatively small departures from low mean precipitation (World Meteorological Organization, 2012).

The SPI is a probabilistic measure of the severity of a dry or wet event. An arbitrary drought classification with specific SPI thresholds was defined by McKee et al. (1993). Recently, an objective method based on percentiles from the United States Drought Monitor (USDM) has been recommended for defining location-specific drought thresholds (Quiring, 2009). For calculating SPI, we used the SPI function in R package SPEI version 1.6 (Beguería and Vicente-Serrano, 2013).

3.2.4 Standardized precipitation–evapotranspiration index

The SPEI is similar to SPI mathematically, but also accounts for the impact of temperature variability on drought through atmospheric water demand, in addition to the water supply from precipitation (Vicente-Serrano et al., 2010). The SPEI is based on a climatological surface water balance, which is calculated as the differences between precipitation and ET_0. In this work, ET_0 was calculated according to the FAO-56 Penman–Monteith equation (Allen et al., 1994; Beguería and Vicente-Serrano, 2013), which is predominately a physical-based method and has been tested over a wide range of climates (Ventura et al., 1999; López-Urrea et al., 2006). The water balance time series is normalized by a log-logistic probability distribution and its parameter fitting is based on the unbiased probability-weighted moments method. For calculating SPEI, we used the SPEI function in R package SPEI version 1.6 (Beguería and Vicente-Serrano, 2013).

3.3 Assessment of JSBACH simulated soil moisture dynamics

In order to evaluate the ability of JSBACH simulated soil moisture to detect drought, the SMI series at the three study sites from the site and the regional (the model grids where the sites are located) JSBACH simulations were compared with the observed soil moisture data over the common data coverage periods (Table 1). SMI based on the Hyytiälä observational data was calculated with θ_{SAT} and θ_{WILT} values measured at the site. Due to the lack of measured soil parameters at the Sodankylä and Kenttärova sites, the volumetric soil moisture measurements were directly used to examine the simulated soil moisture dynamics. An upper limit was set on the presented volumetric soil moisture to exclude abrupt and instantaneous peaks due to heavy snow melting or precipitation.

3.4 Intercomparison of drought indicators

The temporal and spatial coherency between the drought indicators was investigated at the regional level. The time correlations over our study period between the SPI calculated with the observational data set and the SPI calculated with the JSBACH forcing data were derived for the grid boxes in Finland. The same approach was adopted for the SPEIs. Moreover, the time correlations between the meteorological-based drought indicators (SPI, SPEI) calculated with the JSBACH forcing data and the soil moisture-based drought indicators (SMI, SMA) calculated with the JSBACH simulated soil moisture were derived for the grid boxes in Finland, as well as the time correlation between SMI and SMA. Furthermore, the spatial and temporal evolution of drought depicted by indicators was compared through time–latitude transections.

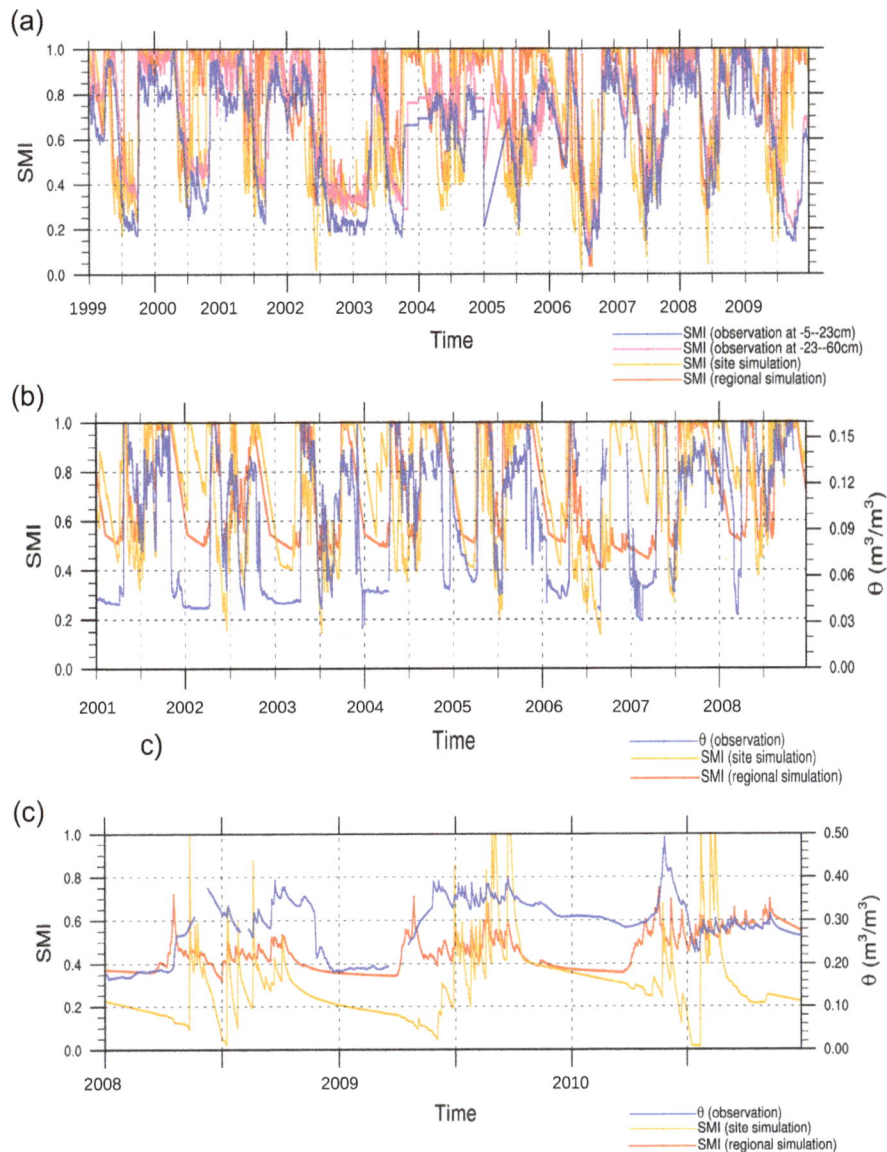

Figure 2. Soil moisture dynamics at the three micrometeorological sites: **(a)** Hyytiälä, **(b)** Sodankylä and **(c)** Kenttärova, comparing results from regional (the model grid boxs where the sites are located) and site JSBACH simulations with observations. The volumetric soil moisture (θ) is shown for the Sodankylä and Kenttärova sites.

3.5 Selection of EDF thresholds for indicators

According to the forest health observation data, we consider the 30 % forest damaged sites in southern Finland as the fraction of the area influenced by the severe drought in 2006, which is a reasonable assumption based on the dense and even distribution of observation sites over southern Finland. Based on this information, we utilized the cumulative area distributions of the SMI and SMA over southern Finland during the driest 28-day period of southern Finland in 2006 (i.e. in the case of SMI, this is the lowest 28-day-running mean value averaged over southern Finland) to derive their thresholds for this kind of extreme drought. Herein, as SMA

was calculated with 28-day-running means for soil moisture, the same time window was adopted for SMI to be consistent with SMA. The SPEI threshold for extreme drought is selected as 2 % of the SPEI data series, according to the recommended percentile classification (Quiring, 2009).

4 Results and discussion

4.1 Comparison of site soil moisture dynamics from JSBACH simulations with observations

In general, the timing of dry spells in summer in most of the years of the simulated soil moisture corresponded well with the observations at the three sites (Fig. 2). There was good agreement between the minimum values reached by the simulated and observed SMIs in summertime at Hyytiälä. The late summer of 2006 was noticeable as being extremely dry in the simulations and observations at Hyytiälä and Sodankylä. At Kenttärova, the extent of the SMI was quite different in the regional and site JSBACH simulations. This was mainly because different soil types are prescribed for this site, which affects not only the soil hydrology but also the values of SMI. In the regional simulation, Kenttärova was situated in a peat soil area, while in reality and in the site simulation the site is classified within a mineral soil area. The soil type in an individual grid for the regional simulation is homogeneous and defined according to the soil type with the highest coverage. The summer of 2010 was the driest among the three years at Kenttärova according to the observation, and the timing of the driest period after mid-summer shown in the observations was successfully captured by the site simulation. Moreover, the soil at Kenttärova was mostly unsaturated during those three years, even in the site simulations where it was realistically represented as a mineral soil. This is related to the small amount of precipitation during those years.

The diverse features of soil moisture among these sites in wintertime were captured by JSBACH. The soil tends to be saturated at Hyytiälä in winter, whereas at Sodankylä and Kenttärova there is a winter recession period of soil moisture when the soil tends to dry out. At Hyytiälä, the difference is due to infiltration of snowmelt water during intermittent periods when air temperature is above 0 °C, while at Sodankylä and Kenttärova, periods when the surface soil is frozen are more persistent and only percolation takes place then. The exceptionally low soil moisture during the winter 2003–2004 was also well simulated for Hyytiälä. This winter dry spell was caused by low rainfall in autumn 2003 and the relatively cold winter afterwards when there was not enough snowmelt water to recharge the deficit volume. This autumn-to-winter drought in 2003 at Hyytiälä was a rain to snow season drought (a precipitation deficit in the rainy season and at the beginning of the snow season) in combination with a cold snow season drought (see Van Loon and Van Lanen (2012) for the drought typology). The winter recession period of soil moisture at Kenttärova is longer than that in Sodankylä, probably because Kenttärova is located at higher latitudes. Large and obvious decreases in soil water immediately after the winter recession periods of soil moisture in 2008 and 2009 were shown by the site simulation, but not by the regional simulation. This is due to less precipita-

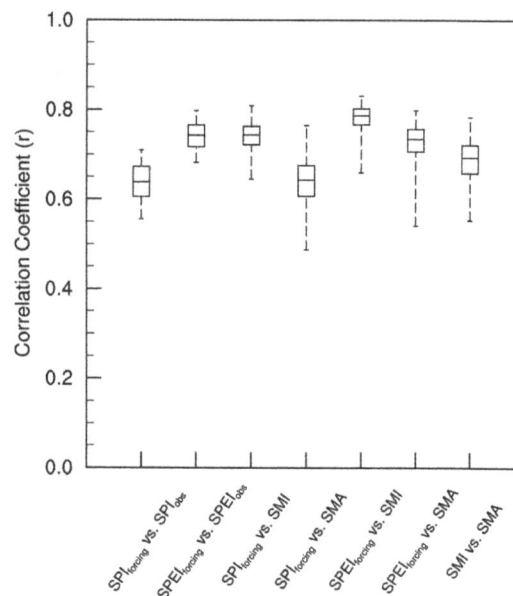

Figure 3. Percentiles of the time correlation coefficients across the grid boxes over Finland. The time correlations over the study period between the SPIs and SPEIs derived from the JSBACH forcing data and the observational data set ($SPI_{forcing}$ vs. SPI_{obs}, $SPEI_{forcing}$ vs. $SPEI_{obs}$), and the time correlations between SPI, SPEI calculated with the JSBACH forcing data and SMI, SMA calculated with the JSBACH simulated soil moisture ($SPI_{forcing}$ vs. SMI, $SPI_{forcing}$ vs. SMA, $SPEI_{forcing}$ vs. SMI, $SPEI_{forcing}$ vs. SMA), as well as the time correlation between SMI and SMA calculated with the JSBACH simulated soil moisture (SMI vs. SMA), are investigated. Dashed lines extend from 5th to 95th percentile of the correlation coefficients over Finland, boxes extend from 25th to 75th percentile and middle horizontal lines within each box are the medians. Those correlation coefficients are statistically significant ($p < 0.01$).

tion during this period in the meteorological forcing data for the site simulation, in comparison to the regional simulation (data not shown). Moreover, the balance between water consumption through evapotranspiration and water gained from snowmelt was more negative in the site simulations. In general, the layer 2 soil moisture in the regional simulation for Kenttärova captures the observed soil moisture dynamics at −10 cm depth better. However, a full evaluation would require observational data from several closely spaced soil layers.

Overall, the timing of summer dry spells and the winter characteristics of the observed soil moisture at the three sites were well captured by the simulated soil moisture, although the simulated soil moisture shows larger amplitudes and a faster response to changes in water inputs. The discrepancies in soil moisture between the site and the regional JSBACH simulations are mainly due to the differences in precipitation in summertime and in surface temperature during winter in the meteorological forcing data, as well as different soil types in specific locations. The latter is related to the difference in scales between the regional grid and the site. Soil characteristics tend to be heterogeneous, so that the characteristics may vary on scales from a metre to a kilometre. While for modelling on the regional grid, effective soil characteristics are chosen that represent the average characteristics of a grid box.

4.2 Intercomparison of drought indicators

The time correlations between the regional results of those drought indicators over our study period showed high correlation coefficients over Finland (Fig. 3). The medians of the time correlation coefficients over the whole of the country were greater than 0.6; with the 5 % percentiles also greater than 0.5, with the exception of the correlation coefficient between SMA and SPI. The agreement between SPEIs calculated with the JSBACH forcing data and the FMI gridded observational data set was better than that for SPIs. Furthermore, the soil moisture-based drought indicators revealed a better correspondence with SPEI than with SPI, which is reasonable as SPEI is based on the water balance. Therefore, in the following, we will focus on SPEI as the climatic driver indicator, and as there was a good correlation between the JSBACH forcing data and the FMI gridded observational data-based SPEIs, we restricted the data set by using the JSBACH forcing data-based SPEI, which was better related to the two soil moisture-based drought indicators from the model. Moreover, the correlation between SPEI and daily SMI was higher than that between SPEI and SMA. This is especially true for peatland areas while the correlations in mineral soil areas are more similar (see regional maps in the Supplement). This results from different soil moisture memory effects in those soil types.

From the time–latitude transections of the selected indicators (Fig. 4), the most exceptional dry years in our study period (e.g. 1994, 2006) can be distinguished, as well as the exceptionally wet years (e.g. 1981, 1998). Although there is generally a good correlation among all three indicators in capturing drought, there are differences among them in depicting drought durations and latitudinal extent at detailed locations and time. First, SPEI and SMA generally show more consistent patterns extending through a wider range of latitudes than SMI. Also, the buffering effect of soil moisture and the associated soil moisture memory can delay and extend dry or wet events as indicated by SMI and SMA, in comparison to those by the SPEI. For instance, the dry period in 1992 over southern Finland in SMI and SMA is longer than that in SPEI, and the wet period in the same year over northern Finland as indicated by SMA starts later in comparison to SPEI; however, this difference is not shown by SMI. Second, SMI exhibits a more distinct south–north gradient than the other two indicators. In particular, SMI describes more frequent droughts in the extreme southern parts of Finland. This is because the shallow soil in those areas is more sensitive to climate drivers. However, there is much less drought indicated by SMI in the extreme northern part of the country (above 68° N). This could be due to the atmospheric water demand at the same SPEI drought level in the north is weaker than that in the south. In other words, the deviation of the multi-year mean value in precipitation surplus (precipitation − evapotranspiration) can lead to a higher change in SPEI values in the very north of Finland than in the south, as the variability of the climate in the north of Finland is lower. Third, SMI between latitudes 66 and 68° N shows an evident narrow range; i.e., the soil is not saturated or deeply dried out. This is due to the abundance of peatland areas with a larger soil moisture buffer than mineral soil areas.

SMI values vary within different ranges for the peatland and mineral soil areas in southern and northern Finland, whereas SMA and SPEI, as they are standardized indicators, show no differences regarding the soil type or location (Fig. 5). The regionally averaged SMI over the peatland areas mainly varies from 0.4 to 0.6 in both the south and north of Finland, while the SPEI averaged over the same area ranges between −2.0 and 2.0. SMI in the mineral soil shows larger variations compared to peat soil under the same climatic conditions. The SMI averaged over the mineral soil areas ranges between 0.1 and 1.0 in the south of Finland and 0.4 to 1.0 in the north of Finland. The higher values associated with the regionally averaged SMI in the north are due to less shallow soils and less meteorological drought in comparison to the south.

4.3 EDFs indicated by drought indicators

Our results showed that the driest 28-day periods of southern Finland in 2006 were the same (from 20 July to 16 August) for SMI and SMA. The SMI and SMA thresholds for the EDF are 0.138 and −2.287, respectively (Fig. 6). Moreover, according to the recommended percentile classification (Quiring, 2009), the SPEI threshold for extreme drought, which is selected as 2 % of the SPEI data series, is −1.85 averaged over the grid boxes in Finland (−1.843 averaged over southern Finland). The averaged SPEI values over the EDF influenced areas in Finland depicted by SMI and SMA for the same period are −1.84 and −1.89, which are very close to the percentile-dependent SPEI threshold for extreme drought. This demonstrates that the degrees of EDF described by the derived SMI and SMA thresholds are consistent with the percentile-based threshold of extreme drought for SPEI, which is taken as the EDF threshold for SPEI.

Furthermore, we compared the regional distributions of the areas influenced by the 2006 EDF in the driest 28-day period indicated by SMI, SMA and SPEI (Fig. 7). The SMI showed that the EDF influenced areas were mainly located in southern Finland, whereas the SMA showed more EDF affected areas located in the middle to northern part of the country (mainly above 64° N). The EDF influenced areas presented by SMA in the north were mainly located in peatland areas, where the porosity of peat is much higher than

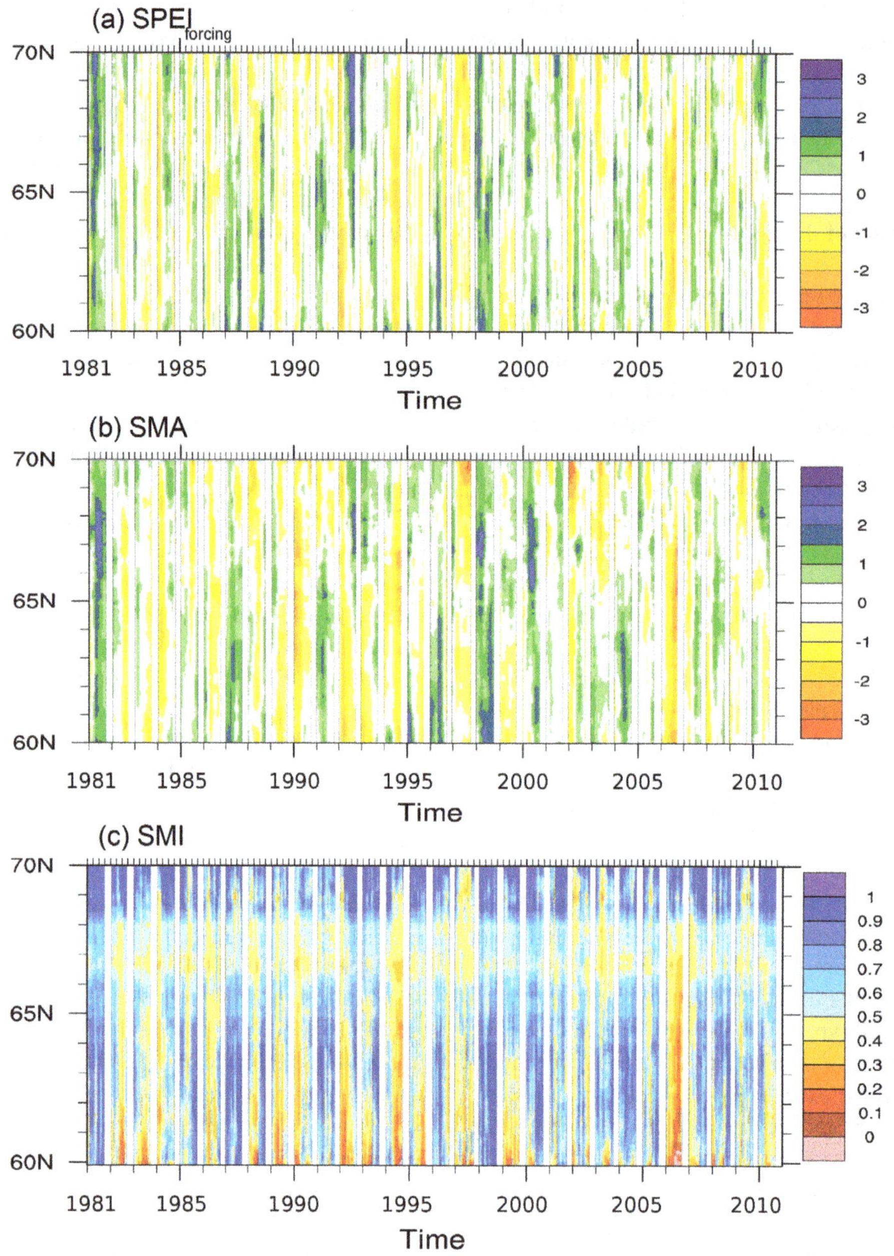

Figure 4. Latitude–time transections of **(a)** SPEI$_{forcing}$, **(b)** SMA, and **(c)** SMI over Finland in the study period (the summer months (June, July, August) in 1981–2010).

that of mineral soils. Although there was a strong decrease in relative soil moisture with respect to the long-term mean value in those areas during this EDF event, the absolute soil moisture was not sufficiently low for those areas to be recognized as EDF in terms of SMI (Fig. 5). Moreover, the EDF influenced areas in the south-eastern part of Finland, as indicated by SMI, were not shown by SMA, although these areas comprise relatively low SMA values. This is because there were more EDF influenced areas indicated by SMA in the middle of Finland compared to SMI. Those areas took up a part of the 30 % influenced area over the entire south-

ern Finland, which has been used for the selection of EDF thresholds by the cumulative area distributions. The areas impacted by EDF as indicated by SPEI, are widespread over Finland, complying with the climate conditions in this period. The extremely dry climate in northern Finland led to the EDF shown by SMA, but was not sufficiently intense for EDF to be captured by SMI. In southern Finland, the EDF areas of SMI generally agree with those of SPEI, except for the shallow soil area along the southern coastline and in the south-eastern part of the country (at 63° N). This more severe drought, which was indicated by SMI rather than by SPEI in

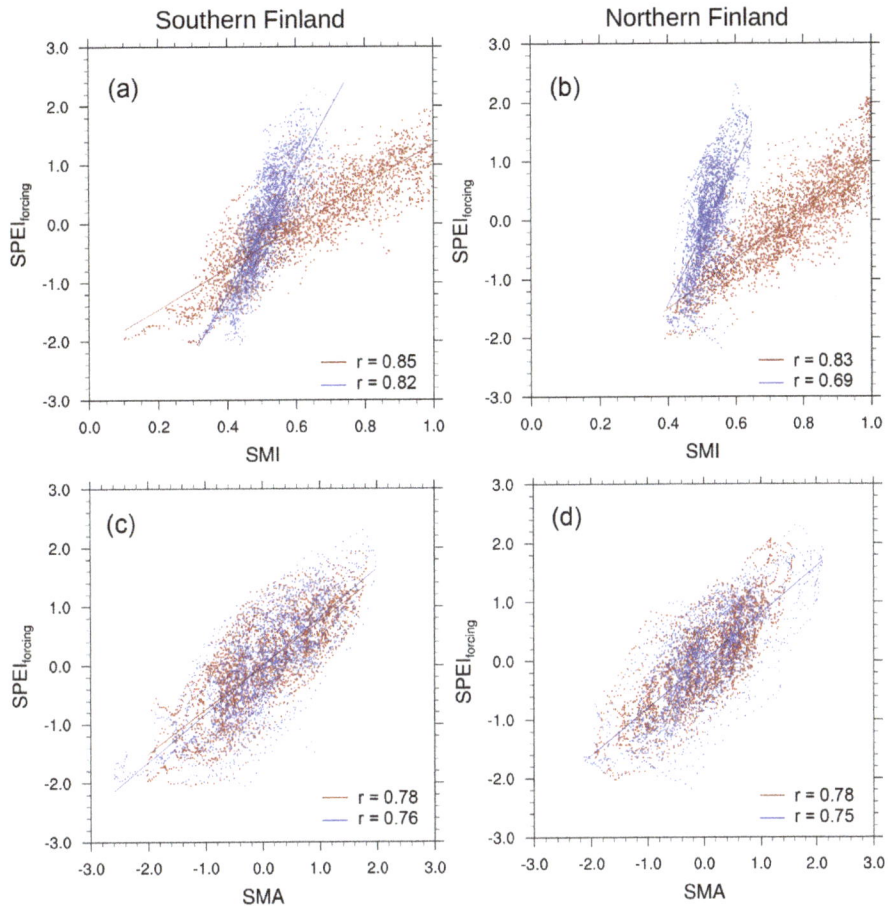

Figure 5. Time correlations over the study period between SPEI$_{forcing}$ and SMI (**a, b**), SPEI$_{forcing}$ and SMA (**c, d**) with the spatial means over the mineral soil areas (brown) and the peat soil areas (blue) in southern Finland (left column panels) and northern Finland (right column panels), respectively.

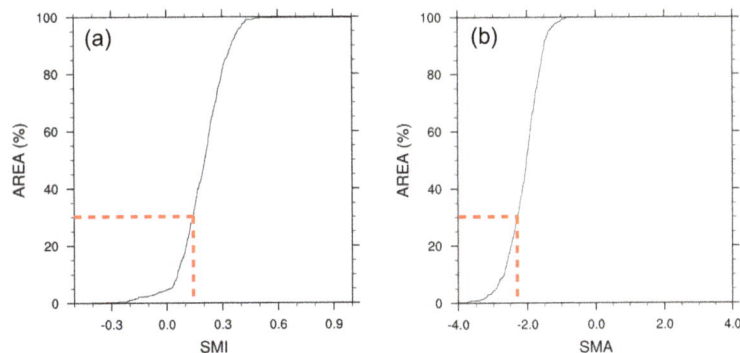

Figure 6. Cumulative area distribution of the (**a**) SMI and (**b**) SMA over southern Finland in the driest 28-day period of southern Finland in 2006 (i.e. the driest day of 28-day-running means of the regionally averaged SMI and SMA over southern Finland). The red dashed lines indicate the corresponding SMI and SMA values at which 30 % of the area is affected by the Extreme Drought that affects Forest health (EDF).

the driest 28-day period, points to the vulnerability of shallow soil to climate variability. Also, it is worth noting that the EDF area in the driest 28-day period as indicated by the SMI, shows a similar spatial pattern to the locations of damaged forest sites in the observation data, where few forest damaged sites are found in northern Finland.

A more comparative analysis of the ability of the three indicators to represent EDF under the derived thresholds was

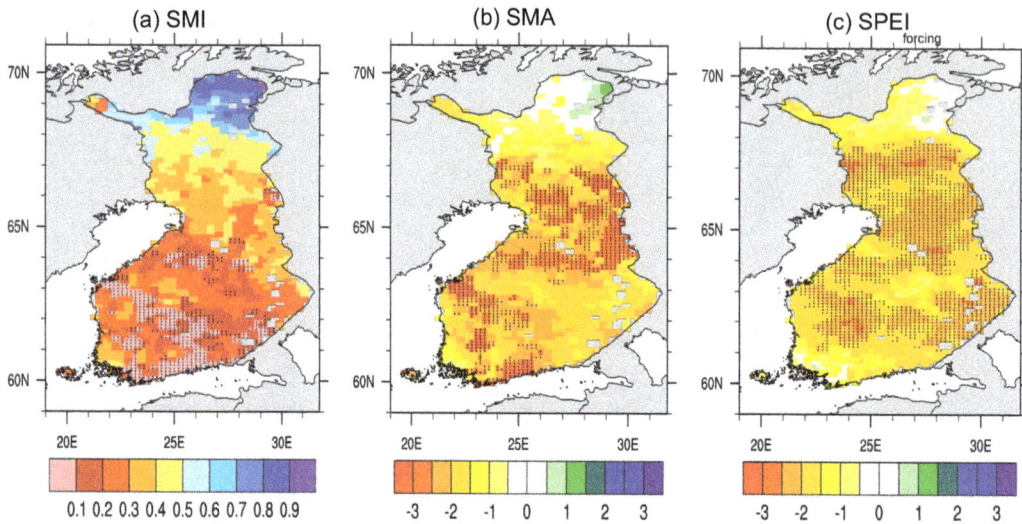

Figure 7. The (**a**) SMI, (**b**) SMA, and (**c**) SPEI$_{forcing}$ in the driest 28-day period of southern Finland in 2006. The dotted areas are under the derived thresholds for EDF. The uncovered grid boxes (grey cells) in Finland represent inland lakes.

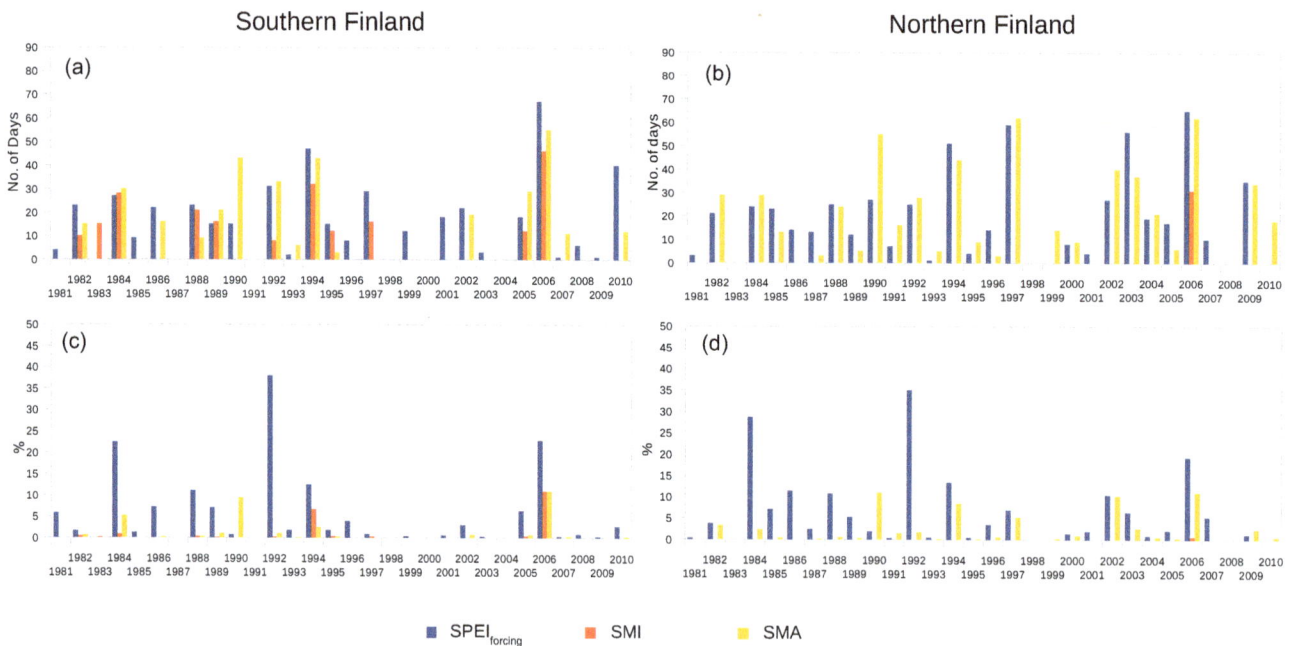

Figure 8. The summer drought periods (**a, b**) and the mean fractional areas affected by drought in these periods (**c, d**) induced by EDF events that are indicated by SMI, SMA, and SPEI$_{forcing}$ for southern Finland (left column panels) and northern Finland (right column panels) in the study period (note that areas with shallow soil (soil depth < 3 m) are excluded).

conducted for the summer months of the 30-year study period (Fig. 8). As the shallow soil is quite sensitive to climate variation, areas with soil depths less than 3 m were excluded to eliminate the influence on drought period by sporadic drought episodes that would have exaggerated the number of drought days. In general, the drought periods (number of days) influenced by EDF show a better consistency among the three indicators than the mean fraction of affected areas.

In general, SMI shows less area under EDFs in both southern and northern Finland than the other two indicators. In particular, the only EDF indicated by SMI in the north was for 2006, but with only a small fractional area of around 1–2 %. In the south, the SMI indicates EDF events in 1994 and 2006, with the mean influenced area larger than 5 % and the period longer than 30 days. In 2006, the mean influenced areas indicated by the SMI and SMA are similar, as are the drought periods. However, the SMA shows less mean influenced areas compared to SMI in 1994, which is related to the longer drought period indicated by SMA than SMI. The SPEI displays higher mean areas influenced by EDFs than the soil moisture drought indicators in all years, except 1990. The reason for this is that the EDF as indicated by SPEI in that year had already commenced before June, which is the first month of summer in our study. The SMA shows a prolonged effect in comparison to meteorological drought, which is not sufficiently strong to allow SMI to reach the EDF threshold due to the high initial soil moisture content.

Overall, the SMI is considered to be more capable in indicating EDFs because it directly reflects the plant available soil moisture. In boreal forests in Finland, EDFs indicated by SPEI and SMA often cannot lead to very low soil moisture that could be indicated as EDFs by SMI, due to the high initial soil moisture or presence of peat.

5   Summary and conclusions

In this study, we assessed the performance of several drought indicators (SPI, SPEI, SMA, and SMI) for their ability to represent the timing and spatial extent of droughts in Finland. The SPI, SPEI and SMA are standardized indicators that describe the degrees of anomalies over a period, whereas SMI is directly related to plant available water. Those standardized indicators were calculated with 28-days-running mean inputs, while SMI is calculated with daily soil moisture. The regional soil moisture is simulated by the land surface model JSBACH with its five layer soil hydrology scheme. The simulated soil moisture can generally capture the timing of dry spells in summer and winter characteristics of the observed soil moisture at the three observation sites in Finland, although inconsistencies exist in the rates of change and amplitudes of variations in soil moisture. The SPEI showed higher time correlation coefficients with the soil moisture-based drought indicators than SPI, as SPEI takes into account the surface water balance rather than precipitation only. Further inspections of the temporal and spatial variability of SPEI, SMA, and SMI revealed that, in general, the SPEI and SMA showed latitudinal-consistent patterns, whereas the SMI described more droughts for the south than the north of Finland. The vulnerable shallow soil area along the coastline in southern Finland and the peat soil area in northern Finland are drought-prone and drought-resistant areas, respectively, as indicated by SMI. Therefore, soil characteristics impact on SMI. In addition, soil moisture buffering effects and the associated soil moisture memory can delay and extend the drought as indicated by soil moisture-based drought indicators, in comparison to those by the SPEI.

Especially, we examined the effectiveness of SPEI, SMA, and SMI to capture the Extreme Drought affecting Forest Health (EDF). The SMI was found to be more capable in spatially representing the EDF in 2006. High discrepancies were found among the indicated EDF periods and the mean fraction of affected areas by the three indicators for the summer months of the 30-year study period. The SPEI was the most sensitive drought indicator and showed the highest amount of EDFs with larger influenced areas, while the SMI showed much less EDF events than the other two indicators.

To conclude, we recommend to use SMI to indicate EDFs in boreal forest because it directly represents the plant available soil moisture, which is a synthesized result of the initial soil moisture content, soil properties, as well as climate conditions. Thus, a land surface model that produces reliable predictions of soil moisture is necessary when assessing EDF risks in boreal areas. To improve the accuracy of soil moisture-based drought indicators (especially SMI) calculated with LSM simulated soil moisture, high-quality soil type distribution and soil parameters data are essential. More sophisticated models are expected to improve simulated soil moisture; for instance, soil layers with different soil types along the soil profile, heterogeneity of soil types in a grid box and thorough consideration of the model formulations and parameters that regulate the rate of evapotranspiration, drainage and run-off. Furthermore, uncertainties associated with the drought indicators may originate from their input data (Naumann et al., 2014), therefore unbiased forcing data are of vital importance for the accurate simulation of soil moisture by a LSM (Maggioni et al., 2012).

The critical points of drought indicators leading to drought damages symptoms of forests are crucial for understanding climate impacts on forest ecosystems. In this study, the EDF thresholds for those indicators were selected only according to the statistics of the forest health observation in 2006. This might induce some uncertainties when they are used for future predictions of EDFs. The method for selecting EDF thresholds for drought indicators could be adopted and the EDF thresholds could be calibrated, when there are more observation data about forest damages induced by drought available. In addition, drought damage on different tree species could be studied. These would require more detailed information and a better monitoring at the forest observation sites. Moreover, satellite data could be explored to monitor the drought effects in boreal forests timely and across large spatial scale (Caccamo et al., 2011).

Acknowledgements. We would like to thank the EMBRACE (EU 7th Framework Programme, Grant Agreement number 282672) and HENVI (Helsinki University Centre for Environment) projects for the financial support. We give our deepest appreciation to Pentti Pirinen from FMI for providing the observational data. The authors acknowledge MPI–MET, MPI–BGC, and CSC (Hamburg) for providing JSBACH and REMO models and assistance in their use, and also the MONIMET project

(LIFE12 ENV/FI/000409) for supporting the drought indicator study. This work was also supported by the Academy of Finland Center of Excellence (no. 272041), ICOS-Finland (no. 281255), and ICOS-ERIC (no. 281250) funded by Academy of Finland.

References

Aalto, J., Pirinen, P., Heikkinen, J., and Venäläinen, A.: Spatial interpolation of monthly climate data for Finland: comparing the performance of kriging and generalized additive models, Theor. Appl. Climatol., 112, 99–111, 2013.

Aalto, T., Peltoniemi, M., Aurela, M., Böttcher, K., Gao, Y., Härkönen, S., Härmä, P., Kilkki, J., Kolari, P., Laurila, T., Lehtonen, A., Manninen, T., Markkanen, T., Mattila, O.-P., Metsämäki, S., Muukkonen, P., Mäkelä, A., Pulliainen, J., Susiluoto, J., Takala, M., Thum, T., Tupek, B., Törmä, M., and Ali, N. A.: Preface to the special issue on monitoring and modelling of carbon-balance-, water- and snow-related phenomena at northern latitudes, Boreal Environ. Res., 20, 145–150, 2015.

Allen, C. D., Macalady, A. K., Chenchouni, H., Bachelet, D., McDowell, N., Vennetier, M., Kitzberger, T., Rigling, A., Breshears, D. D., and Hogg, E. T.: A global overview of drought and heat-induced tree mortality reveals emerging climate change risks for forests, Forest Ecol. Manage., 259, 660–684, 2010.

Allen, R. G., Smith, M., Pereira, L. S., and Perrier, A.: An update for the calculation of reference evapotranspiration, ICID Bulletin, 43, 35–92, 1994.

American Meteorological Society: Meteorological drought-policy statement, B. Am. Meteorol. Soc., 78, 847–849, 1997.

Aurela, M., Lohila, A., Tuovinen, J.-P., Hatakka, J., Penttilä, T., and Laurila, T.: Carbon dioxide and energy flux measurements in four northern-boreal ecosystems at Pallas, Boreal Environ. Res., 20, 455–473, 2015.

Betts, A. K.: Understanding Hydrometeorology Using Global Models, B. Am. Meteorol. Soc., 85, 1673–1688, doi:10.1175/BAMS-85-11-1673, 2004.

Beguería, S. and Vicente-Serrano, S. M.: SPEI: Calculation of the standardised Precipitation–Evapotranspiration Index, R package version 1.6, R Foundation for Statistical Computing, Vienna, Austria, 2013.

Bi, J., Xu, L., Samanta, A., Zhu, Z., and Myneni, R.: Divergent Arctic–Boreal Vegetation Changes between North America and Eurasia over the Past 30 Years, Remote Sens., 5, 2093–2112, doi:10.3390/rs5052093, 2013.

Blauhut, V., Gudmundsson, L., and Stahl, K.: Towards pan-European drought risk maps: quantifying the link between drought indices and reported drought impacts, Environ. Res. Lett., 10, 014008, doi:10.1088/1748-9326/10/1/014008, 2015.

Bonan, G. B.: Forests and Climate Change: Forcings, Feedbacks, and the Climate Benefits of Forests, Science, 320, 1444–1449, doi:10.1126/science.1155121, 2008.

Bréda, N., Cochard, H., Dreyer, E., and Granier, A.: Water transfer in a mature oak stand (Quercuspetraea): seasonal evolution and effects of a severe drought, Can. J. Forest Res., 23, 1136–1143, doi:10.1139/x93-144, 1993.

Bréda, N., Huc, R., Granier, A., and Dreyer, E.: Temperate forest trees and stands under severe drought: a review of ecophysiological responses, adaptation processes and long-term consequences, Ann. For. Sci., 63, 625–644, 2006.

Burke, E. J. and Brown, S. J.: Evaluating Uncertainties in the Projection of Future Drought, J. Hydrometeorol., 9, 292–299, doi:10.1175/2007JHM929.1, 2008.

Caccamo, G., Chisholm, L. A., Bradstock, R. A., and Puotinen, M. L.: Assessing the sensitivity of MODIS to monitor drought in high biomass ecosystems, Remote Sens. Environ., 115, 2626–2639, doi:10.1016/j.rse.2011.05.018, 2011.

Ciais, P., Reichstein, M., Viovy, N., Granier, A., Ogee, J., Allard, V., Aubinet, M., Buchmann, N., Bernhofer, C., Carrara, A., Chevallier, F., De Noblet, N., Friend, A. D., Friedlingstein, P., Grunwald, T., Heinesch, B., Keronen, P., Knohl, A., Krinner, G., Loustau, D., Manca, G., Matteucci, G., Miglietta, F., Ourcival, J. M., Papale, D., Pilegaard, K., Rambal, S., Seufert, G., Soussana, J. F., Sanz, M. J., Schulze, E. D., Vesala, T., and Valentini, R.: Europe-wide reduction in primary productivity caused by the heat and drought in 2003, Nature, 437, 529–533, 2005.

Clenciala, E., Kucera, J., Ryan, M., G., and Lindroth, A.: Water flux in boreal forest during two hydrologically contrasting years; species specific regulation of canopy conductance and transpiration, Ann. For. Sci., 55, 47–61, 1998.

Dale, V. H., Joyce, L. A., McNulty, S., Neilson, R. P., Ayres, M. P., Flannigan, M. D., Hanson, P. J., Irland, L. C., Lugo, A. E., Peterson, C. J., Simberloff, D., Swanson, F. J., Stocks, B. J., and Michael Wotton, B.: Climate change and forest disturbances, BioScience, 51, 723–734, doi:10.1641/0006-3568(2001)051[0723:CCAFD]2.0.CO;2, 2001.

Dee, D. P., Uppala, S. M., Simmons, A. J., Berrisford, P., Poli, P., Kobayashi, S., Andrae, U., Balmaseda, M. A., Balsamo, G., Bauer, P., Bechtold, P., Beljaars, A. C. M., van de Berg, L., Bidlot, J., Bormann, N., Delsol, C., Dragani, R., Fuentes, M., Geer, A. J., Haimberger, L., Healy, S. B., Hersbach, H., Hólm, E. V., Isaksen, L., Kållberg, P., Köhler, M., Matricardi, M., McNally, A. P., Monge-Sanz, B. M., Morcrette, J. J., Park, B. K., Peubey, C., de Rosnay, P., Tavolato, C., Thépaut, J. N., and Vitart, F.: The ERA-Interim reanalysis: configuration and performance of the data assimilation system, Q. J. Roy. Meteorol. Soc., 137, 553–597, doi:10.1002/qj.828, 2011.

Drebs, A., Nordlund, A., Karlsson, P., Helminen, J., and Rissanen, P.: Climatological statistics of Finland 1971–2000, in: Climatological Statistics of Finland 2001, Finnish Meteorological Institute, Helsinki, 2002.

Dümenil, L. and Todini, E.: A rainfall–runoff scheme for use in the Hamburg climate model, in: Advances in Theoretical Hydrology – a Tribute to James Dooge, European Geophysical Society Series on Hydrological Sciences, edited by: O'Kane, J. P., Elsevier Science, Amsterdam, the Netherlands, 129–157, 1992.

Dunne, K. and Willmott, C. J.: Global distribution of plant-extractable water capacity of soil, Int. J. Climatol., 16, 841–859, 1996.

Edwards, D. C. and McKee, T. B.: Characteristics of 20th century drought in the United States at multiple time scales. Climatology Report 97-2, Department of Atmospheric Science, Colorado State University, Fort Collins, Colorado, 1997.

Eichhorn, J., Roskams, P., Ferretti, M., Mues, V., Szepesi, A., and Durrant, D.: Visual Assessment of Crown Condition and Damaging Agents Manual, Part IV in: Manual on methods and criteria for harmonized sampling, assessment, monitoring and analysis of the effects of air pollution on forests, UNECE ICP Forests Programme Co-ordinating Centre, Hamburg, 46 pp., 2010.

European Environment Agency: CLC2006 technical guidelines, EEA Technical Report No. 17, Copenhagen, 2007.

FAO/UNESCO: Soil Map of the World, UNESCO, Paris, 1971–1981.

Finnish Statistical Yearbook of Forestry 2012, Finnish Forest Research Institute, Helsinki, Finland, 454 pp., 2012.

Gao, Y., Weiher, S., Markkanen, T., Pietikäinen, J.-P., Gregow, H., Henttonen, H. M., Jacob, D., and Laaksonen, A.: Implementation of the CORINE land use classification in the regional climate model REMO, Boreal Environ. Res., 20, 261–282, 2015.

Gimbel, K. F., Felsmann, K., Baudis, M., Puhlmann, H., Gessler, A., Bruelheide, H., Kayler, Z., Ellerbrock, R. H., Ulrich, A., Welk, E., and Weiler, M.: Drought in forest understory ecosystems – a novel rainfall reduction experiment, Biogeosciences, 12, 961–975, doi:10.5194/bg-12-961-2015, 2015.

Granier, A., Bréda, N., Biron, P., and Villette, S.: A lumped water balance model to evaluate duration and intensity of drought constraints in forest stands, Ecol. Model., 116, 269–283, doi:10.1016/S0304-3800(98)00205-1, 1999.

Granier, A., Reichstein, M., Bréda, N., Janssens, I. A., Falge, E., Ciais, P., Grünwald, T., Aubinet, M., Berbigier, P., Bernhofer, C., Buchmann, N., Facini, O., Grassi, G., Heinesch, B., Ilvesniemi, H., Keronen, P., Knohl, A., Köstner, B., Lagergren, F., Lindroth, A., Longdoz, B., Loustau, D., Mateus, J., Montagnani, L., Nys, C., Moors, E., Papale, D., Peiffer, M., Pilegaard, K., Pita, G., Pumpanen, J., Rambal, S., Rebmann, C., Rodrigues, A., Seufert, G., Tenhunen, J., Vesala, T., and Wang, Q.: Evidence for soil water control on carbon and water dynamics in European forests during the extremely dry year: 2003, Agr. Forest Meteorol., 143, 123–145, doi:10.1016/j.agrformet.2006.12.004, 2007.

Grossiord, C., Granier, A., Gessler, A., Jucker, T., and Bonal, D.: Does drought influence the relationship between biodiversity and ecosystem functioning in boreal forests?, Ecosystems, 17, 394–404, doi:10.1007/s10021-013-9729-1, 2014.

Grünzweig, J. M., Lin, T., Rotenberg, E., Schwartz, A., and Yakir, D.: Carbon sequestration in arid-land forest, Global Change Biol., 9, 791–799, doi:10.1046/j.1365-2486.2003.00612.x, 2003.

Guttman, N. B.: On the sensitivity of sample L moments to sample size, J. Climate, 7, 1026–1029, doi:10.1175/1520-0442(1994)007<1026:OTSOSL>2.0.CO;2, 1994.

Guttman, N. B.: Accepting the standardized precipitation index: a calculation algorithm, J. Am. Water Resour. As., 35, 311–322, 1999.

Hagemann, S.: An Improved Land Surface Parameter Dataset for Global and Regional Climate Models, Max Planck Institute for Meteorology, Hamburg, 2002.

Hagemann, S. and Stacke, T.: Impact of the soil hydrology scheme on simulated soil moisture memory, Clim. Dynam., 44, 1731–1750, doi:10.1007/s00382-014-2221-6, 2015.

Hain, C. R., Crow, W. T., Mecikalski, J. R., Anderson, M. C., and Holmes, T.: An intercomparison of available soil moisture estimates from thermal infrared and passive microwave remote sensing and land surface modeling, J. Geophys. Res.-Atmos., 116, D15107, doi:10.1029/2011JD015633, 2011.

Hayes, M., Svoboda, M., Wall, N., and Widhalm, M.: The Lincoln declaration on drought indices: universal meteorological drought index recommended, B. Am. Meteorol. Soc., 92, 485–488, doi:10.1175/2010BAMS3103.1, 2011.

Heim, R. R.: A review of twentieth-century drought indices used in the United States, B. Am. Meteorol. Soc., 83, 1149–1165, doi:10.1175/1520-0477(2002)083<1149:AROTDI>2.3.CO;2, 2002.

Hillel, D.: Environmental Soil Physics, Academic Press, San Diego, 1998.

Hirschi, M., Seneviratne, S. I., Alexandrov, V., Boberg, F., Boroneant, C., Christensen, O. B., Formayer, H., Orlowsky, B., and Stepanek, P.: Observational evidence for soil–moisture impact on hot extremes in southeastern Europe, Nat. Geosci., 4, 17–21, doi:10.1038/ngeo1032, 2011.

Irvine, J., Perks, M. P., Magnani, F., and Grace, J.: The response of Pinus sylvestris to drought: stomatal control of transpiration and hydraulic conductance, Tree Physiol., 18, 393–402, doi:10.1093/treephys/18.6.393, 1998.

Jacob, D.: A note to the simulation of the annual and inter-annual variability of the water budget over the Baltic Sea drainage basin, Meteorol. Atmos. Phys., 77, 61–73, doi:10.1007/s007030170017, 2001.

Jacob, D. and Podzun, R.: Sensitivity studies with the regional climate model REMO, Meteorol. Atmos. Phys., 63, 119–129, doi:10.1007/BF01025368, 1997.

Krishnan, P., Black, T. A., Grant, N. J., Barr, A. G., Hogg, E. T. H., Jassal, R. S., and Morgenstern, K.: Impact of changing soil moisture distribution on net ecosystem productivity of a boreal aspen forest during and following drought, Agr. Forest Meteorol., 139, 208–223, 2006.

Lagergren, F. and Lindroth, A.: Transpiration response to soil moisture in pine and spruce trees in Sweden, Agr. Forest Meteorol., 112, 67–85, doi:10.1016/S0168-1923(02)00060-6, 2002.

Law, B. E., Falge, E., Gu, L., Baldocchi, D. D., Bakwin, P., Berbigier, P., Davis, K., Dolman, A. J., Falk, M., Fuentes, J. D., Goldstein, A., Granier, A., Grelle, A., Hollinger, D., Janssens, I. A., Jarvis, P., Jensen, N. O., Katul, G., Mahli, Y., Matteucci, G., Meyers, T., Monson, R., Munger, W., Oechel, W., Olson, R., Pilegaard, K., Paw U, K. T., Thorgeirsson, H., Valentini, R., Verma, S., Vesala, T., Wilson, K., and Wofsy, S.: Environmental controls over carbon dioxide and water vapor exchange of terrestrial vegetation, Agr. Forest Meteorol., 113, 97–120, doi:10.1016/S0168-1923(02)00104-1, 2002.

Lenton, T. M., Held, H., Kriegler, E., Hall, J. W., Lucht, W., Rahmstorf, S., and Schellnhuber, H. J.: Tipping elements in the Earth's climate system, P. Natl. Acad. Sci. USA, 105, 1786–1793, doi:10.1073/pnas.0705414105, 2008.

Lindgren, M., Nevalainen, S., Pouttu, A., Rantanen, H., and Salemaa, M.: Metsäpuiden elinvoimaisuuden arviointi, Forest Focus/ICP Level 1, Finnish Forest Research Institute, Helsinki, Finland, 56 pp., 2005.

Liu, Y. Y., Dorigo, W. A., Parinussa, R. M., de Jeu, R. A. M., Wagner, W., McCabe, M. F., Evans, J. P., and van Dijk, A. I. J. M.: Trend-preserving blending of passive and active microwave soil moisture retrievals, Remote Sens. Environ., 123, 280–297, doi:10.1016/j.rse.2012.03.014, 2012.

López-Urrea, R., Olalla, F. M. d. S., Fabeiro, C., and Moratalla, A.: An evaluation of two hourly reference evapotranspiration equations for semiarid conditions, Agr. Water Manage., 86, 277–282, doi:10.1016/j.agwat.2006.05.017, 2006.

Ma, Z., Peng, C., Zhu, Q., Chen, H., Yu, G., Li, W., Zhou, X., Wang, W., and Zhang, W.: Regional drought-induced reduction in the biomass carbon sink of Canada's boreal forests, P. Natl. Acad. Sci. USA, 109, 2423–2427, doi:10.1073/pnas.1111576109, 2012.

Maggioni, V., Anagnostou, E. N., and Reichle, R. H.: The impact of model and rainfall forcing errors on characterizing soil moisture uncertainty in land surface modeling, Hydrol. Earth Syst. Sci., 16, 3499–3515, doi:10.5194/hess-16-3499-2012, 2012.

McKee, T. B., Doeskin, N. J., and Kleist, J.: The relationship of drought frequency and duration to time scales, in: 8th Conference on Applied Climatology, Anaheim, Califonia, 179–184, 1993.

McKee, T. B., Doesken, N. J., and Kleist, J.: Drought monitoring with multiple time scales, in: Ninth Conference on Applied Climatology, Dallas, Texas, 233–236, 1995.

Mishra, A. K. and Singh, V. P.: A review of drought concepts, J. Hydrol., 391, 202–216, doi:10.1016/j.jhydrol.2010.07.012, 2010.

Muukkonen, P., Nevalainen, S., Lindgren, M., and Peltoniemi, M.: Spatial Occurrence of Drought-Associated Damages in Finnish Boreal Forests: Results from Forest Condition Monitoring and GIS Analysis, Boreal Environ. Res., 20, 172–180, 2015.

Naumann, G., Dutra, E., Barbosa, P., Pappenberger, F., Wetterhall, F., and Vogt, J. V.: Comparison of drought indicators derived from multiple data sets over Africa, Hydrol. Earth Syst. Sci., 18, 1625–1640, doi:10.5194/hess-18-1625-2014, 2014.

Orlowsky, B. and Seneviratne, S. I.: Elusive drought: uncertainty in observed trends and short- and long-term CMIP5 projections, Hydrol. Earth Syst. Sci., 17, 1765–1781, doi:10.5194/hess-17-1765-2013, 2013.

Päivänen, J. and Hånell, B.: Peatland Ecology and Forestry: a Sound Approach, Department of Forest Ecology, University of Helsinki, Helsinki, 2012.

Peng, C., Ma, Z., Lei, X., Zhu, Q., Chen, H., Wang, W., Liu, S., Li, W., Fang, X., and Zhou, X.: A drought-induced pervasive increase in tree mortality across Canada's boreal forests, Nat. Clim. Change, 1, 467–471, 2011.

Quiring, S. M.: Developing objective operational definitions for monitoring drought, J. Appl. Meteorol. Clim., 48, 1217–1229, doi:10.1175/2009JAMC2088.1, 2009.

Raddatz, T. J., Reick, C. H., Knorr, W., Kattge, J., Roeckner, E., Schnur, R., Schnitzler, K. G., Wetzel, P., and Jungclaus, J.: Will the tropical land biosphere dominate the climate–carbon cycle feedback during the twenty-first century?, Clim. Dynam., 29, 565–574, doi:10.1007/s00382-007-0247-8, 2007.

Räisänen, J. and Räty, O.: Projections of daily mean temperature variability in the future: cross-validation tests with ENSEMBLES regional climate simulations, Clim. Dynam., 41, 1553–1568, doi:10.1007/s00382-012-1515-9, 2013.

Räty, O., Räisänen, J., and Ylhäisi, J.: Evaluation of delta change and bias correction methods for future daily precipitation: intermodel cross-validation using ENSEMBLES simulations, Clim. Dynam., 42, 2287–2303, doi:10.1007/s00382-014-2130-8, 2014.

Rebel, K. T., de Jeu, R. A. M., Ciais, P., Viovy, N., Piao, S. L., Kiely, G., and Dolman, A. J.: A global analysis of soil moisture derived from satellite observations and a land surface model,

Hydrol. Earth Syst. Sci., 16, 833–847, doi:10.5194/hess-16-833-2012, 2012.

Reick, C. H., Raddatz, T., Brovkin, V., and Gayler, V.: Representation of natural and anthropogenic land cover change in MPI-ESM, J. Adv. Model. Earth Syst., 5, 459–482, doi:10.1002/jame.20022, 2013.

Roeckner, E., Arpe, K., Bengtsson, L., Christoph, M., Claussen, M., Dümenil, L., Esch, M., Giogetta, M., Schlese, U., and Schultz-Weida, U.: The Atmospheric General Circulation Model ECHAM4: Model Description and Simulation of the Present-Day Climate, Max Planck Institute for Meterology, Hamburg, 90 pp., 1996.

Seneviratne, S. I., Corti, T., Davin, E. L., Hirschi, M., Jaeger, E. B., Lehner, I., Orlowsky, B., and Teuling, A. J.: Investigating soil moisture–climate interactions in a changing climate: a review, Earth-Sci. Rev., 99, 125–161, doi:10.1016/j.earscirev.2010.02.004, 2010.

Sheffield, J. and Wood, E. F.: Drought: Past Problems and Future Scenarios, Earthscan, London, UK and Washington, D.C., USA, 2011.

Simmons, A., Uppala, S., Dee, D., and Kobayashi, S.: ERA-Interim: new ECMWF reanalysis products from 1989 onwards, ECMWF Newsletter, 110, 25–35, 2007.

Sivakumar, M. V. K., Motha, R. P., Wilhite, D. A., and Wood, D. A.: Agricultural drought indices, in: Proceedings of the WMO/UNISDR Expert Group Meeting on Agricultural Drought Indice, 2–4 June 2010, Murcia, Spain, WMO/TD No. 1572; WAOB-2011, World Meteorological Organization, Geneva, Switzerland, AGM-11 197 pp., 2011.

Stevens, B., Giorgetta, M., Esch, M., Mauritsen, T., Crueger, T., Rast, S., Salzmann, M., Schmidt, H., Bader, J., Block, K., Brokopf, R., Fast, I., Kinne, S., Kornblueh, L., Lohmann, U., Pincus, R., Reichler, T., and Roeckner, E.: Atmospheric component of the MPI-M Earth System Model: ECHAM6, J. Adv. Model. Earth Syst., 5, 146–172, doi:10.1002/jame.20015, 2013.

Tallaksen, L. M. and Van Lanen, H. A. J.: Hydrological Drought: Processes and Estimation Methods for Streamflow and Groundwater, Elsevier Science B. V., the Netherlands, 2004.

Thum, T., Aalto, T., Laurila, T., Aurela, M., Lindroth, A., and Vesala, T.: Assessing seasonality of biochemical CO_2 exchange model parameters from micrometeorological flux observations at boreal coniferous forest, Biogeosciences, 5, 1625–1639, doi:10.5194/bg-5-1625-2008, 2008.

Tikkanen, M.: The Physical Geography of Fennoscandia, edited by: Seppälä, M., Oxford University Press, Oxford, 2005.

Törmä, M., Markkanen, T., Hatunen, S., Härmä, P., Mattila, O.-P., and Arslan, A. N.: Assessment of land-cover data for land-surface modelling in regional climate studies, Boreal Environ. Res., 20, 243–260, 2015.

US Geological Survey: Global Land Cover Characteristics Data Base version 2.0, US Geological Survey, Reston, USA, 2001.

Van Loon, A. F. and Van Lanen, H. A. J.: A process-based typology of hydrological drought, Hydrol. Earth Syst. Sci., 16, 1915–1946, doi:10.5194/hess-16-1915-2012, 2012.

Ventura, F., Spano, D., Duce, P., and Snyder, R. L.: An evaluation of common evapotranspiration equations, Irrig. Sci., 18, 163–170, doi:10.1007/s002710050058, 1999.

Vesala, T., Suni, T., Rannik, Ü., Keronen, P., Markkanen, T., Sevanto, S., Grönholm, T., Smolander, S., Kulmala, M., Ilvesniemi, H., Ojansuu, R., Uotila, A., Levula, J., Mäkelä, A., Pumpanen, J., Kolari, P., Kulmala, L., Altimir, N., Berninger, F., Nikinmaa, E., and Hari, P.: Effect of thinning on surface fluxes in a boreal forest, Global Biogeochem. Cy., 19, GB2001, doi:10.1029/2004GB002316, 2005.

Vicente-Serrano, S. M., Beguería, S., and López-Moreno, J. I.: A multiscalar drought index sensitive to global warming: the standardized precipitation evapotranspiration index, J. Climate, 23, 1696–1718, doi:10.1175/2009JCLI2909.1, 2010.

Welp, L. R., Randerson, J. T., and Liu, H. P.: The sensitivity of carbon fluxes to spring warming and summer drought depends on plant functional type in boreal forest ecosystems, Agr. Forest Meteorol., 147, 172–185, doi:10.1016/j.agrformet.2007.07.010, 2007.

World Meteorological Organization: Standardized Precipitation Index User Guide, WMO-No. 1090, edited by: Svoboda, M., Hayes, M., and Wood, D., Geneva, Switzerland, 16 pp., 2012.

Wu, H., Svoboda, M. D., Hayes, M. J., Wilhite, D. A., and Wen, F.: Appropriate application of the standardized precipitation index in arid locations and dry seasons, Int. J. Climatol., 27, 65–79, doi:10.1002/joc.1371, 2007.

10

Spatio-temporal assessment of WRF, TRMM and in situ precipitation data in a tropical mountain environment (Cordillera Blanca, Peru)

L. Mourre[1], T. Condom[1], C. Junquas[1,2], T. Lebel[1], J. E. Sicart[1], R. Figueroa[3], and A. Cochachin[4]

[1]IRD/UGA/CNRS/G-INP, LTHE UMR 5564, Grenoble, France
[2]Instituto Geofísico del Perú (IGP), Lima, Peru
[3]UNASAM, Huaraz, Peru
[4]Glaciology and Water Resources Unit, National Water Authority (ANA-UGRH), Huaraz, Peru

Correspondence to: L. Mourre (lise.mourre@ujf-grenoble.fr)

Abstract. The estimation of precipitation over the broad range of scales of interest for climatologists, meteorologists and hydrologists is challenging at high altitudes of tropical regions, where the spatial variability of precipitation is important while in situ measurements remain scarce largely due to operational constraints. Three different types of rainfall products – ground based (kriging interpolation), satellite derived (TRMM3B42), and atmospheric model outputs (WRF – Weather Research and Forecasting) – are compared for 1 hydrological year in order to retrieve rainfall patterns at timescales ranging from sub-daily to annual over a watershed of approximately $10\,000\,\mathrm{km}^2$ in Peru. An ensemble of three different spatial resolutions is considered for the comparison (27, 9 and 3 km), as long as well as a range of timescales (annual totals, daily rainfall patterns, diurnal cycle). WRF simulations largely overestimate the annual totals, especially at low spatial resolution, while reproducing correctly the diurnal cycle and locating the spots of heavy rainfall more realistically than either the ground-based KED or the Tropical Rainfall Measuring Mission (TRMM) products. The main weakness of kriged products is the production of annual rainfall maxima over the summit rather than on the slopes, mainly due to a lack of in situ data above 3800 m a.s.l. This study also confirms that one limitation of TRMM is its poor performance over ice-covered areas because ice on the ground behaves in a similar way as rain or ice drops in the atmosphere in terms of scattering the microwave energy. While all three products are able to correctly represent the spatial rainfall patterns at the annual scale, it not surprisingly turns out that none of them meets the challenge of representing both accumulated quantities of precipitation and frequency of occurrence at the short timescales (sub-daily and daily) required for glacio-hydrological studies in this region. It is concluded that new methods should be used to merge various rainfall products so as to make the most of their respective strengths.

1 Introduction: the challenge of precipitation estimation in the tropical Peruvian Andes

Located in the north-west of Peru, the Cordillera Blanca is the most glaciated tropical mountain range in the intertropical band. The Cordillera Blanca glaciers are melting at an unprecedented high rate (e.g., Georges, 2004; Silverio and Jaquet, 2005; Vuille et al., 2008a, b; Bury et al., 2011), impacting the whole water cycle of the region. At the seasonal scale, the distribution of water discharge in rivers downstream of glaciers is changing, while at the decadal scale, the mean annual discharge is increasing, with the prospect of decreasing in the long run. Temperature and precipitation are the two major forcing variables most influencing the interannual variability and long-term evolution of the water balance. A proper evaluation of these two variables is consequently a key issue for properly predicting the future of the glaciers and of the associated water resources.

The tropics are thermally characterized by an annual variation less important than the diurnal cycle (e.g., Kaser, 1999; Baraer et al., 2012). This applies to the Cordillera Blanca, where homogeneous thermal conditions are observed throughout the year (Juen et al., 2007). For instance, at Querococha, located in the southern part of the Cordillera Blanca, mean monthly temperature variation is less than 1 °C (Kaser et al., 2003).

By contrast, there is a strong seasonality of precipitation, controlled by the upper air circulation, with easterly wind transporting moisture from the Amazon plain (Aceituno, 1987) and westerly flow causing dry conditions due to the Humboldt Current (Garreaud et al., 2003). This results in two distinct seasons: the wet season from October to April with an average of 80 % of the annual precipitation (Vuille et al., 2008a), and the dry season from May to August. The wet season corresponds to the South American Monsoon System (SAMS) (e.g., Vera et al., 2006; Garreaud, 2009; Marengo et al., 2012), bringing humidity far to the west. The dry season is associated with the North American Monsoon System, the Intertropical Convergence Zone (ITCZ) being located as its northernmost position. The inter-annual variability of rainfall is important in relation to the fluctuations of the sea surface temperature (SST) of the North Atlantic and the El Niño–Southern Oscillation (ENSO) (e.g., Espinoza Villar et al., 2009; Lavado Casimiro et al., 2012; Lavado Casimiro and Espinoza, 2014). According to Lavado Casimiro and Espinoza (2014), the Rio Santa catchment belongs to an area where positive precipitation anomalies are observed during strong Niño as well as during strong Niña events.

The rainfall climatology is also characterized by strong spatial gradients at all temporal scales. First of all, the main annual rainfall pattern between 5 and 30° S is the contrast between the dry and cold conditions on the Pacific coast, stretching to the western slopes of the Andes, and the warm, humid and rainy conditions prevailing on the eastern slopes (Garreaud, 2009). This results in high precipitation amounts on the windward slopes of the Andes in easterly flows situations (up to $6000\,\mathrm{mm\,yr^{-1}}$) and much smaller precipitation amounts on the leeward side, even at high altitudes (under $530\,\mathrm{mm\,yr^{-1}}$) between 5° N and 20° S (Espinoza Villar et al., 2009). Superimposed onto this large-scale spatial pattern, the influence of the topography becomes more and more important when considering smaller temporal scales at which convective and orographic processes have a deep influence. Rainfall hotspots, heavy rainfall gradients over a few kilometers and flash floods (Young and Leon, 2009; Espinoza et al., 2015) are the most prominent hydro-meteorological patterns induced by the rough topography of the region.

Another issue arises from the high altitude, meaning that a significant amount of precipitation falls as snow over 4800 m a.s.l. This requires one to measure reliably both the solid and liquid precipitation all year around, something that is far from granted and that remains a major difficulty in mountain hydrology.

The estimation of precipitation over the broad range of scales of interest for climatologists, meteorologists and hydrologists is thus especially challenging in this region characterized by very uncommon geographical features. And yet socio-economic stakes are high as far as potentially drastic changes of the water cycle related to precipitation variability and long-term changes are concerned, affecting access to drinkable water in urban areas, the yields of agricultural projects and the operation of numerous hydroelectric power plants.

The driving question of this study is to identify and compare the precipitation data sets that can be used for properly characterizing the water balance over catchments of the region from sub-daily to yearly temporal scales. Both the accumulated quantities of precipitation and the frequency of occurrence have to be properly estimated if one is to compute coherent water budgets over this large range of temporal scales, an accomplishment that no single precipitation data set can pretend to achieve on its own.

Each precipitation data set has its own strength and weakness. Starting with ground data, their main shortcoming – beyond their key advantage of being the only direct measurement of rainfall – is a poor sampling of the spatial variability that is especially important in mountainous regions (Scheel et al., 2011). This is compounded by the difficulty of installing and maintaining ground stations in a harsh environment, making whole areas very difficult to access (Salzmann et al., 2013; Schwarb et al., 2011). Rain gauges are thus most often available in the vicinity of villages, meaning that non-habited areas are virtually not sampled, especially at high altitudes, where distinguishing between liquid and solid precipitation is a major issue.

On their side, satellite rainfall products provide the global coverage that is lacking for ground data sets. However, the early satellite rainfall products elaborated in the mid-1980s were solely based on infrared data, affecting their accuracy in the case of convective rainfall and, more generally, in the presence of a strong rainfall gradient. The most recent products now make use of various sources of information, blending infrared and microwave satellite data and often incorporating ground data, which make them more performant in spotting the patches of intense rainfall. It remains that there are still significant differences between the most commonly used satellite rainfall products, especially in the tropics and for orographically forced rainfall (Ward et al., 2011). This means that the ability of these satellite products to fulfill user's expectations must be scrutinized on a case-by-case basis. Note also that satellite products are rather weak in distinguishing between liquid and solid precipitation.

In the perspective of quantifying the spatial and temporal variability of water budgets over catchments, another possibility for providing the required rainfall component is to use the precipitation produced by climate models. This presents two main advantages: (i) the physical coherency of the various elements of the water budgets computed by these models and (ii) the possibility of studying the evolution of the water budgets in the future in a context of global warming. Note, however, that global climate models usually fail to simulate properly the regional processes and their spatial variability, especially for precipitation in mountainous areas, a default particularly critical in the Andes due to their complex topography (Giovannettone and Barros, 2009). To remedy these limitations, downscaling approaches based on the nesting of regional climate models (RCM) into global models is frequently used. The performance of nested regional models depends on the study area, the spatial resolution and the parameterization used (Box and Bromwich, 2004), which means that their added value, as compared to the other sources of rainfall information, should also be considered on a case-by-case basis.

2 Study area and data

2.1 Study area

Draining an area of 11 930 km^2 located between 8 and 10° S and 79 and 77° W, the Rio Santa runs northward, between the Cordillera Negra to the west and the Cordillera Blanca to the east (Mark and Seltzer, 2003), before making its way to the Pacific; 41 % of the catchment area is above 4000 m a.s.l., including the highest point of the cordillera, Huascaran, peaking at 6768 m a.s.l. (Fig. 1). The upper Rio Santa catchment, with an outlet at Condorcerro, drains an area of about 10 000 km^2, and will be our main study area.

Some modeling projections based on the mean of meteorological variables from four GCM grid points predict the disappearance of ice cover for 2080 in some sub-watersheds of the Rio Santa (Juen et al., 2007), which would have a significant impact on the flow regime of the river, since glaciers meltwater regulates its annual flow. For a sub-watershed of the upper Rio Santa watershed (4700 km^2, 8 % glaciated), glacier meltwater currently provides 10–20 % of the annual rate, and up to 40 % in the dry season (Kaser et al., 2003; Mark and Seltzer, 2003; Baraer et al., 2012; Condom et al., 2012).

The larger studied area is a rectangle of 84 000 km^2 (Fig. 1). It can be divided into four hydrological sub-regions from the north-east to the south-west. The Rio Marañon catchment is located on the Amazon side, where the highest yearly precipitated amount was measured in situ during the hydrological year 2012–2013 (> 1100 mm yr^{-1}). The second sub-region is the western side of the Cordillera Blanca, draining into the Pacific. Stations in this area are located inside the Rio Santa catchment. In situ measured precipitation amounts in the Cordillera Blanca area range from 478 to 1000 mm yr^{-1} (Table 1 and Fig. 1). The third region is the Cordillera Negra, which is much drier (from 44 to 434 mm yr^{-1}) and lower in altitude (Table 1 and Fig. 1). This zone includes all stations located west of the Rio Santa

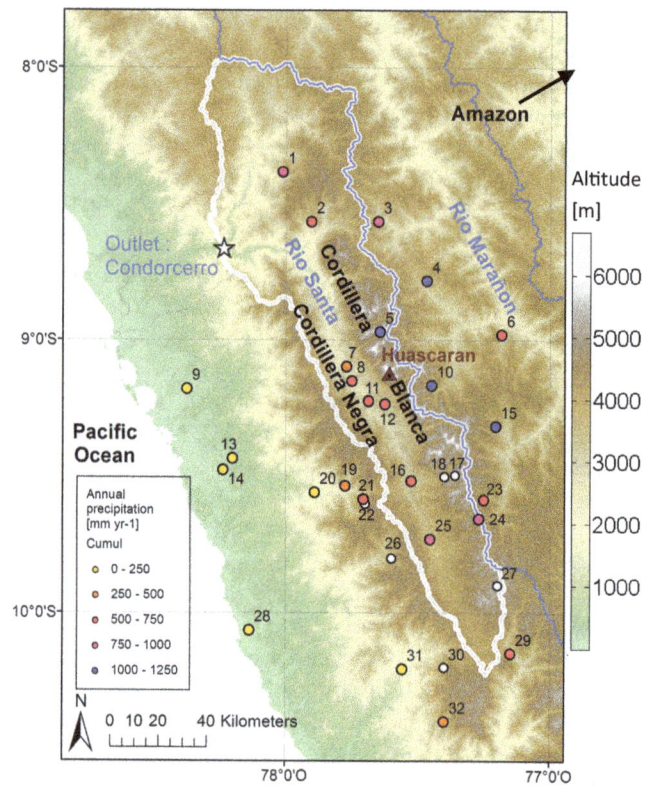

Figure 1. Location of the upper Santa watershed (the star marks outlet Condorcerro). Color dots indicate annual precipitation amounts at in situ stations. White dots correspond to stations with missing data. The Huascaran peak is indicated, as well as the Rio Marañon watershed. Topography is from SRTM (http://srtm.csi.cgiar.org/).

riverbed, up from 3625 m a.s.l. to an altitude of 1000 m a.s.l. Finally, the dry area near the Pacific Ocean, named Costa, is defined as the land area whose altitude ranges from 0 to 1000 m a.s.l. The topography data shown in Fig. 1 are from STRM (90 m resolution). While we will be looking at the entire 84 000 km^2 region, our analysis is focused on the precipitation falling over the upper Santa watershed, because this is our region of interest from a hydrological standpoint and because it is where we have the best ground network coverage.

2.2 In situ data

It was not an easy task to gather data from a sufficiently large number of stations in order to properly document our study area. First of all, there was the need to obtain some background climatological information; 10 stations operated by the Servicio Nacional de Meteorología e Hidrología de Perú (SENAMHI) since 1965 (Table 1) allow computation of monthly and yearly long-term averages. However, their specific location and loose spatial sampling prevent one from estimating correctly the long-term average rainfall either over the upper Rio Santa catchment or over the whole study area. Data from an additional set of eight SENAMHI stations

Table 1. Information on in situ rainfall stations. Under "Location", CB means Cordillera Blanca, CN means Cordillera Negra, M means Marañon, and C means Costa. [NS] indicates stations used for the study along the Rio Santa Valley. [H] indicates stations used for the transect along the Huascaran peak. * indicates stations used to calculate the precipitation index (data from 1965) (Sect. 2.2). Precipitation ($mm\,yr^{-1}$) during the hydrological year 2012–2013 is indicated at each rain gauge station for in situ data (Obs), TRMM and WRF (WRF27, WRF9 and WRF3). Accu indicates the value for glacier accumulation over the year.

UNASAM no.	Lon.	Lat.	Alt. (m)	Situation	Obs	TRMM	WRF27	WRF9	WRF3
2 [NS]	−77.9	−8.6	3172	CB	542	407	2173	1517	1225
6	−77.2	−8.9	2786	M	577	671	3377	1090	1716
7	−77.8	−9.1	2350	CB	478	307	4219	997	796
9	−78.4	−9.2	125	C	31	107	121	214	341
10	−77.4	−9.2	3770	M	1162	271	2758	2421	2821
11 [H]	−77.7	−9.2	2500	CB	598	1596	4219	1000	849
12 [NS]	−77.6	−9.2	3040	CB	738	1596	2758	1073	1145
14	−78.2	−9.5	133	C	14	158	121	182	338
15	−77.5	−9.3	3480	M	1028	558	2558	3472	3948
16 [NS]	−77.5	−9.5	3091	CB	666	434	4625	1663	1025
18	−77.4	−9.5	3850	CB	–	1674	4625	3168	2513
28	−78.1	−10.1	18	C	8	78	49	102	250
29	−77.1	−10.1	3405	CB	624	381	1861	1664	3069
32	−77.4	−10.4	3268	CN	307	523	2203	1990	2860

SENAMHI no.	Lon.	Lat.	Alt. (m)	Situation	Obs	TRMM	WRF27	WRF9	WRF3
1	−78.0	−8.4	3160	CB	972	343	2502	1498	1373
3*	−77.6	−8.6	3375	M	959	437	2173	1651	1483
4 [H]	−77.5	−8.8	3605	M	1030	530	2758	2248	2160
8	−77.7	−9.1	2527	CB	744	307	4219	1000	719
13	−78.2	−9.4	216	C	28	158	121	219	396
16	−77.5	−9.5	3079	CB	634	434	4625	1663	1025
19*	−77.8	−9.5	2285	CN	251	233	1320	797	761
20	−77.9	−9.5	1260	CN	91	234	710	502	528
21*	−77.7	−9.6	3625	CN	668	434	4348	1800	1169
22*	−77.7	−9.6	3325	CN	–	434	4348	1524	1310
23*	−77.2	−10.1	3137	M	687	790	2402	2456	3289
25*	−77.4	−9.7	3444	CB	756	790	4348	1942	1541
26*	−77.6	−9.8	3440	CN	–	358	4348	1705	973
27*	−77.2	−9.9	4400	CB	–	645	3413	2684	3922
29*	−77.2	−10.1	3382	CB	620	381	1861	1678	3069
30	−77.4	−10.2	3200	CN	–	329	1861	837	1867
31	−77.5	−9.6	1221	CN	44	192	499	454	662
32*	−77.4	−10.4	3230	CN	383	271	1861	1255	1586

UGRH no.	Lon.	Lat.	Alt. (m)	Situation	Obs	TRMM	WRF27	WRF9	WRF3
5 [H]	−77.6	−9.0	5100	CB	Accu: 1006	545	4188	3010	2922
17	−77.4	−9.5	4281	CB	–	1674	3215	2691	2479
24	−77.3	−9.6	4955	CB	Accu: 1000	790	3215	2809	3894

cover the period August 2012 to July 2013 at a daily resolution. We also had access to three stations from the Unidad de Glaciología y Recursos Hídricos (UGRH) from the Autoridad Nacional de Agua (ANA). These stations are of a tipping bucket type; they have the double advantage of being located at higher altitudes and of providing data at sub-daily time steps. As compared to previous studies in this region, the key new information used comes from a database of 16 meteorological stations with hourly data located in the An-

cash region of Peru. They were installed in the framework of a project (Centro de Información e Investigación Ambiental de Desarrollo Regional Sostenible – CIIADERS), operated by the Universidad Nacional Santiago Antúñez de Mayolo (UNASAM) of Huaraz. These stations provide essential information for understanding the spatial (increased sampling density) and temporal (hourly resolution) distribution of precipitation within our study area. The SENAMHI data are routinely quality controlled, using standard procedures in use in the Met services worldwide. For the UGRH and UNASAM data, we had to carry out our own quality check, for instance by comparing precipitation amounts reported by stations located in the same area, leading to the removal of errant values.

Unfortunately the CIIADERS network has been in operation since 2012 only, limiting this study to 1 hydrological year (August 2012 to July 2013). The average pluviometric index of this 1-year study period, which corresponds to a reduced centered anomaly, is close to 0 (0.0774), meaning that the annual precipitation is close to the mean precipitation of the 1965–2014 period as calculated from 10 long-term stations among our total of 37. Note also that stations with more than 25 % of missing data during that year have been removed, leaving only 32 stations available to compute our ground-based rainfall grids (Table 1 and Fig. 1).

A weakness of this 32-station network is the lack of data for the dry Cordillera Negra and the high-altitude areas of the Cordillera Blanca (only three stations located above 3800 m a.s.l.). This shortcoming was partly overcome by using accumulation data provided by the UGRH for the Artesonraju and Yanamarey glaciers at near 5000 m a.s.l., which are net accumulations during 1 year, including solid precipitation and melting over the period. Concerning snow, it is important to keep in mind that the rainy season occurs during austral summer, when temperature is slightly higher and consequently few solid precipitations are observed under 4600 m a.s.l. (Condom et al., 2011).

2.3 Gridded precipitation from in situ data

A major problem when comparing precipitation data sets from different sources relates to their different spatial sampling. Satellite and atmospheric model data are provided as gridded products, while rain gauges provide point data. A spatial interpolation procedure is thus required to get each product on the same grid. There is a considerable amount of literature on selecting an appropriate interpolation method for computing rain grids from point data. This is an especially tricky problem in regions of rough topography.

Several studies showed that kriging with external drift (KED), using altitude as an external variable, provides good results over complex terrains (e.g., Masson and Frei, 2014; Tobin et al., 2011; Ochoa et al., 2014). Block kriging with altitude as an external drift was thus chosen here as our reference interpolation method – note however that other types of kriging interpolators were tested, but a cross-validation evaluation showed KED to be the most efficient of all in our case. While accounting for the strong influence of topography on the structure of rain fields is crucial in mountainous regions, another issue arises from the type of variogram to be used and whether it is allowed to vary from day to day. Related to this topic, different concepts of spatio-temporal kriging have been tested in previous studies (Amani and Lebel, 1997; Vischel et al., 2011; Gräler at al., 2012). Daily evolving

variograms assume the hypothesis of a relationship between precipitation amounts of days D and $D-1$, and information from the previous days is considered with a weight chosen by the user (10 % is used in this study). This is the method that was finally chosen to compute daily gridded precipitation at 27 km, 9 km and 3 km spatial resolutions, thus matching the resolution of the satellite and Weather Research and Forecasting (WRF) model products that will be presented below in Sects. 2.4 and 2.5.

2.4 TRMM product

Tropical Rainfall Measuring Mission (TRMM) Multi-Satellite Precipitation Analysis (TMPA) products have been available since 1998. This study makes use of the TRMM3B42 version 7 product, which provides precipitation data at a 3 h time step from a combination of remote sensing observations (microwave imager, precipitation radar, visible and infrared scanner) and monthly in situ observations (Huffman et al., 2007; Huffman and Bolvin, 2012). This product will simply be referred to as TRMM in the rest of the study. The TRMM data set covers a region between 50° S and 50° N, with a spatial resolution of 0.25° (approximately 27 km) (Table 3). This product can be used for hydrological application in regions with scarce in situ data. Even though the TRMM mission was focused on the monitoring of tropical rainfall, it suffers from a number of drawbacks, the main one being its poor time sampling reduced to one or two passages per day depending on the area considered. This causes a significant loss of information for short-duration storms (Roca et al., 2010; Condom et al., 2011; Ward et al., 2011). The effect of these time sampling errors are reduced when aggregating in time (Scheel et al., 2011; Mantas et al., 2014), but TRMM products still show significant biases in monthly values in the tropical Andes (Condom et al., 2011) as well as in solid precipitation (Maussion et al., 2014).

2.5 WRF simulation

In this study we use the high-resolution simulations from the WRF model version 3.4.1 (Skamarock et al., 2008) that had only a few applications in the tropical Andes (Murthi et al., 2011; Ochoa et al., 2014; Sanabria et al., 2014). The WRF is a nonhydrostatic model and uses a terrain-following vertical coordinate (sigma). The limited domain simulations are forced by a boundary condition every 6 h by the National Center for Environmental Prediction (NCEP) Final Analyses (FNL) Global Forecast System (GFS) with 1° of latitude and longitude horizontal resolution. The elevation data set is from the USGS GTOPO30. A large tropical Andes domain was first delimited for simulations at a 27 km resolution (WRF27). Two sub-domains were then used for carrying out simulations at a 9 km (WRF9) and 3 km (WRF3) resolution, respectively, both being centered in the Santa River basin (Tables 2, 3). WRF9 (WRF3) simulations are forced

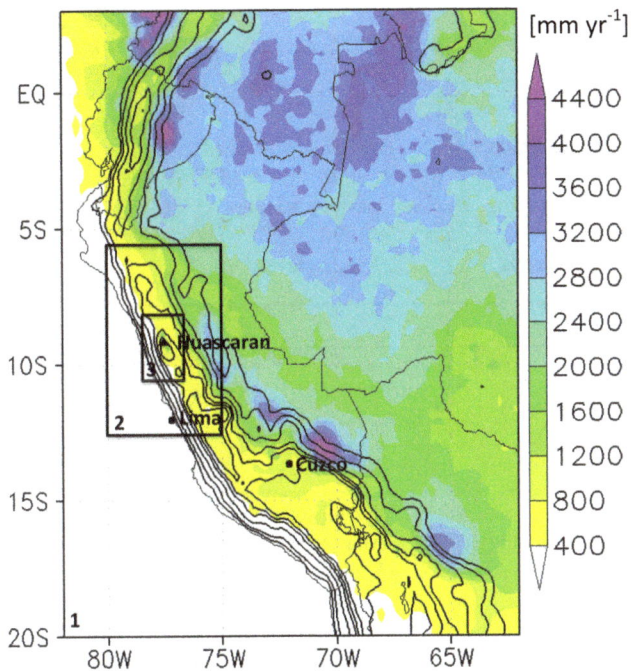

Figure 2. Annual precipitation $(\mathrm{mm\,yr^{-1}})$ from TRMM2B31 for the hydrological year 2012–2013 (August 2012–July 2013). The three boxes indicate the WRF simulation domain: box 1 for $27\,\mathrm{km} \times 27\,\mathrm{km}$; box 2 for $9\,\mathrm{km} \times 9\,\mathrm{km}$; box 3 for $3\,\mathrm{km} \times 3\,\mathrm{km}$. Topography contours are displayed every 500 m.

by the WRF27 (WRF9) simulations using a one-way nesting technique. The simulations begin on April 2012, the first 4 months being used as a spin-up period for producing 1 year of data to be compared to the KED and TRMM products.

Figure 2 shows the boxes corresponding to each simulation domain, and Table 2 lists the resolutions and coordinates of each configuration. Table 3 lists the parameterizations used in the simulations. We use the Thompson microphysical scheme (Thompson et al., 2008) and the Grell–Devenyi ensemble scheme for the cumulus parameterization. We also use a topographic correction for surface winds, previously tested in a complex orographic terrain of the Iberian Peninsula (Jimenez and Dudhia, 2012). The Noah-MP (Multi-Physics) land surface model is used for the representation of land–atmosphere interaction processes (Niu et al., 2011; Yang et al., 2011). Noah-MP is an extended version of the Noah land surface model with enhanced Multi-Physics option to address critical shortcomings in Noah for long-term soil state spin-up and snow modeling. In particular, this version of the Noah model has shown improvements in the representation of surface energy fluxes, snow cover and snow albedo treatment. The partitioning of precipitation into rainfall and snowfall was set to option 2 (opt_snf = 2) using the Biosphere–Atmosphere Transfer Scheme, which assumes all precipitation as snowfall when the air temperature is lower than the freezing point plus 2.2 K, and rainfall otherwise.

The overestimation of precipitation is a frequent bias in numerical models (e.g., Mearns et al., 1995), particularly in complex orographic regions. Preliminary tests of sensitivity with various WRF parameterizations (including different cumulus schemes, cloud microphysics, planetary boundary layer and land surface options) have been done in the tropical Andes at a 27 km horizontal resolution; a clear overestimation of precipitation was observed with all these configurations and over all the domain, including the high mountain areas. The biases found with other configurations were almost similar to those of the configuration selected here in terms of the precipitation spatial distribution, and with quantitative differences more pronounced in the eastern slopes of the Andes and in the Amazon region rather than in high mountain zones like the Cordillera Blanca. The configuration finally retained for this study (Table 2) has been selected because (i) it minimizes the positive precipitation bias in the tropical Andes above 3500 m a.s.l., and (ii) it simulates correctly the spatial distribution of the precipitation in the region, including the zones of maximum precipitation situated in the Amazon basin and on the eastern slopes of the Andes (Fig. 2), when compared with the TRMM2B31 data. At 3 km resolution, the Noah-MP option was found to decrease the precipitation overestimation in the Cordillera Blanca and show a more realistic snow distribution when compared with previous observations.

3 Methods and criteria used for comparing the rainfall products

A total of seven gridded rainfall products are compared here, as described in Table 4. These products differ from one another on two accounts: (i) the type of information used (ground data, satellite data, atmospheric model), (ii) the spatial resolution, ranging from 27 km corresponding to the size of the TRMM satellite product grid mesh, down to 3 km, the finest resolution at which the WRF model was run. These seven gridded products are available at the daily scale, which is the corner scale for the comparison carried out in this paper. While TRMM products and WRF simulations are inherently gridded, in situ data need to be interpolated in order to build grids at the three spatial resolutions: 27, 9 and 3 km.

3.1 Computation of daily precipitation grids from in situ data

The performance of the KED outputs is determined based on a "leave-one-out" cross-validation procedure (Li and Heap, 2008). It consists in leaving aside one measurement point at a time and estimating the value at that point from the remaining 31 stations. The procedure is applied successively to each of the 32 measurement stations, allowing one to compute the bias (Eq. 1), the root mean square error (RMSE) score (Eq. 2)

Table 2. Characteristics of the WRF simulations at the three different spatial scales.

	WRF27	WRF9	WRF3
Horizontal resolution (km)	27	9	3
Domain	Tropical Andes	Rio Santa region	Rio Santa watershed
Domain center coordinates	8°30′ S, 72° W	9°1′4″ S, 77°37′53″ W	9°11′25″ S, 77°43′7″ W
Configuration	Regional simulation	One-way nesting	One-way nesting
Forcing	NCEP_FNL	WRF27	WRF9
Vertical resolution	27 sigma levels	27 sigma levels	27 sigma levels
Run time step (s)	150	50	6
Output time resolution (h)	6	3	1

Table 3. List of the physical parameterizations used in the WRF simulations.

	Parameterizations	References
Clouds microphysics	New Thompson scheme	Thompson et al. (2008)
Radiation	Longwave: Rapid Radiative Transfer Model (RRTM)	Mlawer et al. (1997)
	Shortwave: Dudhia scheme	Dudhia (1989)
Cumulus parameterization	Grell–Devenyi ensemble scheme	Grell and Devenyi (2002)
Planetary boundary layer	Yonsei University scheme	Hong et al. (2006)
	Wind topographic correction (option 1)	Jimenez and Dudhia (2012)
Land surface	Noah-MP (Multi-Physics) partitioning	Niu et al. (2011);
	precipitation option 2	Yang et al. (2011)
Surface layer	MM5 similarity	Paulson (1970)

and the correlation coefficient, as follows:

$$\text{bias} = \sum_{i=1}^{n} \sum_{d=1}^{m} (\hat{P}_{i,d} - P_{i,d}), \tag{1}$$

$$\text{RMSE} = \sum_{d=1}^{m} \sqrt{\frac{1}{n} \sum_{i=1}^{n} (\hat{P}_{i,d} - P_{i,d})^2}, \tag{2}$$

where $\hat{P}_{i,d}$ is the daily precipitation estimated at point i for day d, using all the other gauges, $P_{i,d}$ is the corresponding measured daily rainfall, n is the number of stations (32 when there are no missing data on day d) and m is the number of days studied.

In the following, the gridded daily precipitation product at 27, 9, and 3 km spatial resolutions will, respectively, be referred to as KED27, KED9 and KED3 (see Table 4 and Fig. 3). The daily RMSE value is large (3.41 mm d^{-1}) compared to the mean daily precipitation over all stations (1.85 mm d^{-1}), and errors are reduced with aggregation on a yearly basis (RMSE of 271 mm yr^{-1} for an average in situ amount of 572 mm y^{-1} for the 32 stations and a correlation coefficient of 0.78). In yearly values, kriging products will then be the basis of our comparison to TRMM data and WRF outputs. Despite some bias in the estimation of annual and daily rainfall, it is assumed that the most important spatial pattern is captured by KED.

3.2 Comparing the daily and annual precipitation products

Daily precipitation is defined as the accumulation of rainfall between 00:00:00 LT (local time) and 23:59:00 LT. An important point is that all gridded products suffer from weakness and, thus, the aim of the comparison is to analyze differences between products. The daily products are compared from three different standpoints: the statistical distribution of non-zero rainfall, the grid annual values and the seasonal cycle.

The frequency of daily precipitation at one location (one station or the corresponding grid mesh) was studied through the cumulative distribution function of the non-zero precipitation (Sambou, 2004):

$$f(x) = -\log_{10}(1 - F(x)), \tag{3}$$

where $F(x)$ is the cumulative frequency of the daily precipitation amount above 1 mm d^{-1} and x is the daily precipitation (mm d^{-1}).

To assess the statistical performance of the 3 km resolution products against punctual in situ data at a daily timescale, the contingency table for rainfall/no rainfall was built (Table 5). The bias score (BIAS – ratio of the number of rainy days simulated ($A + B$) over the number of rainy days observed ($A + C$)), false alarm rate (FAR – ratio of the number of rainy days incorrectly simulated (B) over the total number of rainy days simulated ($A + B$)), probability of false detection (POFD – ratio of the number of rainy days incorrectly

Table 4. Precipitation data used in this study, with their spatial and temporal resolutions, and the accumulated amount precipitated over the upper Rio Santa watershed during the hydrological year 2012–2013. WRF and KED (corresponding to kriging data with external drift – daily evolving variogram) are at three different spatial resolutions (27, 9 and 3 km). TRMM is the TRMM3B42 product.

Product	Spatial resolution	Temporal resolution used in this study			Annual precipitation over the watershed (m)
		Hourly	Daily	Yearly	
In situ	Punctual	x	x		–
KED27	27 km × 27 km		x	x	0.83
KED9	9 km × 9 km		x	x	1.01
KED3	3 km × 3 km		x	x	0.95
WRF27	27 km × 27 km		x	x	2.91
WRF9	9 km × 9 km		x	x	1.95
WRF3	3 km × 3 km	x	x	x	1.97
TRMM	27 km × 27 km		x	x	0.57

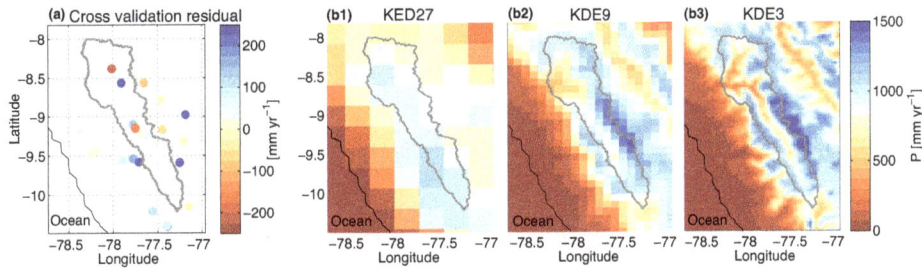

Figure 3. (a) Cross-validation residuals with in situ yearly precipitation amount. (b) Annual precipitation amount from KED interpolations at 27 (b1), 9 (b2) and 3 km (b3) spatial resolutions. Delimitation of the upper Rio Santa watershed is indicated in bold gray lines. The coastline is also indicated in black.

Table 5. Contingency table used to assess the statistical performances of the 3 km resolution products against punctual in situ data at a daily timescale. The B value corresponds for example to a day with no precipitation in the in situ data and precipitation > threshold mm d^{-1} in the 3 km grid product.

		In situ	
	P_j	Yes	No
3 km grid	Yes	A	B
Product	No	C	D

simulated (B) over the number of days without rain in the observations ($B + D$)) and the frequently used Heidke skill score (HSS) (Eqs. 4–6) were calculated.

$$\text{HSS} = \frac{S - S_{\text{ref}}}{1 - S_{\text{ref}}}, \tag{4}$$

$$S = \frac{A + D}{N}, \tag{5}$$

$$S_{\text{ref}} = \frac{(A + B)(A + C) + (B + D)(C + D)}{N^2}, \tag{6}$$

where N is the size of the statistical population, and A, B, C and D values are explained in Table 5.

A perfect product would have a BIAS of 1, a FAR of 0, a POFD of 0 and a HSS of 1.

Annual grids were computed by temporal aggregation of the daily grids. In the aim to study the water balance for the purpose of hydrological applications, each product was evaluated in terms of volume of water precipitated over the area of the upper Rio Santa watershed, corresponding to the watershed limited by the outlet at Condorcerro (Fig. 1).

Finally, to evaluate the seasonal cycle of precipitation in one site, we used the temporal standard score S_t (Eq. 7):

$$S_t = \frac{\overline{P_j}^{10} - \langle P_j \rangle}{\sigma_j}, \tag{7}$$

where $\overline{P_j}^{10}$ is the running means of daily precipitation amounts over 10 days in one location, and $\langle P_j \rangle$ and σ_j are the temporal average and standard deviation of the daily precipitation, respectively.

It is important to mention that when comparing the performances at one location of the KED daily products with those of the TRMM and WRF, use is made of the cross-validation products, so that the local information is not taken into account, which would artificially benefit the ground product with respect to the satellite and model products.

3.3 Assessing the quality of the WRF3 hourly precipitation grids

To facilitate the comparison among all stations, the hourly precipitation amounts were normalized by dividing them by the mean of hourly values during the year. Few studies deal with hourly rainfall amounts from WRF modeling. In this study, we compared the timing of the precipitation peak from hourly rain gauge data and from WRF3 simulation outputs. Studying hourly data allowed us to see whether short time processes governing precipitation in the Rio Santa Valley are well represented in WRF3, considering in situ hourly measurement as the reference.

4 Results

4.1 Frequency and intensities of daily precipitation amounts

In this section, we first analyze the statistics of daily precipitation and the temporal scale for which all eight products are available (Table 4), and present them for the Corongo location (no. 2 in Table 1 and Fig. 1). This station, located in the northern part of the Rio Santa watershed, was selected because it is representative of the 16 stations located inside the upper Rio Santa catchment in terms of the precipitation areal averaging effect, except when comparing the differences between the three different spatial-resolution products of WRF. In a second part, we studied daily precipitation occurrences based on the contingency table indices (see Sect. 3.2, Table 5) for all stations located in the Sierra area.

Figure 4 shows the cumulative frequency of daily precipitation above $1 \, \mathrm{mm \, d^{-1}}$ for the Corongo location comparing (i) the three spatial resolutions of WRF (Fig. 4a), (ii) comparing the three spatial resolutions of KED (Fig. 4b), (iii) comparing TRMM, WRF and KED products at 27 km (Fig. 4c), and (iv) comparing WRF and KED products at 3 km spatial resolution vs. in situ punctual data (Fig. 4d). The number in the box of each graph represents the number of days with precipitation over $1 \, \mathrm{mm \, d^{-1}}$ ($n_{p>1}$) for each product. Regarding KED data, the three spatial resolutions have a few differences that can also be seen in the number of $n_{p>1}$ (Fig. 4b). Concerning the 27 km spatial resolution, KED27 and TRMM are more similar to each other compared to WRF27 (Fig. 4c), despite an underestimation of $n_{p>1}$ for TRMM (108 days) compared to KED27 (183 days). WRF3, as WRF27 (Fig. 4c and d), does not correctly report daily precipitation amounts, with stronger values compared to the other data sets. In this comparison, KED3 seems to underestimate daily precipitation amounts and overestimate $n_{p>1}$ in light of in situ data, but this can be related to a resolution effect between the 3 km resolution grid and punctual measurement.

Noting that WRF products are unrealistic in terms of daily precipitated quantities, we will now evaluate their perfor-

mances in terms of occurrence, a notion that is essential in glacio-hydrological studies. This can be seen in the results of the contingency table and is studied by comparing KED3 and WRF3 with in situ data for different daily precipitation thresholds in Fig. 5. The results are shown for the Sierra region, but are similar for the Cordillera Negra and Marañon areas.

WRF3 largely overestimates the amount of strong daily precipitation, which can be linked to the overestimation of the product (Fig. 4d). The FAR, POFD and HSS show that there is an important improvement considering only precipitation above $1 \, \mathrm{mm \, d^{-1}}$ in KED3 and that the amount of daily precipitation between 0 and $1 \, \mathrm{mm \, d^{-1}}$ is largely overestimated by this product (Fig. 5b–d). POFD can be seen as an inter-comparison indicator as it does not depend on the number of predicted events. Above $1 \, \mathrm{mm \, d^{-1}}$, KED3 is then a better estimator of precipitation occurrence compared to WRF3. However, we faced the same spatial-resolution problem as above when comparing the 3 km mesh grid and in situ data for low-precipitation amounts. HSS indicates that daily precipitation in KED3 is in better accordance with in situ data than WRF3, with a few rainy days well predicted in WRF3. Although we noted a spatial-resolution effect for daily precipitation quantities under $1 \, \mathrm{mm \, d^{-1}}$, KED3 appears to be a good estimate of precipitation in terms of daily average quantities and occurrences, and will be considered later as a basis for comparison between different gridded products.

4.2 Annual amount and seasonal cycle

4.2.1 Annual cumulated precipitation amounts during the hydrological year 2012–2013

The estimations of the annual precipitation over the upper Rio Santa catchment (about $10\,000 \, \mathrm{km^2}$) for the 27 km resolution products, range from $570 \, \mathrm{mm \, yr^{-1}}$ for TRMM to $2910 \, \mathrm{mm \, yr^{-1}}$ for WRF27 (and $830 \, \mathrm{mm \, yr^{-1}}$ for KED27) (Table 4). Thus, even at this large integrative scale, the 27 km products display large discrepancies. KED annual rainfall is 15 % larger at the 3 km resolution ($950 \, \mathrm{mm \, yr^{-1}}$) compared to the 27 km resolution, while it is a diminution of 30 % for WRF (1970 vs. $2910 \, \mathrm{mm \, yr^{-1}}$). Figure 6 shows those annual precipitation amounts for all different products used in this study. Even though the KED3 estimate is certainly not devoid of bias, it is clear that WRF overestimates rainfall. WRF products, compared to KED, show more spatial variability in precipitation amounts at both 3 and 9 km resolutions, with stronger altitudinal gradient. TRMM and KED27 are closer along the Rio Santa Valley, as they both incorporate rain gauge data. However, on the Marañon watershed side, TRMM integrates the tropospheric flows from the Amazonian lowlands compared to KED27, whose ground observations are undersampled over this area, not catching the rainfall effect of the moisture influx from the Amazon basin. Although coarse-resolution products (TRMM and WRF27)

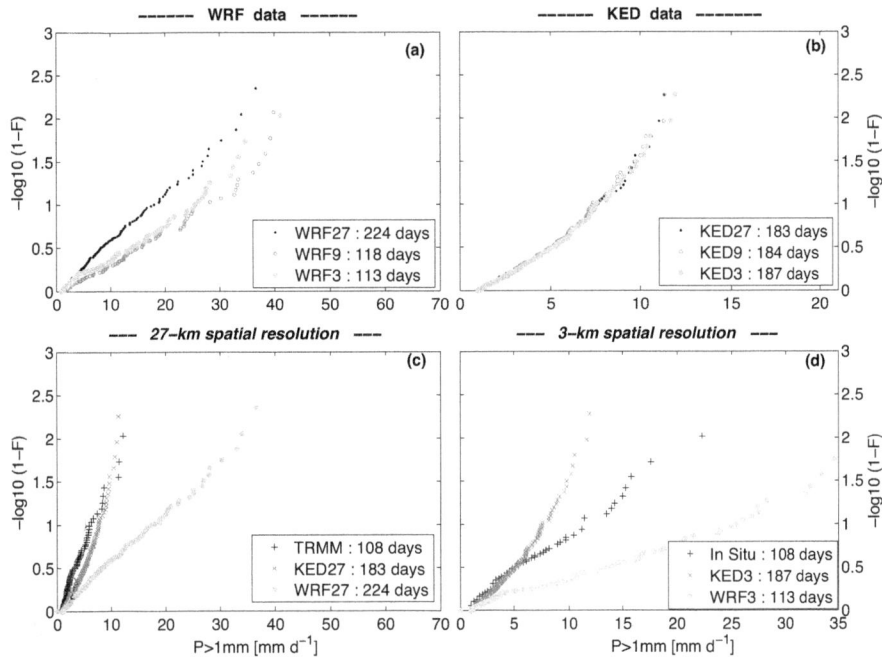

Figure 4. Frequency diagram of Corongo (station n°2) of daily precipitation data $> 1\,\mathrm{mm\,d^{-1}}$ for WRF outputs (**a**) and KED products (**b**) at three different spatial resolutions, and for all products at 27 km (**c**) and 3 km spatial resolution (**d**). Numbers in the bottom right corner indicate the number of days with precipitation $> 1\,\mathrm{mm\,d^{-1}}$ for each data set.

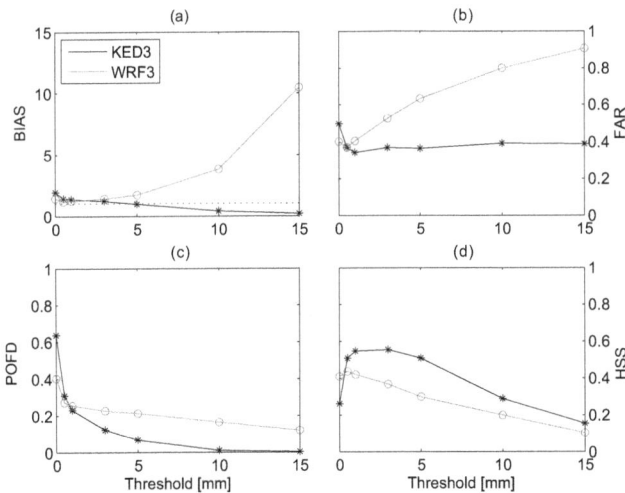

Figure 5. Daily precipitation indices: BIAS (**a**), false alarm rate (**b**), probability of false detection (**c**) and Heidke skill score (**d**). Calculated for KED3 (black) and WRF3 (gray) against rain gauge precipitation data located in the Sierra area. Scores have been evaluated for several daily precipitation thresholds: 0.1, 0.5, 1, 3, 5, 10 and 15 mm.

do not provide acceptable rainfall grids for hydrological applications in complex topography area because of their lack of representation of the finer spatial pattern, they are not totally useless at this annual scale. They correctly represent the longitudinal precipitation gradient between the humid and rainy condition of the Amazon plain, the orographic influence of the Cordillera Blanca and the dry and cold Pacific coast conditions (Fig. 6f and h). Those products may thus be used as indicators of spatial precipitation pattern for the study of long-term trends in precipitation (that are costly to generate with WRF3, and not available with KED, because half of the gauge network was installed only in 2012).

4.2.2 Orographic influence on annual amount at 9 and 3 km spatial resolution

Field data are too remote, with no measurement at high altitude to provide information on the altitudinal gradient of precipitation. On a longitudinal transect near the Huascaran peak, we observed important differences in annual precipitation amount and spatial pattern between KED products and WRF outputs (Fig. 7b and c). At very high altitude, we compared precipitation to accumulation data measured at 5100 m a.s.l. on the Artesonraju glacier (station no. 5 from Table 1 and Fig. 1). We can observe in Fig. 7c that KED3 and KED9 products suffer from one major impediment: in regions of low gauge density, the spatial pattern will be solely driven by the altitude, not taking into account the effect of local slopes and orientation. As a consequence, daily rainfall maxima produced by KED are located over the summits, whereas it is well known that these maxima are rather located on the slopes, as correctly simulated by WRF3 (Fig. 7b). The only area with less precipitation in WRF3 compared to WRF9 is the upper zone of the Cordillera Blanca mountain

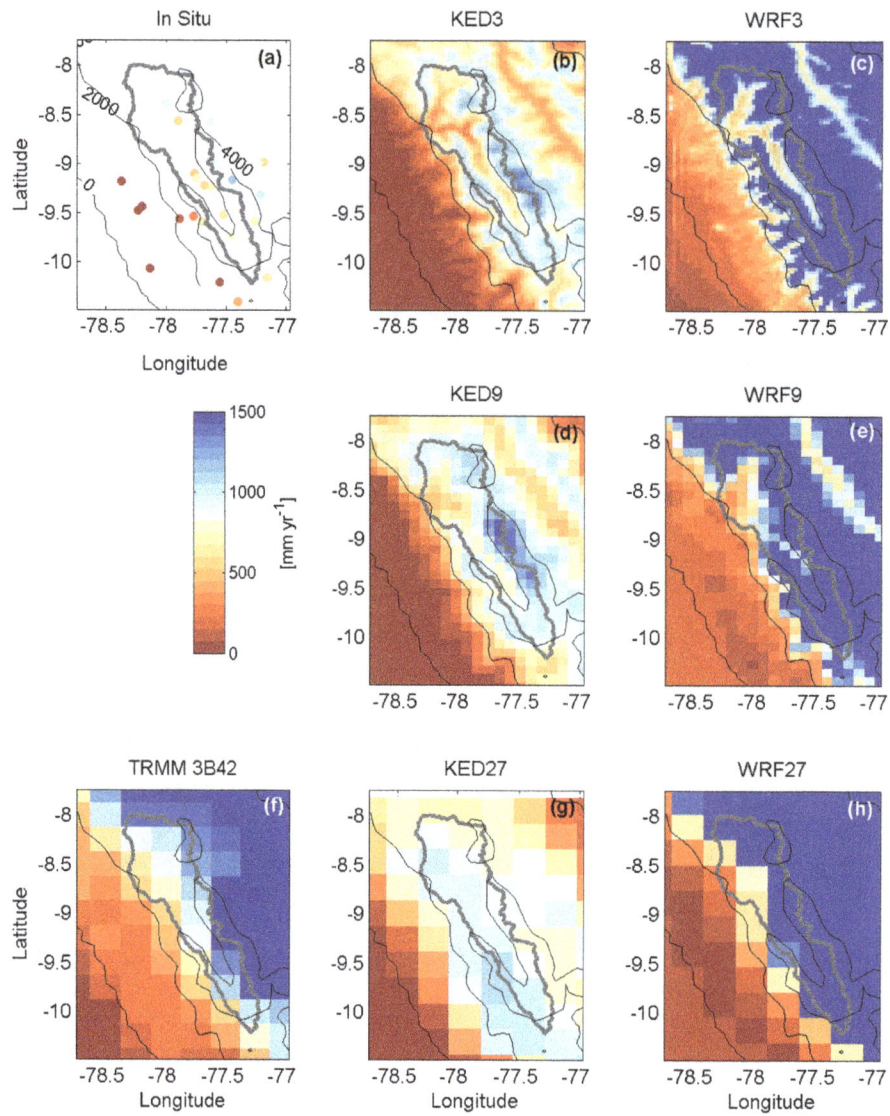

Figure 6. Annual precipitation amounts for all products. Altitudinal contours of WRF9 are drawn every 2000 m (altitudes indicated in **a**). Delimitation of the upper Rio Santa watershed is in bold gray line.

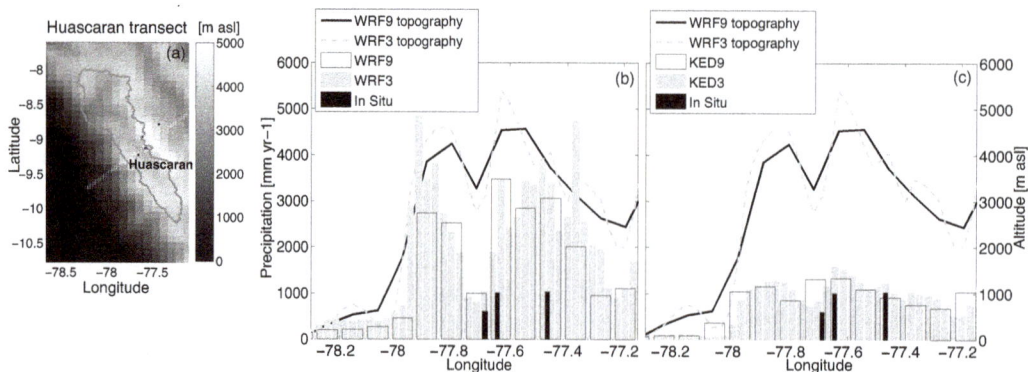

Figure 7. Annual precipitation along a longitudinal transect (white crosses in **a**). Black bars in (**b, c**) correspond to measured precipitation or accumulation. Elevation at 9 km spatial resolution is in solid black line, at 3 km in dotted gray line. WRF9 (empty bars) and WRF3 precipitation (light gray) are plotted in (**b**). KED9 (empty bars) and KED3 precipitation (light gray) are plotted in (**c**).

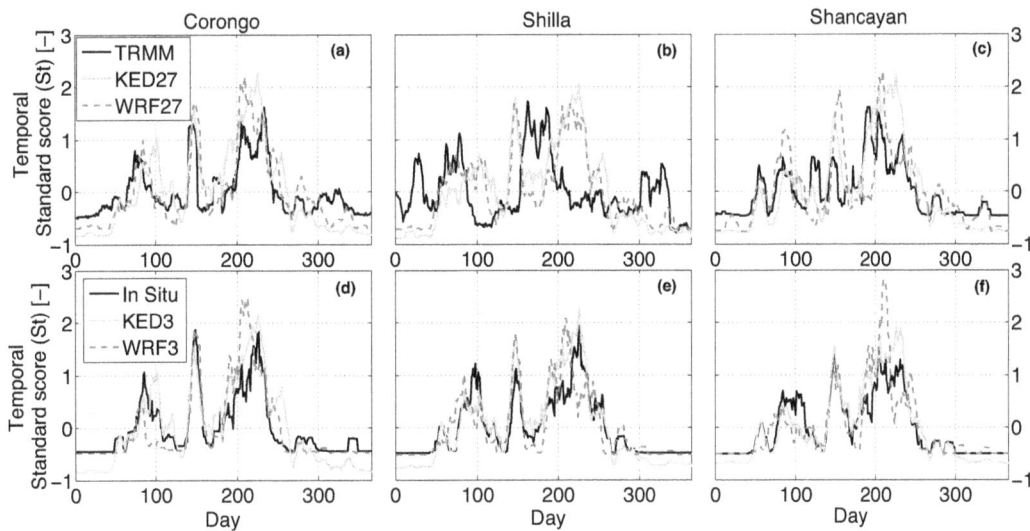

Figure 8. Temporal standard score of running means of daily precipitation amounts over 10 days for three stations along the Rio Santa Valley, for 27 km (**a–c**) and 3 km (**d–f**) spatial resolutions. The gray line is for KED, the dotted line for WRF, and the dark line either for TRMM (upper panels) or in situ (lower panels). Day 1 corresponds to 1 August 2012.

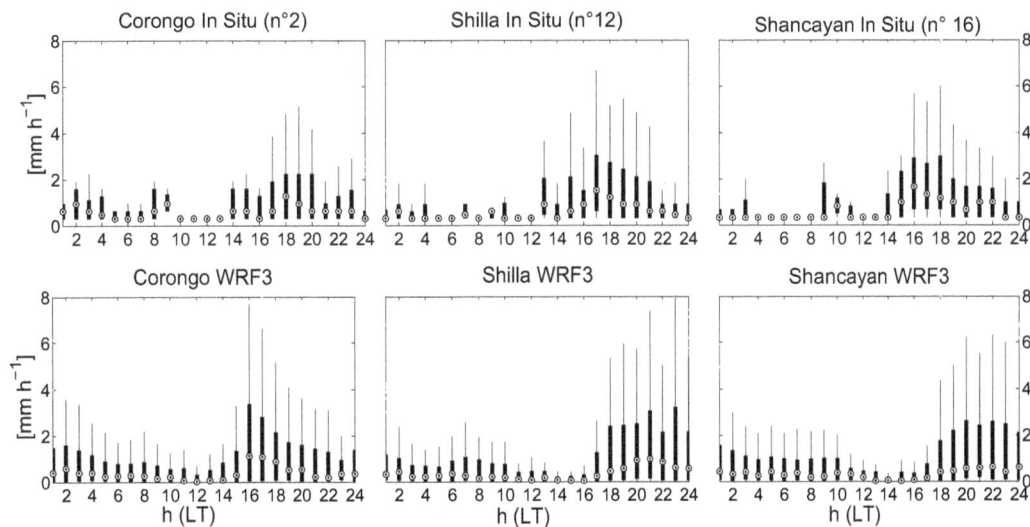

Figure 9. Box plot of hourly precipitation amounts normalized by the mean of hourly data during 1 hydrological year (August 2012–July 2013) for three rain gauges along the Rio Santa Valley (Corongo, Shilla and Shancayan). In situ data are plotted in the upper panel, while WRF3 outputs are plotted in the lower panels.

range, near the highest peaks (Fig. 7b). In WRF3, the altitudinal variation is greater than in WRF9, with the summit reaching 5000 m a.s.l.; the spatial resolution is finer, and in this configuration, the orographic processes on the eastern slopes of the Andes are more pronounced and correctly represented at the 3 km spatial resolution.

4.2.3 Seasonal changes along the Rio Santa Valley

The annual cycle is presented in detail for cells corresponding to three stations located along the Rio Santa Valley, Corongo (station no. 2), Shilla (station no. 12) and Shancayan

(station no. 16) (Fig. 1, Table 1), as these three stations are representative of others located in the Sierra area. Day 1 in Fig. 8 corresponds to the beginning of the hydrological year, 1 August 2012. The upper panels (Fig. 8a–c) correspond to the three products available at the 27 km spatial resolution (TRMM, KED27, WRF27). During the dry period, between days 1 and 50, and 300 to 350, TRMM largely overestimates the precipitation amount for Shilla (Fig. 8b). The percentage of ice-covered area in the mesh corresponding to Shilla station is up to 10 %, while it is less than 0.5 % for the meshes of Corongo and Shancayan. Error in dry season for Shilla can be seen as a poor consideration of ice-covered surface in the TRMM algorithm, as ice on the ground scatters energy in a similar way as precipitation drops in the atmosphere (Yin et al., 2004). Temporal trends of KED27 and WRF27 are similar, with occasional shifts of a few days in heavy rainfall events (for example, between days 200 and 230 for Corongo station; Fig. 8a).

Concerning the finer spatial resolution (Fig. 8d–f), KED3 and in situ data have strong similarities for the three stations, and that confirms the use of the 3 km spatial resolution to compare gridded data with in situ punctual data. Regarding WRF3, intensities of precipitation peaks are false in the heart of the rainy season, but the temporal distribution remains close to that of rain gauge precipitation.

4.3 Diurnal cycle of precipitation along the Rio Santa Valley

Half of the rain gauges available over the region of study are daily reading stations; the network of recording rain gauges is consequently too sparse and too unevenly distributed to permit the computation of relevant rainfall grids at a sub-daily scale. WRF3 thus remains the only product able to account for the diurnal cycle of precipitation by providing hourly rainfall grids (even though TRMM3B42 is available at a 3-hourly time step, the fact that the satellite overpasses the studied area only once or twice daily makes it difficult to trust its accuracy for sub-daily timescales). This is important since the diurnal cycle in a glaciological context controls the precipitation phase and consequently the surface albedo (one strength of WRF is that it produces liquid as well as solid precipitation).

In situ data at Corongo (station no. 2), Shilla (station no. 12) and Shancayan (station no. 16) display a clear precipitation peak in the late afternoon, between 16:00:00 and 19:00:00 LT (Fig. 9). This diurnal cycle is visible in the WRF3 simulations, even though somewhat less pronounced (more rainfall around noon), and with a slight lag at Shilla and Shancayan. Looking at the diurnal cycle of precipitation at a regional scale (Fig. 10), it is noteworthy that the peak hour of precipitation occurs later in the bottom of the Rio Santa Valley (dark green for altitudes below 4000 m a.s.l., and around 19:00:00 LT) than in the surrounding mountains (light green color, around 17:00:00 LT). A lack of hourly information at high altitudes prevents one from validating these hourly scale characteristics with observations, but they correspond to well-documented orographic processes (valley and mountain breezes) (Biasutti et al., 2012; Barros, 2013). In the afternoon, moisture is transported to the peaks by anabatic winds. At the beginning of the night, moisture downs into the valley with katabatic winds. In a physical climate model like WRF, the representation of thermal and orographic circulations theoretically benefits from a finer resolution (Jimenez et al., 2013; Weckwerth et al., 2014), and mountain–valley

breezes seem to be accurately estimated for the 3 km resolution runs.

5 Summary and conclusions

Over the past 40 years, the warming climate of the tropical Andes has led to a significant melting of the glaciers, impacting the hydrological cycle to an extent that remains to be assessed, both for present and for future times. One obstacle to doing so is our limited ability to evaluate properly the precipitation falling over high-altitude catchments, if only because of the difficulties for installing and maintaining sufficiently dense in situ networks. In addition, the rough topography generates strong spatial gradients that are very challenging to sample. In such a context, remote sensing and modeling look to be attractive means for complementing the information provided by in situ measurements. With this in mind, this paper has presented a comparison of rainfall products based on three different sources of information: rain gauge measurements, satellite imagery and atmospheric model outputs. While TRMM3B42 is a widely used standard, making it a natural candidate to represent the family of satellite rainfall products, there is a larger range of possibilities for selecting a ground-based product and an atmospheric model product. Preliminary tests, the results of which are not detailed in this paper, were used to finally select kriging with external drift interpolation (KED) as a typical ground-based product, the external drift being the altitude. As for atmospheric models, the retained product is made of WRF simulations, WRF being run in a configuration minimizing the differences between the observations and the model outputs over the Cordillera Blanca.

The TRMM3B42 product has a resolution of 27 km; the same resolution was thus used for the computation of coarse rainfall grids from gauge measurements (KED27) and for WRF simulations (WRF27). Then gauge rainfall grids and WRF simulations were also produced at the finer resolutions of 9 km (KED9 and WRF9) and 3 km (KED3 and WRF3). This makes a total of seven gridded precipitation products that were computed and inter-compared over the region of the Rio Santa in Peru, a glaciated catchment and the second largest river flowing from the tropical Andes to the Pacific.

Each process leading to the computation of gridded rainfall products has its own weaknesses: interpolation errors for the rain gauge products, indirect measurement of rainfall for the satellite products, sub-mesh parameterization for the WRF model outputs. Therefore none of them can be taken as an indisputable reference, whether it be in terms of quantities or in terms of occurrence. This is why the performances of each product were assessed from a double perspective. A comparison with measured on-site data was carried out when relevant (diurnal and seasonal cycle, statistics of rainfall occurrence), while the ability of each product to reproduce some well-known spatial features of precipitation fields

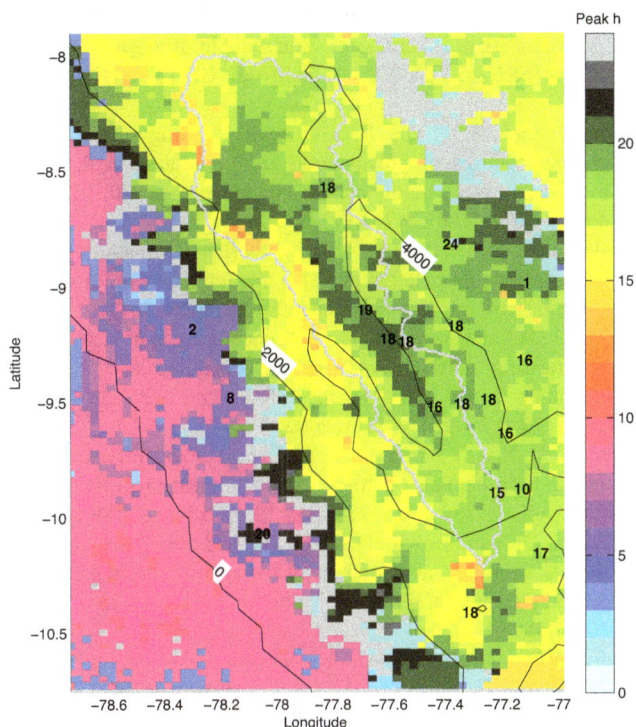

Figure 10. Peak hour of precipitation in WRF3. White numbers correspond to peak hour for the in situ data. The altitude of WRF9 is drawn every 2000 m (black lines). Delimitation of the upper Rio Santa watershed is in gray.

at various timescales (from annual down to daily) was analyzed when no obvious quantitative reference could be used.

In line with the results of other studies, WRF27 simulations are found to be totally unrealistic in terms of annual quantities. WRF9 and WRF3 simulations are better in this respect but still largely overestimate the annual total, with WRF9 being additionally unable to capture properly the details of the spatial pattern that are well restituted by WRF3. This shortcoming of WRF9 can be explained by its resolution that is still too coarse to reproduce correctly the orographic influence, because a number of key features are smoothed out (for instance, grid meshes reaching altitudes above 5000 m a.s.l. are found in the WRF3 topography, which is not the case in the WRF9 topography).

TRMM, with its coarse spatial resolution of 27 km, performs poorly over ice-covered surfaces, because ice on the ground behaves in a similar way as rain or ice drops in the atmosphere in terms of scattering the microwave energy (Yin et al., 2004). Using TRMM in glaciated mountain ranges should thus be avoided, especially at small timescales where spatial error compensation does not occur, as it might do when averaging annual totals over large areas. On the other hand, TRMM might provide some useful information over the lowlands on the Amazonian side of the Andes, as already mentioned by Lavado Casimiro et al. (2009).

Coarse-resolution products (TRMM and WRF27), however, correctly represent the large spatial gradient between the humid Amazonian lowlands and the dry Pacific coast, and their long-term precipitation series can thus be used to study the interannual variability of the spatial patterns at a large regional scale and possible long-term trends linked to climate change.

Comparing the diurnal cycle of the hourly WRF3 simulations with observations in meshes containing one recording rain gauge leads to the conclusion that this diurnal cycle is fairly realistic. Of course the default of the large overestimation of precipitation by WRF3 prevents one from using directly the WRF3 grids as inputs to hydrological models. The challenge is thus to combine the hourly temporal distribution of precipitation in WRF3 with more accurate precipitated amounts. In this respect, one path to explore is to use the WRF3 diurnal cycle for disaggregating the KED daily grids.

A more general conclusion is that the topography and the associated rainfall gradients are too steep in this region for rainfall products at the spatial resolution of either 9 or 27 km to provide good rainfall estimates and good rainfall spatial patterns for glacio-hydrological purposes. Moreover, due to a poor sampling at high altitudes, kriging with external drift does not take into account local slope and orientation effects as the spatial pattern is solely driven by altitude. In summary, combining the daily rain gauge measurements with the spatial patterns generated by WRF3 appears to be promising way for building daily rain fields. There are several techniques to do so, one being to use the WRF3 rain field, instead of the topography, as the external drift when interpolating the in situ measurements with a KED technique.

Acknowledgements. This study was performed thanks to the IRD (French Research Institute for Development) and the LMI-GREATICE IRD program. Field work was carried out in cooperation with the UGRH and the UNASAM. We are grateful to all those who took part in those field campaigns. Gratitude is expressed to the Senamhi (Servicio Nacional de Meteorología e Hidrología del Perú), to Waldo Lavado, to the UGRH and to the UNASAM-CIIADERS project for making station data available. We also warmly thank our Peruvian colleagues Ken Takahashi and Jhan Carlo Espinoza from the IGP (Instituto Geofísico del Perú) who participated in the development and improvement of this work. We are grateful to Dr Maussion and an anonymous reviewer who made constructive and detailed comments, which helped improve the manuscript.

References

Aceituno, P.: On the functioning of the southern oscillation in the South American sector. Part I I. Upper-air circulation, J. Climate, 116, 505–524, doi:10.1175/1520-0442(1989)002<0341:OTFOTS>2.0.CO;2, 1987.

Amani, A. and Lebel, T.: Lagrangian kriging for the estimation of Sahelian rainfall at small time steps. J. Hydrology., 192, 125–157, doi:10.1016/S0022-1694(96)03104-6, 1997.

Baraer, M., Mark, B., McKenzie, J., Condom, T., Bury, J., Huh, K. I., Portocarrero, C., Gomez, J., and Rathay, S.: Glacier recession and water resources in Peru's Cordillera Blanca, J. Glaciol., 58, 134–150, doi:10.3189/2012JoG11J186, 2012.

Barros, A. P.: Orographic precipitation, freshwater resources, and climate vulnerabilities in mountainous regions, in: Climate Vulnerability: Understanding and Addressing Threats to Essential Resources, Elsevier Inc., Academic Press, Waltham, Massachusetts, 57–78, 2013.

Biasutti, M., Yuter, S. E., Burleyson, C. D., and Sobel, A. H.: Very high resolution rainfall patterns measured by TRMM precipitation radar: seasonal and diurnal cycles, Clim. Dynam., 39, 239–258, doi:10.1007/s00382-011-1146-6, 2012.

Box, J. E. and Bromwich, D. H.: Greenland ice sheet surface mass balance 1991–2000: application of Polar MM5 mesoscale model and in situ data, J. Geophys. Res., 109, 1–21, doi:10.1029/2003JD004451, 2004.

Bury, J. T., Mark, B. G., McKenzie, J. M., French, A., Baraer, M., Huh, K. I., Zapata Luyo, M. A., and Gómez López, R. J.: Glacier recession and human vulnerability in the Yanamarey watershed of the Cordillera Blanca, Peru, Climatic Change, 105, 179–206, doi:10.1007/s10584-010-9870-1, 2011.

Condom, T., Rau, P., and Espinoza, J. C.: Correction of TRMM 3B43 monthly precipitation data over the mountainous areas of Peru during the period 1998–2007, Hydrol. Process., 25, 1924–1933, doi:10.1002/hyp.7949, 2011.

Condom, T. Escobar, M., Purkey, D., Pouget, J. C. Suarez, W., Ramos, C., Apaestegui, J., Tacsi, A., and Gomez, J.: Simulating the implications of glaciers' retreat for water management: a case study in the Rio Santa Basin, Peru, Water Int., 37, 442–459, doi:10.1080/02508060.2012.706773, 2012.

Dudhia, J.: Numerical study of convection observed during the winter monsoon experiment using a mesoscale twodimensional model, J. Atmos. Sci., 46, 3077–3107, doi:10.1175/1520-0469(1989)046%3C3077:NSOCOD%3E2.0.CO;2, 1989.

Espinoza, J. C., Chavez, S. P., Ronchail, J., Junquas, C., Takahashi, K., and Lavado, W.: Rainfall hotspots over the southern tropical Andes: spatial distribution, rainfall intensity and relations with large-scale atmospheric circulation, Water Resour. Res., 51, 3459–3475, doi:10.1002/2014WR016273, 2015.

Espinoza Villar, J. C., Ronchail, J., Guyot, J. L., Cochonneau, G., Naziano, F., Lavado, W., De Oliveira, E., Pombosa, R., and Vauchel, P.: Spatio-temporal rainfall variability in the Amazon basin countries (Brazil, Peru, Bolivia, Colombia, and Ecuador), Int. J. Climatol., 29, 1574–1594, doi:10.1002/joc.1791, 2009.

Garreaud, R., Vuille, M., and Clement, A. C.: The climate of the Altiplano: observed current conditions and mechanisms of past changes, Palaeogeogr. Palaeocl., 194, 5–22, doi:10.1016/S0031-0182(03)00269-4, 2003.

Garreaud, R. D.: The Andes climate and weather, Adv. Geosci., 22, 3–11, doi:10.5194/adgeo-22-3-2009, 2009.

Georges, C.: 20th century glacier fluctuations in the tropical Cordillera Blanca, Peru, Arct. Antarct. Alp. Res., 36, 100–107, doi:10.1657/1523-0430(2004)036[0100:TGFITT]2.0.CO;2, 2004.

Giovannettone, J. P. and Barros, A. P.: Probing regional orographic controls of precipitation and cloudiness in the Central Andes using satellite data, J. Hydrometeorol., 10, 167–182, doi:10.1175/2008JHM973.1, 2009.

Gräler, B., Rehr, M., Gerharz, L., and Pebesma, E.: Spatio-temporal analysis and interpolation of PM_{10} measurements in Europe for 2009, ETC/ACM Technical Paper 2012/8, 1–29, 2012.

Grell, G. A. and Devenyi, D.: A generalized approach to parameterizing convection combining ensemble and data assimilation techniques, Geophys. Res. Lett., 29, 38-1–38-4, doi:10.1029/2002GL015311, 2002.

Hong, S. Y., Noh, Y., and Dudhia, J.: A new vertical diffusion package with an explicit treatment of entrainment processes, Mon. Weather Rev., 134, 2318–2341, doi:10.1175/MWR3199.1, 2006.

Huffman, G. J. and Bolvin, D. T.: TRMM and other data precipitation data set documentation, available at: ftp://precip.gsfc.nasa.gov/pub/trmmdocs/3B42_3B43_doc.pdf, last access: 25 May 2014, 2012.

Huffman, G. J., Bolvin, D. T., Nelkin, E. J., Wolff, D. B., Adler, R. F., Gu, G., Hong, Y., Bowman, K. P., and Stocker, E. F.: The TRMM Multisatellite Precipitation Analysis (TMPA): quasi-global, multiyear, combined-sensor precipitation estimates at fine scales, J. Hydrometeorol., 8, 38–55, doi:10.1175/JHM560.1, 2007.

Jimenez, P. A. and Dudhia, J.: Improving the representation of resolved and unresolved topographic effects on surface wind in the WRF model, J. Appl. Meteorol. Clim., 51, 300–316, doi:10.1175/JAMC-D-11-084.1, 2012.

Jimenez, P. A., Dudhia, J., Gonzalez-Rouco, J. F., Montavez, J. P., Garcia-Bustamante, E., Navarro, J., Vila-Guerau de Arellano, J. and Muñoz-Roldan, A.: An evaluation of WRF's ability to reproduce the Surface wind over complex terrain based on typical circulation patterns, J. Geophys. Res., 118, 7651–7669, doi:10.1002/jgrd.50585, 2013.

Juen, I., Kaser, G., and Georges, C.: Modelling observed and future runoff from a glacierized tropical catchment (Cordillera Blanca, Perú), Global Planet. Change, 59, 37–48, doi:10.1016/j.gloplacha.2006.11.038, 2007.

Kaser, G.: A review of the modern fluctuations of tropical Glaciers, Global Planet. Change, 22, 93–103, doi:10.1016/S0921-8181(99)00028-4, 1999.

Kaser, G., Juen, I., Georges, C., Gómez, J., and Tamayo, W. The impact of glaciers on the runoff and the reconstruction of mass balance history from hydrological data in the tropical Cordillera Blanca, Perú, J. Hydrol., 282, 130–144, doi:10.1016/S0022-1694(03)00259-2, 2003.

Lavado Casimiro, W. S. and Espinoza, J. C.: Impactos de El Niño y La Niña en las lluvias del Peru (1965–2007), Revista Brasileira de Meteorologia, 29, 171–182, doi:10.1590/S0102-77862014000200003, 2014.

Lavado Casimiro, W. S., Labat, D., Guyot, J. L., Ronchail, J., and Ordonez, J. J.: TRMM Rainfall Data Estimation over the Peruvian Amazon-Andes Basin and Its Assimilation into a Monthly Water Balance Model. New Approaches to Hydrological Prediction in Data-Sparse Regions, in: Proceedings of Symposium

HS.2 at the Joint IAHS&IAH Convention, September 2009, Hyderabad, India, 245–252, 2009.

Lavado Casimiro, W. S., Ronchail, J., Labat, D., Espinoza, J. C., and Guyot, J. L.: Basin-scale analysis of rainfall and runoff in Peru (1969–2004): Pacific, Titicaca and Amazonas drainages, Hydrol. Sci. J., 57, 625–642, doi:10.1080/02626667.2012.672985, 2012.

Li, J. and Heap, A. D.: A Review of Spatial Interpolation Methods for Environmental Scientists, Geoscience Australia, Canberra, 42–46, 2008.

Mantas, V. M., Liu, Z., Caro, C., and Pereira, A. J. S. C.: Validation of TRMM multisatellite precipitation analysis (TMPA) products in the Peruvian Andes, Atmos. Res., 163, 132–145, doi:10.1016/j.atmosres.2014.11.012, 2014.

Marengo, J. A., Liebmann, B., Grimm, A. M., Misra, V., Silva Dias, P. L., Cavalcanti, I. F. A., Carvalho, L. M. V., Berbery, E. H., Ambrizzi, T., Vera, C. S., Saulo, A. C., Nogues-Paegle, J., Zipser, E., Seth, A., and Alves, L. M.: Recent developments on the South American monsoon system, Int. J. Climatol., 32, 1–21, doi:10.1002/joc.2254, 2012.

Mark, B. G. and Seltzer, G. O.: Tropical glacier meltwater contribution to stream discharge: a case study in the Cordillera Blanca, Peru, J. Glaciol., 49, 271–281, doi:10.3189/172756503781830746, 2003.

Masson, D. and Frei, C.: Spatial analysis of precipitation in a high-mountain region: exploring methods with multi-scale topographic predictors and circulation types, Hydrol. Earth Syst. Sci., 18, 4543–4563, doi:10.5194/hess-18-4543-2014, 2014.

Maussion, F., Scherer, D., Mölg, T., Collier, E., Curio, J., and Finkelnburg, R.: Precipitation seasonality and variability over the Tibetan Plateau as resolved by the high asia reanalysis, J. Climate, 27, 1910–1927, doi:10.1175/JCLI-D-13-00282.1, 2014.

Mearns, L. O., Giorgi, F., McDaniel, L. and Shields, C.: Analysis of daily variability of precipitation in a nested regional climate model: comparison with observations and doubled CO 2 results, Global Planet. Change, 10, 55–78, doi:10.1007/BF00215007, 1995.

Mlawer, E. J., Taubnam, S. J., Brown, P. D., Iacono, M. J., and Clough, S. A.: Radiative transfer for inhomogeneous atmospheres: RRTM, a validated correlated-k model for the longwave, J. Geophys. Res., 102, 663–682, doi:10.1029/97JD00237, 1997.

Murthi, A., Bowman, K. P., and Leung, L. R.: Simulations of precipitation using NRCM and comparisons with satellite observations and CAM: annual cycle, Clim. Dynam., 36, 1659–79, doi:10.1007/s00382-010-0878-z, 2011.

Niu, G. Y., Yang, Z. L., Mitchell, K. E., Chen, F., Ek, M. B., Barlage, M., Kumar, A., Manning, K., Niyogi, D., Rosero, E., Tewari, M., and Xia, Y.: The community Noah land surface model with multiparameterization options (Noah-MP): 1. Model description and evaluation with local-scale measurements, J. Geophys. Res., 116, 1–19, doi:10.1029/2010JD015139, 2011.

Ochoa, A., Pineda, L., Crespo, P., and Willems, P.: Evaluation of TRMM 3B42 precipitation estimates and WRF retrospective precipitation simulation over the Pacific–Andean region of Ecuador and Peru, Hydrol. Earth Syst. Sci., 18, 3179–3193, doi:10.5194/hess-18-3179-2014, 2014.

Paulson, C. A.: The mathematical representation of wind speed and temperature profiles in the unstable atmospheric surface layer, J. Appl. Meteorol., 9, 857–861, doi:10.1175/1520-0450(1970)009<0857:TMROWS>2.0.CO;2, 1970.

Roca, R., Chambon, P., Jobard, I., Kirstetter, P. E., Gosset, M., and Bergès, J. C.: Comparing satellite and surface rainfall products over West Africa at meteorologically relevant scales during the AMMA campaign using error estimates, J. Appl. Meteorol. Clim., 49, 715–731, doi:10.1175/2009JAMC2318.1, 2010.

Sambou, S. Modèle statistique des hauteurs de pluies journalières en zone sahélienne: exemple du bassin amont du fleuve Sénégal/Frequency analysis of daily rainfall in the Sahelian area: case of the upstream basin of the Senegal River, Hydrolog. Sci. J., 49, 115–129, doi:10.1623/hysj.49.1.115.53989, 2004.

Salzmann, N., Huggel, C., Rohrer, M., Silverio, W., Mark, B. G., Burns, P., and Portocarrero, C.: Glacier changes and climate trends derived from multiple sources in the data scarce Cordillera Vilcanota region, southern Peruvian Andes, The Cryosphere, 7, 103–118, doi:10.5194/tc-7-103-2013, 2013.

Sanabria, J., Calanca, P., Alarcón, C., and Canchari, G. Potential impacts of early twenty-first century changes in temperature and precipitation on rainfed annual crops in the Central Andes of Peru, Reg. Environ. Change, 14, 1533–1548, doi:10.1007/s10113-014-0595-y, 2014.

Scheel, M. L. M., Rohrer, M., Huggel, Ch., Santos Villar, D., Silvestre, E., and Huffman, G. J.: Evaluation of TRMM Multisatellite Precipitation Analysis (TMPA) performance in the Central Andes region and its dependency on spatial and temporal resolution, Hydrol. Earth Syst. Sci., 15, 2649–2663, doi:10.5194/hess-15-2649-2011, 2011.

Schwarb, M., Acuña, D., Konzelmann, Th., Rohrer, M., Salzmann, N., Serpa Lopez, B., and Silvestre, E.: A data portal for regional climatic trend analysis in a Peruvian High Andes region, Adv. Sci. Res., 6, 219–226, doi:10.5194/asr-6-219-2011, 2011.

Silverio, W. and Jaquet, J. M.: Glacial cover mapping (1987–1996) of the Cordillera Blanca (Peru) using satellite imagery, Remote Sens. Environ., 95, 342–350, doi:10.1016/j.rse.2004.12.012, 2005.

Skamarock, W. C., Klemp, J. B., Dudhia, J., Gill, D. O., Barker, D. M., Duda, M. G., Huang, X. Y., Wang, W., and Powers, J. G.: A description of the advanced research WRF version 3, NCAR Technical Note, NCAR/TN-475+STR, 113, doi:10.5065/D68S4MVH, 2008.

Thompson, G., Field, P. R., Rasmussen, R. M., and Hall, W. D.: Explicit forecasts of winter precipitation using an improved bulk microphysics scheme. Part II: Implementation of a new snow parameterization, Mon. Weather Rev., 136, 5095–5115, doi:10.1175/2008MWR2387.1, 2008.

Tobin, C., Nicotina, L., Parlange, M. B., Berne, A., and Rinaldo, A.: Improved interpolation of meteorological forcings for hydrologic applications in a Swiss Alpine region, J. Hydrol., 401, 77–89, 2011.

Vera, C., Higgins, W., Amador, J. Ambrizzi, T., Garreaud, R., Gochis, D. Gutzler, D., Lettenmaier, D., Marengo, J., Mechoso, C. R., Nogues-Paegle, J., Silva Dias, P. L., and Zhang, C.: Toward a unified view of the American monsoon systems, J. Climate, 19, 4977–5000, doi:10.1175/JCLI3896.1, 2006.

Vischel, T., Quantin, G., Lebel, T., Viarre, J., Gosset, M., Cazenave, F., and Panthou, G.: Generation of high resolution rainfields in West Africa: evaluation of dynamical interpolation methods, J. Hydrometeorol., 12, 1465–1482, doi:10.1175/JHM-D-10-05015.1, 2011.

Vuille, M., Kaser, G., and Juen, I.: Glacier mass balance variability in the Cordillera Blanca, Peru and its relationship with climate and the large-scale circulation, Global Planet. Change, 62, 14–28, doi:10.1016/j.gloplacha.2007.11.003, 2008a.

Vuille, M., Francou, B., Wagnon, P., Juen, I., Kaser, G., Mark, B. G., and Bradley, R. S.: Climate change and tropical Andean glaciers: past, present and future, Earth-Sci. Rev., 89, 79–96, doi:10.1016/j.earscirev.2008.04.002, 2008b.

Ward, E., Buytaert, W., Peaver, L., and Wheater, H.: Evaluation of precipitation products over complex mountainous terrain: a water resources perspective, Adv. Water Resour., 34, 1222–1231, doi:10.1016/j.advwatres.2011.05.007, 2011.

Weckwerth, T. M., Bennett, L. J., Jay Miller, L., Van Baelen, J., Di Girolamo, P., Blyth, A. M. and Hertneky, T. J.: An Observational and Modeling Study of the Processes Leading to Deep, Moist Convection in Complex Terrain, Mon. Weather Rev., 142, 2687–2708, doi:10.1175/MWR-D-13-00216.1, 2014.

Yang, Z. L., Niu, G. Y., Mitchell, K. E., Chen, F., Ek, M. B., Barlage, M., Manning, K., Niyogi, D., Tewari, M., and Xia, Y. L.: The community Noah land surface model with multiparameterization options (Noah-MP): 2. Evaluation over global river basins, J. Geophys. Res., 116, 1–16, doi:10.1029/2010JD015140, 2011.

Yin, Z. Y., Liu, X., Zhang, X., and Chung, C. F.: Using a geographic information system to improve Special Sensor Microwave Imager precipitation estimates over the Tibetan Plateau, J. Geophys. Res., 109, 1984–2012, doi:10.1029/2003JD003749, 2004.

Young, K. R. and Leon, B.: Natural hazard in Peru: causation and vulnerability, Developments in Earth Surface Processes, 13, 165–180, doi:10.1016/S0928-2025(08)10009-8, 2009.

11

Effects of cultivation and reforestation on suspended sediment concentrations: a case study in a mountainous catchment in China

N. F. Fang[1,2], F. X. Chen[3], H. Y. Zhang[1,2], Y. X. Wang[1,2], and Z. H. Shi[2,3]

[1]State Key Laboratory of Soil Erosion and Dryland Farming on the Loess Plateau, Northwest A & F University, Yangling 712100, People's Republic of China
[2]Institute of Soil and Water Conservation of Chinese Academy of Sciences and Ministry of Water Resources, Yangling 712100, People's Republic of China
[3]College of Resources and Environment, Huazhong Agricultural University, Wuhan 430070, People's Republic of China

Correspondence to: Z. H. Shi (shizhihua70@gmail.com)

Abstract. Understanding how sediment concentrations vary with land use/cover is critical for evaluating the current and future impacts of human activities on river systems. This paper presents suspended sediment concentration (SSC) dynamics and the relationship between SSC and discharge (Q) in the 8973 km^2 Du catchment and its sub-catchment (4635 km^2). In the Du catchment and its sub-catchment, 4235 and 3980 paired SSC–Q samples, respectively, were collected over 30 years. Under the influence of the Household Contract Responsibility System and Grain-for-Green projects in China, three periods were designated, the original period (1980s), cultivation period (1990s) and reforestation period (2000s). The results of a Mann–Kendall test showed that rainfall slightly increased during the study years; however, the annual discharge and sediment load significantly decreased. The annual suspended sediment yield of the Du catchment varied between 1.3×10^8 and 1.0×10^{10} kg, and that of the sub-catchment varied between 6.3×10^7 and 4.3×10^9 kg. The SSCs in the catchment and sub-catchment fluctuated between 1 and 22400 g m^{-3} and between 1 and 31800 g m^{-3}, respectively. The mean SSC of the Du catchment was relatively stable during the three periods (± 83 g m^{-3}). ANOVA (analysis of variance) indicated that the SSC did not significantly change under cultivation for low and moderate flows, but was significantly different under high flow during reforestation of the Du catchment. The SSC in the sub-catchment was more variable, and the mean SSC in the sub-catchment varied from 1058 $\pm$ 2217 g m^{-3} in the 1980s to 1256 $\pm$ 2496 g m^{-3} in the 1990s and 891 $\pm$ 1558 g m^{-3} in the 2000s. Reforestation significantly decreased the SSCs during low and moderate flows, whereas cultivation increased the SSCs during high flow. The sediment rating curves showed a stable relationship between the SSC and Q in the Du catchment during the three periods. However, the SSC–Q of the sub-catchment exhibited scattered relationships during the original and cultivation periods and a more linear relationship during the reforestation period.

1 Introduction

Suspended sediment is conventionally regarded as sediment that is transported by a fluid and is fine enough to remain suspended in turbulent eddies (Parsons et al., 2015). Suspended sediment plays important roles in the hydraulics, hydrology and ecology of rivers (Luo et al., 2013). Land use/cover is thought to affect hydrology and suspended sediment yield (SSY) (Van Rompaey et al., 2002; Casalí et al., 2010). Although many studies have assumed that forest cover is an effective method for controlling sediment yield throughout the world (e.g., Mount et al., 2005; Hopmans and Bren, 2007; Garzía-Ruiz et al., 2008; Stickler et al., 2009; Verbist et al., 2010; Lü et al., 2015; Wei et al., 2015), other studies have disagreed (e.g., Mizugaki et al., 2008; Ide et al., 2009). Additionally, many studies have implicated farmland as a major contributor of sediments (Gafur et al., 2003; Shi et al., 2004; Izaurralde et al., 2007; Cerdan et al., 2010). However, whether changes in land use/cover alter soil loss by changing the runoff volume or by changing the suspended sediment concentration (SSC) has received little attention. The relationships between SSC and discharge (Q have been discussed using sediment rating curves (Walling, 1977), a fuzzy logic model (Kisi et al., 2006), artificial neural networks (Liu et al., 2013) and other multivariate regression methods (Francke et al., 2008). SSCs are highly variable and can vary over many orders of magnitude during storm events (Naden and Cooper, 1999; Cooper, 2002; Fang et al., 2012). The mean annual/monthly SSC fails to capture the highly episodic nature of sediment transport because > 90 % of the sediment load can be transported in < 10 % of the time (Collins et al., 2011). Morehead et al. (2003) indicated that the suspended sediment load carried by rivers varies spatially and temporally and that sediment rating curve parameters can exhibit time-dependent trends. Warrick et al. (2013) concluded that the discharge and sediment relationships from six coastal rivers varied substantially with time in response to land use. In most studies, SSYs were calculated using SSCs and Q. However, little work has focused on the effects of land use/cover change on SSCs.

China contains 22 % of the world's population but only 7 % of the world's croplands (Liu and Diamond, 2005). In China, erosion by water affects an area of $3.6 \times 10^6 \, \text{km}^2$, or approximately 37 % of the country's land area (Ni et al., 2008). Thus, soil erosion has become an important topic for local and national policy makers. In the 1980s, a policy called the Household Contract Responsibility System was implemented in China's rural areas. Consequently, more land was reclaimed for farming. In the late 1999s, the Grain-for-Green project was introduced to increase forest and grassland cover. To combat soil erosion on sloped croplands, farmland with slopes > 25° was restored. The farmers, who agreed to stop cultivating these lands received subsidies to cover their losses (Gao et al., 2012). Before this project, subtropical zones with adequate rainfall were often over-exploited due to economic and demographic pressures. Cultivation of steeply sloping lands in subtropical areas can result in serious soil erosion during intense rainfall (Fang et al., 2012). In this study, a mountainous catchment and its sub-catchment were investigated and analyzed in detail. This catchment is located in the Danjiangkou reservoir area, which is a source area in the Middle Route Project under the South–North Water Transfer Scheme (the largest water transfer project in the world). The study catchment has experienced cultivation and reforestation periods. The first part of this study focuses on how cultivation and reforestation affect Q, SSC and SSY at different timescales. Then, we discuss the dual roles of cultivation and reforestation that affect the relationship between SSC and Q. Finally, the SSC dynamics in the catchment and sub-catchment were determined under land use/cover changes.

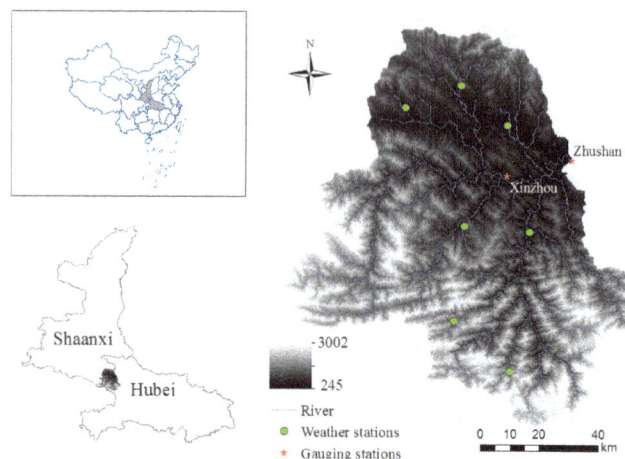

Figure 1. Location of study area.

2   Study area and methods

2.1   Study area

This study was conducted in the Du catchment (31°30′–32°37′ N, 109°11′–110°25′ E), which is located in Hubei Province, China, and covers an area of 8973 km² (Fig. 1). Elevations within the watershed range from 245 to 3002 m. The sub-catchment (Xinzhou catchment) is located in the northwest region of the Du catchment and covers an area of 4635 km². The topography in the Du catchment is undulating and is characterized by mountain ranges, steep slopes and a subtropical climate with a mean temperature of 15 °C. The mean annual precipitation in this region is approximately 1000 mm, with 80 % of the precipitation occurring between May and September. The major soil types include yellow to brown soils, Chao soils and purple soils (National Soil Survey Office, 1992), which correspond to Alfisols, Entisols and Inceptisols, respectively, according to USDA Soil Taxonomy (Soil Survey Staff, 1999). The major crops in this region are corn (*Zea mays L.*) and wheat (*Triticum aestivum L.*). There were 1002 villages with total population of 1.9 × 106 based on the fifth population census of China in 2000.

2.2   Land use/cover change

The land cover was digitized as part of a previous research project. Reconnaissance field surveys were conducted in 2007. A watershed topographic map was used in combination with 1999 ETM (enhanced thematic mapper) photographs and Landsat imagery from 1987 and 2007. The land use/cover units were delineated on the photographs and verified in the field. We assigned the periods of the 1980s, 1990s, and 2000s to original, cultivation, and reforestation periods, respectively. The areas of the various types of land use/cover are presented in Tables 1 and 2. In 1987, forestland, farmland, and shrubland covered areas of 6316 km² (70.4 %),

919 km^2 (10.2 %) and 929 km^2 (10.4 %), respectively. The other land use/cover types covered small areas and included barren land (0.4 %), grassland (7.3 %), urban land (0.9 %), and water bodies (0.4 %) (Table 1). During the 2000s, some steep lands with slopes of more than 25° were converted to forestland. The area of forestland increased to 75.2 % in 2007, whereas the area of farmland decreased to 6.1 % (Fig. 2). The sub-catchment experienced a similar change in farmland, which increased from 11.5 % in 1987 to 14.7 % in 1999 and decreased to 6.7 % in 2007. However, the change in forestland in the sub-catchment was different from that in the Du catchment, in which forestland increased from 66.3 % in 1987 to 67.9 % in 1999 and 74.0 % in 2007 (Table 2).

2.3 Data acquisition

All of the hydrological data were obtained from the Hubei Provincial Water Resources Bureau. Two gauge stations (Zhushan and Xinzhou) and seven weather stations (nearly evenly distributed) are located in the study catchment. The yearly average rainfall measured at three weather stations in Xinzhou was very similar to the mean rainfall measured at the seven weather stations. Therefore, we used the average annual values of rainfall obtained from the seven stations for the Zhushan and Xinzhou stations. A continuously recording water-level stage recorder and a silt sampler (metal type) were used to record discharge and sediment (complemented by manual samples), respectively. The water stage was measured and transformed into discharge by using the calibrated rating curve obtained through periodic flow measurements. SSCs were determined using the gravimetric method, in which water samples were vacuum filtered through a 0.45 μm filter and the residue was oven dried at 105 °C for 24 h. The weight of each dried residue and the initial sample volume were used to obtain the SSC (g m^{-3}). Next, the SSY was calculated from the SSC and Q. During a month, the total SSY was the sum SSY of each event. Monthly SSC was calculated by monthly SSY and Q. During rainfall events, the sampling measurement frequency was increased several times each day. Paired SSC–Q data were obtained during rainfall–runoff events. Because bed-load measurements were not performed in this area, this study does not consider bed-load sediment transport. From 1980 to 2009, 4235 paired SSC–Q samples were collected at the Zhushan station and 3980 samples were collected at the Xinzhou station. This study uses several variables, and their meanings and abbreviations are shown in Table 3. To distinguish between the variables of the two gauges, we used Qd, Dd, SSYd and SSCd for the Zhushan station (Du catchment) and Qx, Dx, SSYx and SSCx for the Xinzhou station (sub-catchment).

The variables for D, SSY$_i$ and SSY are calculated as follows:

$$D = Q/A, \tag{1}$$

$$SSY_i = SSC_i \times Q_i, \tag{2}$$

$$SSY = \int_1^n SSY_i, \tag{3}$$

where A is the area of the catchment and SSY$_i$, SSC$_i$ and Q_i are the suspended sediment yield, suspended sediment concentration and discharge during period i, respectively.

2.4 Statistical analyses

The Mann–Kendall test, which was proposed by Mann (1945) and Kendall (1975), was used to identify trends in P, Q and SSY during the 30-year study period. The S statistic was calculated as follows:

$$S = \sum_{i=1}^{n-1} \sum_{j=i+1}^{n} \text{sgn}\left(x_j - x_i,\right) \tag{4}$$

where n is the number of data points, x_i and x_j are the respective data values in the time series i and j ($j > 1$), and $\text{sgn}(x_j - x_i)$ is the sign function (Gao et al., 2012), which is determined as follows:

$$\text{sgn}\left(x_j - x_i\right) = \begin{cases} +1, & \text{if } x_j - x_i > 0 \\ 0, & \text{if } x_j - x_i = 0 \\ -1, & \text{if } x_j - x_i < 0 \end{cases}. \tag{5}$$

The variance is computed as

$$\text{VAR}(S)$$
$$= \frac{1}{18}\left[n(n-1)(2n+5) - \sum_{i=1}^{q} t_i(t_i-1)(2t_i+5) \right], \tag{6}$$

where n is the number of data points, q is the number of tied groups and t_i is the number of data values in the ith group. The standard test statistic, Z, is computed as follows:

$$Z = \begin{cases} \dfrac{S-1}{\sqrt{\text{VAR}(S)}} & \text{if } S > 0 \\ 0 & \text{if } S = 0 \\ \dfrac{S+1}{\sqrt{\text{VAR}(S)}} & \text{if } S < 0 \end{cases}. \tag{7}$$

A positive value of Z indicates an upward trend, and a negative value of Z indicates a downward trend. We use the threshold of ± 1.96 for significant difference (Gao et al., 2012). The Mann–Kendall statistical test has frequently been used to quantify the significance of trends in hydro-meteorological time series (Gocic and Trajkovic, 2013).

To discuss relationships between SSC and Q, hydrologists often use sediment rating curves. The most common

Table 1. Land use/cover type and change ratio during 1978–2007 in the Du catchment

Land use/cover	Land use/cover (km^2) and ratio			Land use/cover change (km^2) and change ratio		
	1987	1999	2007	1999–1987	2007–1999	2007–1987
Water	35 (0.4 %)	26 (0.3 %)	31 (0.4 %)	−9 (−0.1 %)	5 (0.1 %)	−4 (−0.0 %)
Urban land	81 (0.9 %)	88 (1.0 %)	115 (1.3 %)	8 (0.1 %)	26 (0.3 %)	34 (0.4 %)
Barren land	37 (0.4 %)	38 (0.4 %)	62 (0.7 %)	1 (0.0 %)	24 (0.3 %)	26 (0.3 %)
Forest	6316 (70.4 %)	6232 (69.5 %)	6841 (75.2 %)	−84 (−0.9 %)	609 (6.8 %)	525 (5.9 %)
Shrub	929 (10.4 %)	846 (9.4 %)	851 (9.9 %)	−83 (−0.9 %)	5 (0.1 %)	−78 (−0.9 %)
Grass	657 (7.3 %)	525 (5.8 %)	551 (6.4 %)	−132 (−1.5 %)	26 (0.3 %)	−106 (−1.2 %)
Farmland	919 (10.2 %)	1218 (13.6 %)	522 (6.1 %)	299 (3.3 %)	−695 (−7.7 %)	−397 (−4.4 %)

Table 2. Land use/cover and change ratio during 1978–2007 in the Xinzhou catchment.

Land use/cover	Land use/cover (km^2) and ratio			Land use/cover change (km^2)		
	1987	1999	2007	1999–1987	2007–1999	2007–1987
Water	16 (0.3 %)	15 (0.3 %)	14 (0.3 %)	−1 (0.0 %)	−1 (0.0 %)	−2 (0.0 %)
Urban land	52 (1.1 %)	57 (1.2 %)	51 (1.1 %)	5 (0.1 %)	−6 (−0.1 %)	−1 (0.0 %)
Barren land	20 (0.4 %)	22 (0.5 %)	41 (0.9 %)	2 (0.0 %)	19 (0.4 %)	21(0.5 %)
Forest	3072 (66.3 %)	3148 (67.9 %)	3432 (74.0 %)	76 (1.6 %)	284 (6.1 %)	360 (7.8 %)
Shrub	537 (11.6 %)	422 (9.1 %)	479 (10.3 %)	−115 (−2.5 %)	57 (1.2 %)	−58 (−1.3 %)
Grass	404 (8.7 %)	290 (6.3 %)	307 (6.6 %)	−114 (−2.5 %)	17 (0.4 %)	−97 (−2.1 %)
Farmland	534 (11.5 %)	679 (14.7 %)	312 (6.7 %)	145 (3.1 %)	−367 (−7.9 %)	−222 (−4.8 %)

Table 3. Variables and associated abbreviations used in the statistical analysis.

Abbreviations	Variables	Units
P	Rainfall	mm
Q	Streamflow	m^3 s^{-1}
D	Discharge depth	mm
SSY	Suspended sediment yield	kg or g s^{-1}
SSC	Suspended sediment concentration	kg m^{-3} or g m^{-3}

approach is to fit a power curve to the normal data (Khanchoul et al., 2008) as follows:

$$SSC = \alpha Q^\beta. \tag{8}$$

Here, α and β are constants in the non-linear regression equation. The non-linear model assumes that the dependent variable (SSC) has a constant variance (scatter), which typically does not occur because the scatter around the regression generally increases with increasing Q (Harrington and Harrington, 2013). The Mann–Kendall test was performed in MATLAB 7.0.

3 Results

3.1 Streamflow and sediment yield during different periods

Figure 3 shows the annual P, D and SSY for the hydrological years of 1980–2009 from the Zhushan and Xinzhou gauges. The annual P fluctuated between 665 and 1219 mm. The annual Dd and Dx varied between 253 to 873 mm and 279 to 931 mm, respectively. The annual SSY varied between 1.3×10^8 and 1.0×10^{10} kg yr^{-1} from the Zhushan gauge and between 6.3×10^7 and 4.3×10^9 kg yr^{-1} from the Xinzhou gauge. To identify the relationships between the annual P, Dd, Dx, SSYd and SSYx, we generated a Pearson's correlation matrix, as shown in Fig. 4. The analysis showed significant correlations between all of the variables ($n = 30$, $p < 0.0001$). During the low-flow years (e.g., 1997 or 2001), SSYd was similar to SSYx. However, during the high-flow years (e.g., 1983 or 2005), SSYd was several times greater than SSYx.

The Mann–Kendall test was applied to the annual P, D and SSY data for 1980–2009. The test shows a decreasing but not significant trend for P, a significant (5 % level) decreasing trend for Qd, and highly significant decreasing trends for Qx, SSYx and SSYd (1 % level) (Fig. 5). After 2000, P shows an increasing trend and Q and SSY show decreasing trends.

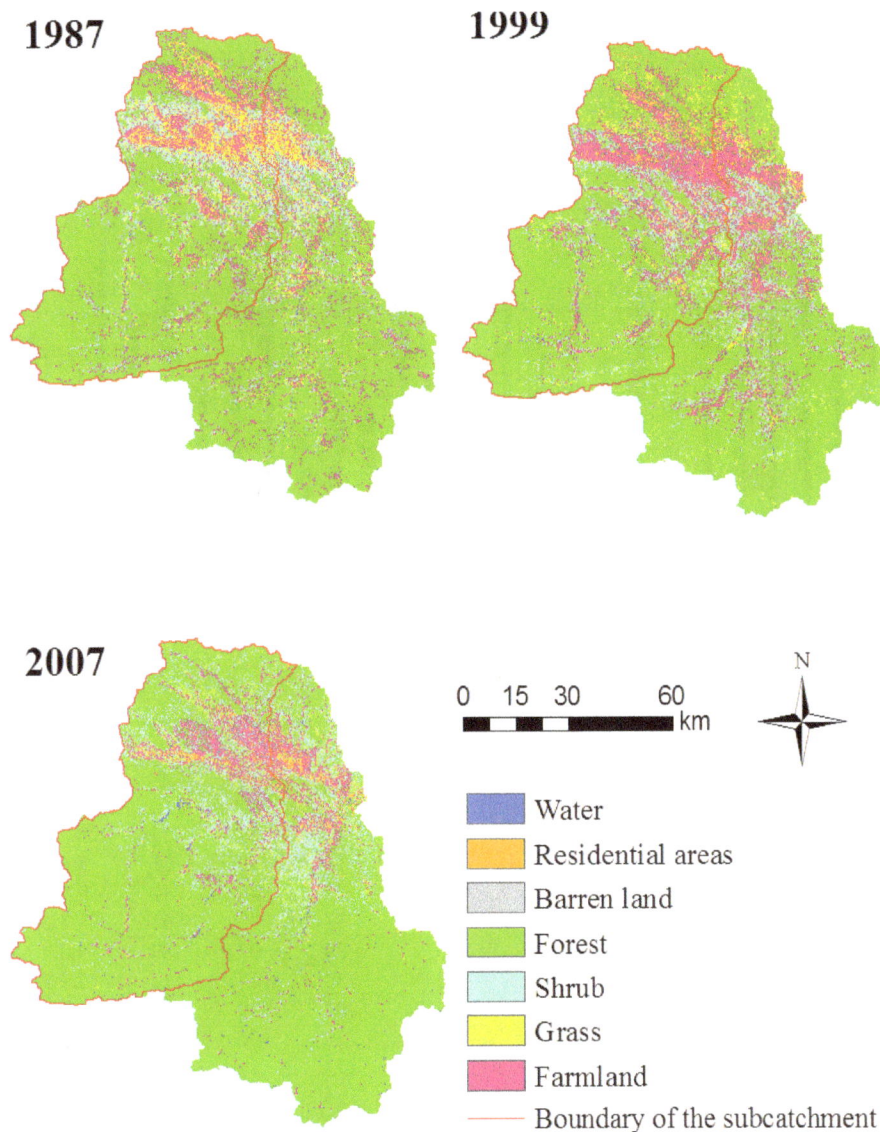

Figure 2. Land use changes during the three periods.

To better understand the dynamics of Q and SSC, Tables 4 and 5 compare the observed average monthly Q and SSC among the three periods monitored at the Zhushan and Xinzhou gauges.

During 1980s–1990s, the annual Qd showed a decreasing trend (Table 4), with only 3 of 12 months showing a slightly increasing trend. The rate of decrease varied from -3.3 to -53.0%. In addition, Qx exhibited a decreasing trend that was similar to that of Qd during the same period. During 1990s–2000s, Qd greatly increased from 1 to 34 % during 9 of 12 months. Meanwhile, Qx increased over 8 months and fluctuated between 10 % and 42 %. During 1990s–2000s, Qd and Qx both exhibited a more obvious increasing trend during the winter than during the flow seasons.

Table 5 shows the monthly mean SSC from the two gauges. SSCd decreased (-1 to -66%) during the flow seasons (May to September), except in August, when it slightly increased (2 %) during 1980s–1990s. The decrease of SSCd did not coincide with that of Qd. During 1990s–2000s, the decrease in SSCd was more obvious than that in 1980s–1990s. In all, 8 of 10 months experienced a decreasing change, and the change over 7 months was $> -40\%$. In addition, the SSCx decreased over 6 months and increased during the other 4 months during 1980s–1990s. During 1990s–2000s, the SSCx decreased over 7 months, and 4 out of 5 months showed a decreasing trend during the flow season. However, the monthly SSC is calculated by SSY and Q and is not the actual SSC. To better understand SSC dynamics,

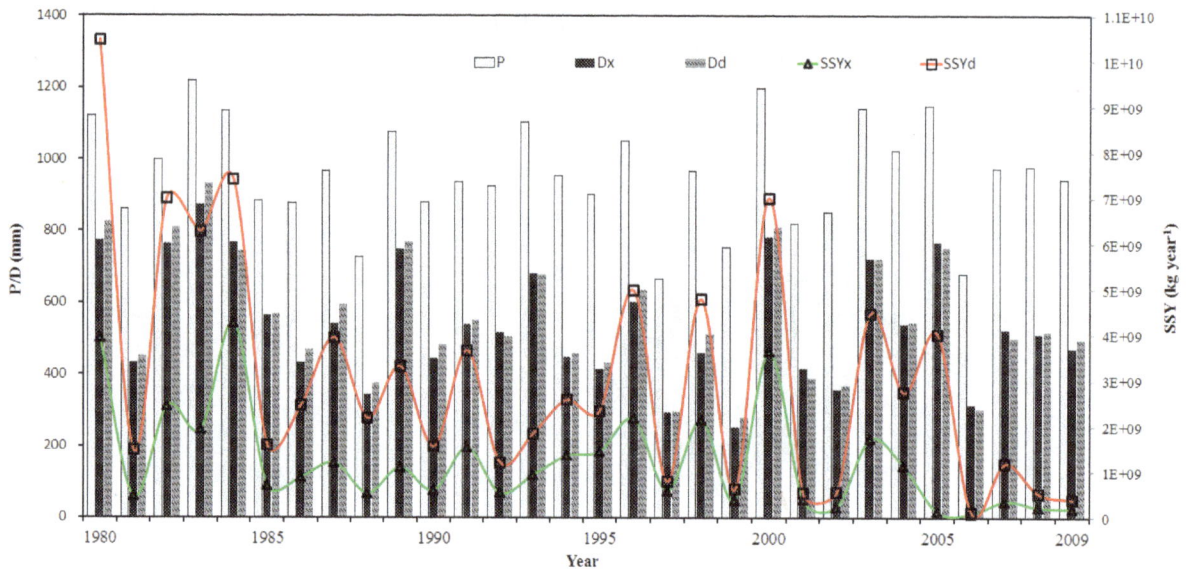

Figure 3. Annual P, D and SSY for the hydrological years of 1980–2009 from the Zhushan and Xinzhou gauges.

Table 4. Monthly mean streamflow from the Xinzhou and Zhushan gauges.

	Qd (m³ s⁻¹)			Change (100 %)		Qx (m³ s⁻¹)			Change (100 %)	
	1980s	1990s	2000s	C1	C2	1980s	1990s	2000s	C1	C2
Jan	35	33	41	−5.7 %	24.2 %	17	13	19	−23.5 %	46.2 %
Feb	37	46	49	24.3 %	6.5 %	18	19	21	5.6 %	10.5 %
Mar	85	96	74	12.9 %	−22.9 %	42	46	31	9.5 %	−32.6 %
Apr	186	146	160	−21.5 %	9.6 %	92	72	61	−21.7 %	−15.3 %
May	185	200	203	8.1 %	1.5 %	89	97	89	9.0 %	−8.2 %
Jun	274	224	192	−18.2 %	−14.3 %	132	115	111	−12.9 %	−3.5 %
Jul	412	223	262	−45.9 %	17.5 %	207	119	173	−42.5 %	45.4 %
Aug	269	260	257	−3.3 %	−1.2 %	129	136	156	5.4 %	14.7 %
Sep	338	159	202	−53.0 %	27.0 %	173	76	109	−56.1 %	43.4 %
Oct	255	136	155	−46.7 %	14.0 %	123	67	103	−45.5 %	53.7 %
Dec	121	94	95	−22.3 %	1.1 %	57	42	47	−26.3 %	11.9 %
Nov	49	41	62	−16.3 %	51.2 %	23	18	30	−21.7 %	66.7 %
Average	187	138	146	−26.2 %	5.8 %	92	68	79	−26.1 %	16.2 %

Note: C1 is the change for 1990–1980; C2 is the change for 2000–1990.

paired SSC–Q data collected by monitoring should be discussed.

3.2 SSC–Q dynamics

Figure 6 shows the statistical characteristics of the SSC and Q during the three periods. The mean SSCd was relatively stable during the three periods (± 83 g m⁻³), and the mean SSCx varied from 1058 g m⁻³ in the 1980s to 1256 g m⁻³ in the 1990s and then decreased to 891 g m⁻³ in the 2000s. In the 1980s, the max SSCd and max SSCx were 22400 and 31800 g m⁻³, respectively. Next, the max SSCd shape decreased to 20000 g m⁻³ during the 1990s and to 17800 g m⁻³

during the 2000s. Meanwhile, the max SSCx decreased to 26900 and 19200 g m⁻³ during the 1990s and 2000s, respectively. The max Qx was more variable than the max Qd and was 12400 g m⁻³ in the 1980s, 3610 g m⁻³ in the 1990s and 3010 g m⁻³ in the 2000s. However, the rate of change of the mean Qx was similar to that of the mean Qd.

Figure 7 shows that the SSCs varied by several orders of magnitude for a given discharge at both gauges. SSCd and SSCx fluctuated between 1 and 22400 g m⁻³ and between 1 and 31800 g m⁻³, respectively. The maximum SSCx (31800 g m⁻³) was larger than the maximum SSCd (21400 g m⁻³). In Fig. 7, SSCd-Qd maintained a stable relationship during the three periods (1980s, 1990s and

Table 5. Monthly mean suspended sediment concentration from the Xinzhou and Zhushan gauges.

	SSCd (g m^{-3})			Change (100 %)		SSCx (g m^{-3})			Change (100 %)	
	1980s	1990s	2000s	C1	C2	1980s	1990s	2000s	C1	C2
Jan	0	0	0	–	–	0	0	0	–	–
Feb	10	1	2	−90 %	100 %	3	0	0	−100 %	–
Mar	7	15	1	114 %	−93 %	3	12	1	300 %	−92 %
Apr	224	147	56	−34 %	−62 %	118	81	28	−31 %	−65 %
May	427	256	139	−40 %	−46 %	298	128	127	−57 %	−1 %
Jun	629	623	321	−1 %	−48 %	471	718	430	52 %	−40 %
Jul	1222	755	686	−38 %	−9 %	929	895	603	−4 %	−33 %
Aug	942	963	364	2 %	−62 %	736	961	411	31 %	−57 %
Sep	674	229	239	−66 %	4 %	409	115	186	−72 %	62 %
Oct	268	146	46	−46 %	−68 %	185	84	84	−55 %	0 %
Dec	26	86	1	231 %	−99 %	18	54	1	200 %	−98 %
Nov	0	0	0	-	-	0	0	0	–	–
Average	369	268	155	−27.4 %	−42.1 %	264	254	156	−3.8 %	−38.6 %

Note: C1 is the change for 1990s–1980s; C2 is the change for 2000s–1990s. Suspended sediment primarily loads during the flow season. Rainfall is rare in the winter (Dec, Nov and Jan), and the streamflow is dominated by a base flow; thus, in most years, there is no suspended sediment load.

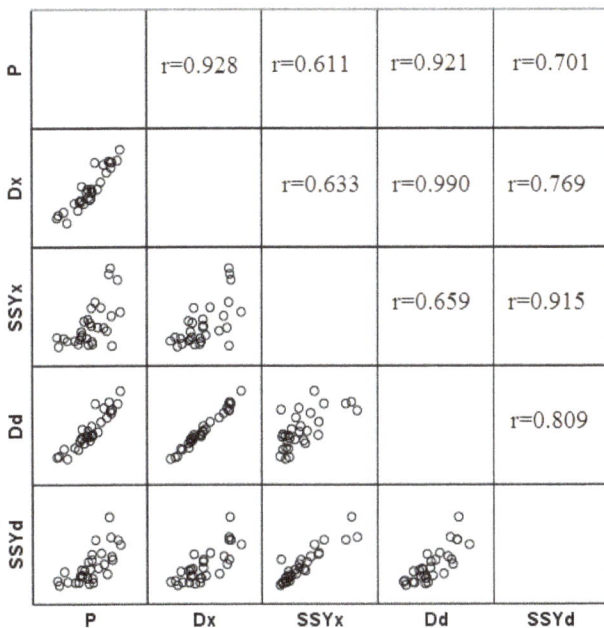

Figure 4. Bivariate scatter-plot matrix of selected variables.

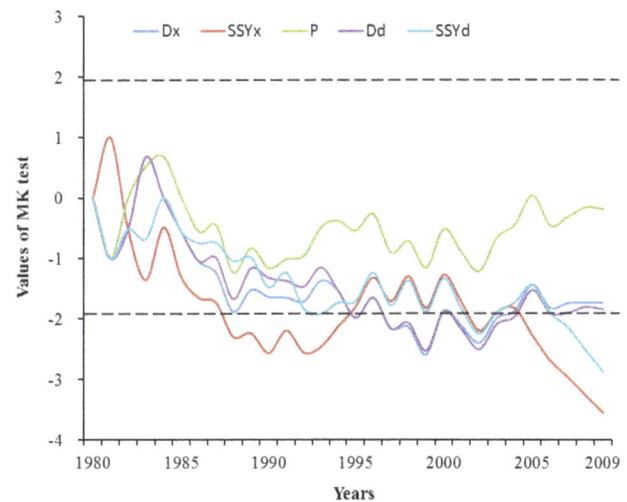

Figure 5. Results of the Mann–Kendall test.

2000s). However, SSCx-Qx showed a scattered relationship from 1980s and 1990s and showed a more liner relationship from 2000s. During the three periods, the max Qd decreased from 9880 to 6140 and 5070 m^{-3} s^{-1}, respectively. Meanwhile, the max Qx was reduced from 5960 to 3580 and 2990 m^{-3} s^{-1}, respectively.

The relationship between SSC and Q is complicated. To better understand the dynamics of SSC, SSC was sorted by ranking the paired Q values, which were classified using a threshold level approach (e.g., low flow ($Q \leq 25 \%$),

moderate flow ($25 < Q < 75 \%$), and high flow ($Q \geq 75 \%$). The SSC dynamics were compared under different flow regimes. For the sub-catchment, the thresholds were 188 and 674 m^3 s^{-1} for the minimum 25 % and maximum 25 %, respectively. For Qd, the thresholds of the minimum and maximum 25 % were 332 and 1100 m^3 s^{-1}, respectively. Figure 8 presents box plots for SSCd and SSCx during the three periods for the three flow grades. The box plots indicate the maximum, 75, 50 and 25 %, and minimum values for each SSC (outliers are excluded). For the sub-catchment, SSCx increased between the original period and the cultivation period for moderate and high flow, but not for low flow. Then, SSCx decreased during the reforestation period for all flows. At the Zhushan station, SSCd was larger during the cultivation period for both moderate and high flows. During the

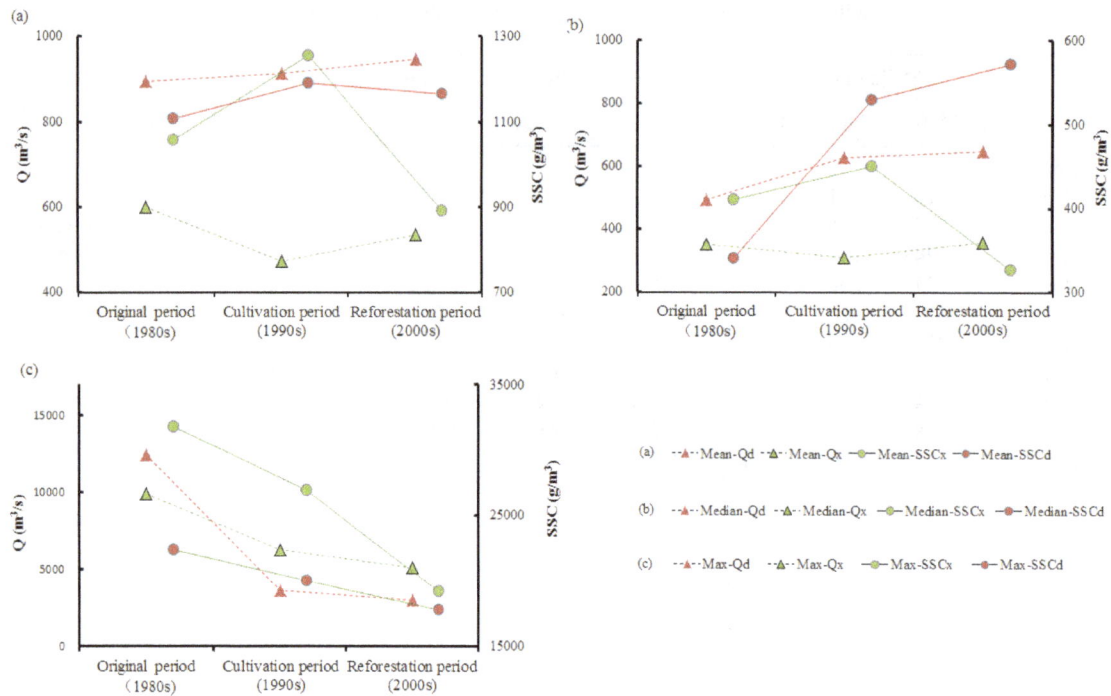

Figure 6. Descriptive statistics of Q and SSC.

reforestation period, the SSCd during low flow was higher than during the other periods.

Six ANOVA tests were performed using SSC as the dependent variable and using the different periods (land use) as independent variables. ANOVA was only conducted for the same flow during different periods. One-way ANOVA (Table 6) revealed that SSCx showed significant differences among the different periods for all three types of flows ($p < 0.001$). However, a significant difference in SSCd was only observed among high flows ($p < 0.001$). No statistically significant differences were observed among the SSCd values during the different periods for low or moderate flows.

4   Discussion

Land use/cover has been widely documented to have dire environmental consequences through their adverse impacts on soil and water qualities (Zhang et al., 2015). Olang et al. (2014) indicated that 40 % and 51 of forest and agriculture land revealed reduced runoff volumes by about 12 %, while 86 % land cover of agriculture increased runoff volumes by about 12 %. Buendia et al. (2015) studied the effects of afforestation on runoff at a Pyrenean Basin (2807 km^2), and the results showed that an increase ranging between 19 % and 57 % in the forest of sub-basins accounted for ~ 40 % of the observed decrease in annual runoff. Liu et al. (2014) demonstrated that afforestation leads to increased runoff in dry seasons in the Yarlung Zangbo River basin. In this study, land use/cover changes significantly affect Q and SSY (Ta-

bles 4 and 5). During the cultivation period, an increase in farmland resulted in an obvious decreasing trend in Q in the Du catchment and its sub-catchment. The sediment concentration in the direct runoff from a slope consists of a combination of the sediment stored on the slope and that generated by flow erosion during the current rainfall event (Aksoy and Kavvas, 2005; Rankinen et al., 2010). Large storms generate sufficient surface runoff to deliver sediment from the uplands to the stream. In forest catchments overflow typically occurs only in a small fraction of the catchment, and it is most likely to occur very close to the stream (Underwood et al., 2015). Reforestation may increased the return period of peak flow and peak sediment yield (Keesstra, 2007). Borrelli et al. (2015) illustrated that a disturbed forest sector could produce about 74 % more net erosion than a 9 times larger, undisturbed forest sector. High SSCs are not detected in the absence of a high-flow velocity to carry the suspended sediment to the outlet of a catchment. SSCs are determined by onsite sediment production and the connectivity of sediment sources to the channel. Sediment delivered to the channel can be deposited (Keesstra et al., 2009). When runoff is decreased, its erodibility is reduced (Bakker et al., 2008; Van Rompaey et al., 2002). Reduced streamflow can reduce the sediment transport capacity and increase the probability for further sediment deposition in the river (Zhu et al., 2015). Human-induced modifications of land use/cover in river basins may cause strong geomorphic responses by disturbing sediment supply, transport and deposition processes (Liébault et al., 2005).

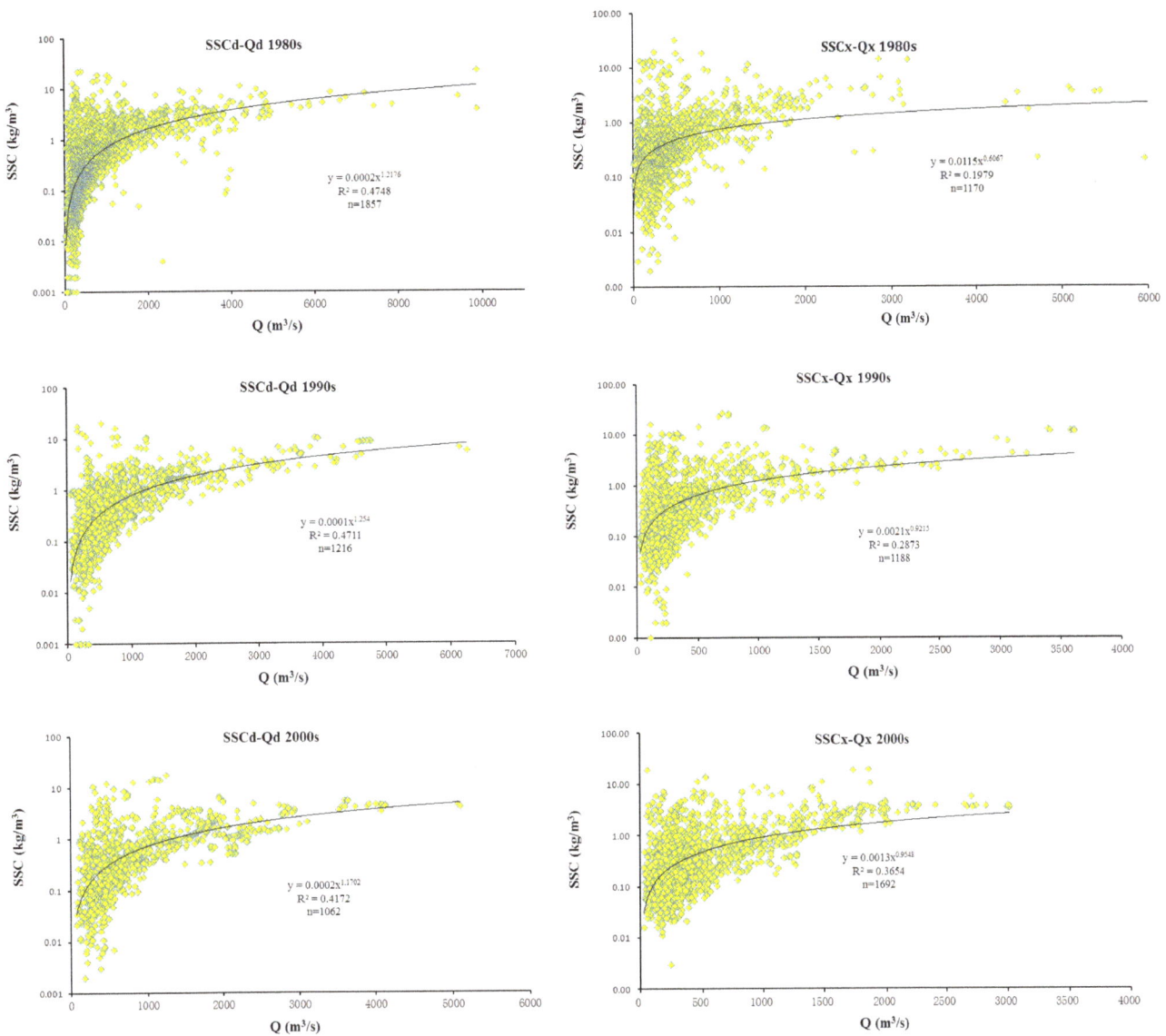

Figure 7. SSC–Q relationships during the three periods for the two gauges.

Table 6. Mean SSC values and one-way ANOVA of SSCs during the different periods.

		Original	Cultivation	Reforestation	p value
Mean SSCd ($g\,m^{-3}$)	Low flow	0.49	0.50	0.44	0.285
	Moderate flow	0.83	0.86	0.97	0.080
	High flow	2.42	2.43	2.02*	0.002
Mean SSCx ($g\,m^{-3}$)	Low flow	0.68	0.66	0.36*	0.000
	Moderate flow	0.87	0.97	0.64*	0.000
	High flow	1.80	2.83*	1.80	0.000

Note: ANOVA was only conducted for the same flow during different periods; * means significant difference at $\alpha = 0.05$.

Hydrological studies rely on the analysis of processes at different spatial scales (García-Ruiz et al., 2008). Sediment yield and watershed areas have been elucidated in many stud-ies (e.g., Renschler and Harbor, 2002; de Vente and Poe-sen, 2013). The mean SSC was stable during the study years in the Du catchment, and the mean SSC varied in the sub-

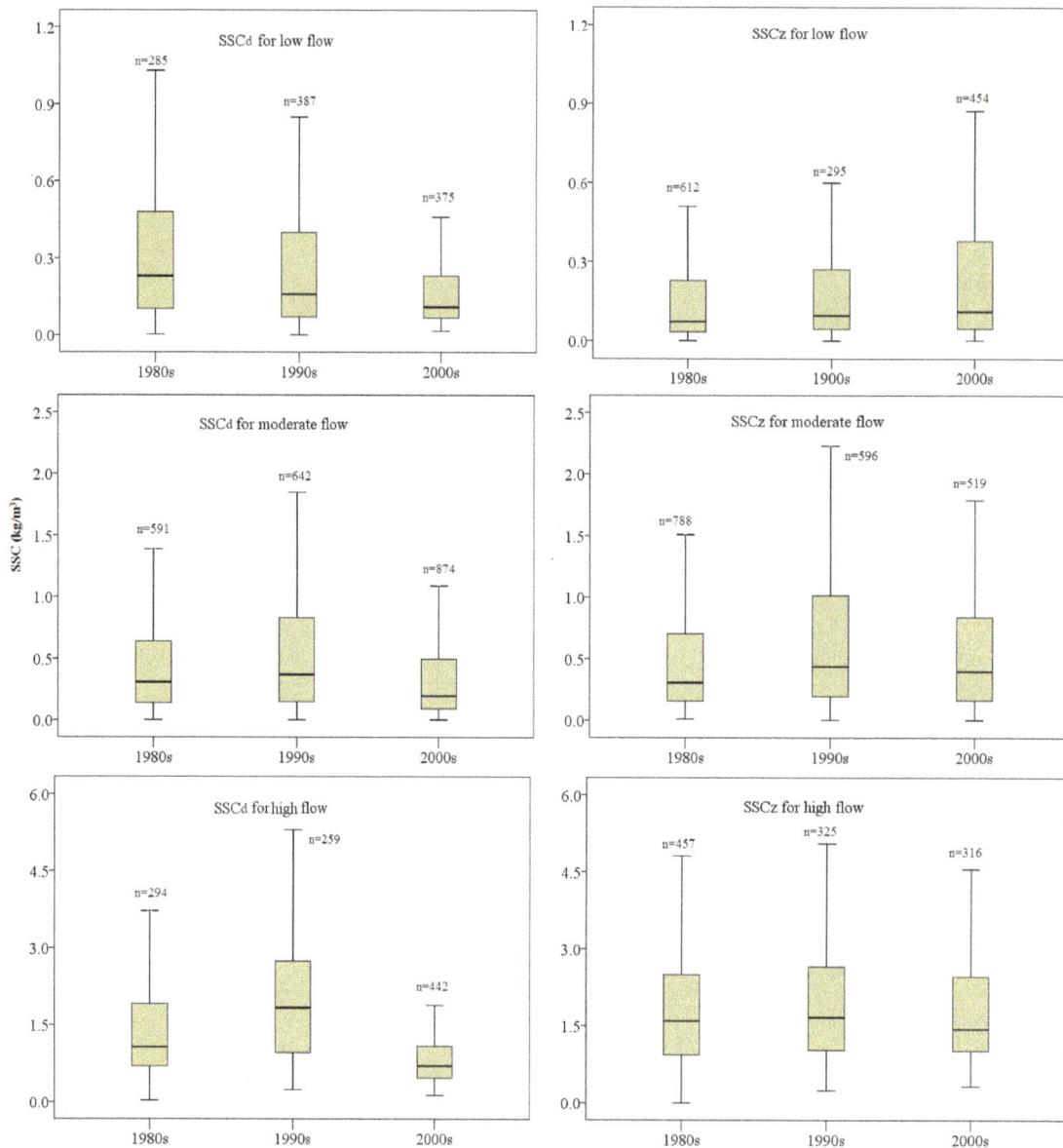

Figure 8. Box plots of SSC.

catchment. The increase in Qx was larger than the increase in Qd. The monitored sub-catchment covered approximately half of the entire catchment. Likewise, the combined mean annual discharge volume of the sub-catchment was nearly half of the total catchment output (i.e., a deficit of approximately 50 % at the outlet). However, the SSC dynamics were more variable. Due to sediment delivery problems, sediment is generated on catchment slopes and is either stored on the surface or removed (Rankinen et al., 2010). Only a fraction of the gross soil erosion within a catchment will reach the outlet and be represented in the sediment yield. In addition, streamflow erodes the sediment directly from the surface or causes channel erosion, which removes the stored surface layer of detached sediment.

Our previous study in Du catchment showed that the area scale dominates the sediment delivery ratio (Shi et al., 2014). The sediment stored in the gullies is flushed to the river when a certain threshold is exceeded, and the deposition of sediment in channels is flushed at higher discharges. The max SSCx is greater than the max SSCd (31 800 vs. 22 400 $g\,m^{-3}$). One possible explanation is that the sediment stock is depleted during a flood; this process does not occur simultaneously within the entire river basin and results in gradually decreasing SSCs downstream (Doomen et al., 2008). Cultivation or reforestation alters the slope surfaces but does not remove gullies and channels. The SSCs in Zhushan were only significantly different during high-flow and the reforestation period when the forest cover greatly increased.

5   Conclusions

This study investigated Q and SSC dynamics for 30 years under cultivation and reforestation. The results of a Mann–Kendall test showed that rainfall slightly increased during the study years; however, the annual discharge and sediment load significantly decreased. The sediment flux is extremely spatially and temporally variable. The relationship between SSC and Q is complicated. Reforestation caused significant differences in the SSC for both low and moderate flows. For low and moderate flow, the changes in SSY primarily resulted from runoff, while the SSC showed little change. For the sub-catchment, the changes in the SSC were more sensitive to land use/cover changes. Meanwhile, cultivation resulted in significant differences in the SSC for high flow. Overall, our results provide useful information regarding SSC dynamics relative to land use/cover changes in mountainous catchments in a subtropical climate, which have largely been undocumented in the literature.

Acknowledgements. Financial support for this research was provided by the National Natural Science Foundation of China (41525003 and 41301294), and the Fundamental Research Funds for the Central Universities (2014YB053).

References

Aksoy, H. and Kavvas, M. L.: A review of hillslope and watershed scale erosion and sediment transport models, Catena, 64, 247–271, doi:10.1016/j.catena.2005.08.008, 2005.

Bakker, M. M., Govers, G., van Doorn, A., Quetier, F., Chouvardas, D., and Rounsevell, M.: The response of soil erosion and sediment export to land-use change in four areas of Europe: The importance of landscape pattern, Geomorphology, 98, 213–226, doi:10.1016/j.geomorph.2006.12.027, 2008.

Borrelli, P., Marker, M., and Schutt, B.: Modelling Post-Tree-Harvesting Soil Erosion and Sediment Deposition Potential in the Turano River Basin (Italian Central Apennine), Land Degrad. Dev., 26, 356–366, doi:10.1002/ldr.2214, 2015.

Buendia, C., Batalla, R. J., Sabater, S., Palau, A., and Marcé, R.: Runoff Trends Driven by Climate and Afforestation in a Pyrenean Basin, Land Degrad. Dev., doi:10.1002/ldr.2384, in press, 2015.

Casalí, J., Giménez, R., Díez, J., Álvarez-Mozos, J., Del Valle de Lersundi, J., Goñi, M., Campo, M. A., Chahor, Y., Gastesi, R., and López, J.: Sediment production and water quality of watersheds with contrasting land use in Navarre (Spain), Agr. Water Manage., 97, 1683–1694, doi:10.1016/j.agwat.2010.05.024, 2010.

Cerdan, O., Govers, G., Le Bissonnais, Y., Van Oost, K., Poesen, J., Saby, N., Gobin, A., Vacca, A., Quinton, J., Auerswald, K., Klik, A., Kwaad, F. J. P. M., Raclot, D., Ionita, I., Rejman, J., Rousseva, S., Muxart, T., Roxo, M. J., and Dostal, T.: Rates and spatial variations of soil erosion in Europe: A study based on erosion plot data, Geomorphology, 122, 167–177, doi:10.1016/j.geomorph.2010.06.011, 2010.

Collins, A. L., Naden, P. S., Sear, D. A., Jones, J. I., Foster, I. D. L., and Morrow, K.: Sediment targets for informing river catchment management: international experience and prospects, Hydrol. Process., 25, 2112–2129, doi:10.1002/hyp.7965, 2011.

Cooper, J. A. G.: The role of extreme floods in estuary-coastal behaviour: contrasts between river- and tide-dominated microtidal estuaries, Sediment. Geol., 150, 123–137, doi:10.1016/S0037-0738(01)00271-8, 2002.

de Vente, J., Poesen, J., Verstraeten, G., Govers, G., Vanmaercke, M., Van Rompaey, A., Arabkhedri, M., and Boix-Fayos, C.: Predicting soil erosion and sediment yield at regional scales: Where do we stand?, Earth-Sci. Rev., 127, 16–29, doi:10.1016/j.earscirev.2013.08.014, 2013.

Doomen, A. M. C., Wijma, E., Zwolsman, J. J. G., and Middelkoop, H.: Predicting suspended sediment concentrations in the Meuse River using a supply-based rating curve, Hydrol. Process., 22, 1846–1856, doi:10.1002/hyp.6767, 2008.

Fang, N. F., Shi, Z. H., Li, L., Guo, Z. L., Liu, Q. J., and Ai, L.: The effects of rainfall regimes and land use changes on runoff and soil loss in a small mountainous watershed, Catena, 99, 1–8, doi:10.1016/j.catena.2012.07.004, 2012.

Francke, T., López-Tarazón, J. A., and Schröder, B.: Estimation of suspended sediment concentration and yield using linear models, random forests and quantile regression forests, Hydrol. Process., 22, 4892–4904, doi:10.1002/hyp.7110, 2008.

Gao, Z. L., Fu, Y. L., Li, Y. H., Liu, J. X., Chen, N., and Zhang, X. P.: Trends of streamflow, sediment load and their dynamic relation for the catchments in the middle reaches of the Yellow River over the past five decades, Hydrol. Earth Syst. Sci., 16, 3219–3231, doi:10.5194/hess-16-3219-2012, 2012.

García-Ruiz, J. M., Reguees, D., Alvera, B., Lana-Renault, N., Serrano-Muela, P., Nadal-Romero, E., Navas, A., Latron, J., Marti-Bono, C., and Arnaez, J.: Flood generation and sediment transport in experimental catchments affected by land use changes in the central Pyrenees, J. Hydrol., 356, 245–260, doi:10.1016/j.jhydrol.2008.04.013, 2008.

Gafur, A., Jensen, J. R., Borggaard, O. K., and Petersen, L.: Runoff and losses of soil and nutrients from small watersheds under shifting cultivation (Jhum) in the Chittagong Hill Tracts of Bangladesh, J. Hydrol., 274, 30–46, doi:10.1016/S0022-1694(03)00262-2, 2003.

Gocic, M. and Trajkovic, S.: Analysis of changes in meteorological variables using Mann-Kendall and Sen's slope estimator statistical tests in Serbia, Global Planet. Change, 100, 172–182, doi:10.1016/j.gloplacha.2012.10.014, 2013.

Harrington, S. T. and Harrington, J. R.: An assessment of the suspended sediment rating curve approach for load estimation on the Rivers Bandon and Owenabue, Ireland, Geomorphology, 185, 27–38, doi:10.1016/j.geomorph.2012.12.002, 2013.

Hopmans, P. and Bren, L. J.: Long-term changes in water quality and solute exports in headwater streams of intensively managed radiata pine and natural eucalypt forest catchments in south-eastern Australia, Forest Ecol. Manag., 253, 244–261, doi:10.1016/j.foreco.2007.07.027, 2007.

Ide, J. i., Kume, T., Wakiyama, Y., Higashi, N., Chiwa, M., and Otsuki, K.: Estimation of annual suspended sediment yield from a Japanese cypress (Chamaecyparis obtusa) plantation considering antecedent rainfalls, Forest Ecol. Manag., 257, 1955–1965, doi:10.1016/j.foreco.2009.02.011, 2009.

Izaurralde, R. C., Williams, J. R., Post, W. M., Thomson, A. M., McGill, W. B., Owens, L. B., and Lal, R.: Long-term modeling of soil C erosion and sequestration at the small watershed scale, Climatic Change, 80, 73–90, doi:10.1007/s10584-006-9167-6, 2007.

Kendall, M. G.: Rank correlation methods, Griffin, London, 1975.

Keesstra, S. D.: Impact of natural reforestation on floodplain sedimentation in the Dragonja basin, SW Slovenia, Earth Surf. Proc. Land., 32, 49–65, doi:10.1002/esp.1360, 2007.

Keesstra, S. D., van Dam, O., Verstraeten, G., and van Huissteden, J.: Changing sediment dynamics due to natural reforestation in the Dragonja catchment, SW Slovenia, Catena, 78, 60–71, doi:10.1016/j.catena.2009.02.021, 2009.

Khanchoul, K. and Jansson, M. B.: Sediment rating curves developed on stage and seasonal means in discharge classes for the Mellah wadi, Algeria, Geogr. Ann. A., 90, 227–236, doi:10.1111/j.1468-0459.2008.341.x, 2008.

Kisi, O., Karahan, M. E., and Şen, Z.: River suspended sediment modelling using a fuzzy logic approach, Hydrol. Process., 20, 4351–4362, doi:10.1002/hyp.6166, 2006.

Liébault, F., Gomez, B., Page, M., Marden, M., Peacock, D., Richard, D., and Trotter, C. M.: Land-use change, sediment production and channel response in upland regions, River Res. Appl., 21, 739–756, doi:10.1002/rra.880, 2005.

Liu, J. and Diamond, J.: China's environment in a globalizing world, Nature, 435, 1179–1186, doi:10.1038/4351179a, 2005.

Liu, Z., Yao, Z., Huang, H., Wu, S., and Liu, G.: Land Use and Climate Changes and Their Impacts on Runoff in the Yarlung Zangbo River Basin, China, Land Degrad. Dev., 25, 203–215, doi:10.1002/ldr.1159, 2014.

Liu, Q. J., Shi, Z. H., Fang, N. F., Zhu, H. D., and Ai, L.: Modeling the daily suspended sediment concentration in a hyperconcentrated river on the Loess Plateau, China, using the Wavelet-ANN approach, Geomorphology, 186, 181–190, doi:10.1016/j.geomorph.2013.01.012, 2013.

Lü, Y. H., Zhang, L. W., Feng, X. M., Zeng, Y., Fu, B. J., Yao, X. L., Li, J. R., and Wu, B. F.: Recent ecological transitions in China: greening, browning, and influential factors, Sci. Rep., 5, 8732, doi:10.1038/srep08732, 2015.

Luo, P. P., He, B., Chaffe, P. L. B., Nover, D., Takara, K., and Mohd Remy Rozainy, M. A. Z.: Statistical analysis and estimation of annual suspended sediments of major rivers in Japan, Environ. Sci. Proc. Imp., 15, 1052–1061, doi:10.1039/c3em30777h, 2013.

Mann, H. B.: Nonparametric Tests Against Trend, Econometrica, 13, 245–259, 1945.

Mizugaki, S., Onda, Y., Fukuyama, T., Koga, S., Asai, H., and Hiramatsu, S.: Estimation of suspended sediment sources using137Cs and210Pbexin unmanaged Japanese cypress plantation watersheds in southern Japan, Hydrol. Process., 22, 4519–4531, doi:10.1002/hyp.7053, 2008.

Morehead, M. D., Syvitski, J. P., Hutton, E. W. H., and Peckham, S. D.: Modeling the temporal variability in the flux of sediment from ungauged river basins, Global Planet. Change, 39, 95–110, doi:10.1016/s0921-8181(03)00019-5, 2003.

Mount, N. J., Sambrook Smith, G. H., and Stott, T. A.: An assessment of the impact of upland afforestation on lowland river reaches: the Afon Trannon, mid-Wales, Geomorphology, 64, 255–269, doi:10.1016/j.geomorph.2004.07.003, 2005.

Naden, P. S. and Cooper, D. M.: Development of a sediment delivery model for application in large river basins, Hydrol. Process., 13, 1011–1034, 1999.

National Soil Survey Office: Soil survey technique in China: Agricultural Press, Beijing, 1992 (in Chinese).

Ni, J. R., Li, X. X., and Borthwick, A. G. L.: Soil erosion assessment based on minimum polygons in the Yellow River basin, China, Geomorphology, 93, 233–252, doi:10.1016/j.geomorph.2007.02.015, 2008.

Olang, L. O., Kundu, P. M., Ouma, G., and Furst, J.: Impacts of Land Cover Change Scenarios on Storm Runoff Generation: A Basis for Management of the Nyando Basin, Kenya, Land Degrad. Dev., 25, 267–277, doi:10.1002/ldr.2140, 2014.

Parsons, A. J., Cooper, J., and Wainwright, J.: What is suspended sediment?, Earth Surf. Proc. Land., 40, 1417–1420, doi:10.1002/esp.3730, 2015.

Rankinen, K., Thouvenot-Korppoo, M., Lazar, A., Lawrence, D. S. L., Butterfield, D., Veijalainen, N., Huttunen, I., and Lepistö, A.: Application of catchment scale sediment delivery model INCA-Sed to four small study catchments in Finland, Catena, 83, 64–75, doi:10.1016/j.catena.2010.07.005, 2010.

Renschler, C. S. and Harbor, J.: Soil erosion assessment tools from point to regional scales-the role of geomorphologists in land management research and implementation, Geomorphology, 47, 189–209, doi:10.1016/S0169-555x(02)00082-X, 2002.

Shi, Z. H., Cai, C. F., Ding, S. W., Wang, T. W., and Chow, T. L.: Soil conservation planning at the small watershed level using RUSLE with GIS: a case study in the Three Gorge Area of China, Catena, 55, 33–48, doi:10.1016/s0341-8162(03)00088-2, 2004.

Shi, Z. H., Huang, X. D., Ai, L., Fang, N. F., and Wu, G. L.: Quantitative analysis of factors controlling sediment yield in mountainous watersheds, Geomorphology, 226, 193–201, doi:10.1016/j.geomorph.2014.08.012, 2014.

Soil Survey Staff: Soil Taxonomy, A basic system of soil classification for making and interpreting soil surveys , 2nd Edn., Agricultural Handbook 436, Natural Resources Conservation Service, USDA, Washington DC, USA, 869 pp., 1999.

Stickler, C. M., Nepstad, D. C., Coe, M. T., McGrath, D. G., Rodrigues, H. O., Walker, W. S., Soares-Filho, B. S., and Davidson, E. A.: The potential ecological costs and cobenefits of REDD: a critical review and case study from the Amazon region, Global Change Biol., 15, 2803–2824, doi:10.1111/j.1365-2486.2009.02109.x, 2009.

Underwood, J. W., Renshaw, C. E., Magilligan, F. J., Dade, W. B., and Landis, J. D.: Joint isotopic mass balance: a novel approach to quantifying channel bed to channel margins sediment transfer during storm events, Earth Surf. Proc. Land., 40, 1563–1573, doi:10.1002/esp.3734, 2015.

Van Rompaey, A. J. J., Govers, G., and Puttemans, C.: Modelling land use changes and their impact on soil erosion and sediment supply to rivers, Earth Surf. Proc. Land., 27, 481–494, doi:10.1002/esp.335, 2002.

Verbist, B., Poesen, J., van Noordwijk, M., Widianto, Suprayogo, D., Agus, F., and Deckers, J.: Factors affecting soil loss at plot scale and sediment yield at catchment scale in a tropical volcanic agroforestry landscape, Catena, 80, 34–46, doi:10.1016/j.catena.2009.08.007, 2010.

Walling, D. E.: Assessing the accuracy of suspended sediment rating curves for a small basin, Water Resour. Res., 13, 531–538, 1977.

Warrick, J. A., Madej, M. A., Goni, M. A., and Wheatcroft, R. A.: Trends in the suspended-sediment yields of coastal rivers of northern California, 1955–2010, J. Hydrol., 489, 108–123, doi:10.1016/j.jhydrol.2013.02.041, 2013.

Wei, W., Chen, L. D., Zhang, H. D., and Chen, J.: Effect of rainfall variation and landscape change on runoff and sediment yield from a loess hilly catchment in China, Environ. Earth Sci., 73, 1005–1016, doi:10.1007/s12665-014-3451-y, 2015.

Zhang, F., Tiyip, T., Feng, Z. D., Kung, H. T., Johnson, V. C., Ding, J. L., Tashpolat, N., Sawut, M., and Gui, D. W.: Spatio-Temporal Patterns of Land Use/Cover Changes over the Past 20 Years in the Middle Reaches of the Tarim River, Xinjiang, China, Land Degrad. Dev., 26, 284–299, doi:10.1002/ldr.2206, 2015.

Zhu, J. L., Gao, P., Geissen, V., Maroulis, J., Ritsema, C. J., Mu, X. M., and Zhao, G.-J.: Impacts of Rainfall and Land Use on Sediment Regime in a Semi-Arid Region: Case Study of the Wuqi Catchment in the Upper Beiluo River Basin, China, Arid Land Res. Manag., 29, 1–16, doi:10.1080/15324982.2014.919041, 2015.

12

Diagnosing hydrological limitations of a land surface model: application of JULES to a deep-groundwater chalk basin

N. Le Vine[1], A. Butler[1], N. McIntyre[1,2], and C. Jackson[3]

[1]Department of Civil and Environmental Engineering, Imperial College London, London, UK
[2]Centre for Water in the Minerals Industry, the University of Queensland, St. Lucia, Australia
[3]British Geological Survey, Keyworth, UK

Correspondence to: N. Le Vine (nlevine@imperial.ac.uk)

Abstract. Land surface models (LSMs) are prospective starting points to develop a global hyper-resolution model of the terrestrial water, energy, and biogeochemical cycles. However, there are some fundamental limitations of LSMs related to how meaningfully hydrological fluxes and stores are represented. A diagnostic approach to model evaluation and improvement is taken here that exploits hydrological expert knowledge to detect LSM inadequacies through consideration of the major behavioural functions of a hydrological system: overall water balance, vertical water redistribution in the unsaturated zone, temporal water redistribution, and spatial water redistribution over the catchment's groundwater and surface-water systems. Three types of information are utilized to improve the model's hydrology: (a) observations, (b) information about expected response from regionalized data, and (c) information from an independent physics-based model. The study considers the JULES (Joint UK Land Environmental Simulator) LSM applied to a deep-groundwater chalk catchment in the UK. The diagnosed hydrological limitations and the proposed ways to address them are indicative of the challenges faced while transitioning to a global high resolution model of the water cycle.

1 Introduction

Guidance to support adaptation to the changing water cycle is urgently required, yet the ability of water cycle models to represent the hydrological impacts of climate change is limited in several important respects. Climate models are an essential tool in scenario development, but suffer from fundamental weaknesses in the simulation of hydrology. Hydrology (as well as other soil–vegetation–atmosphere interactions) in climate models is represented via land surface models (LSMs) that partition water between evapotranspiration, surface runoff, drainage, and soil moisture storage. The deficiencies in hydrological processes representation lead to incorrect energy and water partitioning at the land surface (Oleson et al., 2008) that propagates into precipitation and near-surface air temperature biases in climate model predictions (Lawrence and Chase, 2008). Furthermore, improving the representation of hydrology is a step towards the development of a global hyper-resolution model for monitoring the terrestrial water, energy, and biogeochemical cycles that is considered as one of the *grand challenges* to the community (Wood et al., 2011).

The most recent third generation LSMs operate in a continuous time and distributed space mode, and simulate exchanges of energy, water, and carbon between the land surface and the atmosphere using physics-based process descriptions (Pitman, 2003). The physics-based nature of third generation LSMs allows widely available global data sets, such as soil properties, land use, weather states, etc., to be used as model parameters and inputs, thus making predictive modelling with LSMs very appealing.

A significant body of literature exists on LSM hydrology assessment and inter-comparison, including comparison with observed point scale evapotranspiration fluxes, soil moisture, observed river flow rates and depths to groundwater (Balsamo et al., 2009; Blyth et al., 2011; Boone et al., 2004; Lohmann et al., 2004; Maxwell and Miller, 2005). Blyth et al. (2011) used point-scale evapotranspiration fluxes from

10 FLUXNET observation sites covering the major global biomes as well as river flows from seven large rivers to assess the performance of the JULES (Joint UK Land Environmental Simulator) model. The evaluation used monthly average fluxes, over a period of 10 years, and demonstrated a number of model weaknesses in energy partitioning as well as in water partitioning and routing, thus providing a direction for further model improvements. Balsamo et al. (2009) revised the soil representation in the TESSEL LSM (used by the European Centre for Medium-Range Weather Forecasts) and showed better agreement of the new H-TESSEL model with soil moisture point observations from the Global Soil Moisture Bank. Lohmann et al. (2004) evaluated four LSMs coupled to a surface runoff routing model over 1145 small and medium size basins in the USA, and found that "the modeled mean values of the water balance terms are of the same magnitude as the spread of the models around them". The authors name both parameter selection and model structure improvements as the key factors to achieve better model performance for hydrological predictions.

LSMs focus on modelling processes in the near-surface layer (typically, the top three metres). Typically, a unit gradient (free drainage) or other simple lower boundary condition is generally assumed in place of explicitly representing the groundwater boundary (e.g. Best et al., 2011; Kriner et al., 2005; Yang and Niu, 2003). However, in permeable basins the depth to the water table is often much deeper, for example in the Kennet case study, introduced below, this can be as much as 100 m (Jackson et al., 2008), calling into question the adequacy of a relatively shallow lower boundary condition. This can result in unrealistically dry lower soil layers (e.g. Li et al., 2008). To address this problem, the NASA Seasonal-to-Interannual Prediction Project Catchment Model (NSIPP) model relies on an approximate relationship (derived from detailed simulations) to estimate the soil moisture transfer rate between the root zone and water table at a catchment scale (Koster et al., 2000). In contrast, Community Land Model (CLM) uses the hydraulic gradient between the bottom of the soil column and the water table to approximate the drainage rate from the soil column (Oleson et al., 2008). Another approach is to use the location of the water table as a lower boundary condition. The soil, water, atmosphere and plant (SWAP) model uses a variable depth soil column, whose base is located at the water table (Gusev and Nasonova, 2003). Maxwell and Miller (2005) developed this further by coupling CLM to a physics-based 3-D groundwater model ParFlow (Kollet and Maxwell, 2008) at the land surface, replacing the soil column/root-zone soil moisture formulation in CLM with the ParFlow formulation. They concluded that the resulting model provided reasonable predictions for runoff rates and shallow groundwater levels on monthly time steps. However, the explicit inclusion of the deep unsaturated zone requires the estimation of hydraulic properties that are generally not included in existing soil databases.

The tendency for LSMs to use relatively shallow soil column depths and a simplistic or non-existent representation of groundwater also questions their applicability to catchments with deep-groundwater systems (where an average water table is tens of metres deep). Such systems represent a major storage of water and their interaction with the unsaturated zone can influence river flows, soil moisture, and evapotranspiration rates (Maxwell and Miller, 2005). Consequently, the addition of a groundwater modelling capability into LSMs will not only address these issues, but also be a step forward for multi-purpose modelling (e.g. representing groundwater levels for water resources) (Wood et al., 2011).

Most LSMs assume a 1-D vertical flow in a soil column neglecting lateral flow (e.g. Kriner et al., 2005; Gusev and Nasonova, 2003). Although this assumption is sufficiently accurate only for soils that are relatively homogeneous in horizontal and vertical directions (Protopapas and Bras, 1991), it is a fairly common feature for LSMs that employ a gridded surface representation. A further complicating factor is that 1-D flow is usually described in physics-based LSMs using Richards' equation, which was derived at the point scale and used to represent single permeability, single porosity soils. The validity of this is questionable for a wide variety of soils, particularly at larger scales (Beven and Germann, 2013). Chalk is an example of a soil–rock system that consists of both matrix and fractures, whose properties are significantly distinct from each other, forming a dual porosity, dual-permeability system (Price et al., 1993). Therefore, a traditional single domain soil water representation is unsuitable to adequately characterize its properties (Ireson et al., 2009). To the best of our knowledge, there is no currently operational LSM that is capable of realistically representing such dual porosity, dual porosity-dual permeability.

Another important challenge in improving hydrological fluxes in LSMs is the representation of surface and near-surface heterogeneity, in particular how it affects partitioning between surface runoff, evaporation and infiltration. For example, 15 LSMs and a two-layer conceptual hydrological model were used to represent river discharge in the Rhone, one of Europe's major basins, in the Rhone-Aggregation Land Surface Scheme Inter-comparison Project (Boone et al., 2004); it was concluded that an LSM's ability to provide a good performance for daily discharge simulation is linked to their ability to generate sub-grid runoff, that is, to the representation of top-soil heterogeneity.

In light of these concerns, the scope of the study is to assess the hydrological behaviour of a typical third-generation LSM, the JULES, in a comprehensive and consistent way and adapt the model accordingly. For this, an evaluation strategy focusses on the primary functions of a hydrological system in a hierarchical way. While other alternatives exist (Black, 1997; Wagener et al., 2007), the following four hydrological functions are considered (Yilmaz et al., 2008): (1) to maintain an overall water balance (i.e. water partitioning between different water cycle components), (2) to redistribute water vertically through the soil, (3) to redistribute water in time, and (4) to redistribute water spatially over the catchment's groundwater and surface-water systems. The hierarchical evaluation strategy (or diagnosis) allows for inferences to be made about the specific aspects of the model structure that are causing the problems via targeted evaluations of the model response. The diagnostic evaluation makes use of multiple measures of model performance that are relevant for each of the four functions evaluated. When model performance is poor in a particular hydrological aspect, model modifications are based on hydrological expert knowledge that, whilst subjective, is the only currently available way to adjust the model. The Kennet catchment in southern England is chosen as a complex case study that represents a number of the modelling challenges; however, the methodology and the results are of interest beyond this study due to the similarities across the hydrological modules of different third-generation LSMs, and also the broad importance of chalk aquifers and deep-groundwater systems (Brouyere et al., 2004; Downing et al., 1993; Kloppmann et al., 1998; Pinault et al., 2005; Dahan et al., 1998, 1999; Nativ and Nissim, 1992; Nativ et al., 1995).

2 Case study

2.1 The Joint UK Land Environmental Simulator

JULES is a community land surface model, based upon the established UK Met Office Surface Exchange Scheme (MOSES) (Cox et al., 1999). In addition to representing the exchange of fluxes of heat and moisture between the land surface and the atmosphere, the model also represents fluxes of carbon and some other gases, such as ozone and methane (Clark et al., 2011). It includes linked processes of photosynthesis and evaporation, soil and snow physics as well as plant growth and soil microbial activity. These processes are all linked through a series of equations that quantify how soil moisture and temperature govern evapotranspiration, energy balance, respiration, photosynthesis, and carbon assimilation (Best et al., 2011; Clark et al., 2011). JULES includes multi-layer, finite-difference models of subsurface heat and water fluxes, as described in Cox et al. (1999). There are options for the specification of the hydraulic and thermal characteristics, the representation of soil moisture and the subsurface heterogeneity of soil properties (for more details see Best et al., 2011). JULES can be used as a stand alone land surface model driven by observed forcing data, or can be coupled to an atmospheric model (for example, the UK Met Office Unified Model). The model runs at a sub-daily time step, using meteorological drivers of rainfall, incoming radiation, temperature, humidity, and wind speed. When meteorological data have coarser temporal resolution than required by the model, the standard model (version 2.2) disaggregates the data as constant values.

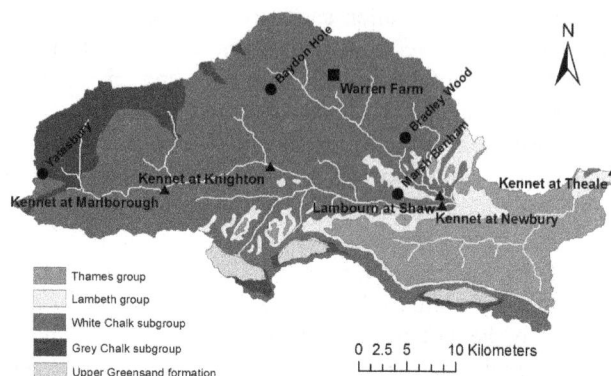

Figure 1. Hydrogeological map of the Kennet catchment. The square indicates the Warren Farm site, the triangles are flow gauging stations, and the circles are observational boreholes.

JULES is typically employed with a 3 m fixed depth of soil, a unit hydraulic head gradient lower boundary condition, and no groundwater component. Shallow groundwater can be optionally represented via the (topography-based model) TOPMODEL approach (Beven and Kirkby, 1979; Clark and Gedney, 2008). Further, top-soil heterogeneity can be optionally represented via a probability distributed model (PDM) (Moore, 2007; Clark and Gedney, 2008). Both options require specification of parameters that are conceptual in nature and are not directly related to the existing data on soil/vegetation properties. JULES is able to generate an infiltration-excess (when PDM is used, or when rainfall intensity exceeds the near-surface infiltration rate) as well as saturation-excess (when the TOPMODEL is used, or through the upward movement of water from saturated soil layers) surface runoff.

The study uses and implements modifications to JULES version 2.2, termed the standard JULES. The standard set-up is used with a 3 m depth of soil, four soil layers: 0.1, 0.25, 0.65, and 2 m deep, starting from the surface. The model is spun-up over 3 years, repeating the weather inputs for the first year of available data 3 times (one of the model warming-up options provided), and initialising soils with saturated conditions.

2.2 Case study catchment

The Kennet is a groundwater-dominated catchment in southern England (Fig. 1). The topographic catchment has an area of 1030 km^2 with an annual average rainfall of 759 mm (1961–1990). It is predominantly a permeable catchment (Upper Cretaceous Chalk). The western and northern parts of the catchment have exposed bedrock with only a thin, permeable soil. However, in the southern and eastern parts of the catchment there is significant drift cover, and, in its lowest quarter, it is largely impermeable due to overlying Palaeogene deposits (Fig. 1). It is a primarily rural catchment with scattered settlements. The flow regime is domi-

Table 1. Data used for JULES set-up and performance evaluation.

JULES input type	Data description	Source
Catchment grid	(1) 50 m resolution raster file (2) catchment outlet coordinates	(1) http://digimap.edina.ac.uk/ (2) http://www.environment-agency.gov.uk/hiflows/station.aspx?39016
Land use	50 m IGBP 2007 reclassified from 17 IGPB classes to 9 JULES classes (Smith et al., 2006)	MODIS land cover product: http://webmap.ornl.gov/wcsdown/dataset.jsp?ds_id=10004
Soil properties	1 km NSRI soil maps (Brooks and Corey parameterization)	The Cranfield Soil and AgriFood Institute: http://www.landis.org.uk/data/
Meteorological inputs	Daily, 1 km CHESS data, 1971–2007	E. Blyth, personal communication, 2012 with CEH, UK
Observations	(1) Soil moisture and soil matric potential measurements at Warren Farm, 2003–2006 (2) automatic weather station data at Warren Farm, 2002–2009 (3) Daily river flow data (4) Groundwater levels at observation boreholes	(1) N. Hewitt, personal communication, 2011 with CEH, UK, and LO-CAR project data (2) N. Hewitt, personal communication, 2011 with CEH, UK, and LO-CAR project data (3) http://www.ceh.ac.uk/data/nrfa/data/search.html (4) National Groundwater Level Archive: http://www.bgs.ac.uk/research/groundwater/datainfo/levels/ngla.html

nated by the slow response of the groundwater held within the chalk aquifer (the base flow index, that is, the proportion of total flow as base flow, is 0.87). Where the chalk outcrops, there is generally little surface runoff. At the Chalk–Palaeogene boundary, surface runoff from low permeability deposits gives rise to focussed recharge into the chalk. As a consequence, there are a number of swallow holes in the area (West and Dumbleton, 1972) that serve as surface-water sinks. The flow at the catchment outlet at Theale is monitored using a crump profile weir, where bypassing of the weir occurs above 29 m^3 s^{-1}. The unsaturated zone of the chalk has two characteristic behaviours: slow drainage over summer, and bypass flow during rainfall events (Ireson and Butler, 2013). Both behaviours are important under extreme conditions (i.e. droughts or extreme rainfall) for sustaining river flows and rapid water table response.

2.3 Case study data sets

A number of gridded data types are required for JULES parameterization and forcing (Table 1), including land cover and soil profile data, and meteorological drivers. Using a 50 m resolution topographic map, the Kennet catchment is discretized into 1 km^2 grids, which matches the resolution of the soil and meteorological data. Soil property data are provided by the National Soil Resources Institute (NSRI). Most soil profiles from the NSRI database extend as deep as 1.5 m for the basin (about 70 % of the profiles) and are provided with vertically variable Brooks and Corey soil moisture retention parameters. At the surface, the NSRI database differentiates between soil hydraulic parameters depending on land use (arable, permanent grassland, ley grassland, and other). Land use cover is provided from data collected by the

International Geosphere-Biosphere Programme (IGBP). The IGBP 2007 data are utilized to determine land cover types from 17 IGBP classes. These are re-classified to the 9 JULES land use types (Smith et al., 2006). The outcome is that cropland and mosaic/natural land use are the dominant land use types in the area (97 %).

Meteorological inputs to JULES were provided by the data from the Climate, Hydrology and Ecology research Support System (CHESS) project. The data set, produced by the Centre of Ecology and Hydrology (CEH), UK (E. Blyth, personal communication, 2012), includes 1 km gridded daily rainfall amounts derived from the UK rain gauge network measurements for the period 1971–2008 (Keller et al., 2006). In addition, air temperature, vapour pressure, long and short wave downward radiation, and wind speed, derived by downscaling the observed meteorology from the Meteorological Office Rainfall and Evaporation Calculation System (MORECS) 40 km × 40 km data set (Hough and Jones, 1997) accounting for the effects of topography, are also included. Daily observations of river flow at a number of gauging stations, along with groundwater levels at various observation frequencies (daily to monthly) from boreholes in the catchment, are used to evaluate model performance (Fig. 1). Groundwater levels at the same observational boreholes were previously examined by Jackson (2012), who used a conceptual model to estimate recharge rates to groundwater.

Chalk hydraulic properties are not available from standard national/global soil data sets (in the NSRI data set it is classified as a rock). Instead, these properties are estimated using soil moisture and matric potential observations at Warren Farm in the Kennet along with data from an on-site automatic weather station (Ireson et al., 2006) (Fig. 1). Soil moisture was measured between May 2003 and February 2006 using neutron probe measurements at different depths between 0.1 and 4.1 m taken fortnightly, on average. Either pressure transducer tensiometer (wet conditions) or equitensiometer (dry conditions) readings were taken for the same period of time to measure soil matric potential at 1 m depth every 15 min (Ireson et al., 2006; Ireson, 2008). Weather data include hourly observations of rainfall, downward short-wave solar and downward long-wave radiation, air temperature, specific humidity, and wind speed for the period between October 2002 and January 2009. The sub-daily weather data are used to account for any soil moisture sensitivity to the rainfall timing and intensity.

3 Method

The hydrological process representation in JULES is assessed with respect to the four primary functions of a hydrological system (Yilmaz et al., 2008): (1) overall water balance, (2) vertical redistribution, (3) temporal redistribution, and (4) horizontal spatial redistribution. Table 2 lists the assessment metrics for each of the four functions: the examined model assumptions/simplifications, the implemented model modifications, and the information sources used to inform the model modifications. Each of these information sources is described in the following sub-sections. The implemented model modifications considered below consist of a sub-daily weather generator, representation of sub-grid scale heterogeneity, dual Brooks and Corey curve representation of chalk hydraulic properties, change of the lower boundary condition, and coupling to a groundwater model.

3.1 Sub-daily weather generator

The daily CHESS weather data are downscaled in time (15 min) by a weather generator (WG) (D. Clark, personal communication, 2013). The code provided by CEH uses a cosine variation for sub-daily temperature defined by the average daily temperature and temperature variation range (defined as 7 °C based on the local automatic weather station – AWS). Sub-daily incoming long-wave radiation is calculated using the same phase of the cosine function as that used for the temperature disaggregation. Sub-daily downward short-wave radiation is calculated as a product of the daily average downward shortwave radiation and a normalized fraction of a daily total solar radiation defined by a geographical location, time of year and day. Sub-daily specific humidity is assumed to be equal to the minimum of the saturated specific humidity (for a given sub-daily temperature) and the average daily specific humidity. Wind speed and air pressure are assumed to be constant throughout the day. Sub-daily precipitation is divided into large-scale rainfall, convective rainfall and large-scale snow. This differentiation is based on the mean daily temperature. Precipitation is defined as snow if the temperature is below 0 °C; convective if the temperature is above

20 °C; and large-scale rainfall, otherwise. It is set to start at a random time during a day and to continue for a specified number of hours over the entire corresponding model grid: 2 h for a convective storm, and 5 h for large-scale precipitation. The model configuration that includes the weather generator is referred to as JULES + WG (see Table 2).

3.2 Representation of sub-grid scale heterogeneity of near-surface soil hydraulic properties

A statistical approach is chosen to represent sub-grid scale heterogeneity of soil hydraulic properties; therefore, the upper soil layer storage capacity is assumed to be heterogeneous and to have a Pareto probability distribution with shape parameter b and upper soil layer depth dz (Bell et al., 2009). This representation is available in the standard version of JULES, but there is no guidance on selection of the two parameters. This approach limits the amount of water available for infiltration according to the soil moisture state, with the rest of the rainwater becoming surface runoff. The infiltrated water is then routed vertically through the soil using Richards' equation.

Since there is limited information to constrain both parameters, the effective upper layer soil depth dz is fixed to the JULES default value of 1 m. A regionalized base flow index (BFI) from the HOST soil classification (BFIHOST) (Boorman et al., 1995; Bulygina et al., 2009) is used to specify the Pareto distribution shape parameter b for each soil type in the catchment. The parameter is calibrated using water partitioning between surface runoff and drainage by JULES. The parameter value that results in the drainage-to-total-runoff ratio closest to the expected BFIHOST for that soil classification is chosen to be representative of the soil heterogeneity. Due to the high computational requirement of JULES, only 21 regularly spaced values between 0 and 2 are considered. The considered parameter b range is found to provide suitable drainage-to-total-runoff ratios for the catchment soils and meteorological conditions. The model configuration that includes both the weather generator and the PDM model is referred to as JULES + WG + PDM (Table 2).

3.3 Chalk hydraulic properties estimation

Modelling vertical soil water flow in JULES using Richards' equation requires the following descriptors: air entry pressure head, Brooks and Corey exponent (Brooks and Corey, 1964), saturated hydraulic conductivity, soil moisture at saturation, residual soil moisture, soil moisture at the critical point when transpiration starts to decrease, and soil moisture at wilting point. Due to the two distinct flow domains in chalk–matrix and fractures, two intersecting Brooks and Corey curves are employed when fitting a chalk soil moisture retention curve. The effective soil moisture at the curves' intersection is estimated using available observations. This leads to a double curve representation of hydraulic conduc-

Table 2. Metrics and information used for model diagnosis and modification.

Metrics	Examined model assumptions/simplifications	Implemented model modifications	Resulting model configuration	Information sources
I. Water balance				
Relative bias in total runoff[1]: $$RBias Q = \frac{\sum_t Q_t^{mod} - \sum_t Q_t^{obs}}{\sum_t Q_t^{obs}}$$ Relative bias in subsurface runoff[2]: $$RBias SR = \frac{\sum_t SR_t^{mod} - \sum_t SR_t^{obs}}{\sum_t SR_t^{obs}}$$	Weather inputs are constant when disaggregated to finer timescales; soil properties are constant over relatively large spatial scales (1 km)	Sub-daily weather generator (Sect. 3.1); PDM model and its parameterization (Sect. 3.2)	JULES + WG; JULES + WG + PDM	Observed flows at the catchment outlet, precipitation from CHESS data set, regionalized base flow index HOST (BFI-HOST)
II. Vertical redistribution				
Mean square relative error for soil moisture[3]: $$MSRelE = \frac{1}{N}\sum_{n=1}^{N}\frac{1}{T_n}\sum_{t=1}^{T_n}\frac{(\theta_{n,t}^{mod} - \theta_{n,t}^{obs})^2}{var(\theta_n^{obs})}$$	A single permeability Richards' equation is capable of representing chalk soil moisture behaviour	Chalk representation via a double Brooks–Corey soil moisture retention curve and calibrated K_{sat} (Sect. 3.3)	JULES + CHALK	Observations of soil matric potential and soil moisture
III. Temporal redistribution				
Nash–Sutcliffe efficiency for flow: $$NS = 1 - \frac{\sum_t (Q_t^{mod} - Q_t^{obs})^2}{\sum_t (Q_t^{obs} - E|Q^{obs}|)^2}$$ and log-transformed flow: $$NS = 1 - \frac{\sum_t (\log(Q_t^{mod}) - \log(Q_t^{obs}))^2}{\sum_t (\log(Q_t^{obs}) - E|\log(Q^{obs})|)^2}$$	Horizontal unsaturated zone disconnection; no root uptake from deep saturated zone; unit gradient lower boundary condition; no surface/subsurface routing	Change of the lower boundary condition; approximation of unsaturated water travel through the deep unsaturated zone (Sect. 3.4); and coupling to the groundwater model ZOOMQ3D (Sect. 3.5) routing	JULES + WG + PDM; JULES + WG + PDM + CHALK + GW	States/fluxes from a detailed physics-based model of a chalk hillslope, observed flows at the catchment outlet; Observed flows and groundwater level hydrographs at internal catchment points
V. Spatial redistribution				
Nash–Sutcliffe efficiency for raw and log-transformed flow, relative bias in total runoff, visual inspection of groundwater levels at selected boreholes	Groundwater model parameterization	Change of specific yield parameters in the groundwater model	JULES + WG + PDM + CHALK + GWadj	Observed flows and groundwater level hydrographs at selected boreholes

[1] mod refers to modelled values; obs refers to the observed values; Q_t denotes runoff value at time t; SR_t denotes subsurface runoff value at time t. [2] Subsurface runoff is calculated via a hydrograph separation for observations. [3] $\theta_{n,t}$ denotes effective soil moisture at time t at the nth soil depth.

tivity dependence on soil moisture. The JULES soil module is modified accordingly to allow for a dual curve soil moisture retention representation. Although preferential *bypass* flow can occur in chalk (Ireson et al., 2012), it is considered to be relatively rare in the Kennet (Ireson and Butler, 2011). Consequently, it is not a major component in groundwater recharge and JULES has not been modified to include this effect.

The residual soil moisture content cannot be readily observed in the field, as chalk never dries out sufficiently to reach this state (Ireson, 2008). Therefore, the residual soil moisture content is estimated as a difference between maximum observed soil moisture and the effective porosity. The effective porosity (which includes matrix and fractures) is fixed at 0.36 (i.e. matrix porosity of 0.35 and fracture porosity of 0.01) (Bloomfield, 1997; Price et al., 1993). While fracture porosity tends to be higher at the soil surface due to the chalk weathering process (Ireson, 2008), this is not represented here due to the lack of comprehensive observations of soil moisture dynamics at multiple vertical levels; and the effects of the assumption are discussed in Sect. 4. Two sets of Brooks and Corey parameters are estimated by fitting the dual curve soil moisture retention representation to measurements of soil moisture and matric potential at 1 m depth obtained from field data collected at Warren Farm, Berkshire (Lowland Catchment Research – LOCAR – experiment data described in Ireson, 2008). Mean Square Error (MSE) is used to measure goodness of fit. Then, using the derived soil moisture retention curve, soil moisture at the critical point is calculated using the *wet* end curve at −40 kPa matric potential, while soil moisture at wilting point is calculated using the *dry* end curve at −1500 kPa.

Chalk saturated hydraulic conductivity is estimated by fitting the simulated soil moisture profiles to the available soil moisture neutron probes at multiple depths down to 4 m. For calibration purposes, 100 random values of saturated hydraulic conductivity are sampled logarithmically between 0.001 and 10 m day^{-1}. Mean square relative error (MSRelE; see definition in Table 2) for soil moisture between modelled and observed soil moisture for all observation depths is used as an objective function. The objective function increases error weights for the deeper layers that have less variable soil moisture, which is deemed to be important for drainage evaluation purposes. The model configuration that includes the weather generator, the PDM model and the chalk representation is referred to as JULES + WG + PDM + CHALK (Table 2).

3.4 A detailed physics-based model of a chalk hillslope

A physics-based model for 2-D flow in chalk (Ireson and Butler, 2013) represents a hillslope transect through unconfined chalk in the Pang catchment, located in close proximity to the river Kennet. While this model is an approximation itself and can only represent one set of hillslope properties,

it is built upon the best current knowledge of the hydrology of chalk hillslopes and is the best available test bed for simpler approximations. Flows in the 2-D model are governed by Richards' equation in both the saturated and unsaturated zones; and the properties of the chalk matrix and fractures are represented using an equivalent continuum approach (Peters and Klavetter, 1988; Doughty, 1999; Ireson et al., 2009). The Richards' equation is solved using a finite volume method.

Fluxes and states of the chalk hillslope model for the period 1970–2000 are examined to assess the following two assumptions underlying the JULES hydrology: (a) there is no hydrological interaction between neighbouring vertical soil columns, and (b) a unit gradient flow is a satisfactory approximation of the lower boundary condition at the 3 m base of the soil column on a hillslope location with a typically deep unsaturated zone. Further, the hillslope model is used to evaluate the nature of coupling between the unsaturated zone and groundwater, as well as the nature of water transport in the deep unsaturated zone located between the base of the JULES soil column and the water table. For these purposes, lateral fluxes in the unsaturated and saturated zones, hydraulic gradients and drainage rates at the soil column base, transpiration volumes extracted from the saturated and unsaturated zones by plants, and recharge rates at the groundwater table are extracted from the model. To reduce boundary condition effects at the upper and lower ends of the hillslope, the above variables are considered in the middle of the hillslope.

3.5 ZOOMQ3D distributed groundwater model

Groundwater flow in the Kennet is simulated using the ZOOMQ3D finite difference code (Jackson and Spink, 2004). The groundwater model is set up to simulate fluctuations in groundwater level, river baseflow, and spring discharge on a daily time step. The model uses gridded catchment representations at two scales; a 2 km base grid is locally refined to 500 m over the central part of the catchment. Rivers are simulated using an interconnected set of river reaches that exchange water with the aquifer according to a Darcian type flux equation. The vertical variations in rock hydraulic properties are represented using a three-layer model based on geological models of the hydrostratigraphy within the London Basin. The model is assessed to be a relatively good representation of the processes in the region in comparison with other chalk modelling examples (Jackson et al., 2011; Power and Soley, 2004).

ZOOMQ3D requires a significant number of parameters including horizontally and vertically distributed hydraulic conductivity and storage coefficient values. The parameters were zonally regularized and calibrated to approximate regional water table elevations (Jackson et al., 2011). For parameter estimation purposes, recharge has been modelled using a distributed recharge model ZOODRM (Mansour and Hughes, 2004) based on a conceptual Penman–Grindley soil moisture deficit model (Penman, 1948; Gridley, 1967). As a result, it needs to be understood that the calibrated groundwater parameters are only representative for the ZOODRM recharge field and are not, therefore, adjusted for recharge fluxes obtained using JULES.

The model configuration JULES + WG + PDM + CHALK is coupled to ZOOMQ3D based on the findings from the detailed 2-D model (Sect. 3.4) and is referred to as JULES + WG + PDM + CHALK + GW (Table 2). When groundwater model parameters are adjusted to examine sensitivity of the model response, the configuration is referred to as JULES + WG + PDM + CHALK + GWadj.

3.6 Surface runoff routing

The standard JULES configuration (version 2.2) does not have a surface-water routing option. Therefore, given the catchment size, flows are averaged over 10-day intervals to reduce the impact of routing effects. For the chosen flow averaging interval, any inaccuracy in the estimated river discharge due to the lack of surface routing is believed to be minor when compared to the total flow magnitude (groundwater contributes 87 % of the flow, on average), and inaccuracies in both actual baseflow index estimation (when BFIHOST is used) and in groundwater routing representation. Further, swallow holes in the catchment (West and Dumbleton, 1972) and river-soil water exchange for surface runoff (i.e. infiltration of surface runoff into the river bed) are not represented in the model, and possible consequences of this are discussed later in the Results and discussion, Sect. 4.

3.7 Other JULES parameters

The remaining JULES parameters are assigned as follows. Land use fractions are taken from the IGBP 2007 data set, and re-classified into the nine land use types commonly used for JULES applications (Smith et al., 2006). Soil hydraulic parameters are taken from the NSRI soil database with the exception of soil layers that are classified as chalk. Soil hydraulic properties below the deepest NSRI horizon, typically at 1.5 m, are assigned the deepest horizon properties. Chalk hydraulic properties derived in this study are assigned to soil horizons that are classified as chalk in the NSRI database. The dominant agricultural crop for the area is spring barley (Limbrick et al., 2000). The root depth for the crop was chosen as 1 m (average value based on Breuer et al., 2003) and canopy height was chosen as 0.8 m (Hough and Jones, 1997; Mauser and Schadlich, 1998). Leaf area index (LAI) changes seasonally, with maximum of LAI = 3 (Mauser and Schadlich, 1998; Petr et al., 2002). The maximum interception capacity per unit leaf area is fixed at 0.2 mm, so that the upper limit to interception is $0.2 \times$ LAI (Hough and Jones, 1997). Other vegetation parameters are set at their recommended default value for JULES (Cox et al., 1999).

4 Results and discussion

Observations of water fluxes, soil moisture and groundwater levels in the Kennet catchment are compared with the simulated values derived using the sequentially modified JULES model structure to represent the four hydrological functions of a catchment.

4.1 Water balance

The long-term water balance is calculated for the period 1972–2007 from observations and various model configurations, and three metrics are calculated – relative bias for total runoff ($RBias_Q$) and surface runoff ($RBias_{SR}$), and MSRelE (Table 3). The unmodified JULES (version 2.2) is found to overestimate the total runoff by 24 % and, correspondingly, underestimate the evapotranspiration (ET) by 15 %. This is attributed to the constant temporal disaggregation of weather variables that is hard-coded into the model. When weather variables are temporally disaggregated using the WG, described in Sect. 3.1, the total runoff is only 2 % lower than the observed value. However, neither configuration is capable of producing any surface runoff. This is because the hydraulic conductivities of the catchment soils (derived from the NSRI parameter database), even for relatively clayey soils, are sufficiently high to enable virtually all the instantaneous rainfall rates obtained using temporal disaggregation to infiltrate into the soil.

The parameter b of the PDM model is selected based on regionalized information from BFIHOST (Sect. 3.2) and ranges, mainly, between 0 and 0.4, except for two grids where b is set as 0.7 and 1. Further, the parameter b is assigned 0 value over approximately 60 % of the catchment for the locations with permeable (chalky) top-soils. Addition of the PDM model (JULES + WG + PDM configuration) with parameters selected based on regionalized information from BFIHOST (Sect. 3.2) generated, on average, 70 mm yr^{-1} of surface runoff (compared to 39 mm yr^{-1} derived by baseflow separation at the catchment outlet). This is likely to originate from the regionalization error – the catchment average regionalized BFIHOST value (0.78) is lower than the BFI value calculated from observed flow at the catchment outlet (0.87). This difference may arise from a number of locally relevant soil properties and processes that are not represented in the regionalized BFIHOST, for example there is a focussed recharge into sink or swallow holes of runoff from the Palaeogene deposits in the lower reaches of the Kennet catchment (West and Dumbleton, 1972). Such localized processes could, in principle, be explicitly represented in the land surface model, but this would be difficult in practice due to the scales involved; for example, representing the sink holes would require fine-scale data (at 0.1 to 1 m resolution) describing the land surface features.

It is to be noted that the proposed model modification with PDM and its parameterization is not the only possible model

Table 3. Observed and simulated water balance and metrics of model performance.

Source/model configuration	Rainfall, mm yr^{-1}	ET, mm yr^{-1}	Total runoff, mm yr^{-1}		RBias$_Q$	RBias$_{SR}$	MSRelE
			Surface runoff, mm yr^{-1}	Subsurface runoff, mm yr^{-1}			
Obs[1]	784	485	299		–	–	–
			39	260			
Standard JULES (version 2.2)[2]	784	410	370		0.24	0.42	–[3]
			0	370			
JULES + WG	784	489	292		−0.02	0.12	–
			1	291			
JULES + WG + PDM	784	489	299		0.00	−0.12	3.64
			70	229			
JULES + WG + PDM + CHALK	784	495	293		−0.02	−0.13	1.12
			67	226			
JULES + WG + PDM + CHALK + GW	784	496	293		−0.02	−0.13	1.07
			67	226			
JULES + WG + PDM + CHALK + GWadj	784	496	293		−0.02	−0.13	1.07
			67	226			

[1] For observations, ET is calculated as a residual between the long-term precipitation and runoff; surface and subsurface runoff are calculated based on the hydrograph separation. [2] For model configurations, surface and subsurface runoff are taken as surface runoff and drainage fluxes, respectively, produced by a model. [3] MSRelE is calculated starting from the JULES + WG + PDM configuration.

modification. An alternative, which potentially leads to increased surface runoff production, includes spatial and, perhaps, further temporal downscaling of rainfall to produce more intense events over parts of the 1 km discretization grids. Table 3 shows that the model modifications used to improve the representation of the additional processes observed in the catchment (and outlined in Table 2) do not compromise the simulated water balance.

4.2 Vertical redistribution through the soil

Both JULES + WG + PDM and JULES + WG + PDM + CHALK configurations use 4.5 m long soil columns with 0.1 m thick soil layers to facilitate the comparison with the observed soil moisture. The JULES + WG + PDM configuration results in overly dry soils between 1 and 4.1 m depth when compared to the observations (Fig. 2; a representative subset of the soil moisture time series is shown); and the corresponding MSRelE metric equals 3.64. This soil dryness is attributed to incorrect representation of chalk soil hydraulic properties. Figure 3 shows two Brooks and Corey soil moisture retention curves fitted to the pairs of soil moisture and matric potential observations at 1 m depth in chalk; the curves intersect at an effective soil moisture of 0.31 (effective soil moisture equals soil moisture with subtracted residual soil moisture). The figure illustrates a threshold change

in the chalk soil moisture retention curve and consequently, through the Brooks–Corey–Mualem model, the unsaturated hydraulic conductivity relationship. This change in properties is related to the dual porosity–dual permeability nature of the chalk soil (Ireson et al., 2009). Estimated chalk hydraulic properties are given in Table 4. Further, the time varying vertical distribution of soil moisture estimated by the JULES + WG + PDM + CHALK configuration is shown in Fig. 2; this corresponds to an MSRelE metric of 1.12. This value stays approximately the same throughout further model modifications to include additional functions of the hydrological system. The inclusion of chalk properties into the model produces a better simulation of soil moisture content at the Warren Farm site than that from the JULES + WG + PDM configuration. This corresponds well with the observed soil moisture below ∼1 m depth. However, the upper soil tends to be wetter than the observed moisture levels. This is attributed to the chalk's vertical heterogeneity; fractures appear more frequently and are larger in the upper chalk. Depth-variable soil hydraulic properties are required to capture the phenomenon. This is not attempted here due to the lack of soil moisture–matric potential observational pairs at multiple vertical levels to define entire soil moisture retention curves.

Table 4. Hydraulic double Brooks and Corey curve parameters for chalk.

Parameter	Description	Wet end	Dry end	Source
b	Exponent	30.2	1.3	Calibration to s oil moisture and matric potential at 1 m
α, m	Soil matric potential at saturation	0.15	12.2	Calibration to soil moisture and matric potential at 1 m
K_{sat}^*, m day^{-1}	Saturated hydraulic conductivity	0.016 (0.02)		JULES calibration to soil moisture at multiple depths down to 4.1 m
θ_s^{eff}	Effective saturated soil moisture	0.36		Price et al. (1993), Bloomfield (1997)
θ_r	Residual soil moisture	0.11		Soil moisture observations and θ_s^{eff} value
θ_{cr}^{eff}	Effective saturated soil moisture at critical point	0.32		Brooks and Corey equation at -40 kPa
θ_{wilt}^{eff}	Effective saturated soil moisture at wilting point	0.05		Brooks and Corey equation at -1500 kPa
θ_{inter}^{eff}	Effective soil moisture at the two curves intersection	0.31		Calibration to soil moisture and matric potential at 1 m

* K_{sat} is fitted using JULES + WG + PDM + CHALK as well as JULES + WG + PDM + CHALK + GW configurations. The value for the latter is shown in the parenthesis.

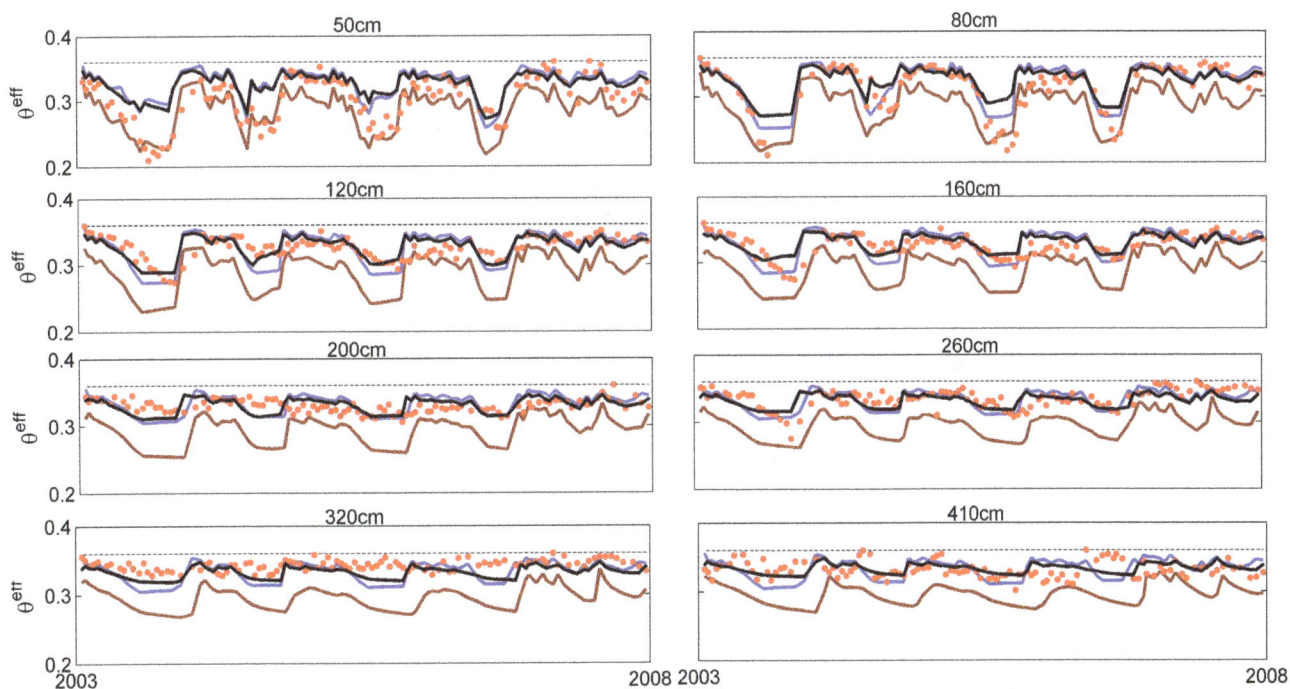

Figure 2. Comparison of the optimized effective soil moisture (θ^{eff}) time series with the observed soil moisture (red dots) at various depths at Warren Farm, Berkshire, UK. Brown lines show soil moisture estimated by JULES + WG + PDM; blue lines show soil moisture estimated by JULES + WG + PDM + CHALK; and black lines show soil moisture estimated by JULES + WG + PDM + CHALK + GW. Grey horizontal line shows the effective soil moisture at saturation 0.36. Note, only a representative subset of soil moisture time series utilized in the analysis is shown in the figure.

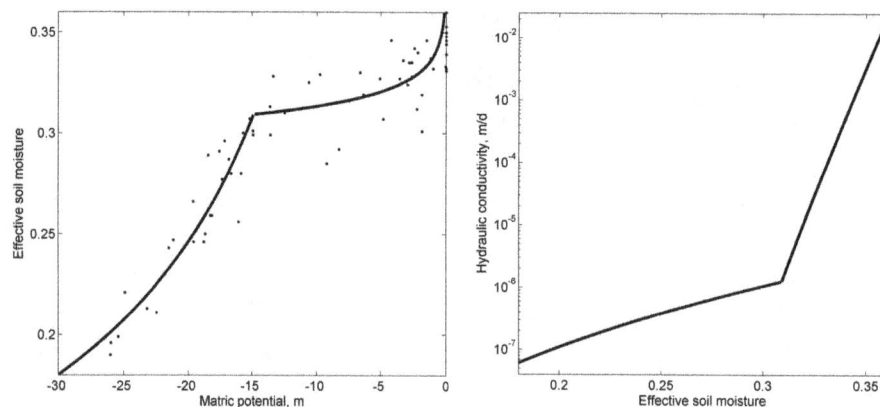

Figure 3. Chalk soil moisture retention fit using a dual Brooks and Corey curve and the corresponding hydraulic conductivity dependence on effective soil moisture. Black dots are observed data.

4.3 Temporal redistribution

The original as well as the modified model configurations are only capable of partitioning water fluxes at the point/grid scale, and do not have a mechanism for further routing to provide temporal water redistribution at the catchment outlet. Here, some assumptions about the nature of such water redistribution (Sect. III in Table 2) are assessed, and lateral routing through the saturated zone is achieved through coupling the model to a distributed groundwater model.

Fluxes extracted from the physics-based 2-D model of a chalk hillslope imply that there are two simplifications that can be made with regards to the 2-D nature of hillslope hydrological processes. First, lateral fluxes in the chalk unsaturated zone are found to be insignificant when compared to the net (vertical and lateral) fluxes in the unsaturated zone. Hence, the simplifying assumption about inter-soil-column-independent hydrological behaviour is a reasonable and sufficiently accurate approximation for the area. Second, evapotranspiration losses from the chalk saturated zone are found to be negligible compared to those from the unsaturated zone. It is therefore assumed that evapotranspiration processes can be restricted to the unsaturated zone when coupling the unsaturated zone to groundwater for the study area investigated herein.

Extracted vertical hydraulic gradients from the 2-D hillslope model are compared to the unit gradient lower boundary condition along with a number of alternative lower boundary conditions (using mean absolute difference as an objective function). Of these, it is found that a *persistent gradient* condition is the most consistent and accurate approximation of the lower boundary condition for the area. The persistent gradient condition assumes that hydraulic gradient is time varying but almost constant with depth at the soil column base. The condition can be approximated using the hydraulic gradient between soil column nodes just above the column base, requiring a relatively fine node mesh at the column base. The persistent gradient condition can be seen as a general conse-

quence of the following lower boundary condition $\frac{\partial^2 h}{\partial z^2} = 0$, where the unit gradient lower boundary condition $\frac{\partial h}{\partial z} = 1$ is a special case. Note that only the hydraulic gradient at the soil column base is approximated herein. This gradient is used to substitute the unit gradient in the formula for the drainage flux in JULES. This implies that hydraulic conductivity at the base of the soil column is based on the nearest to the bottom node state.

Further, the persistent gradient approximation is evaluated for multiple soil column depths to optimize its applicability. The mean absolute difference between the persistent hydraulic gradient at the lower boundary and hydraulic gradients extracted from the hillslope model at a number of depths is used as an objective function. It is found that the objective function improves with increasing depth of soil column but less significantly after 6 m. As a trade-off between the soil column depth and the lower boundary approximation accuracy, an optimal depth to apply the persistent gradient lower boundary condition is chosen to be 6 m. Figure 4 compares hydraulic gradients at a 6 m column base extracted from the 2-D model to the unit gradient as well as to the gradient just above 6 m (approximately at 5.5 m depth, based on the model mesh), representing the persistent boundary condition approximation.

Lastly, to draw a connection between the modelled potential recharge at 6 m depth and the modelled actual recharge at the water table, temporally averaged vertical fluxes extracted from the 2-D hillslope model are considered for 6 hourly (the model step), daily, weekly and 30-day periods. The correlations between the time series of actual and potential recharges for the averaging periods are 0.75, 0.8, 0.89, and 0.94, respectively. Total actual and total potential recharges for the 1970–2000 period are found to be less than 1 % different. Average daily (the regional groundwater time step) potential simulated recharge at 6 m and actual simulated recharge at the water table extracted from the 2-D model are shown in Fig. 5. It can be seen that the potential recharge

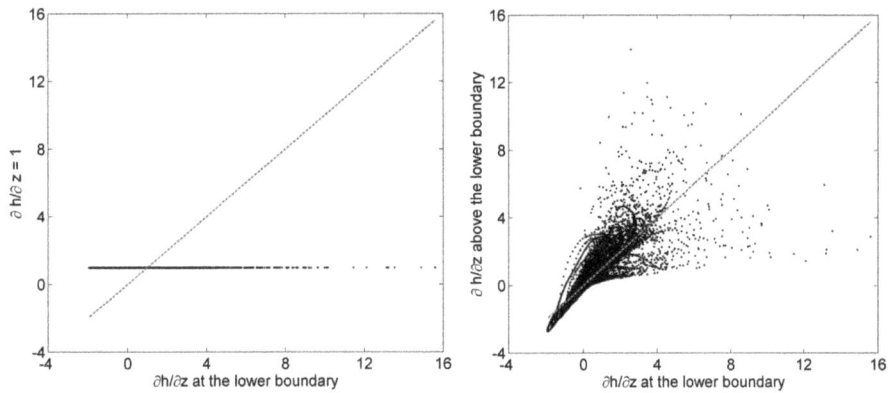

Figure 4. Comparison of hydraulic head gradient $\partial h/\partial z$ at a 6 m depth extracted from the 2-D model with a unit gradient condition (left panel), and with a persistent gradient condition extracted at a 5.5 m depth (right panel).

is widely spread at low actual recharge rates (below about 2 mm day^{-1}). However, the potential recharge becomes quite a consistent predictor for the actual recharge at mid- to high actual recharge rates.

Based on the above findings, JULES + WG + PDM + CHALK is used with a 6 m deep soil column and 0.1 m thick soil layers, and is coupled via a weak two-way coupling to the groundwater model ZOOMQ3D implemented through the lower boundary condition (persistent gradient). The *weak* coupling assumes that the drainage flux from JULES is used as an upper boundary condition by ZOOMQ3D, and any upward water fluxes from the saturated zone to the upper unsaturated zone are calculated based on the (persistent gradient) lower boundary condition. Note, the saturated hydraulic conductivity for chalk soil is re-calibrated following the procedure given in Sect. 3.3 (Table 4) as the new persistent gradient lower boundary condition impacts the soil moisture dynamics.

The resulting 10-day-averaged river flow at the catchment outlet (Theale) for the period 1994–2006 is shown on Fig. 6. The period includes two droughts in the region (1995–1998 and 2003–2006) as well as substantially wet 1999–2001 period that led to groundwater flooding. Figure 6 also shows model performance measures for the total simulation period of 1972–2007, with a Nash–Sutcliffe efficiency for simulated flow (NS = 0.82) and log-transformed simulated flow (NS$_{\log}$ = 0.81), as well as a relative bias for the total flow (RBias$_Q$ = 0.01). Note, the relative bias is calculated using the flow at the catchment outlet, not the sum of surface and drainage fluxes produced by the land surface model component of the configuration. This explains the slight difference between the RBias$_Q$ values in Table 3 and Fig. 6 for the model configuration.

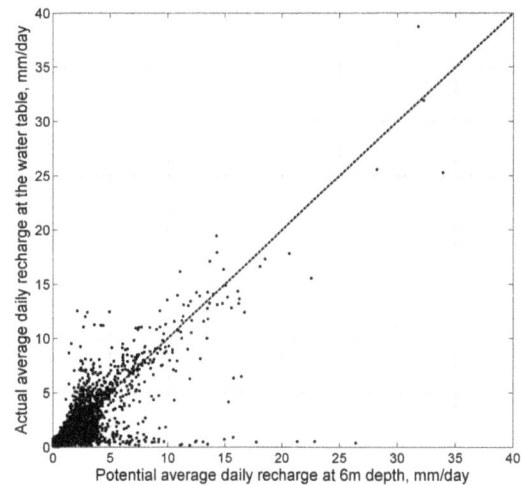

Figure 5. Correspondence between potential and actual daily recharge rates extracted from the 2-D model.

4.4 Spatial redistribution over the groundwater and surface-water systems

Due to the distributed nature of the coupled model configuration, flows (Fig. 6) and groundwater levels (Fig. 7) can be examined at the internal catchment points shown in Fig. 1. It can be seen that total flow tends to be underestimated in the smaller sub-catchments such as the Kennet at Marlborough and the Lambourn at Shaw. Inspection of water movement patterns inside the groundwater model ZOOMQ3D offers a possible explanation. The Lambourn groundwater catchment area is found to be underestimated by ZOOMQ3D when compared to the groundwater catchment area extracted from local observational boreholes and spring head data (Parker, 2011; Parker et al., 2015). Further, the model tends to direct some water from the Lambourn to the middle part of the Kennet (Parker, 2011, S. Parker, personal communication, 2013), which helps to explain the total flow overestimation

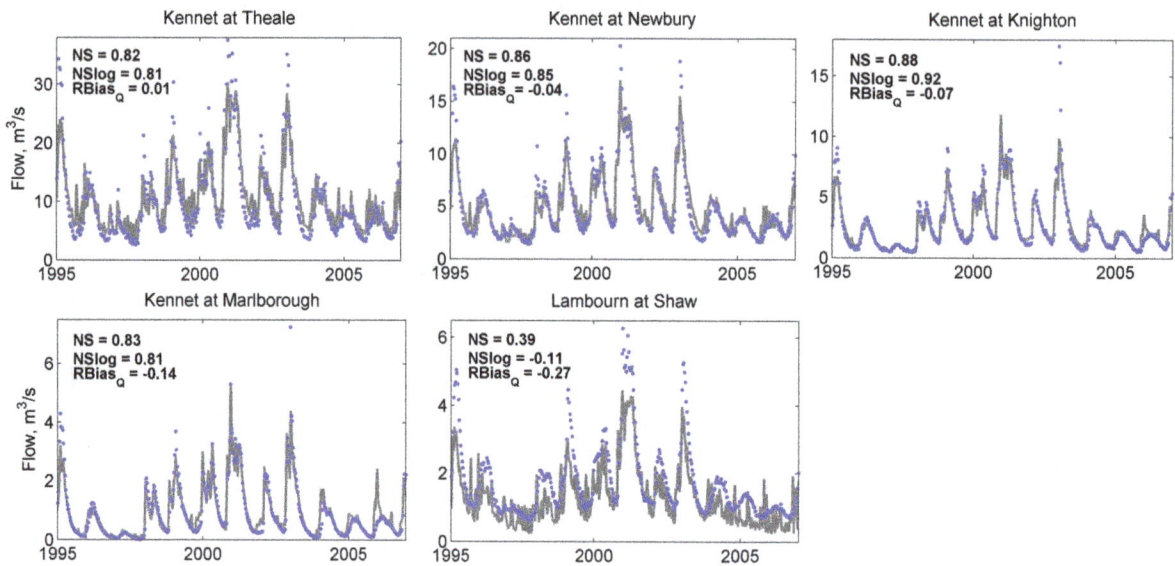

Figure 6. 10-day average flows at five gauging stations in the Kennet generated by the JULES + WG + PDM + CHALK + GW model configuration. Grey lines denote simulated flows, and blue dots are observations.

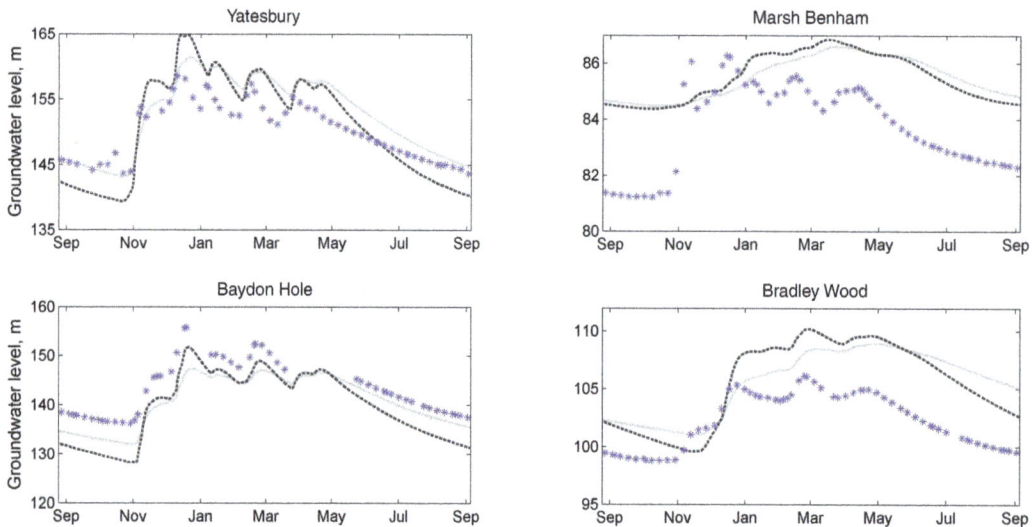

Figure 7. Water levels at four observational boreholes in the Kennet. Blue stars are observed levels, grey lines represent groundwater levels generated by JULES + WG + PDM + CHALK + GW configuration; and black dotted lines represent groundwater levels generated by JULES + WG + PDM + CHALK + GWadj configuration.

at the Marlborough, Newbury, and Knighton gauges. Further, during wet years, peak flows appear to be underestimated at all gauging stations; meanwhile, low flows are slightly over-estimated for the Kennet at Theale and the Kennet at Newbury, and underestimated for the Lambourn at Shaw. Treating potential recharge from the land surface model as actual recharge to ZOOMQ3D might partly explain the low flow overestimation.

Because of the mismatch of scales between an observation borehole (order of 1 m) and JULES and ZOOMQ3D grid scales (1 km and 500 m, respectively), only a visual assess-

ment of the predicted water levels is attempted. Figure 7 illustrates simulated water levels at four selected boreholes for September 2000–August 2001 representing an unusually wet year leading to a groundwater flooding in the area. Similar to the results from Jackson (2012), who considered the same period and boreholes, water levels are mainly overestimated at the Marsh Benham and Bradley Wood boreholes. Moreover, the modelled response at Marsh Benham and Bradley Wood is more attenuated than the observed response. At the model scale (1 km), the estimated groundwater levels are indicative of the boreholes partly due to soil heterogeneity. For exam-

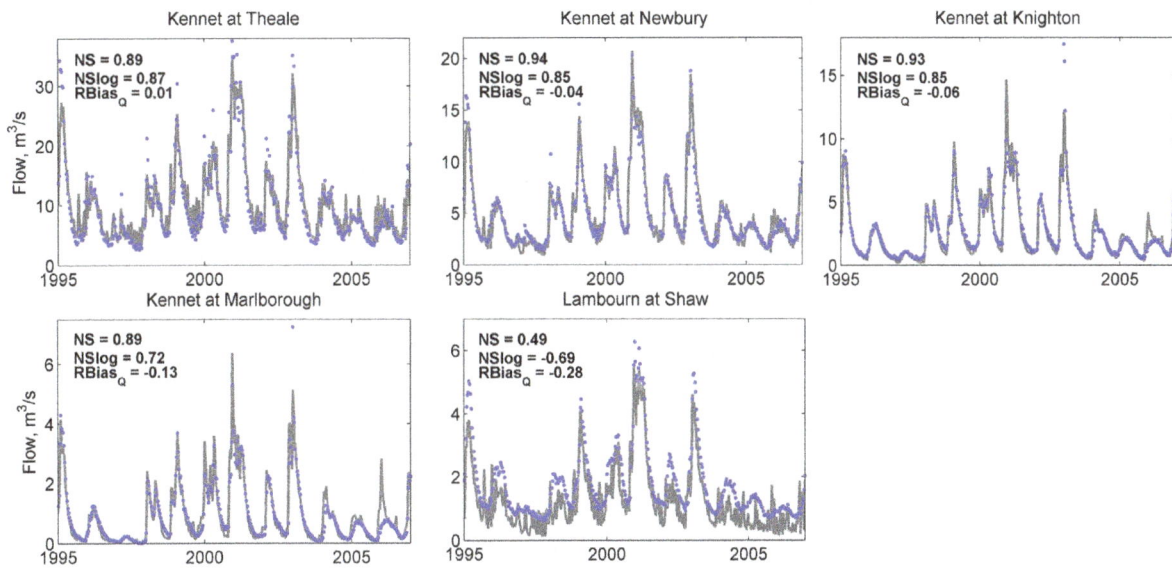

Figure 8. 10-day average flows at five gauging stations in the Kennet generated by the JULES + WG + PDM + CHALK + GWadj model configuration. Grey lines denote simulated flows, and blue dots are observations.

ple, the PDM model parameter *b* (as well as soil hydraulic properties) is chosen based on a dominant soil type; therefore, the recharge to total runoff ratios are 0.52 and 0.17 for Bradley Wood and Marsh Benham, correspondingly. However, other soils present in the model grids have very different recharge to total runoff ratios, for example, 0.98 and 0.88 for Bradley Wood and Marsh Benham, respectively. Incorporating the hydrology of these soils can potentially lead to more responsive water level behaviour at the boreholes as more water will infiltrate.

As indicated in Sect. 3.5, the parameters of the ZOOMQ3D groundwater model are derived using recharge from a different near-surface model, and thus are likely to be sub-optimal when recharge produced by JULES is used. A manual sensitivity analysis of model parameters showed that tuning values of specific yield or hydraulic conductivity leads to better agreement with the observed data. For example, Fig. 8 shows flows generated by the coupled model when ZOOMQ3D specific yield parameters are halved over the whole Kennet area. This results in a better representation of high flows, but mixed outcomes for low flows (according to the NS and NS_{log} performance measures). Groundwater levels at the selected boreholes become slightly more responsive, but do not change significantly (Fig. 7). As the primary research objective is to diagnose the hydrological limitations of a land surface model, a formal recalibration of an auxiliary groundwater model is not pursued here.

5 Conclusions

The paper is motivated by the goals of using land surface models as a basis for global hyper-resolution modelling of

the terrestrial water, energy, and biogeochemical cycles, including application to a range of complex hydrological prediction problems. This comes alongside the recognition that there are significant limitations in the accuracy with which hydrological fluxes and storages are represented in general in LSMs due to their focus on supporting large-scale climate modelling problems (Oleson et al., 2008; Wood et al., 2011). The paper uses a case study of the JULES LSM model applied to the Kennet catchment in southern England, which represents the challenging problem of hydrological modelling in a chalk dominated catchment with a predominantly deep unsaturated zone. A diagnostic approach is taken to identify the model inadequacies with respect to the four functions of a hydrological system: overall water balance, vertical redistribution of water through the soil, temporal redistribution of water, and spatial redistribution over the catchment's surface-water and groundwater systems. The approach facilitates a sequential model improvement using hydrological expert knowledge about model assumptions and simplifications relevant for each hydrological aspect considered. The following model modifications are presented and assessed in the paper:

- overall water balance: introduction of a weather generator and statistical description of top-soil heterogeneity via regionalized information;

- vertical redistribution through the soil: approximation of the dual permeability – dual porosity hydraulic soil behaviour;

- temporal redistribution: change of the lower boundary condition and approximation of coupling to a groundwater model;

– spatial redistribution over the catchment: alteration of groundwater model parameters.

It is noted that improving the model physics in sequence preserves model performance quality with respect to the other previously considered functions. For example, improving vertical distribution does not corrupt the water balance achieved at a previous model modifications stage. This might be explained by the physical basis of both the model and reasoning for model modifications. The improvements are illustrative of the potential outcomes of a diagnosis approach, and alternative or additional improvements are possible. These include: the representation of the temporal and spatial distribution of precipitation; inclusions of point/small-scale features such as sink holes; and more physics-based inclusion of the vertical and horizontal distribution of soil hydraulic properties. As a procedural improvement, uncertainty analysis could be used to indicate if output errors can be explained by estimates of particular input uncertainties.

For some applications, the intermediate model configurations might be sufficient. For example, while JULES + WG + PDM configuration cannot provide flow/groundwater level hydrographs as it lacks surface and subsurface water routing, the configuration can still be used to represent the water balance over an area. This is useful for regions where no groundwater model and/or detailed geology data are available. It is to be noted that the findings are catchment specific and result from a weak surface-groundwater coupling, and as such cannot be readily generalized to other environments with shallower water tables.

Diverse sources of information were used to guide the model assessment and include remotely sensed data (topography, land use), spatially extrapolated point data (soils, weather conditions), point measurements (soil moisture and matric potential, flow, groundwater level), regionalized hydrological information (BFIHOST), and states/fluxes extracted from an auxiliary physics based hillslope model (Ireson and Butler, 2013). Fewer data might result in a less detailed representation of the water cycle depending on the specifics the hydrological system being investigated.

Whilst this application of JULES to the Kennet catchment is highly specific, it conveniently illustrates the type of challenges and parameterization of complex and distributed hydrological processes, model coupling using simplified boundary conditions, and assimilation of different sources of information to model identification, which will be encountered in almost any attempt to improve the utility of LSMs for catchment-scale water cycle modelling, arising due to the *uniqueness of place* problem. The paper has demonstrated the considerable accuracy gains that can be achieved using a sequential model error diagnosis strategy and expert-lead model adjustments. These can be taken forward to develop a general comprehensive guidance for transitioning to high resolution land surface modelling.

Acknowledgements. The study was made possible through code and data provided by Andrew Ireson (2-D hillslope model and soil moisture data, University of Saskatchewan, Canada), Doug Clark (weather generator, CEH, UK), Ned Hewitt (AWS and neutron probes, CEH, UK), and Eleanor Blyth (CHESS, CEH, UK). Further, the authors thank Zed Zulkafi (Imperial College London, UK) for help with the initial JULES set-up and Simon Parker (Imperial College London, UK) for specific information on chalk groundwater catchment behaviour. The author Jackson publishes with the permission of the Executive Director of the British Geological Survey. The work was funded under the UK Natural Environment Research Council's Changing Water Cycle program, grant NE/I022558/1.

References

Balsamo, G., Viterbo, P., Beljaars, A., van den Hurk, B., Hirschi, M., Betts, A., and Scipal, K.: A revised hydrology for the ECMWF model: verification from field site to terrestrial water storage and impact in the integrated forecast System, J. Hydrometeorol., 10, 623–643, 2009.

Bell, V., Kay, A., Jones, R., Moore, R., and Reynard, N.: Use of soil data in a grid-based hydrological model to estimate spatial variation in changing flood risk across the UK, J. Hydrol., 377, 335–350, 2009.

Best, M. J., Pryor, M., Clark, D. B., Rooney, G. G., Essery, R. L. H., Ménard, C. B., Edwards, J. M., Hendry, M. A., Porson, A., Gedney, N., Mercado, L. M., Sitch, S., Blyth, E., Boucher, O., Cox, P. M., Grimmond, C. S. B., and Harding, R. J.: The Joint UK Land Environment Simulator (JULES), model description – Part 1: Energy and water fluxes, Geosci. Model Dev., 4, 677–699, doi:10.5194/gmd-4-677-2011, 2011.

Beven, K. and Germann, P.: Macropores and water flow in soils revisited, Water Resour. Res., 49, 3071–3092, doi:10.1002/wrcr.20156, 2013.

Beven, K. and Kirkby, M.: A physically based, variably contributing model of basin hydrology, Hydrol. Sci. Bull., 24, 2415–2433, 1979.

Black, P. E.: Watershed functions, J. Am. Water Resour. As., 33, 1–11, 1997.

Bloomfield, J.: The role of diagenisis in the hydrogeological stratification of carbonate aquifers: an example from the chalk at Fair Cross, Berkshire, UK, Hydrol. Earth Syst. Sci., 1, 19–33, doi:10.5194/hess-1-19-1997, 1997.

Blyth, E., Clark, D. B., Ellis, R., Huntingford, C., Los, S., Pryor, M., Best, M., and Sitch, S.: A comprehensive set of benchmark tests for a land surface model of simultaneous fluxes of water and carbon at both the global and seasonal scale, Geosci. Model Dev., 4, 255–269, doi:10.5194/gmd-4-255-2011, 2011.

Boone, A., Habets, F., Noilhan, J., Clark, D., Dirmeyer, P., Fox, S., Gusev, Y., Haddeland, I., Koster, R., Lohmann, D., Mahanama, S., Mitchell, K., Nasonova, O., Niu, G., Pitman, A., Polcher, J., Shmakin, A., Tanaka, K., van den Hurk, B., Verant, S., Verseghy, D., Viterbo, P., and Yang, Z.: The Rhone-aggregation land surface scheme intercomparison project: an overview, J. Climate, 17, 187–208, 2004.

Boorman, D., Hollis, J., and Lilly, A.: Hydrology of Soil Types: A Hydrologically-based Classification of the Soils of the UK, Institute of Hydrology, Wallingford, 1995.

Breuer, L., Eckhardt, K., and Frede, H.-G.: Plant parameter values for models in temperate climates, Ecol. Model., 169, 237–293, 2003.

Brooks, R. and Corey, A.: Hydraulic Properties of Porous Media, Hydrology Papers, Colorado State University, Fort Collins, Colorado, 1964.

Brouyere, S., Dassargues, A., Hallet, V.: Migration of contaminants through the unsaturated zone overlying the Hesbaye chalky aquifer in Belgium: a field investigation, J. Contam. Hydrol. 72, 135–164, 2004.

Bulygina, N., McIntyre, N., and Wheater, H.: Conditioning rainfall-runoff model parameters for ungauged catchments and land management impacts analysis, Hydrol. Earth Syst. Sci., 13, 893–904, doi:10.5194/hess-13-893-2009, 2009.

Clark, D. B. and Gedney, N: Representing the effects of subgrid variability of soil moisture on runoff generation in a land surface model, J. Geophys. Res., 113, D10111, doi:10.1029/2007JD008940, 2008.

Clark, D. B., Mercado, L. M., Sitch, S., Jones, C. D., Gedney, N., Best, M. J., Pryor, M., Rooney, G. G., Essery, R. L. H., Blyth, E., Boucher, O., Harding, R. J., Huntingford, C., and Cox, P. M.: The Joint UK Land Environment Simulator (JULES), model description – Part 2: Carbon fluxes and vegetation dynamics, Geosci. Model Dev., 4, 701–722, doi:10.5194/gmd-4-701-2011, 2011.

Cox, P., Bett, R., Bunton, C., Essery, R., Rowntree, P., and Smith, J.: The impact of new land surface physics on the GCM simulation of climate and climate sensitivity, Clim. Dynam., 15, 183–203, 1999.

Dahan, O., Nativ, R., Adar, E., Berkowitz, B.: A measurement system to determine water flux and solute transport through fractures in the unsaturated zone, Groundwater, 36, 444–449, 1998.

Dahan, O., Nativ, R., Adar, E., Berkowitz, B., Ronen, Z.: Field observation of flow in a fracture intersecting unsaturated chalk, Water Resour. Res., 35, 3315–3326, 1999.

Doughty, C.: Investigation of conceptual and numerical approaches for evaluating moisture, gas, chemical and heat transport in fractured unsaturated rock, J. Contam. Hydrol. 38, 69–106, 1999.

Downing, R. A., Price, M., and Jones, G.: The Hydrogeology of the Chalk of Northwest Europe, Oxford University Press, Oxford, 1993.

Grindley, J.: The estimation of soil moisture deficits, Meteorol. Mag., 96, 97–108, 1967.

Gusev, Y. and Nasonova, O.: The simulation of heat and water exchange in the boreal spruce forest by the land-surface model SWAP, J. Hydrol., 280, 162–191, 2003.

Hough, M. N. and Jones, R. J. A.: The United Kingdom Meteorological Office rainfall and evaporation calculation system: MORECS version 2.0 – an overview, Hydrol. Earth Syst. Sci., 1, 227–239, doi:10.5194/hess-1-227-1997, 1997.

Ireson, A.: Quantifying the hydrological processes governing flow in the unsaturated Chalk, PhD thesis, Imperial College London, 270 pp., available at: http://www3.imperial.ac.uk/pls/portallive/docs/1/38553696.PDF (last access: 1 March 2015), 2008.

Ireson, A. M. and Butler, A.: Controls on preferential recharge to Chalk aquifers, J. Hydrol., 398, 109–123, 2011.

Ireson, A. M. and Butler, A. P.: A critical assessment of simple recharge models: application to the UK Chalk, Hydrol. Earth Syst. Sci., 17, 2083–2096, doi:10.5194/hess-17-2083-2013, 2013.

Ireson, A. M., Wheater, H., Butler, A., Mathias, S., and Finch, J.: Hydrological processes in the chalk unsaturated zone – insights from an intensive field monitoring programme, J. Hydrol., 330, 29–43, 2006.

Ireson, A. M., Mathias, S., Wheater, H., Butler, A., and Finch, J.: A model for flow in the Chalk unsaturated zone incorporating progressive weathering, J. Hydrol., 365, 244–260, 2009.

Ireson, A. M., Butler, A., and Wheater, H.: Evidence for the onset and persistence with depth of preferential flow in unsaturated fractured porous media, Hydrol. Res., 43, 707–719, 2012.

Jackson, B., Browne, C., Butler, A., Peach, D., Wade, A., and Wheater, H.: Nitrate transport in Chalk catchments: monitoring, modeling and policy implications, Environ. Sci. Policy, 11, 125–135, 2008.

Jackson, C.: Simple automatic time-stepping for improved simulation of groundwater hydrographs, Ground Water, 50, 736–745, 2012.

Jackson, C. and Spink, A.: User's Manual for the Groundwater Flow Model ZOOMQ3D, IR/04/140, British Geological Survey, Nottingham, UK, 107 pp., 2004.

Jackson, C., Meister, R., and Prudhomme, C.: Modeling the effects of climate change and its uncertainty on UK chalk groundwater resources from an ensemble of global climate model projections, J. Hydrol., 399, 12–28, 2011.

Keller, V., Young, A., Morris, D., and Davies, H.: Task 1.1: Estimation of Precipitation Inputs, Environment Agency R&D Project w6-101, Centre for Ecology and Hydrology, Wallingford, UK, 1–35, 2006.

Kloppmann, W., Dever, L., and Edmunds, W.: Residence time of Chalk groundwaters in the Paris Basin and the North German Basin: a geochemical approach, Appl. Geochem., 13, 593–606, 1998.

Kollet, S. and Maxwell, R.: Capturing the influence of groundwater dynamics on land surface processes using an integrated, distributed watershed model, Water Resour. Res., 44, W02402, doi:10.1029/2007WR006004, 2008.

Koster, R., Suarez, M., Ducharne, A., Stieglitz, M., and Kumar, P.: A catchment-based approach to modeling land surface processes in a general circulation model 1. Model structure, J. Geophys. Res., 105, 24809–24822, 2000.

Krinner, G., Viovy, N., de Noblet-Ducoudré, N., Ogée, J., Polcher, J., Friedlingstein, P., Ciais, P., Sitch, S., and Prentice, I. C.: A dynamic global vegetation model for studies of coupled atmosphere-biosphere system, Global Biogeochem. Cy., 19, GB1015, doi:10.1029/2003GB002199, 2005.

Lawrence, P. and Chase, T.: Climate impacts of making evapotranspiration in the Community Land Model (CLM3) consistent with the Simple Biosphere Model (SiB), J. Hydrometeorol., 10, 374–394, 2008.

Li, B., Peters-Lidard, C., Kumar, S., Rheingrover, S., and Anantharaj, V.: Free drainage or not: an evaluation of simulated soil moisture profiles by the Noah land surface model in the Mississippi region, 22nd Conference on Hydrology, 20–24 January 2008, New Orleans, USA, paper 134829, available at: https://ams.confex.com/ams/88Annual/

techprogram/paper_134829.htm (last access: 4 December 2015), 2008.

Limbrick, K., Whitehead, P., Butterfield, D., and Reynard, N.: Assessing the potential impacts of various climate change scenarios on the hydrological regime of the River Kennet at Theale, Berkshire, south-central England, UK: an application and evaluation of the new semi-distributed model, INCA, Sci. Total Environ., 251/252, 539–555, 2000.

Lohmann, D., Mitchell, K., Houser, P., Wood, E., Schaake, J., Robock, A., Osgrove, B., Sheffield, J., Duan, Q., Luo, L., Higgins, W., Pinker, R., and Tarpley, J.: Streamflow and water balance intercomparisons of four land surface models in the North American Land Data Assimilation System project, J. Geophys. Res., 109, D07S91, doi:10.1029/2003JD003823, 2004.

Mansour, M. and Hughes, A.: User's manual for the distributed recharge model ZOODRM, British Geological Survey Internal Report IR/04/150, British Geological Survey, Keyworth, 2004.

Mauser, W. and Schadlich, S.: Modeling the spatial distribution of evapotranspiration on different scales using remote sensing data, J. Hydrol., 212/213, 250–267, 1998.

Maxwell, R. and Miller, N.: Development of a coupled land surface and groundwater model, J. Hydrometeorol., 6, 233–247, 2005.

Moore, R. J.: The PDM rainfall-runoff model, Hydrol. Earth Syst. Sci., 11, 483–499, doi:10.5194/hess-11-483-2007, 2007.

Nativ, R. and Nissim, I.: Characterization of a desert aquitard – hydrologic and hydrochemical considerations, Groundwater, 30, 598–606, 1992.

Nativ, R., Adar, E., Dahan, O., and Geyh, M.: Water recharge and solute transport through the vadose zone of fractured Chalk under desert conditions, Water Resour. Res., 31, 253–261, 1995.

Oleson, K., Niu, G.-Y., Yang, Z.-L., Lawrence, D. M., Thornton, P. E., Lawrence, P. J., Stöckli, R., Dickinson, R. E., Bonan, G. B., Levis, S., Dai, A., and Qian, T.: Improvements to the community land model and their impact on the hydrological cycle, J. Geophys. Res.-Biogeo., 113, G01021, doi:10.1029/2007JG000562, 2008.

Parker, S.: Chalk Regional Groundwater Models and their Applicability to Site Scale Processes, PhD thesis, Imperial College London, 301 pp., available at: https://workspace.imperial.ac.uk/ewre/Public/Parker_PhD.pdf (last access: 1 March 2015), 2011.

Parker, S., Butler, A., and Jackson, C.: Seasonal and interannual behaviour of groundwater catchment boundaries in a Chalk aquifer, Hydrol. Process., doi:10.1002/hyp.10540, 2015.

Penman, H. L.: Natural evaporation from open water, bare soil and grass, P. Roy. Soc. Lond. A, 193, 120–145, 1948.

Peters, R. and Klavetter, E.: A continuum model for water movement in an unsaturated fractured rock mass, Water Resour. Res. 24, 416–430, 1988.

Petr, J., Lipavsky, J., and Hradecka, D.: Production process in old and modern spring barley varieties, Bodenkultur, 53, 19–27, 2002.

Pinault, J., Amraoui, N., and Golaz, C.: Groundwater-induced flooding in macropore-dominated hydrological system in the context of climate changes, Water Resour. Res., 41, W05001, doi:10.1029/2004WR003169, 2005.

Pitman, A.: The evolution of, and revolution in, land surface schemes designed for climate models, Int. J. Climatol., 23, 479–510, 2003.

Power, T. and Soley, R.: A comparison of chalk groundwater models in and around the River Test catchment, NC/03/05, Environment Agency of England and Wales Report, Environment Agency, Science Group, Olton, 2004.

Price, M., Downing, R., and Edmunds, W.: The chalk as an aquifer, in: The Hydrogeology of the Chalk of North-West Europe, edited by: Downing, R., Price, M., and Jones, G., Clarendon Press, Oxford, 1993.

Protopapas, A. and Bras, R.: The one-dimension approximation for infiltration in heterogeneous soils, Water Resour. Res., 27, 1019–1027, 1991.

Smith, R., Blyth, E., Finch, J., Goodchild, S., Hall, R., and Madry, S.: Soil state and surface hydrology diagnosis based on MOSES in the Met Office Nimrod nowcasting system, Meteorol. Appl., 13, 89–109, 2006.

Wagener, T., Sivapalan, M., Troch, P., and Woods, R.: Catchment classification and hydrologic similarity, Geography Compass, 1, 901–931, 2007.

West, G. and Dumbleton, M.: Some observations on swallow holes and mines in the chalk, Q. J. Eng. Geol., 5, 171–177, 1972.

Wood, E., Roundy, J., Troy, T., van Beek, L., Bierkens, M., Blyth, E., de Roo, A., Döll, P., Ek, M., Famiglietti, J., Gochis, D., van de Giesen, N., Houser, P., Jaffé, P., Kollet, S., Lehner, B., Lettenmaier, D., Peters-Lidard, C., Sivapalan, M., Sheffield, J., Wade, A., and Whitehead, P.: Hyperresolution global land surface modeling: meeting a grand challenge for monitoring Earth's terrestrial water, Water Resour. Res., 47, W05301, doi:10.1029/2010WR010090, 2011.

Yang, Z. and Niu, G.: The versatile integrator of surface and atmosphere processes – Part 1. Model description, Global Planet. Change, 38, 175–189, 2003.

Yilmaz, K., Gupta, H., and Wagener, T.: A process-based diagnostic approach to model evaluation: application to the NWS distributed hydrological model, Water Resour. Res., 44, W09417, doi:10.1029/2007WR006716, 2008.

13

Hydroclimatological influences on recently increased droughts in China's largest freshwater lake

Y. Liu and G. Wu

Nanjing Institute of Geography & Limnology, Chinese Academy of Sciences, 73 East Beijing Road, Nanjing 210008, China

Correspondence to: Y. Liu (ybliu@niglas.ac.cn; yb218@yahoo.com)

Abstract. Lake droughts are the consequence of climatic, hydrologic and anthropogenic influences. Quantification of droughts and estimation of the contributions from the individual factors are essential for understanding drought features and their causation structure. This is also important for policymakers to make effective adaption decisions, especially under changing climate. This study examines Poyang Lake, China's largest freshwater lake, which has been undergoing drastic hydrological alternation in the past decade. Standardized lake stage is used to identify and quantify the lake droughts, and hydroclimatic contributions are determined with a water budget analysis, in which absolute deficiency is defined in reference to normal hydrologic conditions. Our analyses demonstrate that in the past decade the lake droughts worsened in terms of duration, frequency, intensity and severity. Hydroclimatic contributions to each individual drought varied between droughts, and the overall contribution to the lake droughts in the past decade came from decreased inflow, increased outflow, and reduced precipitation and increased evapotranspiration in the lake region. The decreased inflow resulted mainly from reduced precipitation and less from increased evapotranspiration over the Poyang Lake basin. The increased outflow was attributable to the weakened blocking effects of the Yangtze River, which the Three Gorges Dam (TGD) established upstream. The TGD impoundments were not responsible for the increased number of drought events, but they may have intensified the droughts and changed the frequency of classified droughts. However, the TGD contribution is limited in comparison with hydroclimatic influences. Hence, the recently increased droughts were due to hydroclimatic effects, with a less important contribution from anthropogenic influences.

1 Introduction

A drought is a temporary lack of water caused by abnormal climatic or environmental influences, among other factors (Kallis, 2008; Mishra and Singh, 2010; and references therein). There are meteorological droughts (abnormal precipitation deficits), hydrological droughts (abnormal streamflow, groundwater, or lake deficits), agricultural droughts (abnormal soil moisture deficits), ecological droughts (abnormal water deficits causing stress on ecosystems) and socio-economic droughts (abnormal failures of water supply to meet economic and social demands) (Tallaksen and van Lanen, 2004; Kallis, 2008; Mishra and Singh, 2010). The drought phenomena may have different temporal features and causation structures (Kallis, 2008; Mishra and Singh, 2010). It is anticipated that droughts would likely increase owing to global climate change (Kallis, 2008; Mishra and Singh, 2010).

Hydrological droughts occur when land-water resources decrease significantly below normal conditions, represented by low water levels in streams, lakes, reservoirs and groundwater as well (Nalbantis and Tsakiris, 2009; Keskin and Sorman, 2010). Streamflow droughts may occur with basin-scale precipitation deficiency and/or excessive evapotranspiration (Zelenhasic and Salvai, 1987; Tallaksen et al., 1997; Kingston et al., 2013). In addition to local precipitation and evaporation, lake droughts involve other hydrological components, including inflows from streams surrounding the lake and outflows out of the lake. Hence, lake droughts can be more complicated than streamflow droughts in causation structure. Furthermore, both inflows and outflows may be affected by human activities, for example, groundwater pumping, reservoir construction or land cover change (Wilcox et al., 2010). Therefore, lake droughts are the consequence of combined climatic, hydrologic and anthropogenic influences. In contrast to floods that have received a great deal of attention in hydrology, droughts are not yet comprehensively understood (Kallis, 2008; Mishra and Singh, 2011). Quantification of lake droughts and clarification of contributions from individual factors are essential for understanding drought features and their causation structure. This is important for policymakers to make effective adaption decisions, especially under changing climate. Site-based drought analysis is a starting point towards integrated theories of drought (Kallis, 2008).

Poyang Lake is China's largest freshwater lake, which has been undergoing hydrological alterations in recent decades (Jiao, 2009; Finlayson et al., 2010; Hervé et al., 2011; Liu et al., 2013; Zhang et al., 2014). The lake is located at the south bank of the Yangtze River, which is a humid monsoon climatic region. Although the region historically experiences significant floods (Shankman and Liang, 2003; Shankman et al., 2006), severe lake droughts have occurred frequently in the past decade, resulting in tremendous hydrological, biological, ecological and economic consequences (Feng et al., 2012; Environment News Service, 2012; Wu and Liu, 2014). Because the lake is the primary part of the well-known Poyang Lake wetland and the lake region serves as an important food base for China, the frequently occurring lake droughts have also received increasing international attention (Jiao, 2009; Finlayson et al., 2010; Liu et al., 2011; Environment News Service, 2012; The Ramsar Convention, 2012; Zhang et al., 2012, 2014).

Lake droughts are usually defined as an abnormal decline in lake stage or lake size. A number of studies have documented this decline in Poyang Lake and its influencing factors (Guo et al., 2012; Zhang et al., 2012; Liu et al., 2013; Lai et al., 2014a; Zhang et al., 2014). Feng et al. (2012) used satellite images with a 250 m spatial resolution and reported that the lake size had a decreasing trend between 2000 and 2010. Liu et al. (2013) revealed an abrupt decrease in the lake size in 2006, mainly in October and November. Zhang et al. (2014) demonstrated that the lake stage fell to its lowest level during the 2000s compared to previous decades, in particular in the autumn recession periods. Since Poyang Lake receives inflows from its surrounding basin and discharges into the Yangtze River via a narrow outlet at the Hukou (Fig. 1), the strong lake–river interaction makes it complex to separate relative impacts of the inflow and outflow on the lake stage (Hu et al., 2007; Lai et al., 2014a). Zhang et al. (2014) employed a hydrodynamic model for Poyang Lake for the separation and declared that the lake decline in the 2000s was primarily ascribed to the weakened blocking effect of the Yangtze River. Compared to climate variability on the lake basin, modifications to the Yangtze River flows have had a much greater influence on the seasonal (September–October) dryness of the lake (Zhang et al., 2014). The modification was largely attributable to the operation of the Three Gorges

Dam (TGD), established upstream of the Yangtze River in 2003. Water impoundments of the TGD incurred water level drops with an average estimate of 2 m at the outlet of Poyang Lake in mid-September to November for the period 2003–2008 (Guo et al., 2012; Zhang et al., 2012). Alternatively, Lai et al. (2013) developed a hydrodynamic model for the middle Yangtze River region (CHAM-Yangtze), in which they coupled both Poyang Lake and the Yangtze River to account for the lake–river interactions explicitly. They demonstrated that the lake stage was more sensitive to the alternation in lake inflow compared to the same discharge modification in the Yangtze River (Lai et al., 2014a). The recent extremely low water levels in the Yangtze River resulted mainly from remarkable declines in inflows to the river, rather than solely from the TGD impoundments (Lai et al., 2014b). These studies highlighted the complexity of the multiple influences on Poyang Lake's decline in the complex basin–lake–river system.

Drought differs from low water level and persistent dryness. Water level can be low in seasonal dry seasons, but this does not necessarily constitute a drought (Smakhtin, 2001). Persistent dryness refers to water decrease in a long run, which is usually unrecoverable in the short term (Zhang et al., 2012). Droughts are complex events that have a recurrent feature, and may occur in any season and last several months or longer (Todd et al., 2013). Feng et al. (2012) quantified the drought severity of Poyang Lake in 2011 and showed that the drought was primarily due to low basin-scale precipitation, rather than discharge differences between the lake and the Yangtze River with TGD impoundments. Very recently, Wu and Liu (2014) used satellite-delineated inundation area to quantify two lake droughts in 2006 and 2011. The results indicated that the 2006 drought was mainly attributable to abnormal decrease of water flow in the Yangtze River and the 2011 drought was due to the combined influences of the Poyang Lake basin and the Yangtze River. Although these were two extreme drought events, it is not certain if they explain the more frequently occurring droughts in Poyang Lake as well.

In principle, drought identification, quantification or characterization with a consistent standard is a prerequisite for drought analysis. However, few studies have comprehensively quantified and addressed the Poyang Lake droughts in the 2000s. The current understanding of the Lake's decline in autumn cannot provide a complete explanation to the lake droughts spanning non-autumn seasons. It remains unknown to what extent the climatic, hydrologic and anthropogenic influences have contributed to the lake droughts, which is one of the key issues for developing integrated, interdisciplinary theories on droughts (Kallis, 2008). Especially for practice, clarification of the multiple influences on the recently increased droughts is essential for the effective prevention of droughts.

The complicated causality of lake droughts requires a robust approach for determining the contributions from multiple influences. Analogous to standardized precipitation index, this study utilizes standardized lake index to quantify lake droughts. With the principle of lake water balance, it proposes to define an absolute deficiency for each water component and determine their relative contributions to lake droughts. The approach is applicable to basin-scale water balance, quantifying regional hydroclimatic influences on lake inflow, and subsequently on lake droughts (Sect. 2). Poyang Lake droughts are examined with the proposed approach, in combination with 5-decade hydroclimatic data including the latest satellite products (Sect. 3). The drought features in the 2000s and their causes are subsequently addressed (Sect. 4). Our findings should be valuable for improving our understanding of lake droughts under changing climate conditions and be useful for local water resources management and climate change adaptation.

2 Methodology

The main properties of a drought are time of initiation and termination, duration, severity, magnitude and intensity, as well as spatial extent in the case of meteorological or agricultural droughts (Yevjevich, 1967; Dracup et al., 1980; Wilhite and Glantz, 1985; McKee et al., 1993; Mishra and Singh, 2010; Spinoni et al., 2014). Drought initiation time is the beginning of the drought. Termination time is the end, i.e., when the drought ceases. Drought duration is the period between the initiation and the termination (Yevjevich, 1967; Mishra and Singh, 2010). Drought severity is the total, cumulative water deficiency for the duration of the drought. Drought magnitude is a derivative of drought severity, defined as the average water deficit in the drought period (Dracup et al., 1980; Wilhite and Glantz, 1985). Drought intensity usually refers to the largest departure from the normal conditions (McKee et al., 1993; Spinoni et al., 2014). For a given historical period, another important drought statistic is drought frequency, which refers to the number of drought events that have occurred (Mishra and Singh, 2010; Spinoni et al., 2014).

2.1 Quantification of lake droughts

Various indices have been proposed to characterize and quantify the complex features of droughts (Dracup et al., 1980; Keyantash and Dracup, 2002; Mishra and Singh, 2010, 2011). Among these, the standardized precipitation index (SPI) is most commonly used (McKee et al., 1993; Mishra and Singh, 2010). It is a normalized dimensionless index, defined as the difference of precipitation from the mean divided by the standard deviation for a given period, in which a gamma distribution is generally fitted to the long-term precipitation records for each calendar month to account for seasonal differences (McKee et al., 1993). The SPI is simple but capable of quantifying drought features, and has been re-

cently recommended by the World Meteorological Organization (WMO) to characterize meteorological droughts (Hayes et al., 2011). Nevertheless, it was proposed to quantify precipitation deficiency, the SPI methodology has been applied in a similar manner to other hydroclimatic variables, for example, streamflow discharge, soil moisture, reservoir storage and groundwater level (McKee et al., 1993; Sheffield et al., 2004; Vicente-Serrano and López-Moreno, 2005; Mendicino et al., 2008; Shukla and Wood, 2008).

In the case of lake drought, it can be described with lake stage, lake area or water storage. Of these variables, lake stage is usually continuously measured and is suitable for drought analysis. In comparison to SPI, the standardized lake index (SLI) is described as follows

$$\text{SLI}_{ij} = \frac{L_{ij} - \overline{L}_j}{\sigma_j}, \tag{1}$$

where L_{ij} is the monthly average lake stage (unit in meters) of year i and month j ($j = 1, 2, ..., 12$), which is transformed from gamma distribution into the normal distribution (McKee et al., 1993). $\overline{L}_j$ is the multi-year mean of monthly average stage for month j, and σ_j is the standard deviation of monthly average stage for month j. Since SLI uses $\overline{L}_j$ and σ_j, both of which are monthly dependent, it removes seasonal differences in the lake stage.

A drought event is discernible with SLI. While a negative SLI indicates the lake stage is lower than the normal, not all the negatives can be classified into a drought event. Only when SLI deviates away from the normal by more than 1 standard deviation (SLI < −1), an event can be established (McKee et al., 1995). Furthermore, a drought initializes when SLI becomes negative and terminates before SLI becomes positive in SLI time series (McKee et al., 1995). The initialization and the termination time yield drought duration (unit in day, month or year). Once all the drought events are identified, drought frequency can be determined for a given period.

In accordance with the definition of SPI by McKee et al. (1993), SLI represents a departure of lake stage from its normal conditions. The departure corresponds to a probability of drought intensity, useful for drought risk analysis, namely, SLI = −1 denotes an occurrence probability of 15.9 % (Lloyd-Hughes and Saunders, 2002). Positive (negative) value indicates lake stage higher (lower) than the normal condition for the period. For an individual drought event, its lowest SLI value indicates the intensity of the event (McKee et al., 1993; Spinoni et al., 2014). Accordingly, a drought event can be classified into four categories with its lowest SLI: extreme drought (−∞, −2.0], severe drought (−2.0, −1.5], moderate drought (−1.5, −1.0] and mild drought (−1.0, 0.0) (Dracup et al., 1980; McKee et al., 1993).

In addition, drought severity may be calculated as follows

$$\text{severity} = \sum_{k=m}^{k=n} \text{SLI}_k, \tag{2}$$

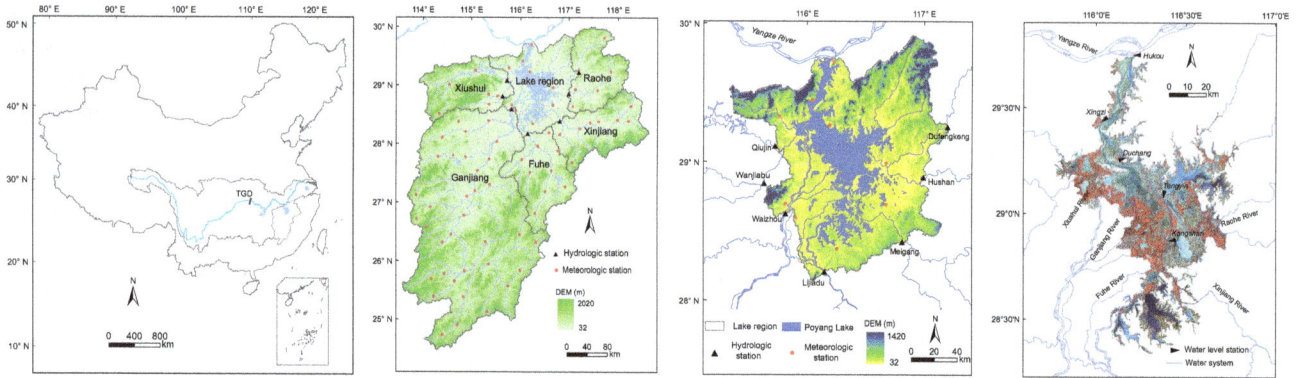

Figure 1. Geographic location of Poyang Lake, China. The lake is principally fed by five river systems of the Poyang Lake basin. Lake water flows into the Yangtze River via a single outlet at the Hukou. Jiujiang is located 25 km upstream of the Hukou on the Yangtze River. The Three Gorges Dam (TGD) is upstream of the river.

where m denotes the initialization time of a drought and n represents the termination time (Keyantash and Dracup, 2002; Mishra and Singh, 2010). Drought magnitude is then calculated as (Keyantash and Dracup, 2002)

$$\text{Magnitude} = \text{Severity}/\text{duration.} \qquad (3)$$

2.2 Contribution of water deficiency to lake droughts

A lake drought results directly from an abnormal change in the lake water budget. A general water balance for lake in a period can be described as follows

$$\Delta_L = P_L - E_L + G_L + I_L - O_L, \qquad (4)$$

where Δ_L is the lake water budget or change in lake storage (millimeters or cubic meters, hereafter mm or m^3), P_L is precipitation (mm or m^3), E_L is lake evaporation (mm or m^3), G_L is groundwater net inflow to the lake (mm or m^3), I_L is inflow (mm or m^3) and O_L is outflow (mm or m^3) from the lake.

Once the water budget appears abnormal, it suffers from the anomalies of some or all the water components, namely, low precipitation, high evapotranspiration, low inflow and/or high outflow. At the monthly scale, for a water component X, being Δ_L, P_L, E_L, G_L, I_L or O_L, its anomaly is described as follows

$$\widetilde{X}_{ij} = X_{ij} - \overline{X}_j, \qquad (5)$$

where $\widetilde{X}_{ij}$ denotes the anomaly of the water component (mm or m^3) for year i and month j. $\overline{X}_j$ is the multi-year mean of X_{ij} in month. Notably, Eq. (5) defines an absolute water deficiency from its normal amount, different from Eq. (1) that defines a relative deficiency for drought identification. The equation offers a baseline to quantify contributions from individual water components to a drought.

During a drought event, the anomaly of lake water storage in month $l(m < l \leq n)$, $\widetilde{S}_l$, results from the consecutive

anomalies of the lake water budget (Seneviratne et al., 2012; Teuling et al., 2013), which can be described as

$$\widetilde{S}_l = \sum_{k=m}^{k=l} \widetilde{\Delta}_k. \qquad (6)$$

The contribution from an individual water component to the water deficiency of the lake water storage is quantifiable with a ratio defined as follows

$$C_{X_l} = \frac{\sum\limits_{k=m}^{k=l} \widetilde{X}_k}{\sum\limits_{k=m}^{k=l} \widetilde{\Delta}_k}, \qquad (7)$$

where C_{X_l} denotes the contribution, the numerator is the sum of the monthly anomalies of the water component from month m to l. $\sum\limits_{k=m}^{k=l} \widetilde{\Delta}_k$ is generally negative, but $\sum\limits_{k=m}^{k=l} \widetilde{X}_k$ may vary with hydroclimatic conditions. For example, precipitation deficiency leads to a negative $\sum\limits_{k=m}^{k=l} \widetilde{X}_k$ value and produces a positive C_{X_l}. Low evapotranspiration may lessen water deficiency and generate a negative C_{X_l}. Therefore, C_{X_l} can be either positive or negative.

Notably, Eq. (7) requires all the involved water components to be independent from each other, which is the general case for P_L, E_L, G_L and I_L, but not for O_L. Given O_L is largely dependent on the sum of $(P_L - E_L + G_L + I_L)$, its net contribution to the lake water budget can be described with $\Delta_{Lk} = (P_{Lk} - E_{Lk} + G_{Lk} + I_{Lk}) - O_{Lk}$. The anomaly of the net contribution is $\widetilde{\Delta}_k = (\widetilde{P}_k - \widetilde{E}_k + \widetilde{G}_k + \widetilde{I}_k) - \widetilde{O}_k$, and it is used to replace $\widetilde{O}_k = O_k - \overline{O}_k$ in Eq. (7) for the quantification of the relative contribution of the outflow.

Equation (7) is useful to quantify hydroclimatological influences on drought, and applicable to any single month in a drought period. However, it may not be meaningful for an entire drought, because the storage anomaly will return back to zero at the end of the drought. In contrast, when a drought event reaches its highest storage deficit, it has the highest intensity, the main criterion for drought classification. Therefore, quantification of the hydroclimatological contribution for the month with the highest intensity (lowest SLI) is fundamental to clarify drought causes.

2.3 Contribution of basin-scale hydroclimatic influences on lake droughts

In addition to the quantification of water deficiency in inputs and outputs to the lake, it is also important to understand the causes of inflow deficiency for a complete understanding of hydroclimatic influences on lake droughts. Lake inflow originates from precipitation on its contributing basin. Given the water balance for the basin in a period, lake inflow is described as

$$I_L = P_B - E_B + \Delta_B, \tag{8}$$

where P_B is the precipitation (mm or m^3), E_B is the evapotranspiration (mm or m^3) and Δ_B is the change of water storage (mm or m^3), including soil moisture and groundwater in the basin.

In practice, there are often areas ungauged downstream from hydrological stations. In this case, the lake inflow includes two parts, one from gauged areas and another from ungauged areas ($I_L = I_U + I_G$):

$$I_U = P_U - E_U + \Delta_U, \tag{9a}$$

$$I_G = P_G - E_G + \Delta_G, \tag{9b}$$

where the subscript G represents the components for the gauged areas and the subscript U for the ungauged areas.

In combination with Eqs. (4) and (9a), the lake water budget can be expressed as

$$\Delta_L = \underbrace{P_L - E_L + G_L}_{\text{lake}} + \underbrace{P_U - E_U + \Delta_U}_{\text{ungauged\_area}} + I_G - O_L. \tag{10a}$$

Or in parallel,

$$\Delta_L = \underbrace{P_R - E_R + \Delta_R}_{\text{lake\_region}} + I_G - O_L, \tag{10b}$$

where the subscript R represents the components for the lake region. It shows that the lake change involves water budgets in the lake and the ungauged areas, in addition to gauged inflow and outflow.

Further incorporated with Eq. (9b), the lake water budget can be expressed as

$$\Delta_L = \underbrace{P_L - E_L + G_L}_{\text{lake}} + \underbrace{P_U - E_U + \Delta_U}_{\text{ungauged\_area}}$$
$$+ \underbrace{P_G - E_G + \Delta_G}_{\text{gauged\_area}} - O_L. \tag{11a}$$

Or in parallel,

$$\Delta_L = \underbrace{P_R - E_R + \Delta_R}_{\text{lake\_region}} + \underbrace{P_G - E_G + \Delta_G}_{\text{gauged\_area}} - O_L. \tag{11b}$$

The water anomaly of each component in Eqs. (10) and (11) can be defined with Eq. (5). Their contributions to the abnormal change of the lake storage (Eq. 6) can be determined for water balance at different spatial scales, namely, the lake, the lake region and the lake basin.

3 Study area and data processing

3.1 Study area and data

Poyang Lake is located at the northern part of the Poyang Lake basin, a sub-basin of the Yangtze River basin of China (Fig. 1a). The lake has a maximum area of 3860 km^2 with an average depth of 8 m at the lake stage of 22 m (Shankman et al., 2003). It varies remarkably from several thousand km^2 in summer to less than one thousand km^2 in winter (Liu et al., 2013). There are five stations (Kangshan, Tangyin, Duchang, Xingzi and Hukou) to measure lake stage across the lake from the south to the north (Fig. 1d). Lake water flows out into the Yangtze River via the Hukou outlet. The lake water principally comes from five major river systems including Xiushui, Ganjiang, Fuhe, Raohe and Xinjiang. Seven hydrological control stations (Qiujin, Wanjiabu, Waizhou, Lijiadu, Meigang, Dufengkeng and Shizhenjie) are located downstream to measure the discharge of the five rivers (Fig. 1b). The lake region (Fig. 1c) downstream from the stations is ungauged, with an area of 23 089 km^2, approximately 6 times the maximum lake size. The Poyang Lake basin has an area of 162 225 km^2 and belongs to a humid subtropical climate zone with an annual mean surface air temperature of 17.5° and a mean annual precipitation of 1640 mm for the years 1960–2010 (Liu et al., 2012). Forestlands, agricultural fields, grasslands, bare land and water surfaces are the dominant land cover types (Liu et al., 2012).

Daily lake stage data from five hydrological stations and daily discharge data from seven control stations were obtained from the Hydrological Bureau of Poyang Lake. Lake stage data from Xingzi and Hukou were available for the period 1961–2010, but the data from the other three stations were available only until 2008. Daily discharge data for the Hukou outlet are available from the Hydrological Bureau of the Yangtze River Water Resources Commission.

Daily precipitation data from 73 national weather stations within the Poyang Lake basin are available from the National Meteorological Information Center of China for the period 1961–2010. Regional evapotranspiration estimates were extracted from the latest satellite products (MOD16) of the Moderate resolution Imaging Spectroradiometer (MODIS) (http://www.ntsg.umt.edu/project/mod16) (Mu et al., 2011) for the lake region and the gauged basin in 2000–2010. In addition, the lake stage at Hukou is available for the case without the TGD for the period 2006–2010, which is the output of the CHAM-Yangtze model (Lai et al., 2013).

3.2 Drought quantification

To identify lake droughts, SLI was calculated with Eq. (1) from monthly lake stage. Since it is approximately 110 km from the north to the south of Poyang Lake, a representative gauge was sought. Among five stations to measure the lake stage, the SLI values of Xingzi station had the highest correlation with that calculated from averaged lake stage of the five stations using all the available data for the period 1960–2008 ($y = 0.9953x$, $R^2 = 0.9901$, $p < 0.0001$). Thus, the Xingzi station was selected for drought quantification.

Prior to drought quantification with SLI, the monthly lake stage was evaluated for its fit with the gamma distribution in each calendar month (McKee et al., 1993). The statistical evaluation demonstrated the goodness-of-fit at a significant level of 1 % with the Kolmogorov–Smirnov test (Lloyd-Hughes and Saunders, 2002) for all 12 months (Fig. 2). Subsequently, drought initialization, termination, duration, intensity, severity and frequency were determined from the SLI values of Xingzi, with the criteria described in Sect. 2.1. Finally, all the lake droughts were identified and classified into extreme, severe or moderate drought (McKee et al., 1993).

In addition, for the case without the TGD impoundments, the lake stage at Xingzi was estimated from its highly correlated relationship with Hukou, $y = 0.9594x + 0.8034$ ($R^2 = 0.9949$) (Min and Zhan, 2013), for the period 2006–2010. Consequently, the SLI values of Xingzi were recalculated for the case without TGD. This serves as a reference to evaluate the TGD effect on the lake droughts.

3.3 Water budget analysis

To quantify hydroclimatic influences on lake droughts, the water budget analysis was designed at multi-spatial scales: the lake, lake region and gauged basin (Fig. 1). At the lake scale, water components include precipitation, evaporation, groundwater net inflow, inflows from gauged and ungauged areas, and outflow (Eq. 4). It has been difficult to perform water balance analysis with a high accuracy for the lake. First, evaporation data are unavailable for the lake in monthly time series. Second, the lake inundation area shows remarkable variation, which significantly regulates wetland evapotranspiration (Zhao and Liu, 2014). Third, there are many small

rivers and brooks downstream from the hydrological control stations. It is impractical to measure all of the surface runoff into Poyang Lake. Given the hydrological data, the Poyang Lake region is thus the minimum closure entity directly available for the water budget analysis. Furthermore, for complete understanding of climatic, hydrologic and anthropogenic influences on lake droughts, water budget analysis should be performed for the lake basin, with a focus on the causes of inflow deficiency. Besides, the boundary effect of the Yangtze River is taken into consideration to account for the anomaly of lake outflow.

Specification of normal hydrologic conditions is a prerequisite for determining water deficiency. First, precipitation data were grouped for the Poyang lake region and the gauged basin. Multi-year means of monthly precipitation were obtained from the observation data for the period 1961–2010. Second, multi-year means of monthly discharge were calculated from the data for inflows and outflow in the period. Third, since evapotranspiration data prior to 2000 were unavailable, the annual evapotranspiration was calculated from the difference between annual precipitation and discharge, respectively, for the lake region and the gauged basin in 1961–2010. The multi-year mean of monthly evapotranspiration was then obtained by distributing the annual value with a monthly weighting factor calculated from the MOD16 time series, with an assumption that the seasonal variation is relatively similar for the period 1961–2010.

Once the normal hydrologic condition was established, the water deficiency of a water component (Eq. 5) and its contribution to the lake drought were determined (Eq. 7). It was applied to the water budget for the lake region (Eq. 10b) and the gauged basin (Eq. 11b), respectively. For the basin, a 1-month lag was determined with correlation analysis between peak rainfall and peak discharge, and it was applied to account for the peak difference (Senay et al., 2011; Liu et al., 2013). In addition, there are three points addressed here. First, Δ_R in Eq. (10b) for the lake region is only 1.3 % of the total water balance and is neglected in the present study (Wan and Xu, 2010; Zhang et al., 2014). Second, Δ_G in Eq. (9b) for the gauged basin is generally unavailable. According to Feng and Liu (2014), it is roughly 5 % of the total water balance and is neglected here. Third, while the MOD16 data sets have been extensively evaluated and applied worldwide, e.g., in Australia, Brazil, Asia and the United States (Loarie et al., 2011; Kim et al., 2012; Velpuri et al., 2013; Wang et al., 2014), our recent assessment showed that it had an error of approximately 10 % for the study area (Wu et al., 2013). This error, together with the neglected Δ_R and Δ_G, may introduce uncertainties. Given that the water anomaly (Eq. 5) and relative contribution (Eq. 7) are used, the uncertainties would not be enlarged but minimized in water balance calculation and water budget analysis.

It should be emphasized that, for the sake of addressing water contribution at the lake region and the gauged basin, the water amounts (unit in m^3) of all the water components

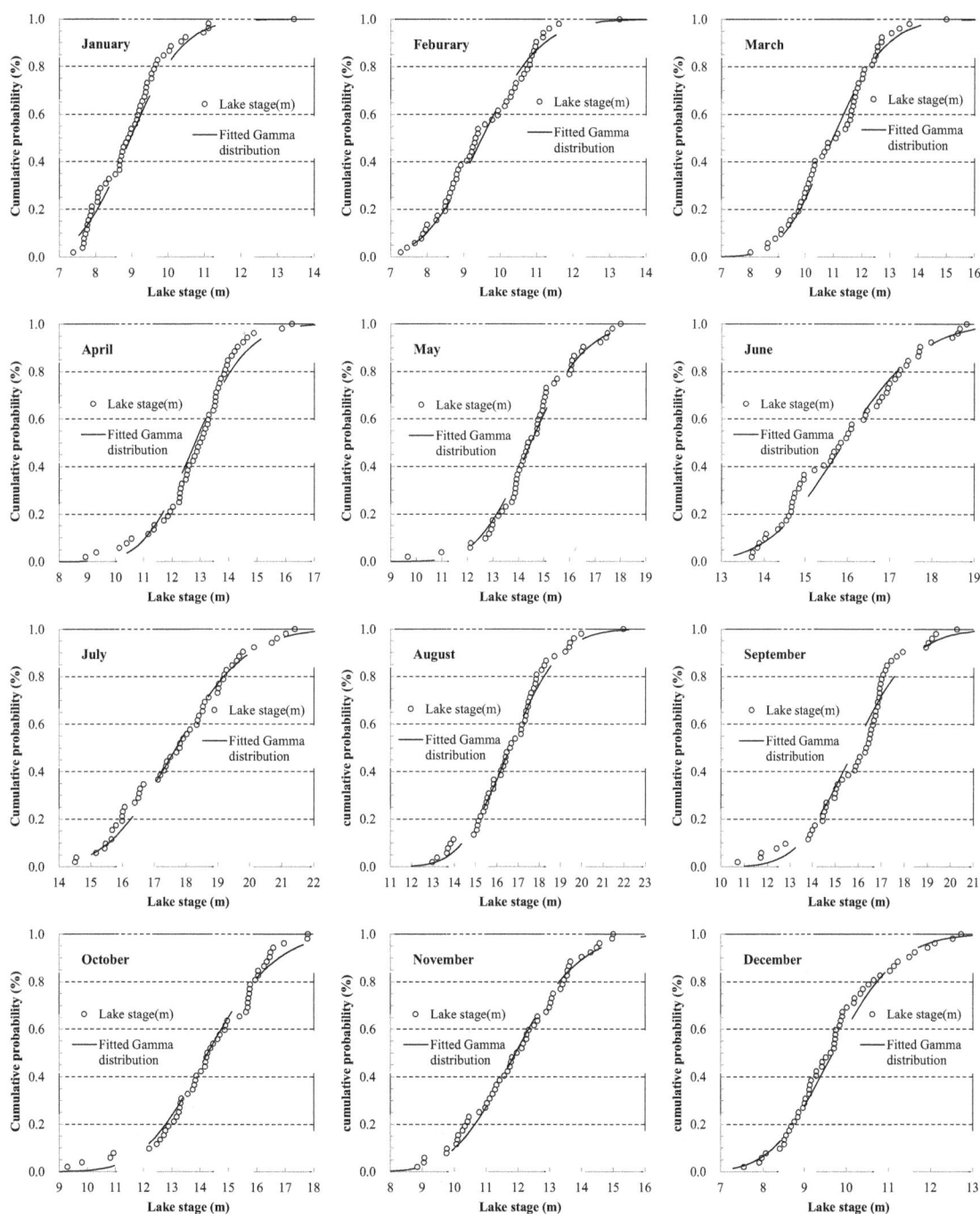

Figure 2. Statistical distribution of monthly average lake stages with a fitted gamma distribution for each calendar month at Xingzi of Poyang Lake in 1961–2010.

were normalized to equivalent water height (unit in mm) of the whole basin (unit in km²). Besides, statistical approaches were adopted in the present analysis (Lomax, 2001), in which paired F test (and t test) were used to examine the variance (and mean) difference between the statistics for the period 1961–2000 and that for the period 2001–2010.

4 Results and discussion

4.1 Poyang Lake droughts in the past decade

Figure 3a illustrates the SLI variation for Poyang Lake in the past decade. Negative values prevail over positive values, indicating the dry phase dominates the lake for the period. Three extreme, two severe and four moderate droughts, ac-

(a)

(b)

(c)

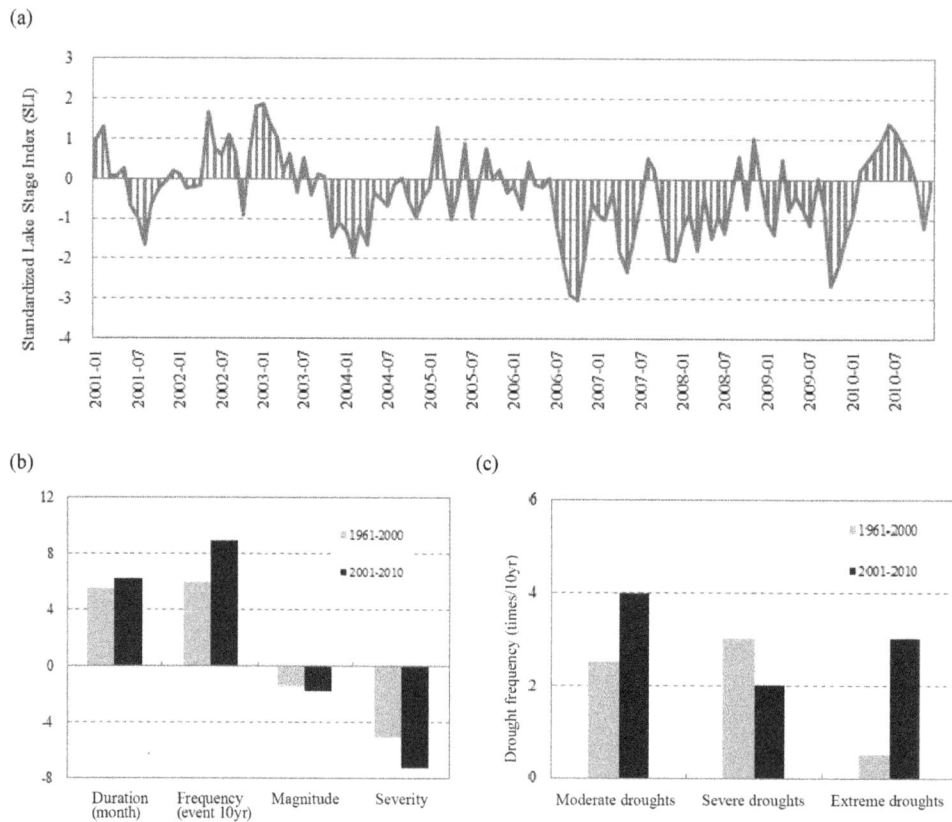

Figure 3. Poyang Lake droughts in 2001–2010. (a) Variation in the standardized lake stage index (SLI). (b) Drought duration, frequency, intensity and severity, and (c) Drought frequency for moderate, severe and extreme droughts, compared to 1961–2000.

Table 1. Drought events occurring during the 2001–2010 time period.

No.	Period	Duration (month)	Intensity (probability)	Severity	Drought classification
1	Jun 2001–Nov 2001	6	−1.64 (0.0505)	−4.33	severe
2	Nov 2003– Aug 2004	10	−1.96 (0.0250)	−10.23	severe
3	Apr 2005–May 2005	2	−1.03 (0.1515)	−1.39	moderate
4	Jul 2006–Jul 2007	13	−3.03 (0.0012)	−20.22	extreme
5	Oct 2007–Aug 2008	11	−2.01 (0.0222)	−13.21	extreme
6	Jan 2009–Feb 2009	2	−1.39 (0.0823)	−2.47	moderate
7	Apr 2009–Jul 2009	4	−1.14 (0.1271)	−3.11	moderate
8	Sep 2009–Jan 2010	5	−2.66 (0.0039)	−7.90	extreme
9	Oct 2010–Dec 2010	3	−1.21 (0.1131)	−1.56	moderate
Totals		7.3 ± 3.8	-1.79 ± 0.70	-7.16 ± 6.41	

cording to the drought classification criteria (McKee et al., 1993), occurred in this period. Among the nine cases, three droughts started in spring, two in summer and four in autumn (Table 1). Drought duration varied from 2 to 13 months with a mean of 7.3 months and a standard deviation (SD) of 3.8 months, which demonstrated that the lake droughts could occur in any month. Drought intensity ranged from −1.03 to −3.03 with a mean of -1.79 ± 0.70. The top three lowest SLI values were −3.03, −2.66 and −2.01, corresponding to

probabilities of 0.12, 0.39 and 2.22 %, respectively, for each occurrence. Drought severity varied from −1.39 to −20.22 with a mean of -7.16 ± 6.41. More specifically, in the category of *extreme drought*, the drought event that ranked first in both intensity and severity occurred from July 2006 to July 2007, lasting 13 months. The 2006 drought was addressed in Feng et al. (2012) and Wu and Liu (2014) in terms of inundated area, whereas the present study quantified its probability of occurrence and revealed that the drought lasted longer than the previous reports. The second most severe drought event emerged in September 2009–January 2010, persisting 5 months. The third most severe drought took place from October 2007 to August 2008, lasting 11 months. The two droughts in the *severe drought* category spanned 6 and 10 months, respectively. The four droughts in the *moderate drought* category lasted 2–4 months. It appears that a drought with a lower SLI is usually more severe and lasts for a longer time.

In comparison to the years 1961–2000, the lake droughts changed in terms of duration, frequency, intensity and severity in the past decade (Fig. 3b). On average, drought duration extended from 5.6 to 6.2 months. Drought frequency increased from 6.0 to 9.0 events per decade. Drought intensity intensified from -1.38 to -1.79 and drought severity increased from -5.02 to -7.16. With regard to the intensification, further analysis revealed that the moderate drought events increased, severe droughts decreased, but the extreme droughts increased from 0.5 to 3.0 events per decade (Fig. 3c). Overall, the lake droughts have worsened in terms of duration, frequency, intensity and severity over the last decade.

4.2 Hydroclimatic change at Poyang Lake region

Normal variation of water components is a baseline for the quantitative analysis of drought occurrence as an abnormal change. Figure 4a shows the multi-year mean of monthly precipitation (P_R) and evapotranspiration (E_R) for the lake region, lake inflow (I) from five major rivers and outflow (O) into the Yangtze River. The monthly precipitation varied with a peak in June followed by a sharp decrease. Inflow had a similar seasonal pattern. Outflow had the maximum value in June and the minimum in January. Maximum evapotranspiration appeared in August and the minimum in December. From a perspective of water balance, the water budget was positive from January to June with a peak in June. It became negative from July to December, and the minimum value appeared in October. These results indicate a shift in water budget from a surplus phase in the first half of the year to a deficit phase in the second half of the year. The deficit phase is a part of the normal hydrologic condition, and thus it does not necessarily mean a drought occurrence. Furthermore, in the annual water budget, the equivalent water supply from local precipitation was 312.0 mm, and that from inflow was 714.4 mm, more than 2 times that of the local precipitation. The water loss from local evapotranspiration was 118.3 mm, and that from outflow was 908.1 mm, approximately 7.7 times that of the local evapotranspiration. The fact that the inflow and outflow were much higher than local precipitation and evapotranspiration implies the dominant role of hydrologic components over meteorological components in regulating Poyang Lake within the lake region.

Lake droughts occur when abnormal change appears in the water budget. Table 2 lists the water components for the lake

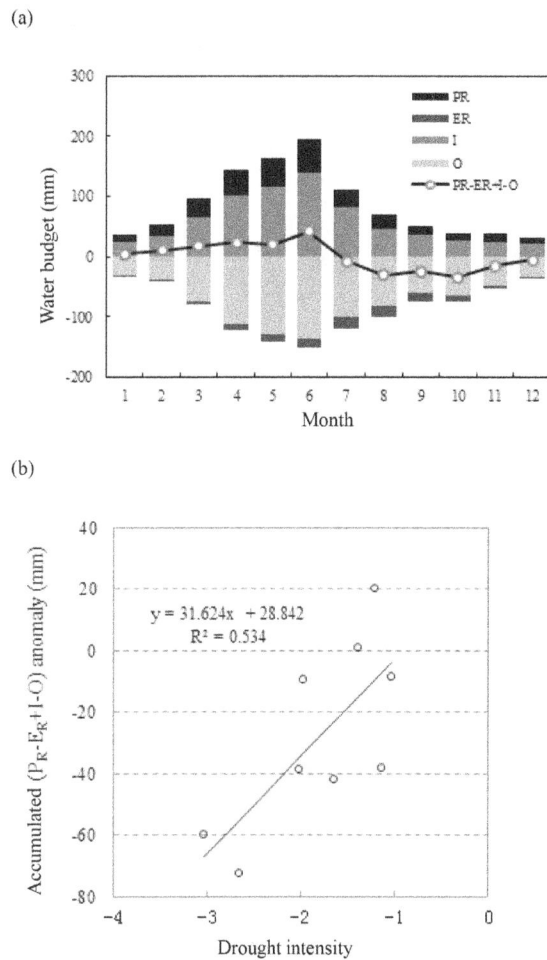

Figure 4. (a) Multi-year mean of monthly precipitation (P_R) and evapotranspiration (E_R) for the Poyang Lake region, lake inflow (I) from five major rivers of the Poyang Lake basin, and outflow (O) into the Yangtze River. All the water amounts are normalized to equivalent water height of the whole Poyang Lake basin. (b) The relationship between drought intensity and accumulated anomaly of ($P_R - E_R + I$) of each event for nine cases of Poyang Lake droughts.

region during the periods of lake droughts. The water budgets ($P_R - E_R + I - O$) were -67.1, -12.6 and -69.3 mm for three extreme lake droughts, -71.6 and 27.7 mm for two severe lake droughts and 72.0, 5.4, 18.5 and -66.1 mm for four moderate lake droughts. In sum, the budget was negative (deficit) for five cases and positive (surplus) for four cases. Despite the positive water budgets, the large negative anomalies of $P_R - E_R + I$ for the cases illustrated that the lake water income ($P_R - E_R + I$) was exceptionally lower than normal. The low water income resulted from largely decreased inflow and precipitation, as well as increased evapotranspiration. The positive water budgets were attributed to the water surplus period in the first half of the year. In this sense, the definition of drought is a water anomaly referenced to a normal state of either water surplus or deficit phase. It indicated

Table 2. Water components (unit in mm) of the Poyang Lake region and of the gauged basin for the drought periods. The values in parenthesis are the anomaly of a component against the multi-year mean in 1961–2010. All the water amounts are normalized to equivalent water height of the whole Poyang Lake basin.

	Lake region						Gauged basin			
	P_R	E_R	I	O	$P_R - E_R + I$	$P_R - E_R + I - O$	P_G	E_G	$P_G - E_G$	$P_G - E_G - O$
Jun 2001–Nov 2001	133.6	106.6	333.9	432.5	360.9	−71.6	755.9	575.3	180.6	−251.9
	(−15.9)	(24.8)	(−19.6)	(−58.5)	(−60.2)	(−1.7)	(−23.3)	(55.1)	(−78.4)	(−19.9)
Nov 2003–Aug 2004	246.0	122.1	367.0	463.2	490.9	27.7	1282.4	666.7	615.7	152.5
	(−39.9)	(28.1)	(−284.2)	(−320.5)	(−352.2)	(−31.7)	(−210.3)	(85.1)	(−295.4)	(24.7)
Apr 2005–May 2005	81.2	31.2	225.2	203.3	275.2	72.0	489.5	149.9	339.6	136.3
	(−8.3)	(9.1)	(6.9)	(−39.1)	(−10.4)	(28.7)	(19.3)	(14.6)	(4.7)	(43.8)
Jul 2006–Jul 2007	241.9	182.0	584.3	711.3	644.2	−67.1	1469.4	1010.3	459.1	−252.2
	(−100.9)	(44.8)	(−210.7)	(−296.3)	(−356.4)	(−60.1)	(−321.2)	(152.1)	(−473.3)	(−177.0)
Oct 2007–Aug 2008	258.6	134.3	536.3	673.1	660.6	−12.6	1312.0	717.1	594.9	−78.2
	(−39.0)	(31.6)	(−141.9)	(−175.1)	(−212.5)	(−37.4)	(−248.0)	(72.5)	(−320.5)	(−146.6)
Jan 2009–Feb 2009	24.1	7.7	26.3	37.3	42.7	5.4	101.4	42.8	58.6	21.3
	(−8.2)	(1.2)	(−30.1)	(−30.1)	(−39.5)	(−9.4)	(−74.7)	(4.9)	(−79.6)	(−49.5)
Apr 2009–Jul 2009	129.9	72.5	268.1	307.0	325.5	18.5	698.8	401.2	297.6	−9.4
	(−44.9)	(16.3)	(−169.9)	(−171.8)	(−231.1)	(−59.3)	(−194.4)	(51.8)	(−246.1)	(−74.3)
Sep 2009–Jan 2010	49.9	45.1	93.4	167.5	98.2	−69.3	328.2	256.5	71.7	−95.8
	(−12.0)	(11.2)	(−39.6)	(−68.6)	(−62.8)	(5.8)	(−11.8)	(35.4)	(−47.2)	(21.4)
Oct 2010–Dec 2010	46.0	20.9	79.7	170.9	104.8	−66.1	228.9	123.1	105.8	−65.1
	(12.0)	(15.5)	(6.0)	(24.2)	(12.6)	(−11.6)	(48.6)	(13.4)	(35.1)	(10.9)
Totals	1211.1	722.3	2514.2	3166.1	3003.0	−163.1	6666.4	3942.9	2723.5	−442.5
	(−257.1)	(172.5)	(−883.1)	(−1135.9)	(−1312.5)	(−176.7)	(−1016.0)	(484.8)	(−1500.8)	(−366.5)

that a drought occurrence was more closely related to the water deficiency (negative anomaly) of water budget than the net water budget. For example, the net water budget did not show statistically significant relationships with drought intensity. On the contrary, the total water anomaly of a drought event showed a significant relationship with drought intensity (x) ($y = 31.624x + 28.842$, $n = 9$, $R^2 = 0.534$, $p < 0.05$) (Fig. 4b). In general, the water budget analysis highlighted the importance of water deficiency in reference to a normal condition of either water surplus or deficit.

Drought causes can be traced from the relative contribution of individual water components. Table 3 shows the ratios of the total water anomaly of a component to that of the water budget up the time of peak drought (maximum intensity) for each event (Eq. 7). In the lake region, the ratio for inflow is largest for most cases, followed by O, P_R and E_R, indicating the dominant role of inflow in drought formation. Meanwhile, hydroclimatological contributions to each individual drought varied greatly between droughts. For example, O was larger than P_R for June 2001–November 2001, and lower for September 2009–January 2010. In addition to the positive contribution, a water component may contribute negatively. For example, one negative value appeared for inflow (Table 3), which was attributable to the normal inputs ($I + P_R - E_R$) accompanied by excessive O for July 2006–October 2006. Since inflow reduction is the major contribution to drought formation in the lake region, it is vital to trace how precipitation and evapotranspiration have changed at the basin scale.

4.3   Hydroclimatic change at Poyang Lake basin

Likewise, prior to performing a water budget analysis, it is necessary to clarify the normal hydrologic condition. Generally, precipitation (P_G) and evapotranspiration (E_G) had similar seasonal patterns in the gauged basin as its counterpart in the lake region (Fig. 5a). Monthly precipitation varied seasonably with a peak in June, followed by peaks in May and April. Major precipitation appeared in the first half of the year. Monthly evapotranspiration was generally less than precipitation and its top three highest values appeared from June–August. Monthly outflow was approximately half of the precipitation with a similar seasonal pattern. Consequently, the monthly water budget was positive (surplus) from December to June and negative (deficit) from July to November. The highest water surpluses appeared in March, April and May, and the lowest water deficits in July, August and September. On an annual scale, runoff was approximately 55 % of precipitation, 10 % higher than evapotranspiration, which is one of the climate features of this humid subtropical region.

The water budget was −252.2, −79.4 and −95.8 mm for the three extreme droughts, −251.9 and 152.0 mm for the two severe droughts, and 136.3, 21.3, −9.5 and −65.1 mm for the moderate droughts (Table 2). For six negative cases, the water budget featured less precipitation (negative anomaly) and more evapotranspiration (positive anomaly). For three positive cases, $P_G - E_G$ had large negative anomalies over 100 mm, but the water budgets became positive due to the largely reduced outflow. For the three extreme droughts, $P_G - E_G$ was much lower than the normal, sug-

Table 3. Contribution of each water component (unit in 100 %) to the total water anomaly up the time of peak drought (highest drought intensity) for the lake region and gauged basin.

Lake droughts	Lake region				Gauged basin to I	
	P_R	E_R	I	O	P_G	E_G
Jun 2001–Nov 2001	0.23	0.13	0.20	0.44	0.61	0.39
Nov 2003–Aug 2004	0.12	0.04	0.59	0.25	0.84	0.16
Apr 2005–May 2005	0.19	0.03	0.41	0.38	0.98	0.02
Jul 2006–Jul 2007	0.68	0.57	−1.25	0.99	1.12	−0.12
Oct 2007–Aug 2008	0.25	0.14	0.32	0.29	0.75	0.25
Jan 2009–Feb 2009	0.17	0.03	0.41	0.40	0.93	0.07
Apr 2009–Jul 2009	0.11	0.07	0.60	0.22	0.73	0.27
Sep 2009–Jan 2010	0.34	0.14	0.36	0.16	0.73	0.27
Sep 2010–Dec 2010	0.01	0.13	0.24	0.62	0.56	0.44

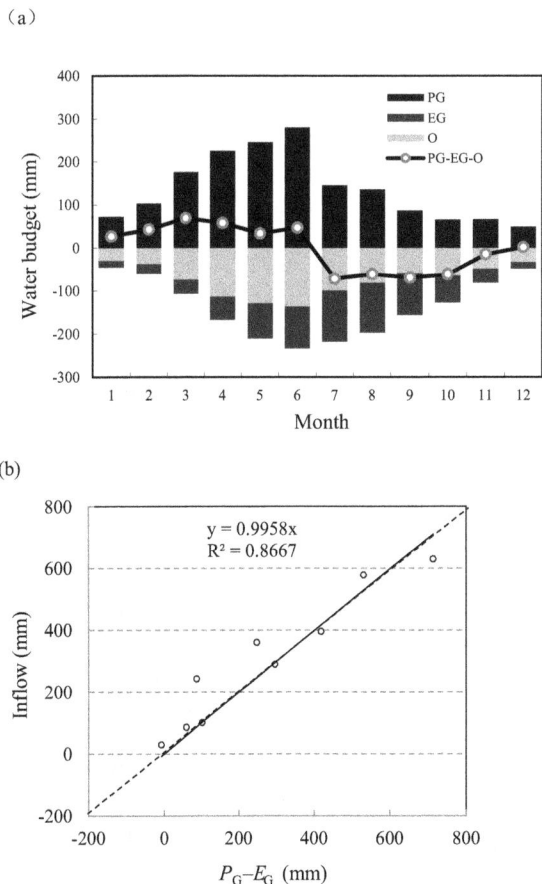

(a)

(b)

Figure 5. (a) Multi-year mean of monthly precipitation (P_G) and evapotranspiration (E_G) for the Poyang Lake basin, and outflow (O) into the Yangtze River. **(b)** The relationship between $P_G - E_G$ and inflow for nine cases of Poyang Lake droughts. All the water amounts are normalized to equivalent water height of the whole Poyang Lake basin.

gesting that meteorological droughts have made significant effects on the drought formation. For all the cases, the water anomalies of P_G and $P_G - E_G$ had positive relationships with drought severity, which was consistent with the water budget for the lake region. Nevertheless, the basin-scale precipitation is the most important water source to the lake, as confirmed by a correlation between $P_G - E_G$ (x) and I (y) ($y = 0.9958x$, $R^2 = 0.8667$, $n = 10$, $p < 0.005$) (Fig. 5b). The $P_G - E_G$-to-I difference was −28.3 mm, approximately 10 % of I, in agreement with our previous study (Wu et al., 2013). The high correlation and relatively small difference also confirmed the suitability of the satellite evapotranspiration data for the study area.

While the inflow reduction resulted from combined hydroclimatic change, precipitation and evapotranspiration may have made different contributions to the drought formation. Table 3 shows that the relative contribution for P_G varied from 0.56 to 1.12, suggesting that precipitation deficiency is the main driver to reduce the lake inflows during the drought development. Alternatively, the contribution for E_G ranged from −0.12 to 0.44, highlighting the importance of evapotranspiration in amplifying droughts, in agreement with the conclusion that reduced precipitation can coincide with increased evaporation (Teuling et al., 2013).

4.4 Mechanisms accounting for recent lake droughts

The above sections detail the lake droughts as abnormal phenomena and the hydroclimatic contribution to individual drought events. Yet, it remains unclear why the droughts strengthened in the past decade, and whether the droughts resulted from a long-term change of hydroclimatic influences or a seasonal combination of these influences.

Figure 6a shows the accumulated anomalies of the water budget from 2001 to 2010. At the lake region, the water budget ($P_R - E_R + I - O$) declined from mid-2002 to a low value in September 2009, and then increased yet remained in a negative phase. Obviously, the decrease in the water budget is a hydroclimatic setting for the recent drought increase.

(a)

(b)

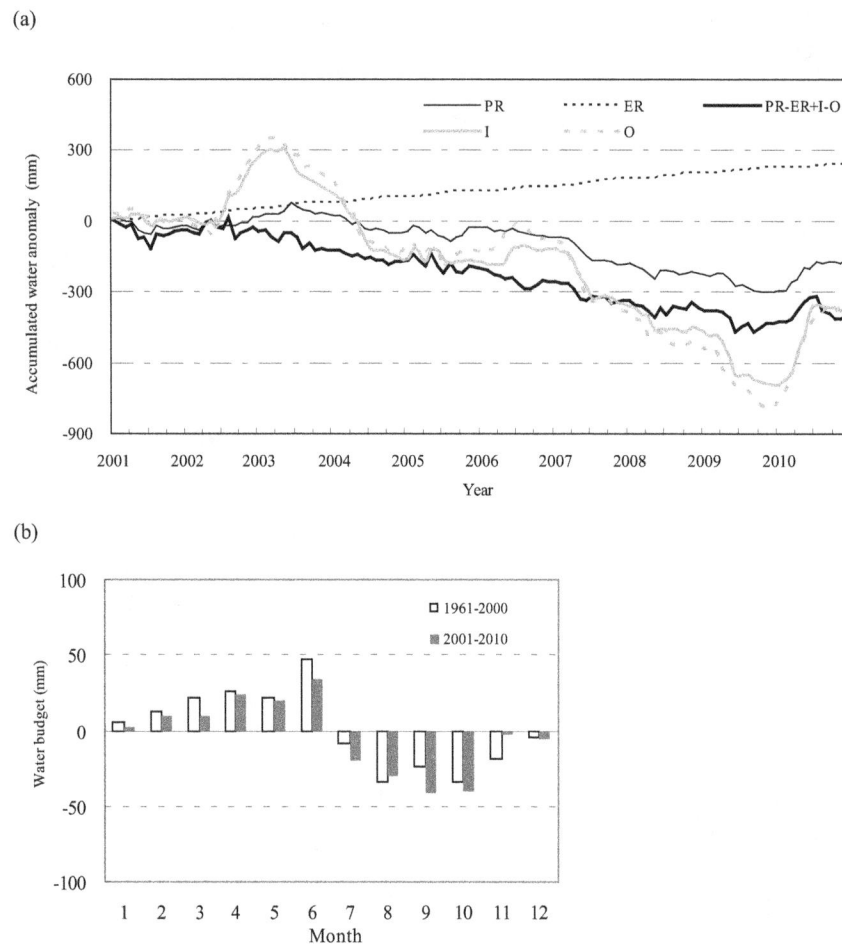

Figure 6. (a) Accumulated anomaly of water components and **(b)** water budget at the Poyang Lake region for period 2001–2010 compared to 1961–2000. All the water amounts are normalized to equivalent water height of the whole Poyang Lake basin.

The water deficits involve local precipitation and evapotranspiration, lake inflow and outflow, but each of these components has different courses. The accumulated P_R showed a decreasing trend after mid-2003. The accumulated E_R increased gradually but steadily, which was consistent with the rapid increase of surface temperature in the Poyang Lake basin since 1998 (Liu et al., 2012). The E_R exceeded the P_R after April 2010, exhibiting an increasing effect on the water budget. Comparatively, the accumulated I or O had a relatively large variation, consistent with their dominance over P_R and E_R at seasonal scale. They displayed a similar behavior with a peak in spring 2003, and then declined by the end of 2009. In the entire period, precipitation decreased by 5 %, evapotranspiration increased by 19 %, inflow declined by 5 % and outflow declined by 4 %, accounting for the negative water budget in the lake region.

Figure 6b displays seasonal variation of the water budget during 2001–2010. In comparison to 1961–2000, the water surplus reduced for the first half of the year, and the water deficit enhanced for the second half of the year except for

August and November. The large reduced surplus includes March and June, and the enhanced deficit includes July and September. The reduced surplus and the enhanced deficit would increase the possibility of drought occurrence and intensify the drought intensity. In the enlarged water deficit, the $P_R - E_R$ and the $I - O$ contributed to 43 and 57 %, respectively. In the $I - O$ deficit, inflow decreased but outflow increased. Usually, the outflow decreases with reduced inflow and $P_R - E_R$. Since the Yangtze River serves as a boundary condition of Poyang Lake, the increased outflow is generally a result of weakened blocking effects of the lake–river interactions (Guo et al., 2012; Zhang et al., 2012; Lai et al., 2014).

The weakened effects involve climate change in the upper reaches of the Yangtze River, and water impoundments of the TGD (Guo et al. 2012). Routinely, the TGD impoundment begins in mid-September and spans 1 to 2 months. Among all the drought events, none occurred during exactly the same time span. Accordingly, the TGD impoundments should not be responsible for the increased drought events. However, the impoundments lowered the lake stage at the Hukou outlet by 1–2 m for September–October (Guo et al. 2012; Zhang et al., 2012, 2014). Our analysis indicated that the impoundments led to a change in SLI from −2.70 to −3.03 for the extreme drought in July 2006–July 2007, and from −1.81 to −2.66 for the extreme drought in September 2009–January 2010. The change has two implications. First, the droughts were intensified with the TGD impoundments. Second, a severe drought ($-2.0 < \mathrm{LI} \leq -1.5$) was intensified to an extreme drought ($-\infty < \mathrm{SLI} \leq -2.0$), which changed the frequency of classified droughts. This is a reasonable explanation for the decrease in the number of severe droughts but increase in extreme droughts (Fig. 3c). Furthermore, according to the latest lake storage curve described in Tan et al. (2013), the lowered lake stages would result in a water loss of 7.1×10^8 and $24.1 \times 10^8 \, \mathrm{m}^3$ for each event, respectively. The losses occupied approximately 11.3 and 24.1 % of the total anomalies of outflow at the time of peak drought, corresponding to 11.3 and 3.7 % of the contribution for each drought. In comparison with the hydroclimatic influences, the TGD contribution is limited.

In general, the recently increased droughts were principally attributed to decreased inflow, increased outflow, and reduced local precipitation and increased evapotranspiration at the lake region. First, use of satellite-retrieved evapotranspiration data makes it possible to analyze drought causes from a perspective of water budget with independent measures of major water components. Given that measurement errors are quality controlled, the independent observations are more faithful than model simulations that are susceptible to model uncertainty and empirical parameterization. Indeed, the MOD16 products have been used effectively in water balance studies in the past. Second, in addition to drought quantification, absolute deficiency was defined for water components and water budget in reference to the normal hydrologic state. Individual hydroclimatic contributions were isolated from the total anomalies of the water budget, and the drought causation structure was subsequently distinguished. The quantification approach is straightforward and applicable to separate hydroclimatic influences on droughts, a key issue identified for developing integrated theories of droughts (Kallis, 2008). Third, it is the first time that all the drought events and their causation structure in Poyang Lake were quantified. Most existing studies did not explicitly quantify the droughts but focused on low water levels mainly in autumn seasons. A few studies addressed one-to-two extreme droughts with statistical regression analysis (Feng et al., 2012; Wu and Liu, 2014), without a systematic water balance analysis of the droughts and their statistics in the past decade. The present study completed drought quantification, water budget analysis, isolation of hydroclimatic contributions and clarification of causation structure for the recently increased droughts in Poyang Lake. The results demonstrated that the droughts were due to hydroclimatic factors, with a less important contribution from the TGD influences. Yet, it should be noted that the present study did not address some potential influences, for example, land cover/use change, agricultural water use, soil moisture variation and vegetation dynamics. These factors may affect the hydrological processes at seasonal and annual scales, and subsequently affect lake stage and droughts, which should be taken into consideration in future studies.

5 Conclusions

This paper used standardized lake stage to identify and quantify droughts on the case of Poyang Lake in China. From a perspective of water budget, it defined an absolute deficiency of the water components and water budget in reference to normal hydrologic conditions to determine hydroclimatic contributions to drought formation. Given 5 decades of hydroclimatic observations and latest satellite products, the water budget analysis was performed in the study area.

Our analyses demonstrated that the lake droughts had strengthened in the past decade, in terms of duration, frequency, intensity and severity. The overall contribution to the lake droughts came from decreased inflow, increased outflow, and reduced precipitation and increased evapotranspiration at the lake region. The decreased inflow resulted mainly from lower basin-scale precipitation and less from increases in evapotranspiration. The TGD impoundments were not responsible for the increased drought events, but they did intensify the droughts and change the frequency of classified droughts. Overall, the TGD contribution was limited, compared with the hydroclimatic influences.

The findings of this study provide an example of intensified lake droughts, and offer an insightful view into the droughts under the hydroclimatic and anthropogenic influences. The methodology proposed for quantification of lake droughts and isolation of hydroclimatic contributions has potential applications to other lakes. Moreover, the results of the study should be useful for local water resources management under climate change.

Acknowledgements. This work is supported by the 973 Program of National Basic Research Program of China (2012CB417003), the State Key Program of National Natural Science of China (41430855), and the Key Program of Nanjing Institute of Geography and Limnology of the Chinese Academy of Sciences (CAS) (NIGLAS2012135001). We thank David Shankman for his constructive comments on an earlier version of the manuscript, R. Guo for data pre-processing, Y. Chen for providing hydrological data, and X. Lai for sharing simulation data from the CHAM-Yangtze model. We are very grateful to Peter Molnar for his inexhaustible patience, critical but constructive comments, and language editing that have substantially improved the manuscript. The anonymous reviewers are acknowledged for their constructive comments on the early version of the manuscript.

References

Dracup, J., Lee, K., and Paulson Jr., E.: On the definition of droughts, Water Resour. Res., 16, 297–302, doi:10.1029/WR016i002p00297, 1980.

Environment News Service: China's Largest Freshwater Lake Shrinks in Record Drought, available at: http://ens-newswire.com/2012/01/06/chinas-largest-freshwater-lake-shrinks-in-record-drought/ (last access: 7 January 2016), 2012.

Feng, H. and Liu, Y.: Trajectory based detection of forest-change impacts on surface soil moisture at a basin scale [Poyang Lake Basin, China], J. Hydrol., 514, 337–346, 2014.

Feng, L., Hu, C., Chen, X.: Satellites capture the drought severity around China's largest freshwater lake, IEEE J-STARS, 5, 1266–1271, 2012.

Finlayson, M., Harris, J., McCartney, M., Lew, Y., and Zhang, C.: Report on Ramsar visit to Poyang Lake Ramsar site, P. R. China, available at: http://archive.ramsar.org/pdf/Poyang_lake_report_v8.pdf (last access: 7 January 2016), 2010.

Guo, H., Hu, Q., Zhang, Q., and Feng, S.: Impacts of the Three Gorges Dam on Yangtze River flow and river interaction with Poyang Lake, China: 2003–2008, J. Hydrol., 416–417, doi:10.1016/j.jhydrol.2011.11.027, 2012.

Hayes, M., Svoboda, M., Wall, N., and Widhalm, M.: The Lincoln declaration on drought indices: universal meteorological drought index recommended. B. Am. Meteorol. Soc., 92, 485–488, 2011.

Hervé, Y., Claire, H., Lai, X., Stéphane, A., Li, J., Sylviane, D., Muriel, B.-N., Chen, X., Huang, S., Burnham, J., Jean-François, C., Tiphanie, M., Li, J., Rmié, A., and Carlos, U.: Nine years of water resources monitoring over the middle reaches of the Yangtze River, with ENVISAT, MODIS, Beijing-1 time series, altimetric data and field measurements, Lake Reserv. Manage., 16, 231–247, doi:10.1111/j.1440-1770.2011.00481.x, 2011.

Hu, Q., Feng, S., Guo, H., Chen, G., and Jiang, T.: Interactions of the Yangtze River flow and hydrologic processes of the Poyang Lake, China. J. Hydrol., 347, 90–100, doi:10.1016/j.jhydrol.2007.09.005, 2007.

Jiao, L.: Scientists line up against dam that would alter protected wetlands, Science 326, 508–509, doi:10.1126/science.326_508, 2009.

Kallis, G.: Droughts, Annu. Rev. Environ. Resour., 33, 85–118, doi:10.1146/annurev.environ.33.081307.123117, 2008.

Keskin, F. and Sorman, A.: Assessment of the dry and wet period severity with hydrometeorological index, Int. J. Water Resour. Environ. Engineer., 2, 29–139, 2010.

Keyantash, J. and Dracup, J. A.: The quantification of drought: an evaluation of drought indices, B. Am. Meteorol. Soc., 83, 1167–1180, 2002.

Kim, H. W., Hwang, K., Mu, Q., Lee, S. O., and Choi, M.: Validation of MODIS 16 global terrestrial Evapotranspiration products in various climates and land cover types in Asia, KSCE J. Civ. Eng., 16, 229–238, 2012.

Kingston, D., Fleig, A., Tallaksen, L., and Hannah, D.: Ocean–atmosphere forcing of summer streamflow drought in Great Britain, J. Hydrometeor., 14, 331–344, doi:10.1175/JHM-D-11-0100.1, 2013.

Lai, X., Jiang, J., Liang, Q., and Huang, Q.: Large-scale hydrodynamic modeling of the middle Yangtze River Basin with complex river-lake interactions, J. Hydrol., 492, 228–243, 2013.

Lai, X., Huang, Q., Zhang, Y., and Jiang, J.: Impact of lake inflow and the Yangtze River flow alternations on water levels in Poyang Lake, China, Lake and Reservoir Management. 30, 321–330, 2014a.

Lai, X., Jiang, J., Yang, G., and Lu, X.: Should the Three Gorges Dam be blamed for the extremely low water levels in the middle–lower Yangtze River?, Hydrol Process., 28, 150–160, 2014b.

Liu, Y., Song, P., Peng, J., Fu, Q., and Dou, C.: Recent increased frequency of drought events in Poyang Lake Basin, China: climate change or anthropogenic effects? Hydro-climatology: Variability and Change (IAHS Publ.), 344, 99–104, 2011.

Liu, Y., Zhang, Q., Liu, J., and Li, H.: Climatic, Hydrologic and Environmental Change in Poyang Lake Basin, Science Press, Beijing, 262 pp., 2012 (in Chinese).

Liu, Y., Wu, G., and Zhao, X.: Recent declines of the China's largest freshwater lake: trend or regime shift?, Environ. Res. Lett., 8, 014010, doi:10.1088/1748-9326/8/1/014010, 2013.

Lloyd-Hughes, B. and Saunders, M.: A drought climatology for Europe, Int. J. Climato., 22, 1571–1592, doi:10.1002/joc.846, 2002.

Loarie, S. R., Lobell, D. B., Asner, G. P., Mu, Q., and Field, C. B.: Direct impacts on local climate of sugar-cane expansion in Brazil, Nature Climate Change, 1, 105–9, 2011.

Lomax, R. G.: An Introduction to Statistical Concepts for Education and Behavioral Sciences, Lawrence Erlbaum Associates, Inc. Mahwah, 519 pp., 2001.

McKee, T. B., Doesken, N. J., and Kliest, J.: The relationship of drought frequency and duration to time scales. Proceedings of the 8th Conference of Applied Climatology, 17–22 January, Anaheim, CA, American Meteorological Society, Boston, MA, 179–184, 1993.

McKee, T. B., Doesken, N. J., and Kliest, J.: Drought monitoring with multiple time scales. Ninth Conference on Applied Climatology, 15–20 January 1995, Dallas, TX, American Meteorological Society, Boston, MA, 233–236, 1995.

Mendicino, G., Senatore, A., and Versace, P.: A groundwater resource index (GRI) from drought monitoring and forecasting in a Mediterranean climate, J. Hydrol., 357, 282–302, 2008.

Min, Q. and Zhan, L.: Analysis of lake stage relationships between different locations in Poyang Lake, The Yangtze River, 44, 5–10, 2013 (in Chinese).

Mishra, A. and Singh, V.: Review of drought concepts, J. Hydrol., 391, 202–216, doi:10.1016/j.jhydrol.2010.07.012, 2010.

Mishra, A. and Singh, V.: Drought modeling – A review, J. Hydrol., 403, 157–175, doi:10.1016/j.jhydrol.2011.03.049, 2011.

Mu, Q., Zhao, M., and Running, S. W.: Improvements to a MODIS global terrestrial evapotranspiration algorithm, Remote Sens. Environ., 115, 1781–1800, doi:10.1016/j.rse.2011.02.019, 2011.

Nalbantis, I. and Tsakiris, G.: Assessment of hydrological drought revisited, Water Resour. Manage., 23, 881–897, doi:10.1007/s11269-008-9305-1, 2009.

The Ramsar Convention: The List of Wetlands of International Importance, 25 April 2012, available at: http://www.ramsar.org/pdf/sitelist.pdf (last access: 7 January 2016), 2012.

Senay, G. B., Leake, S., Nagler, P. L., Artan, G., Dickinson, J., Cordova, J. T., and Glenn, E. P.: Estimating basin scale evapotranspiration (ET) by water balance and remote sensing methods, Hydrol. Process., 25, 4037–4049, doi:10.1002/hyp.8379, 2011.

Shankman, D. and Liang, Q.: Landscape changes and increasing flood frequency in China's Poyang Lake region,

The Professional Geographer, 55, 434–445, doi:10.1111/0033-0124.5504003, 2003.

Shankman, D., Keim, B. D., and Song, J.: Flood frequency in China's Poyang Lake region: trends and teleconnections, Int. J. Climatol., 26, 1255–1266, doi:10.1002/joc.1307, 2006.

Sheffield, J., Goteti, G., Wen, F., and Wood, E. F.: A simulated soil moisture based drought analysis for the United States, J. Geophy. Res., 109, D24108, doi:10.1029/2004JD005182, 2004.

Seneviratne, S. L., Lehner, I., Gurtz, J., Teuling, A. J., Lang, H., Moser, U., Grebner, D., Menzel, L., Schroff, K., Vitvar, T., and Zappa, M.: Swiss prealpine Rietholzbach research catchment and lysimeter: 32 year time series and 2003 drought event, Water Resour. Res., 48, W06526, doi:10.1029/2011WR011749, 2012.

Shukla, S. and Wood, A. W.: Use of a standardized runoff index for characterizing hydrologic drought, Geophys. Res. Lett., 35, L02405, doi:10.1029/2007GL032487, 2008.

Smakhtin, V. U.: Low flow hydrology: a review, J. Hydrol., 240, 147–186, 2001.

Spinoni, J., Naumann, G., Carrao, H., Barbosa, P., and Vogt, J.: World drought frequency, duration, and severity for 1951–2010, Int. J. Climato., 34, 2792–2804, 2014.

Tallaksen, L. M., Madsen, H., and Clausen, B.: On the definition and modelling of streamflow drought duration and deficit volume, Hydrol. Sci. J., 42, 15–33, 1997.

Tallaksen, L. M. and van Lanen, H. A. J.: Hydrological Droughts: Processes and Estimation Methods for Streamflow and Groundwater, Elsevier Science, Amsterdam, the Netherlands, 576 pp., 2004.

Tan, G., Guo, S., Wang, J., and Lv, S.: Hydrologic Change and Water Resources in Poyang Eco-economic Zone, China Water & Power Press, Beijing, China, 277 pp., 2013.

Teuling, A. J., Van Loon, A. F., Seneviratne, S. I., Lehner, I., Aubinet, M., Heinesch, B., Bernhofer, C., Grünwald, T., Prasse, H., and Spank, U.: Evapotranspiation amplifies European summer drought, Geophys. Res. Lett., 40, 2071–2075, doi:10.1002/grl.50495, 2013.

Todd, R., Macdonald, N., Chiverrell, R. C., Caminade, C., and Hooke, J. M.: Severity, duration and frequency of drought in SE England from 1697 to 2011, Clim. Chang., 121, 673–687, 2013.

Velpuri, N. M., Senay, G. B., Singh, R. K., Bohms, S., and Verdinb, J. P.: A comprehensive evaluation of two MODIS evapotranspiration products over the conterminous United States: Using point and gridded FLUXNET and water balance ET, Remote Sensing of Environment, 139, 35–49, 2013.

Vicente-Serrano, S. M. and López-Moreno, J. I.: Hydrological response to different time scales of climatological drought: an evaluation of the Standardized Precipitation Index in a mountainous Mediterranean basin, Hydrol. Earth Syst. Sci., 9, 523–533, doi:10.5194/hess-9-523-2005, 2005.

Wan, X. and Xu, X.: Analysis of supply and demand balance of water resources around Poyang Lake, The Yangtze River, 41, 43–47, 2010.

Wang, H., Guan, H., Gutiérrez-Jurado, H., and Simmons, C.: Examination of water budget using satellite products over Australia, J. Hydrol., 511, 546–554, 2014.

Wilhite, D. A. and Glantz, M. H.: Understanding the drought phenomenon: the role of definitions, Water Int., 10, 111–120, 1985.

Wilcox, B. P., Huang, Y., and Walker, J. W.: Long-term trends in streamflow from semiarid rangelands: uncovering drivers of change, Glob. Change Biol., 14, 1676–1689, 2010.

Wu, G., Liu, Y., and Zhao, X.: Analysis of spatio-temporal variations of evapotranspiration in Poyang Lake Basin using MOD16 products, Geophysical Research, 32, 617–627, 2013 (in Chinese with English abstract).

Wu, G. and Liu, Y.: Satellite-based detection of water surface variation in China's largest freshwater lake in response to hydroclimatic drought, International Journal of Remote Sensing, 35, 4544–4558, doi:10.1080/01431161.2014.916444, 2014.

Yevjevich, V.: An objective approach to definitions and investigations of continental hydrologic droughts, Hydrology Paper No. 23, Colorado State University, Fort Collins, CO, 18 pp., 1967.

Zelenhasic, E. and Salvai, A.: A method of streamflow drought analysis, Wat. Resour. Res., 23, 156–168, 1987.

Zhang, Q., Li, L., Wang, Y.-G., Werner, A., Xin, P., Jiang, T., and Barry, D.: Has the Three-Gorges Dam made the Poyang Lake wetlands wetter and drier?, Geophys. Res. Lett., 39, L20402, doi:10.1029/2012GL053431, 2012.

Zhang, Q., Ye, X., Werner, A., Li, Y, Yao, J., Li, X., and Xu, C.: An investigation of enhanced recessions in Poyang Lake: Comparison of Yangtze River and local catchment impacts, J. Hydrol., 517, 425–434, 2014.

Zhao, X. and Liu, Y.: Lake fluctuation effectively regulates wetland evapotranspiration: A case study of the largest freshwater lake in China, Water, 6, 2482–2500, 2014.

PERMISSIONS

LIST OF CONTRIBUTORS

D. Kurtzman
Institute of Soil, Water and Environmental Sciences, The Volcani Center, Agricultural Research Organization, P.O. Box 6, Bet Dagan 50250, Israel

S. Baram
Dept. of Land, Air and Water Resources, University of California Davis, CA 95616, USA

O. Dahan
Dept. of Hydrology and Microbiology, Zuckerberg Institute for Water Research, Blaustein Institutes for Desert Research, Ben Gurion University of the Negev, Sde Boker Campus, Negev 84990, Israel

S. Biskop and M. Fink
Department of Geography, Friedrich Schiller University Jena, Germany

F. Maussion
Institute of Atmospheric and Cryospheric Sciences, University of Innsbruck, Austria

P. Krause
Thuringian State Institute for Environment and Geology, Jena, Germany

L. Gill and P. Johnston
Department of Civil, Structural & Environmental Engineering, Trinity College Dublin, Dublin 2, Ireland

M. M. R. Jahangir
Department of Civil, Structural & Environmental Engineering, Trinity College Dublin, Dublin 2, Ireland
Department of Environment, Soils & Land Use, Teagasc Environment Research Centre, Johnstown Castle, Co. Wexford, Ireland

O. Fenton and K. G. Richards
Department of Environment, Soils & Land Use, Teagasc Environment Research Centre, Johnstown Castle, Co. Wexford, Ireland

M. G. Healy
Civil Engineering, National University of Ireland, Galway, Co. Galway, Ireland

C. Müller
School of Biology and Environmental Science, University College Dublin, Belfield, Dublin, Ireland
Department of Plant Ecology (IFZ), Justus-Liebig University Giessen, Giessen, Germany

M. Boudou, B. Danière and M. Lang
Irstea, UR HHLY, Hydrology-Hydraulics, 5 rue de la Doua, 69626 Villeurbanne, France

S.-H. Suh
Department of Environmental Atmospheric Sciences, Pukyong National University, Daeyeon campus 45, Yongso-ro, Namgu, Busan 608-737, Republic of Korea

C.-H. You
Atmospheric Environmental Research Institute, Daeyeon campus 45, Yongso-ro, Namgu, Busan 608-737, Republic of Korea

D.-I. Lee
Department of Environmental Atmospheric Sciences, Pukyong National University, Daeyeon campus 45, Yongso-ro, Namgu, Busan 608-737, Republic of Korea
Atmospheric Environmental Research Institute, Daeyeon campus 45, Yongso-ro, Namgu, Busan 608-737, Republic of Korea

C.-S. Huang, J.-J. Chen and H.-D. Yeh
Institute of Environmental Engineering, National Chiao Tung University, Hsinchu, Taiwan

J. W. Kirchner
ETH Zürich, Zurich, Switzerland
Swiss Federal Research Institute WSL, Birmensdorf, Switzerland

Y. Gao, T. Markkanen, T. Thum, M. Aurela, A. Lohila, M. Kämäräinen and T. Aalto
Finnish Meteorological Institute, P.O. Box 503, 00101 Helsinki, Finland

I. Mammarella
University of Helsinki, Department of Physics, P.O. Box 48, 00014 Helsinki, Finland

S. Hagemann
Max Planck Institute for Meteorology, Bundesstr. 53, 20146 Hamburg, Germany

L. Mourre, T. Condom, T. Lebel and J. E. Sicart
IRD/UGA/CNRS/G-INP, LTHE UMR 5564, Grenoble, France

C. Junquas
IRD/UGA/CNRS/G-INP, LTHE UMR 5564, Grenoble, France
Instituto Geofísico del Perú (IGP), Lima, Peru

R. Figueroa
UNASAM, Huaraz, Peru

A. Cochachin
Glaciology and Water Resources Unit, National Water Authority (ANA-UGRH), Huaraz, Peru

N. F. Fang, H. Y. Zhang and Y. X. Wang
State Key Laboratory of Soil Erosion and Dryland Farming on the Loess Plateau, Northwest A & F University, Yangling 712100, People's Republic of China
Institute of Soil and Water Conservation of Chinese Academy of Sciences and Ministry of Water Resources, Yangling 712100, People's Republic of China

F. X. Chen
College of Resources and Environment, Huazhong Agricultural University, Wuhan 430070, People's Republic of China

Z. H. Shi
Institute of Soil and Water Conservation of Chinese Academy of Sciences and Ministry of Water Resources, Yangling 712100, People's Republic of China

College of Resources and Environment, Huazhong Agricultural University, Wuhan 430070, People's Republic of China

N. Le Vine and A. Butler
Department of Civil and Environmental Engineering, Imperial College London, London, UK

N. McIntyre
Department of Civil and Environmental Engineering, Imperial College London, London, UK
Centre for Water in the Minerals Industry, the University of Queensland, St. Lucia, Australia

C. Jackson
British Geological Survey, Keyworth, UK

Y. Liu and G. Wu
Nanjing Institute of Geography & Limnology, Chinese Academy of Sciences, 73 East Beijing Road, Nanjing 210008, China

Index